Biology for the Global Citizen

Biology for the Global Citizen

GEORGE W. SHIFLET, ANGELA B. SHIFLET,
AND MURRAY P. PENDARVIS

First Edition

cognella
SAN DIEGO

Bassim Hamadeh, CEO and Publisher
Mieka Portier, Senior Acquisitions Editor
Tony Paese, Project Editor
Casey Hands, Production Editor
Emely Villavicencio, Senior Graphic Designer
Greg Isales, Licensing Coordinator
Natalie Piccotti, Director of Marketing
Kassie Graves, Vice President of Editorial
Jamie Giganti, Director of Academic Publishing

Printed in the United States of America.

cognella® | ACADEMIC PUBLISHING
3970 Sorrento Valley Blvd., Ste. 500, San Diego, CA 92121

BRIEF CONTENTS

DETAILED CONTENTS

PREFACE

Biology for the Global Citizen is intended to be an easily accessible introductory learning resource for students with varying backgrounds in biology and various life goals. As we wrote this text, we interweaved applications and active learning experiences that are meant to encourage and motivate our audience to study and appreciate biological systems. We are optimistic that most students, through this course, will find biology interesting and relevant. While the students learn about biology, we hope to foster habits and skills to help them sort out information that is supported by evidence from speculation, opinion, or "conspiracy theories." We intend that most of those who complete this course will be better prepared to make informed decisions and to reject misleading and contrived forms of communication.

The authors became scientists because we were curious about the natural world. There were so many questions, and science was there to further our interests and to help our search, even if it could not immediately answer all of our questions. There is clearly much left to learn. If we can encourage students to be inquisitive, sensible, and attentive, they may also seek and demand truth in all areas of life.

In the right atmosphere, students do ask questions. For instance, the following questions are just a sampling of many interesting questions students might ask:

What is the difference between weather and climate, and why does it matter?
How does a cancer cell differ from a normal cell?
What is a healthy body weight, and how can I maintain it?
With the overwhelming information available, how do I determine what to believe?
How is it possible that I am taller than either of my parents?

In this book, we try to help students answer these questions while facilitating their understanding of how science, and biology in particular, is pertinent to their lives. We need students who become citizens who are concerned and responsible. Most will not become scientists, of course, but we are all accountable for ensuring that we have a natural environment that supports a healthy life for human beings, as well as for all other living species on our planet.

With these goals in mind, we have developed this text as a series of modules, including related laboratories. The structure and content of this text is intended to maximize flexibility for individual courses. Thus, each module contains the following *features*:

- **Applying Science to Life**. Engaging, short *applications* or *case studies* with content using current or historical events, experiments, or issues that are relatable to students' lives. Such accounts motivate students to explore and learn. Most applications tell a story and contain questions for further exploration.
- Brief, practical, and engaging overview of the content:
  - **Quick Review Questions** embedded in the text. Moreover, Cognella's Active Learning site can automatically grade answers, allowing even large sections to use the questions conveniently.
  - **Points of Inquiry** are questions that promote critical thinking and more in-depth exploration. Suggested answers and references are available for instructors.
  - **Glossary** of terms, available through Active Learning, with definitions of most terms being inline as well.
  - **Active learning assignments** for in-class and out-of-class learning, including references, links to online resources and videos, a database of original biophotographs with paragraphs related to the biology, and other material to support teaching and student learning.
  - **References** for each module, including others that might provide further insight into the topics.
- **Laboratory exercises using a computer simulation** that reinforce the material, employ the scientific method, enhance *quantitative literacy* through interpretation of graphs and basic statistics, employ data to support conclusions, emphasize critical thinking and analysis, and include *automatically graded online questions*. Guided by the laboratory exercises, students consider various scenarios, make hypotheses, adjust variables to test the effect of each scenario, run the simulation, generate and interpret graphs and tables of data, make observations, draw conclusions, apply their conclusions to decision-making, and gain a deeper understanding of the science that the model simulates. For courses that do not include a laboratory, the simulations could be used for in-class demonstrations or outside project assignments.

  The computer simulations use the software tool NetLogo, which "is used by many tens of thousands of students, teachers and researchers worldwide" and can be downloaded free of charge for Macintosh, Windows, and UNIX computers. NetLogo is very easy to use, and a short video on its use is available in Active Learning. NetLogo includes many scientific models. Moreover, additional models have been developed for topics where NetLogo models do not exist. Some of the laboratories also employ spreadsheets to generate simple statistics, such as mean and standard deviation, and a laboratory tutorial on spreadsheets is available for various application software packages in Active Learning.

  Each laboratory has an associated introductory video, accessible through Active Learning, that illustrates how to use the model, points out significant aspects of the model and laboratory, and indicates the important biological concepts. Several laboratories have associated documents, also on Active Learning, illustrating chemical formulas or mathematical calculations.

As with traditional laboratories, the exercises can be completed in teams, reducing the number of required computers for in-class assignments. For flexibility, each laboratory contains several multipart questions, and the instructor can decide which to assign. Moreover, the instructor can have students complete exercises in class, outside of class, or during a designated laboratory period. Most of the answers to laboratory questions can be quickly graded by hand or automatically graded in Active Learning, while other questions are open-ended and appropriate for classroom discussion. Thus, classes of a variety of sizes and pedagogies can conveniently use the materials.

- **Suggestions for various approaches** for instructors, including how to employ polling and flipped classes. This course could be taught using traditional pedagogical methods, or our materials could be used very effectively in what is termed a "flipped classroom." Instructors can assign study material (including modules and videos) prior to class and quiz students either in or outside class with written answers or through Active Learning. The flipped-class approach would allow for a more interactive classroom and would foster student participation through discussions, presentations, or other in-class exercises.

With the material and the support of these features, we aspire for users to understand what science is, how it integrates into our lives, and how we can apply it to everyday problems. In short, we would like students to learn how to become good, thoughtful citizens who make decisions based on validated information.

ACKNOWLEDGMENTS

The entire staff at Cognella Publishing has done a splendid job of support and accommodation for this project. If we attempted to thank everyone, we would run the risk of leaving out someone. We thank them enthusiastically for doing such a terrific job. However, in particular, we would like to thank Mieka Portier, who understood our project and advocated effectively for its publication; our fine project editor, Tony Paese; Haley Brown, who has been so responsive and resourceful in development of the Active Learning resources; Casey Hands, who shepherded the production process; Sabrina Baeta, who managed the instructor resources; and the whole Cognella team, which has been excellent. Our sincere thanks to all!

We would also like to thank the faculty (listed below) and students who class-tested our material during the 2020–2021 academic year. This "pandemic" year will be a very memorable one for most college/university faculty and students.

Le Paliulis and Kate Toner, Bucknell University (PA)
Maria Oehler, Florida State College at Jacksonville
Mike Wolyniak, Hampden-Sydney College (VA)
Payal Ray, Presbyterian College (SC)
Julie Smoak, University of South Carolina-Upstate

Otherwise, we would like to thank our family and friends who have tolerated our need to work and who supported us so steadfastly through the development of this book.

UNIT I

Setting the Stage

MODULE

1 Science for Citizens

1.1. Introduction

APPLYING SCIENCE TO LIFE 1 Science Can Be a Lifesaver

Justine Henning feels strongly about science, and she has excellent reasons. She comes from a family where many of the women have died of breast cancer. In 1945, her paternal grandmother died of it at the age of 36. Eventually, all of her grandmother's sisters also died of breast cancer. The only treatments for this cancer at that time were mastectomy and radiation. She doesn't know if it was just the inevitable progression of the disease that killed them, the treatments, or both.

In 2009, her mother was diagnosed with the disease. Fortunately, a University of Chicago biochemist, Eugene DeSombre, and his colleagues had previously discovered through their research that the hormone estrogen promoted the growth of some breast tumors. As a consequence of DeSombre's and other researchers' work, treatments had been developed that were available for her mother, and her mother was still cancer-free as of nine years later. When Justine was diagnosed with estrogen-receptor-positive breast cancer in her 50s, she knew it wasn't a death sentence.

There are many breast cancer types. In the estrogen-receptor-positive form, which occurs frequently in inherited forms of breast cancer, the cells have receptors to which estrogen (released by the ovaries) will bind and cause the cancer cells to divide uncontrollably. There are medications for such cancers that either inhibit the production of estrogen (after menopause) or block the binding of estrogen to the cancer cells (Mayo Clinic 2019).

Two genes, BRCA1 and BRCA2, are expressed in breast tissue. These genes normally code for proteins important for the repair of damaged DNA and thereby suppress tumor formation. Any of the many types of mutations in these genes can lead to unreliable repair of DNA and an amplified breast cancer risk in both women and men (Barrois M. et al. 2004; Ewald et al. 2009). There is also increased risk for carriers of a mutation to develop ovarian or prostate cancer. Unfortunately, you only need one copy of a mutated gene to have an increased risk for cancer.

Either of the genes can be hormone-sensitive or hormone-negative, meaning the breast cancer cells do or do not, respectively, have receptors for estrogen to promote cancer growth. About 12 percent of women in the general population will develop breast cancer during their lifetimes. In contrast, by the age of 80, 72 percent of women with BRCA1 mutations and 69 percent of those with BRCA2 mutations will develop breast cancer (Kuchenbaecker et al. 2017; Howlader et al. 2014; NCI 2018). In Justine's case, science was a lifesaver.

Impressed with what scientists had accomplished and grateful for the lifesaving medicine that helped both her and her mother to beat breast cancer, Justine Henning sought a way to encourage young people to pursue careers in science, engineering, and mathematics. To accomplish this goal, she created a remarkable organization that promotes math education—math taught in an effective way. Math is extremely important in those three areas, and if students obtain skills, knowledge, and confidence in math, they are more likely to pursue such careers. Her privately funded organization is called Math4Science (math-4science.org) (Henning 2017).

Science is difficult to ignore but easy to take for granted. We are all surrounded by evidence of scientific achievements, but sometimes we fail to appreciate the time, creativity, and work that went into many of these advances. The preceding story is real and offers you but one example of how science is applied to life. Of course, not everyone can or wants to be a scientist, but to be a good citizen, parent, and so on you ought to understand enough about science to read or listen to the news or help your children with their homework and questions. Hopefully, learning and studying about some of the breakthroughs in our understanding of the world will interest and excite you. You can and should influence decisions about the applications of science in your family, community, state, and country. To do that, you should be informed and able to evaluate what the issues are and what their outcomes may be. Voting is a privilege and an important responsibility for citizens of any country. Sometimes your vote for a candidate or on a ballot initiative/proposition concerning the environment, health care, technology, utilities, and so on will significantly affect your life and the lives of others.

In this text, we will consider a wide array of biological topics. As you study biology, you study living things—how they are constructed and function, how they interact with other living things, and how the environment (physical, chemical, etc.) affects life. We will present to you many examples of the remarkable adaptations and life forms on the Earth. In every module, we try to show you how biology and science apply to your life. Hopefully, as you study life through this course, you will gain knowledge, build skills, and learn of applications that will make you a better citizen, not just of your own country, but of the world.

1.2. The Process of Science—Understanding How Science Is Done

1.2.1. Making Results Matter: The Scientific Method

Experience and evidence validate the *scientific method* as the most reliable approach for determining what in the natural world is factual and authentic. To understand the organization and functions of the earth and the universe, authentic information (facts) provides the groundwork for establishing scientific principles and increasing our knowledge. Results from scientific research must be *repeatable* and undergo *peer review* (evaluation of scientific work by others who are experts in the same field). All research is verified by what might be called "organized scrutiny" by other scientists, those who are experts in the field. The results of this procedure provide us with the collective wisdom of the scientific community. No matter the brilliance of one investigator, unless their ideas and results are verified by the scientific community, they'll need to refine those ideas and test them further to gain acceptance.

The importance of using serious scientific evidence is demonstrated by the relationship between smoking tobacco and lung cancer. Considerable research by many scientists has established links between smoking and respiratory diseases, including lung cancer. The links are and have been widely understood, even by the executives of the purveyors of tobacco products. Interestingly, during the 1930s, scientists working under the Nazi regimes knew of such connections. Unfortunately, for many years, the lawyers, consultants, and boards of tobacco companies in the United States and elsewhere played a game of "smoke and mirrors" by denying the facts with which they were familiar and filling the air with misleading and untruthful information. Their obfuscations delayed political policy changes for decades, and the confused and deceived public suffered and died. This strategy of misleading people continues today. Science is not a political entity. It is not a religion or belief system. However, it is an asset to policymaking, because science provides elemental, verifiable information about the world and how it functions. It is not the only factor for consideration while seeking solutions, but science ought to be utilized when making policies and laws. Political decisions affect the lives of all citizens, and scientific evidence is the most objective and reliable information we have to ensure that the political and social decisions are reasonable and fair.

The process of scientific research follows a general pattern that is referred to as the scientific method. Doubtless, if you have had a course in science, you are familiar with it. Although the way it is generally taught makes the process appear to be very rigid, in practice, the approach to any science problem may be modulated to fit the question. For example, if we look at the methods used by Charles Darwin, he did not start out with a hypothesis. He started out with wide-ranging and detailed observations, from which he developed his generalized ideas on evolution by natural selection.

Generally, the steps associated with the scientific method include the following:

1. **Observation.** Making direct observations of natural phenomena is a dependable approach to understanding these phenomena. From these observations, a scientist is able to generate questions and potential explanations that can be tested. Also, researchers will do a thorough examination of the literature to find out what is already known about the object(s) of their study.

2. **Hypothesis.** A hypothesis is a functional explanation. Actually, scientists may come up with a number of hypotheses to explain their observations. Hypotheses, and predictions derived from those hypotheses, can be tested. If the hypothesis is supported by the results of the observations or experiments (data), it is still not proven. The experimental design and results of these tests must be repeatable by others. The more support found for the hypothesis, the more likely the hypothesis is to be accepted. On the other hand, a scientific hypothesis can be found to be false by lack of observational or experimental support. Even if the hypothesis is unsupported, the results may be useful for developing other hypotheses. With supportive

evidence, there will still be more questions, and related hypotheses can be generated for a deeper understanding.

3. **Modeling.** Today, with the increased capabilities of computers and results from mathematical research, models can be developed, and simulations performed. Hypotheses may be tested by computer models and simulations. Simulations may help scientists to eliminate unlikely hypotheses and to design more robust experiments. Scientists can now use computers to analyze enormous amounts of data and to conduct virtual experiments that are too dangerous or too complex to perform in the lab or field. Computer models and simulations are becoming integral to the scientific method.
4. **Experiments and Testing.** Many hypotheses can be tested using controlled experiments, the designs of which are crucial to collecting meaningful results. An experiment includes both an *experimental group* and a *control group*, and each group includes replicates. Within each group, replicates for each treatment will be exposed to the same conditions. Only one factor in the experimental group actually varies from that factor for the control group. Thus, in an experiment involving plant growth, factors like temperature, humidity, light regime, and so on would be the same for both experimental and control groups. One factor, such as soil type or fertilizer treatment, would be the experimental variable.
5. **Systematic Observations.** For hypotheses that involve systems that are too complicated to be handled experimentally, scientists must rely on systematic observations. For example, let's say we are studying the hunting strategies of different species of mammalian predators on African savannas. It would be unfeasible to run a laboratory experiment with cheetahs, lions, and so on and their prey, such as impalas, wildebeest, and so forth.
6. **Results.** From the experiments, researchers must collect and analyze the results. Data gathered must be analyzed statistically, which helps the researchers determine how much confidence they have that the results are accurate and significant. For large populations, it is more feasible to take representative samples, and researchers must plan sampling procedures and numbers carefully in preparation for the study.
7. **Conclusions.** Following analysis of the data, conclusions may be drawn. Also, the researchers may propose new hypotheses, whether or not the data supports the hypothesis in the original study.
8. **Publication and/or Presentation.** Researchers will submit a formal paper to a pertinent, professional journal, summarizing the experiments, detailing their methods, presenting their results, and discussing their conclusions and the significance of their findings. Reviewers, who are acknowledged experts in that field, will scrutinize the paper, recommend any needed changes, and urge publication or rejection. In some cases, researchers may present their results at an appropriate professional meeting. Once the paper is published or the presentation is made, the researchers' scientific peers must be able to replicate their work.

As you can see from the outline of this method, consideration of any scientific question requires a careful and systematic approach. However, every approach varies with the problem. Let's look at a historical example.

APPLYING SCIENCE TO LIFE 2 **Serendipitous Scientific Sleuthing**

Lucy Wills was born in 1888 to a relatively well-off family near Birmingham, England. She came from a family that was interested in science. It would be an understatement to say that women of that time had very limited opportunities for education, much less for entry into a professional career. As a girl, she attended a local private school but went on to Cheltenham Ladies College. Considered progressive, this college provided her with an intellectually challenging education, with emphasis in mathematics and science. Such an education for women was far from common. In 1907, she began studies at Newnham College, one of the two colleges associated with Cambridge University that educated women. In 1911, she was permitted to take her qualifying exams, but she would receive no degree—Cambridge did not give women degrees until 1948. Following her years at Cambridge, she traveled to South Africa, serving briefly as a nurse when World War I broke out. It was likely that her experience nursing inspired her to attend medical school. She graduated from the London School of Medicine for Women in 1920, but she was more interested in research than in practicing clinical medicine. She was particularly interested in hematology (blood studies).

Through contacts from the medical school, Wills travelled to India and began studies of a particularly severe form of *anemia* that especially affected poor women textile workers. It caused fever and diarrhea and could be fatal. She *observed* that the disease was typical of poor but not of rich women, and she began to investigate specific living conditions and other factors associated with impoverished women. She followed up on the two most likely causes: disease from food, water, and so on or dietary factors.

The *first hypothesis*: Because the disease is characterized by fever and diarrhea, the illness is caused by typhoid or some other sort of disease-causing (*pathogenic*) *organism* (*bacteria*). Obtaining and examining stool samples from the women, she searched for pathogens and conducted Widal tests (test for presence of typhoid bacteria). Despite tedious work, she was unable to find any evidence for a bacterial disease.

Dietary deficiency had been found to cause at least one other form of anemia, so a *second hypothesis* was made: A dietary deficiency is causing the anemia. She began to collect data on the women's diets, which were quite poor in quality and quantity. At first, she suspected a deficiency of vitamin A or C, and she decided to try experimentation using white rats. She had some limited success with the rats, trying various supplements, but her rat colony was infested by lice, which can also cause anemia in those animals. Furthermore, supplementing the women's diets with vitamins A and C did not cure the anemia.

So Wills turned to monkeys. She placed the monkeys on a diet similar to the women, and she gave supplements to some of them. One of the monkeys who was fed a diet without the supplement began to grow very sick. Liver extract had been shown to be effective against another type of anemia, and a colleague told her about marmite (food spread made from yeast extract), which supposedly contained generous amounts of vitamins. The *third hypothesis*: The dietary deficiency is caused by a vitamin deficiency other than those of vitamins A, C, or the vitamins in liver extract. She tried the marmite on the sick monkey, and it recovered. Some factor (vitamin) in the marmite had cured the anemia in the monkey. She then tried the marmite on 22 pregnant women with anemia, and they all recovered.

Conclusion: Despite the small sample size, she concluded that macrocytic anemia was caused by a deficiency of a factor found in marmite. It wasn't until 1941 that biochemists extracted the factor from spinach. They called it *folate*.

Until 1941, folate, isolated, had been known as the "Wills factor." This factor turned out to be a B vitamin, B9, and it was responsible for two health conditions associated with pregnancy, *macrocytic anemia* and *neural tube defects*. Later, research found folate (B9) to be essential for the replication of DNA and for the synthesis of several amino acids.

A deficiency in folate impairs these processes, and so it is not surprising that low levels of folate manifests in tissues that are rapidly replicating. Macrocytic anemia results from an impairment in the production of red blood cells (RBCs); characteristically, the number of RBCs is low, and the cells are fairly large (*macro* + *cytic* means "large cell"). This condition often afflicts women who are pregnant.

Neural tube defects result from the incomplete formation of the neural tube, which develops into your brain and spinal cord. If the tube does not close properly, parts of the brain may not form, or the spinal cord is left open. *Anencephaly* is an example of failure of the tube to close at the head of the infant, leaving parts of the brain and skull unformed. It is always fatal. *Spina bifida*, where the tail end of the spinal cord does not close, can result in a variety of functional losses, some fairly benign, but others, such as paralysis, less so.

Now, folate (synthetic form called *folic acid*) is considered essential for all human diets, particularly for women who are pregnant or are planning to get pregnant. The *Recommended Daily Allowance* (*RDA*) for pregnant women is 600 μg (micrograms) and 500 μg during lactation. For all people over the age of 14, the RDA is 400 μg. Foods like leafy green vegetables, beans, and beef liver contain folate. Although you can obtain folic acid by consuming fortified foods (e.g., cereals) or taking dietary supplements, the best sources are folate-rich foods (Institute of Medicine 1998).

1.2.2. Collaboration

Although the study of life is endlessly fascinating, it is very complicated, as evidenced by the work of Dr. Lucy Wills. This complexity requires many different skills, which make it a multidisciplinary science. Like all of the sciences, biology is composed of many subdisciplines—cell biology, molecular biology, ecology, and so forth. Besides the many biological variables, there are physical factors that affect living organisms. Biological scientists therefore must acquire knowledge and experience from other fields, such as chemistry, physics, mathematics, computer science, and geology. Certainly, it is impossible to know all the areas in pursuing research in any biological specialty. Consequently, collaboration with scientists that specialize in these other areas is commonplace.

Getting the Message (GtM) 1. Let's compare Lucy Wills's work with the standard scientific method:

a. Does her work include observations and asking questions?
b. Does she formulate and reformulate hypotheses based on observations or experimental results?
c. Did her experiments use control groups?
d. What was her conclusion?
e. How did she use her results to come to her conclusion?
f. What applications did her findings have?

Quick Review Questions, indicated by QRQ, appear in the text to help you learn the material.

QRQ 1. Select all that must be done by peers concerning a researcher's research and results: make hypotheses about, praise results, present results, repeat research, review and critique results, talk to the researcher(s)

QRQ 2. (T/F) The scientific method is the most reliable approach for determining what in the natural world is factual and authentic.

QRQ 3. (T/F) In the scientific method, a hypothesis is a true statement about the results.

QRQ 4. In an experiment, an experimental group has the intervention or treatment, while the ______ group does not. Otherwise, the groups are exposed to the same conditions.

QRQ 5. Select methods of testing various hypotheses: arithmetic, computer models/simulations, conclusions, laboratory/field experiments, observations, publications, suppositions, systematic observations

QRQ 6. (T/F) Scientific investigation is a solitary endeavor.

1.2.3. Practice in Experimental Design

You are going to have a chance to design your own experiment. The organisms you will use are sea urchins (Figure 1.1), which are a type of echinoderm, related to sea stars (starfish). Sea urchins are *radially symmetrical* (symmetrical around a central axis) and have a hollow, rounded shell, from which spines and tube feet project. Both structures help in urchin movements, and the spines are somewhat protective. Their mouth is in the center, on the underside. Urchins are primarily herbivores, feeding on algae. There are male and female urchins, and their gametes (sperm and eggs) are released on the topside. In nature, they synchronize the release of their gametes, and fertilization takes place in the seawater. In the laboratory, we can inject the animals with potassium chloride, which causes them to release their gametes, which we can collect in beakers of sea water. We can maintain adults in saltwater aquaria, kept at about 10°C. After fertilization, the zygotes (single-cell, fertilized eggs) begin dividing and eventually metamorphose into larvae, four-armed, then eight-armed (Figure 1.2).

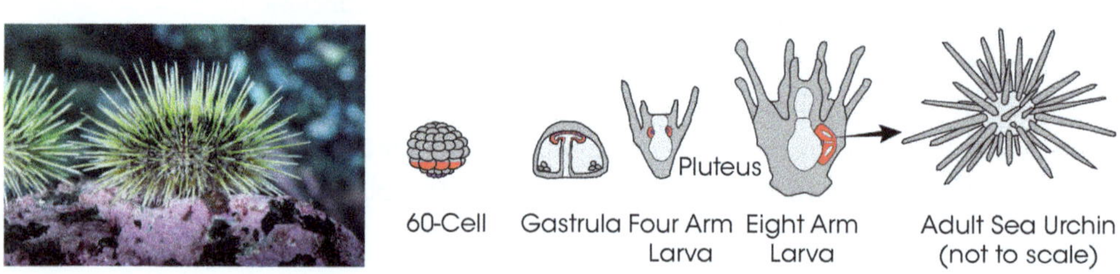

FIGURES 1.1 AND 1.2 Generalized development of larva to an adult sea urchin

Observation: With climate change, seawater temperatures are rising. Organisms that are adapted to cold water may not survive and reproduce as well with the temperature increase. Specifically, we are interested in the development and survival of the immature urchin stages.

Hypothesis: Higher seawater temperatures will reduce survival of the embryos that develop from fertilized eggs to the pluteus stage at temperatures higher than 10°C.

Experimental Design

Materials available: fertilized eggs, small containers in which to rear embryos and larvae, seawater, chambers that can be used to regulate temperatures, microscope, counting chambers to make your counts. If you think you need other equipment or materials, you can add them to this list.

You may assume that embryos develop from fertilized egg to pluteus in six weeks. Also, assume you have a way to produce a flow of fresh seawater, containing food for the embryos, through the chambers.

1. Describe your experimental setup.
2. Describe the steps of your experiment.

1.3. Misleading, Untrue, or Confusing Information

Science depends on reliable, repeatable, and authentic evidence gathered from carefully planned and conducted research. When this evidence is reported in a serious scientific journal, it has been scrutinized by other experienced and knowledgeable experts in the same field. The same cannot be said about news, social media posts, blogs, websites, entertainment venues, and so forth. Very little of that information has been verified by any reliable expert. With the flood of information that arrives on your television, phone, tablet, and computer, it can sometimes make you feel a bit overwhelmed; and much of that information is bewildering, misleading, or just not true. It is difficult to live in such an environment, and it puts the responsibility on us to try to filter out the garbage and to learn what is actually true. There is a huge quantity of *misinformation* (information that is not true, but the sender or poster did not deliberately mean to mislead or lie), but there is also information that is *deliberately* misleading, often to evoke anger, fear, and so on; to sow discord; or to bring you to the senders' side of an issue. We call this sort of false information *disinformation* (intentional misinformation, often malicious in intent). Misinformation is not new, but there is so much more of it now, and it is easily distributed and more accessible.

In an interesting study of the spreading of true and false information online, researchers analyzed more than 125,000 true and false news stories on Twitter. These stories were tweeted by about three million people more than 4.5 million times. What scientists found was that false news spread much farther, faster, deeper, and more broadly. False political news spread much faster than other evocative false news (e.g., natural disasters, terrorism, etc.). Moreover, false news stories stirred emotions like fear and disgust, whereas true news stories tended to stimulate joy, sadness, or trust.

They also concluded that users were more likely to share false news because it was more novel than true stories. Apparently, people prefer to share unusual information (Vosoughi, Roy, and Aral 2018).

There is no one way to approach "fact- or truth-finding," but, in this section, we'll explore some examples and provide pointers.

1.3.1. Autism Spectrum Disorder: Complex Problem Embroiled in Misunderstanding

Autism is not a new condition, but it appears to be an increasingly diagnosed one. Now called *autism spectrum disorder (ASD)*, the condition is considered a developmental syndrome because symptoms tend to appear before the age of two years. Although the symptoms and severity vary among individuals affected with ASD, the National Institute of Mental Health describes it as "a developmental disorder that affects communication and behavior." Despite being a developmental disorder, diagnosis may be made at an age well beyond early childhood (NIH 2018). Although individuals with ASD display a wide array of symptoms (therefore, a spectrum), the *Diagnostic and Statistical Manual of Mental Disorders* lists the following central diagnostic characteristics (APA 2013):

- persistent deficits in social communication and social interaction across multiple contexts
- restricted, repetitive patterns of behavior, interests, or activities

This disorder has no cure and occurs in all ethnic, racial, and economic groups. There are a number of therapies and behavioral mediations that may improve some symptoms. Also, there are some medications that can moderate specific accompanying symptoms (e.g., depression, anxiety) (NIH 2018).

Researchers in the field are finding a broad set of genetic influences associated with autism. Some of these genetic factors are present in the parents, to be inherited, while other factors arise from mutations that form after conception. Most of these genetic factors increase susceptibility for a person to develop ASD. A large study recently reported that researchers have now identified 102 genes that influence ASD (Satterstrom et al. 2020). In fact, there are so many genetic factors and interactions involved, it may take decades of research to sort them all out. We do know that many of the susceptibility genes code for two wide-ranging classes of proteins: those important for the development of *synapses* (junctions between nerve cells) and those important for regulating gene expression.

Scientists who study the causes of autism are aware of quite a few environmental factors that may influence the development of autism, and many studies seek to determine how such environmental factors interact with the genes. Out of hundreds of suggested environmental factors, two appear to increase risk: parental age (particularly in the father) and being male. Researchers are looking to see if there is some protective effect for females or a vulnerability effect for males. There is some thought that female ASD patients are characterized by more internalized types of symptoms (e.g., depression), whereas more outward symptoms (e.g., aggressiveness) are displayed by males. Researchers are also looking at potential environmental factors that might influence development during pregnancy (e.g., medications, infections, smoking, etc.) (Rylaarsdam and Guemez-Gamboa 2019).

According to the CDC (Centers for Disease Control and Prevention) the *prevalence* (meaning proportion of individuals in a population having a disease) of autism for children born in 1992 was about 1/150. For children born in 2006, the prevalence increased to 1/59 (Maenner et al. 2020). Some of this increase has been a result of expanding diagnostic criteria, with therefore more people fitting into this classification. It may be that doctors, therapists, and parents are recognizing and reporting the signs and symptoms more frequently as ASD. And of course, the numbers of cases may really be increasing. The question of why ASD rates are rising may eventually be answered, but certain events that took place in the 1990s diverted attention and years of research trying to respond to those events.

In 1998, the esteemed medical journal *Lancet* published a paper submitted by a British physician, Andrew Wakefield, and 12 colleagues. The authors reported finding the measles virus in the digestive system of 12 patients who exhibited symptoms associated with autism. The patients had all received the MMR vaccine (vaccine given to protect against measles, mumps, and rubella). The paper itself did not suggest that there might be a link between the MMR vaccine and autism, but in an interview after the paper's publication, Wakefield did so. He went further to propose suspension of the MMR, replacing it with a single viral vaccine. As you might imagine, these revelations and suggestions ignited panic in parents whose young children had recently been vaccinated and anger in parents whose children manifested autism. Subsequently, MMR vaccination rates fell below the rates needed to achieve *herd immunity* (resistance to spread of an infection when sufficient numbers of people are immunized [93–95 percent for measles]). In Britain the number of measles cases and measles-related complications and deaths increased dramatically.

The human body is the natural host for the measles virus, and it is the only known *reservoir* (where an infectious agent, such as a virus, resides) for measles. However, measles is an airborne disease that is typically spread through coughs or sneezes containing the virus. Some children might acquire it from contact with an infected surface and infect themselves by putting their hands to their nose, eyes, and so forth. Symptoms include high fever, cough, inflammation of eyes, and rash. Rarely, the infection may result in pneumonia or *encephalitis* (brain inflammation), and some measles victims die from the disease. The virus is very contagious, but with historically massive immunizations in children, it had virtually been eliminated from the United States and other developed countries (CDC 2019, "Measles").

Wakefield was "jumping the gun" when he suggested a link between the vaccine and a disorder like autism. A well-trained, conscientious scientist should not do something so intentionally irresponsible. What could he have been thinking? Based on this rather spurious claim, the medical and scientific community began a long, rigorous investigation of the possibility that the vaccine contributed to the development of ASD. For 12 years, no reputable, relevant study supported his claim, nor have any confirmations developed since that time. The conclusion in 2010 was and now is that there was *no causal relationship* between the MMR vaccine and autism (Godlee, Smith, and Marcovitch 2011; AAP 2018).

In 2004, Wakefield was notified by the editor of *Lancet* that he had not revealed to the journal that some of the patients for his study were recruited by lawyers seeking to sue vaccine manufacturers. Furthermore, Wakefield had received funding from this same group. Most of his coauthors withdrew from his interpretation, and in 2010 the *Lancet* retracted the article. Three months after this retraction, the General Medical Council of Great Britain banned Wakefield from practicing medicine in Britain.

Other revelations about Wakefield and his study have appeared too. First, Wakefield had applied for a patent on the single-virus vaccine he recommended to replace the MMR vaccine. Furthermore, a British journalist, examining research records and interviewing parents of Wakefield's study patients, concluded that Wakefield committed research fraud by falsifying data, as listed in Table 1.1 (Deer 2011).

TABLE 1.1. Summary of falsified data, according to Deer 2011, in Wakefield's paper from *Lancet*.

Paper's Claim	Actual Result
8/12 patients showed GI or autism-like symptoms within days after the MMR vaccination.	A maximum of two (2) patients experienced symptoms within days.
All patients were normal prior to vaccination.	2/12 had developmental delays noted prior to vaccination.
Six of the patients had a combination of regressive autism, nonspecific colitis (colon inflammation), and symptoms appearing within days of MMR vaccination.	None of the children met these criteria.

Thus, most of the subsequent confusion, anxiety, and furor has been the result of a very small study that was poorly designed and based on parental recall and used no control group. The results have been irreparable damage to public health and the misery of victims of the measles disease. Worst of all, some of these victims died. Sadly, the MMR vaccine might have averted most or all of this disaster.

Points of Inquiry, indicated by PoI and appearing throughout the text, emphasize critical thinking and more in-depth exploration.

PoI 1. Why is the measles virus especially dangerous? (Hint: Research its effects on immunity.)

1.3.2. Deluded by Misinformation—"Don't Confuse Me with the Evidence"

Although we do not know when the smallpox virus emerged, we do know it has been around for a long time. According to the World Health Organization, smallpox *pustules* (sharply raised spots, round and firm to the touch, containing pus) have been found on the head of a 3,000-year-old Egyptian mummy. In Asia, the practice of *variolation* (blowing smallpox scabs up the nose) would give the recipient a mild form of smallpox but would leave him/her with *immunity* (the ability to resist a disease or toxin by the action of specific antibodies or sensitized white blood cells) against a more virulent outbreak of the disease.

The English physician *Edward Jenner* likely never anticipated the public opposition to using his variolation (inoculation of a well individual with material from a pustule of cowpox, a disease related to smallpox) to protect people from smallpox. The cowpox virus was similar enough to that of smallpox that the patient would develop a mild disease but, upon recovery, would be immune to smallpox. The practice of variolation in Asia preceded Edward Jenner's version, but he was unaware of this earlier practice. Both versions of variolation resulted in many fewer people dying from the disease.

Smallpox is caused by the variola virus, which is passed from one person to another by direct contact or prolonged face-to-face encounters. After about a two-week incubation period, there is high fever and flu-like aching, followed by a rash, which appears initially on the tongue and in the mouth. The rash spreads over the skin, and the spots develop into pustules. Once the rash appears, the patient is very contagious, and the rash may last about three weeks. The pustules usually leave permanent scars. At the height of smallpox epidemics, about one-third of the patients died.

The World Health Organization (WHO) organized efforts in 1967 to vaccinate broadly in order to combat the disease, and gradually the disease was eliminated. The last person to die from smallpox was a medical photographer who died in 1978. Official world eradication was declared in 1980 (CDC 2017). It is safe to say that smallpox's eradication was a result of implementing aggressive vaccination policies.

The practice of variolation became more widespread in England during the 19th century. Many people did not take easily to the practice for various reasons. New ideas often encounter fear and resistance. In 1853, the Vaccination Act in England mandated vaccination for smallpox of infants and was later extended to 14 years of age. All of this legislation brought on a fury of opposition, and in 1896, the penalties were removed, and many people were excused from participating.

APPLYING SCIENCE TO LIFE 3 Independence: The Importance of Variolation

When George Washington was a young man, he acquired the dread disease of smallpox. Although he was very sick, he did survive, becoming the commander of the Continental Army and later the president of the newly formed United States of America. Having suffered smallpox, he was very concerned about the toll the disease might take on his very meager army during the Revolutionary War. As smallpox began to infect his men, he ordered isolation and other measures we would call *quarantining*. He was also aware that men who had had the disease were resistant to contracting it again. In fact, he would use these soldiers for the siege of Boston. His northern army suffered a severe outbreak, which led to their defeat by the British in Canada. These losses in battle and numbers convinced him that he must begin variolation of new recruits.

Variolation (introducing pus from a victim's sores into a small surgical incision of a healthy individual) was held with suspicion and was banned by the Continental Congress. Although variolation of new recruits was a sound move, he realized he also needed to variolate all of his soldiers who had not already had smallpox. Within a year, thousands of colonial soldiers had been variolated. Losses of men to smallpox plummeted, prompting the Continental Congress to legalize variolation. Moreover, variolation allowed Washington and his army to concentrate on defeating the British with a much weaker force. So in many ways, independence of the colonies from the British and the subsequent establishment of the United States were made possible by variolation.

Misinformation—Pneumonia, Brain Damage, Death

As you can see, resistance to vaccinations is not new, and it continues with new energy today. Let us consider the opposition to the *MMR vaccine*. Despite the disgrace of Andrew Wakefield, some people cling to the notion that the MMR vaccine is the cause of autism. After all these years, wherein scientists have labored to determine if there is any relationship between MMR and autism, they have *not* found a shred of any real evidence that the vaccine and the disorder are related.

Vaccine activists have also targeted vaccine components. One such component was *thimerosal*, which was a mercury-containing preservative used in very small amounts since the 1930s, primarily in multidose vials of certain vaccines, to prevent the overgrowth of bacteria or fungi in the vaccine. The chemical form of mercury released by metabolism of thimerosal is not the same form as that of the mercury obtained by consuming certain kinds of seafood and is cleared from the body faster. In the United States, *no* childhood vaccines contain thimerosal. MMR, inactivated polio, and chickenpox have *never* contained thimerosal. In 2001, thimerosal was removed from other vaccines given to children, and autism rates did not drop in response but rather continued to rise. Thimerosal is currently only used in multiuse vials of influenza vaccines (CDC 2015). Currently, there is *no known causative relationship* between thimerosal and autism or any other neurological disorder.

Because vaccines recommended for children increased after the 1980s, critics of vaccinations proposed another hypothesis: The increased exposure to vaccines was causing autism. *No evidence* to support this hypothesis has been found.

In some vaccines, an aluminum salt *adjuvant* (additive that promotes a stronger immune response to *antigens*, or foreign substances) is added. Vaccine critics proposed that it was this aluminum additive that causes autism. First of all, the amount of aluminum is much smaller than that from other sources, including breast milk or infant formulas. Furthermore, *no study to date implicates* these vaccine adjuvants in autism or any other childhood disease (AAP 2019).

No parent wants to have an autistic child. Although there is now more therapeutic support for this disorder and scientists are gaining new understanding of its causes, autism spectrum disorder (ASD) certainly affects children and their families in deleterious ways. There are many well-meaning people who are confusing *correlation* (variables that change together) with a *causal relationship* (in which change in the value of one variable changes the other variable). Being a celebrity does not make you an expert on autism or any other type of medical disorder. And for any responsible celebrity, politician, or billionaire to broadcast information that has no basis in reality is misusing their position. A child having a vaccination and manifesting autism does not mean that the vaccination caused the disorder. The two variables may not even show any correlation, unless they always vary together. By analogy, the child might have eaten a new cereal for breakfast, but it doesn't mean the cereal is the cause of her developing a cold a few days later.

APPLYING SCIENCE TO LIFE 4 Do Storks Bring Human Babies? What Numbers Do and Don't Say.

After World War II (WWII) ended in Europe, for years cities like Copenhagen, Denmark, saw an increase in the numbers of immigrants from rural areas. The number of people in Copenhagen increased, as did the number of births. With greater population, there was also more construction (resulting in more chimneys). European white storks had long taken advantage of chimneys, towers, and roofs in Copenhagen for their nests. The Danes thought the storks were harbingers of good fortune and actively encouraged these spring arrivals. An increase in human babies came simultaneously with an increase in storks and their babies. So if you graph the number of human births, the number of human adults, and the number of storks for the 12 years following WWII, all populations increased. Would you say that the increase in human births in Copenhagen was better related to the increase of human adults or the increasing number of storks? Human births could be correlated positively with both. However, would you conclude that the increase in numbers of adult human beings or the increase in number of storks *caused* the increase in human babies? Although both human adult numbers and stork numbers are *correlated* with increased human births, the increase in storks is clearly not the *cause* of the increase.

GtM 2. After WWII, did the increasing number of storks cause the increase in human births?

GtM 3. After WWII, did the increasing number of adult humans cause the increase in the number of storks? Why or why not?

Vaccines have been one of the most important and effective developments from biomedical research. In Figure 1.3, note the extraordinary effects of administering the MMR vaccine on the incidence of measles in the United States between 1938 and 2019. Vaccines thwart various illnesses and deaths, and the relationships between the vaccine and the disease prevented are more than correlations. These results have *causal* relationships.

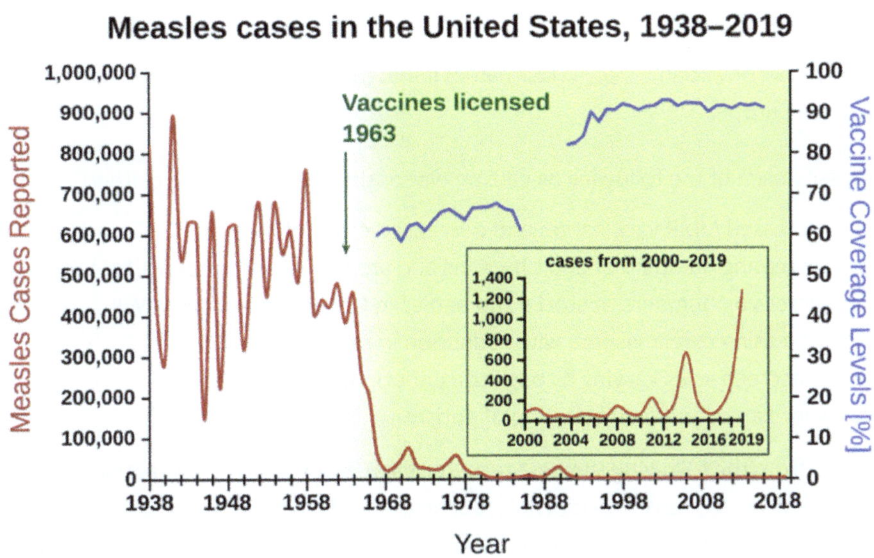

FIGURE 1.3 Graph showing the impact of the MMR vaccine on the incidence of measles in the United States. The insert shows recent increases in measles cases in 2000-2019.

PoI 2. How can the increased incidence of measles in the United States in 2019 be explained?

PoI 3. In 2015, in California there was an outbreak of measles (159 cases), which apparently started from an infected foreign teenager's visit to Disneyland. Why was this single person able to set off this outbreak in California?

PoI 4. How might this individual also have started an outbreak in Quebec, Canada?

PoI 5. What was California Senate Bill 277? What impact did it have on the incidence of measles in California?

PoI 6. A child in England who had just turned one year old acquired measles before she was old enough to be vaccinated. She was very ill but managed to survive. How might she have acquired this disease? How could this near tragedy have been prevented?

PoI 7. For what reasons are people resistant to vaccinations?

PoI 8. What can public health workers do to encourage vaccination?

QRQ 7. (T/F) Generally, false news spreads much farther, faster, deeper, and more broadly than true news.

QRQ 8. Select factors that may influence the development of autism: adjuvants; aluminum additive in vaccines; being male; bowel disease; genetic factors present in the parents; increased exposure to vaccines; MMR vaccine; mutations after conception; parental age (particularly, father); single-virus vaccine; smallpox vaccination; thimerosal

QRQ 9. The English physician ______ (last name) initiated the use of variolation to protect people from smallpox in Europe.

QRQ 10. Classify each of the following as causal relationship, correlation, or neither:

a. increased MMR vaccination and decreased cases of measles
b. increasing numbers of adult humans and storks in Copenhagen after WWII
c. increasing numbers of storks and babies in Copenhagen after WWII
d. increasing resort visitors with reduction in rental fees
e. sunscreen sales varying as beer consumption varies
f. vaccination and manifestation of autism

QRQ 11. Select from the following the best term for each description below: antibody, antigen, autism spectrum disorder, causal relationship, correlation, disinformation, herd immunity, immunity, incidence, misinformation, pool, prevalence, relationship, reservoir, vaccination, variolation

a. ability to resist disease or toxin by action of specific antibodies or sensitized white blood cells
b. blood protein generated when immune system is stimulated by a foreign substance or microbe
c. change in the value of one variable changes the other variable
d. developmental disorder affecting communication and behavior
e. foreign substance that stimulates an immune response
f. false information, but the sender/poster did not deliberately mean to mislead or lie
g. inoculation of well individual with material from cowpox pustule to induce immunity
h. intentional misinformation, often malicious in intent
i. proportion of individuals in a population having a disease
j. resistance to spread of an infection when enough people are immunized
k. variables change together
l. where an infectious agent like a virus resides

1.3.3. Nutrition and Dietary Information: Sorting Out the Confusion

APPLYING SCIENCE TO LIFE 5 Nutritional Recommendations: It Depends

In 2014, a *New York Times* writer proclaimed "Butter Is Back" in an opinion piece that was obviously the writer's opinion. It was catalyzed by a review article published earlier that claimed that from a cluster of studies they could find no positive effects of high polyunsaturated fatty acids and low saturated fatty acids on cardiovascular health (Chowdhury et al. 2014). In 2017, another group of researchers wrote a very interesting, but controversial, article claiming that saturated fat was not the villain in cardiovascular disease, but that it was inflammation, which could be reduced by healthier lifestyle choices (staying active, losing weight, stress reduction) (Malhotra, Redburg, and Meier 2017).

Reaction to these claims was swift and firm, citing omissions of research that did not conform to their findings. A presidential advisory, published by the American Heart Association, counters both these sets of claims that undermined recommendations about the value of reducing saturated fats (animal products) and replacing them with polyunsaturated oils (plant products). The advisory cited randomized, controlled trials where a dietary change in fat and oil intake reduced *cardiovascular disease* (*CVD*) by about 30 percent, roughly the same as level of reduction achieved using statin medications. Other observational studies supported this claim in reducing CVD and other causes of death. Furthermore, clinical trials replacing saturated fats with carbohydrates (mostly refined sugars) did not accomplish this reduction. The reduction in CVD incidence from replacing saturated with unsaturated fats is accomplished by reducing levels of *low-density lipoproteins* (*LDLs*) (a lipid-transport protein that transports cholesterol; LDL cholesterol has a role in formation of arterial plaque, which leads to CVD), a major factor associated with atherosclerosis (Sacks et al. 2017).

Dietary and nutritional information is almost always a bit befuddling. As you can see from the above story, there are differences in opinion among the researchers, although most researchers, public health specialists, and nutrition experts still back the current dietary guidelines for kinds of fats and oils that should be consumed. The reason for these differences exposes one of the *strengths* of science. For the most part, scientists are skeptics. They will not endorse any hypothesis until they are convinced by the evidence. In this case, the majority of support still holds with the hypothesis that lowering saturated fat intake and replacing saturated fat with unsaturated fat sources lowers the risk of CVD. For many science problems, there is rarely universal acceptance of a hypothesis, there is only consensus. Consensus in the scientific community is extraordinarily important and compelling. Only if this hypothesis is demonstrated to be false will it be discarded.

Generally, scientists test hypotheses exhaustively, modifying their research to fit better previous results and to understand the object of their research. Human genetics, physiology, and nutrition are very complex to unravel, especially with the myriad ways in which these components interact. Although estimates still vary, there are somewhere between 20,000 and 25,000 protein-coding genes, 250 different cell types, and 80,000 to 400,000 proteins in the human body. Proteins in your cells are being recycled, replaced, and modified at differing rates. Everyone has many variants for the proteins in their cells. So even though all obese people react by producing inflammation-related proteins, each person put on the same diet has differing rates of decline in these proteins and respond differently to the same diet (Schroder 2017). It is little wonder that there is so much variation in results from studies of causes for CVD or any other pathology that is influenced by lifestyle.

Because government funding for nutritional and dietary studies varies and is often reduced, scientists need to find other funding sources for their research. Unfortunately, the result is that much of it is funded by the food and beverage industry. Accepting industry money for dietary research is so commonplace, most nutrition investigators accept it as the norm. Probably every one of them entered such partnerships believing that they could not be influenced by the support they receive. Dr. Marion Nestle, a renowned nutrition scientist, was interviewed by Bonnie Lieberman of the Center for Science in the Public Interest in 2017 about the influence of industry on nutrition science. Dr. Nestle likened the influence of the food industry to that of the tobacco industry on research into the health effects of tobacco or of the pharmaceutical industry's influence on medical research and prescribing practices. For one year, she gathered a collection of 168 studies that were funded by industry. Almost 93 percent of these studies gave favorable interpretations to their sponsors (Nestle 2017). Researchers in Spain found that studies of links between sugar-sweetened beverages and weight gain that were sponsored by beverage interests were five times more likely to find no link between consumption of these beverages and obesity than studies lacking beverage company support (Bes-Rastrollo et al. 2013). Nestle's conclusion is that industry's funding of nutrition research is a way to generate doubt in the science and promote the food's benefits (Dow 2017).

If you combine the food industry's influence in research and add to that the misinformation of websites, social media, television, and so on, it is no wonder we are often baffled by what to do. Also, human beings have an unfortunate tendency to believe or like recommendations that correspond to what they are doing or are likely to do.

Don't be discouraged by the confusion, but persist in finding out what the best, current science is. The science is complex and will evolve as we understand more.

APPLYING SCIENCE TO LIFE 6 Better Living from an Analytical Chemist

Dr. Harvey Wiley was not an easily discouraged man. Born on a rural farm in Indiana in 1844, he received a medical degree from Indiana Medical College and a science degree from Harvard University. Dr. Wiley taught for a number of years, becoming a professor of chemistry at Purdue University. In 1883, Wiley took a position as chief of chemistry at the US Department of Agriculture in Washington, DC, where he remained until 1912.

Wiley's passion was analysis of foods and "patent medicines," wherein he documented the chemical composition. *Patent medicines* are products that were marketed as cures for various ailments without regard for their effectiveness. Wiley was more interested in the adulteration of foods, which was quite a common practice for food manufacturers. Such *adulteration* often was accomplished by adding or substituting some ingredient that would often lower the cost of manufacturing or preserve the food. These adulterants were usually not acknowledged by the seller. Almost always the additives lowered the quality and may have been toxic. Wiley discovered borax, salicylic acid, and copper sulfate in canned foods, chalk and formaldehyde in milk, corn syrup in honey, and so on. Wiley thought this practice was not only fraudulent to consumers but also quite dangerous.

In 1902, Wiley began experimenting on healthy young male volunteers by administering various chemical additives, one chemical for each experiment, to meals he had prepared for them. Some of the food additives made the men ill. When he tested formaldehyde, all the men got very sick quickly, and Wiley terminated the experiments. The men in his tests began to be known as the "Poison Squad."

For years Wiley tried to persuade politicians in Washington to establish some standards for food manufacturers to meet. Then, as now, many politicians were greatly influenced by paid lobbyists and wealthy owners of commercial enterprises. Any laws that were passed were often severely weakened, with no means for enforcement. However, in 1906, enough evidence had piled up and women's groups had been enlisted by Wiley to help influence lawmakers. *The Pure Food and Drug Act* was passed and was initially overseen by Wiley. Enforcement of the act was difficult and unpopular with influential men, and political opposition became strong enough to force Wiley to leave his post in 1912.

Undaunted, Wiley went to work for *Good Housekeeping Magazine*, where he could continue to push for pure foods and government regulation. The magazine had its own research facilities, and Wiley had his own chemistry laboratory there. For 18 years, he directed the Bureau for Foods, Sanitation, and Health for *Good Housekeeping*, advocating for such things as more rigorous inspection of meat production and exposure of peddlers of fraudulent or misleading diet products. His editorials forced the government to fund better infant care, and in 1927, he published his suspicion that tobacco products were unhealthy and might be linked to cancer. As increasing evidence accumulated, *Good Housekeeping* refused to accept cigarette ads. *The Good Housekeeping Seal* must be earned, and it still has clout.

Relentlessly, Wiley spent virtually all of his professional life promoting disclosure of fraudulent food products and regulation of food manufacturing, both of which would be in the best interest of consumers. We all owe Dr. Wiley our gratitude. Although there are still problems and fierce resistance to regulation of food and beverages, our food supply is healthier for us all. He was buried with honor in Arlington Cemetery in 1930 (PBS 2020).

PoI 9. What effects does borax have on the human body?

PoI 10. What effects does formaldehyde have on the human body?

APPLYING SCIENCE TO LIFE 7 What to Believe

For several months, Arthur, a retired librarian, had often felt very fatigued, had numbness in his hands, and periodically felt light-headed. Then, while watching *The Doctor Oz Show*, he heard that high doses of vitamin B12 would amplify his energy levels. Dr. Oz seemed to think it was a miracle cure, so Arthur went to a health food store and bought some.

After a short time, he did feel like his old energy level was back. But then he read an article published in *The Atlantic*, a highly respected news source, where the author discussed the risks of taking vitamin B6 and B12. First, the article humorously ridiculed *The Doctor Oz Show* and the notion of taking B12 as an energizer. He checked the NIH website, which states that there is no evidence that vitamin B12 increases energy or athletic performance. Although the vitamin might help you regain normal energy levels, from all his reading, he knew now that myriad people accept such a nonevidential notion because B12 deficiency (fairly common in older people, vegetarians, and vegans) causes *anemia* (a condition caused by low production or high destruction of red blood cells), among many other symptoms. With an insufficient number of red blood cells to carry oxygen to the tissues, it is little wonder you feel tired, weak, and so on. B12 is normally absorbed from foods (e.g., eggs, poultry, dairy, meat, fortified grains). If diagnosed with B12 deficiency, doctors normally prescribe supplements or even injections.

Arthur thought the information on the television program was a bit overblown and perhaps misleading. What really bothered him is that *The Atlantic* article went on to say that scientists had found evidence that taking high dosages of vitamin B12 (and folate, or vitamin B9) was associated with increased risk of cancer (primarily lung cancer in men who smoked) and mortality (Ebbing et al. 2009). Alarmed, Arthur started looking into this claim, and there were some studies that supported the idea that high doses of these B vitamins, given long-term, may be linked to colorectal cancer (Araghi et al. 2019; Brasky et al. 2017). However, researchers are still uncertain whether both or which of the vitamins might be linked to causation (NIH 2020).

PoI 11. What do you think Arthur should do, given this somewhat conflicting information?

PoI 12. Arthur utilized three sources of health information. Rank the three sources for reliability and indicate how you justify your ranking.

Answer the following questions through research on the internet, but use only sources that end in .org. Avoid .com websites, because they often are attempting to sell a product.

PoI 13. What agencies regulate food and drug safety in the United States? Dietary supplements?

PoI 14. How is the regulation of drugs different from the regulation of dietary supplements?

PoI 15. Why is there a regulatory difference for drugs versus dietary supplements?

Dietary Supplements—Irrational Exuberance for Often Untested Dietary Ingredients

Dietary supplement sales have burgeoned over the past 20 years. In late 1990s, the value of supplements used in the United States was about $12 billion, and the estimated value for 2020 is $180 billion (Binns, Lee, and Lee 2018). So dietary supplements are big business, and their manufacturers and distributors can afford to hire the best lobbyists. Consequently, dietary supplement companies have great clout within the US government.

The FDA, under the DSHEA (Dietary Supplement Health and Education Act 1994), recognizes components of these supplements as "dietary ingredients." The ingredient classifications included are the following: vitamin, mineral, herb or other botanical, amino acid, dietary substance (e.g., enzyme, organ/gland tissue) supplementing dietary intake, concentrate, metabolite, constituent, or extract.

Perusing the shelves in supermarkets and health food stores, you will find an astounding array of supplement products. Even multiple vitamin products have become specialized for age, gender, level of physical activity, and so on. Although there may be only a few hundred on the shelves, thousands are available. The proliferation of products seems to result from marketing, consumer demand, tantalizing profits, lack of strict regulation, and no enforced quality control. And worst of all, these ingredients have *not* been subjected to testing that might clarify their efficacy and safety.

Public health officials and researchers are very concerned about the potential risks that surround the current supplement culture. Each individual's distinctive physiological and genetic makeup mixed with untested and unregulated supplements makes responses to ingredients highly unpredictable. Also, supplements can have harmful interactions with the medicines people are taking. Some of the risks supplement takers face are the following:

PoI 16. What are people's motivations for this increased supplement use?

- **Contamination or hidden ingredients.** For example, certain weight-loss or obesity treatment products used to contain *sibutramine*, supposedly an appetite suppressant. It was originally the active ingredient in the drug *Meridia*, but the drug maker removed *Meridia* from the market because clinical trials indicated that the chemical increased the risk of heart attacks and strokes (US Food and Drug Administration 2019).
- **Quality concerns.** Studies of fortified foods and food supplements found that vitamin D varied between 8 percent and 177 percent of the amount of the labeled value (Verkaik-Kloosterman, Seves, and Ocké 2017).
- **Lack of efficacy.** Large trials of the effect of vitamin E and selenium in reducing the risk of developing prostate cancer refuted health benefit claims for either or a combination of the two (Lippman et al. 2009).

As you can see, the use of dietary supplements is fraught with peril. Untested and unregulated, many of these products are at best a waste of money and do no damage, or, at worst, they can do physical harm to the user. Current policy set forth in the 1994 DSHEA limits what the FDA can do. Highlights (or "lowlights") of the policy are as follows:

- Manufacturers are not required to demonstrate efficacy or safety.
- It appears the law emphasized the importance the supplement industry played in the economy over the safety, effectiveness, and personal economy of American consumers.
- The FDA is saddled with the responsibility of proving a supplement is unsafe before they can withdraw the product from the market (Binns et al. 2018).

If, as desired by some in the government, the budget of the Office of Dietary Supplements (ODS) at the FDA is cut, the ability of the ODS to oversee public health concerning supplements will be further diminished. Currently, we are relying on the manufacturers to make sure a dietary supplement is safe before marketing it and to make sure that the information on the label is truthful and not misleading. No proof is required that an ingredient does what is touted in the marketing (Binns et al. 2018). So we might question the wisdom of reducing regulation and quality assurance when supplement variety and the demand for supplements keeps rising.

APPLYING SCIENCE TO LIFE 8 Identifying Misinformation and Misleading Claims

There is so much misinformation on websites, newspapers, and social media, it can be very confusing. To help you recognize a misleading source, the Health and Nutrition Letter of Tufts University (below) provides us with a list of cues to misinformation about nutrition. This list is similar to others from a number of highly reputable nutrition-related organizations.

Ten Red Flags of Misinformation and Misleading Nutrition Claims (Flaherty 2015)

1. Recommendations promise a "quick fix."
   - Many are associated with weight-loss supplements.
   - These are generally advertised in tabloids or late-night infomercials.
   - They may refer to a conspiracy meant to keep you from finding out about the product.
2. Focus on a danger from a product or regimen.
   - "Don't eat this or that kind of food."
   - For example, fruit vilified because it contains sugar; however, the scare tactics neglect that fruit also contains vitamins, micronutrients, and so on.
3. Claims that sound too good to be true.
   - For example, "eat as much of this food as you like—it has no unhealthy sugars, and it will help you lose weight!"
4. Simplistic conclusions based on a complex study.
   - Mice fed enormous amounts of any ingredient might be more likely to gain weight and develop cancer.
5. Recommendations made from only one study.
   - For example, the paper that claimed saturated fats did not promote CVD.
   - See if other researchers back up the claim.
6. Statements that are refuted by reputable scientific organizations.
   - Many organizations, such as the National Institutes of Health, American Heart Association, and so on, provide science-based information; questions can be answered from their websites.
7. "Good" and "bad" food lists.
   - It is true that we should try to eat foods that are unprocessed.
   - It is also OK to eat an occasional brownie or cookie.
8. Recommendations that are intended to sell a product.
   - These usually will be for a supplement or diet book.
   - Some nutrition research is now funded by the food and beverage industry, which wants to persuade you to buy more of its products; check other studies not funded by industry.
9. Recommendations that are based on studies that are not peer-reviewed.
   - Usually you can check the source for indications of peer review.
10. Recommendations from studies that ignore differences among individuals or groups.
    - Some studies include just adult males, so such a study might not be as applicable to women and children.

QRQ 12. (T/F) By law in the United States, dietary supplements are regulated and required to be tested for efficacy and safety.

QRQ 13. Select the following that are risks of dietary supplements: contamination, different ingredient amount than labeled, hidden ingredients, improper refrigeration, might not do what label says, no country of origin, no prescription needed, not age-specific

1.3.4. Purveyors of Doubt and Confusion

Have you ever heard of the "tobacco strategy"? Well, if you replace "tobacco" with "ozone hole," "acid rain," "secondhand smoke," or "climate change," the opposition to each has used a strategy modeled on the "tobacco strategy"—deny scientific evidence by *casting uncertainty* and *doubt* about that evidence. The interesting thing about the success of this strategy is that essentially the same group of scientists employed this argument in each case. Moreover, most of these were scientists whose main scientific achievements came during the time of the Soviet Union and the early development of rockets. By the time these men became involved in the issues above, most of their contributions to science were part of the past. More importantly, these scientists were *not* experts in the areas they were seeking to discredit. Though well-known physicists, there was not one climate scientist, biomedical scientist, or environmental scientist in the bunch. Would you want an astrophysicist to perform surgery on your brain?

For each of the above issues, the science is solid. If you can get 90+ percent of scientists to agree that the evidence clearly supports a claim or hypothesis in question, it shows that the consensus is very compelling. For instance, consider the issue of climate change. Scientists warned then President Lyndon Johnson in the mid-1960s that they thought the climate was warming. At that time, even scientists believed there was time to make adjustments that would stop the warming, but there was not yet a consensus about global warming.

However, in the 1980s, the changes in the climate were moving at such a fast pace, many scientists became more alarmed. In 1989, various governments formed the *International Panel on Climate Change (IPCC)*, and this body began a comprehensive study of the accumulating scientific evidence. In their second report, issued in 1995, the panel indicated that human impact on climate change was apparent. Now, as this story unfolds, pay attention to the strategies used by people who deny human involvement in the changing climate.

The response that these findings ignited harkened to the days when the tobacco companies and their confederates undermined the growing evidence supporting the link between smoking and various health risks, including lung cancer. Some of the same players (mostly physicists) that delayed acknowledgment of the links of tobacco to health issues were up to their old tricks. Their mission was to undermine the findings of the IPCC, agreed to by hundreds of climate scientists and official representatives from many countries.

Two of the physicists (Frederick Seitz and Fred Singer) had been heavily involved in the tobacco-cancer fight. They worked with and for the tobacco companies to cultivate *uncertainty* and *doubt* about the data linking smoking and cancer. At that time, both of these men were very concerned about the threat of the Soviet Union and worried about any type of governmental regulation. They feared new regulations would be put on the tobacco industry if the evidence regarding smoking and

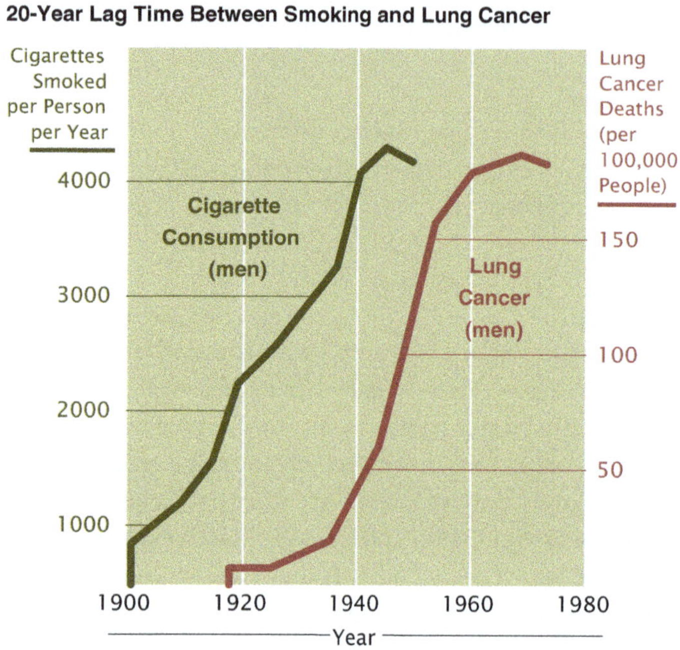

FIGURE 1.4 Correlation between smoking and lung cancer deaths 20 years later. Carcinogens act through a multistep process that produce mutations, promoting cancer.

lung cancer were released. Figure 1.4 shows graphs that compare cigarette smoking habits among men from the 1900s (most cigarette smokers were men), showing the correlation of smoking with deaths per 100,000 from lung cancer after a 20-year lag. The correlation became causation with more research. This data was known by the tobacco industry well before the tobacco settlement (1998) but withheld from the public until much later.

Following the thinking of those within the tobacco defense team, people should have a right to smoke (without government regulation) whether or not it makes them sick and/or kills them. Amazingly, these scientists were so worried about regulations they seemed oblivious to the health effects of smoking—chronic disease and early death. They also seemed unconcerned about the costs to US taxpayers, who pay about 60 percent of the health care costs of smokers through publicly funded programs like Medicare. In 2014, direct medical care costs for smoking-related disease in the United States were estimated to be in excess of $170 billion. Including health effects from secondhand smoke, just about an equal amount comes from lost productivity (CDC 2019, "Fast Facts").

Returning to the climate change issue, a small group of scientists, supported by antiregulation think tanks, the fossil fuel industry, and other private corporations, relentlessly attacked the findings of the IPCC. They savaged the American scientist who, in the 1995 report from the IPCC, had authored chapter 8, "Detection of Climate Change and Attribution of Causes." These men attacked this eminent atmospheric scientist through the press and also wrote members of Congress, Department of Energy officials, and editors of scientific journals, trying to get him fired from his job at a national laboratory. And what had this scientist done? Accused by this group of physicists of altering the data, he had only revised the report's chapter based on peer review by other climate scientists. Under close scrutiny by his peers, the accusation of his altering data was proven to be impossible. Fortunately, he did keep his job, but he certainly suffered personally.

Moreover, the purveyors of misleading accusations and information had great clout in Washington. Falsely presenting themselves as climate authorities and with their well-known names, these men had a strong influence on politicians and the media. However, these men presented no evidence with their allegations. Most disappointing of all, their deceptive and disingenuous story is still being promulgated, expanding to various media outlets. The effort has paid off, at least for the fossil fuel industry. Many Americans have been confused by their false claims, leading to a disparity of public opinion. However, with an extraordinary strength of the consensus ($\geq$ 99 percent), there are *not* two sides to the issue from the perspective of climate scientists—warming of the earth is a reality, and the rapid warming is primarily a result of the emissions of greenhouse gases, derived from the burning of fossil fuels and the loss of forested areas and replaced by agricultural activities (Oreskes 2004). Alarmingly, the efforts begun by some of the same characters involved in promoting doubt about the effects of smoking on health are now being employed to delay action on climate change. In the case of the organized opposition to revealing tobacco's dangers, they had an internal slogan, "Doubt Is Our Product." It still is.

Some of the tactics initially used by climate deniers include falsely claiming the following:

1. **There is no consensus.** In the media, they get a few scientists (most are not climate scientists) who just don't accept global warming, period, and others that are not convinced human impact is responsible to issue statements or answer interview questions. All of this public debate signals to the public that climate scientists widely disagree. They do not! However, in 2015, an analysis of the peer-reviewed literature on anthropogenic global warming (AGW), published in 2013 and 2014, found that out of 69,406 scientists, only 4 rejected the idea of AGW. The consensus was 99.99 percent (Powell 2015). It could hardly have been stronger.
2. **The science is not settled.** Researchers are still skeptical of some aspects of climate research (such as warming causing extreme weather events). However, the science on human impact *is* settled. Of course, research continues for many issues related to climate change and related questions, which is how science operates.
3. **The computer models are not reliable.** Deniers confuse weather and climate. NASA defines *weather* as "atmospheric conditions that occur locally *over short periods* of time—from minutes to hours or days" and defines *climate* as the "*long-term* regional or even global average of temperature, humidity and rainfall patterns over seasons, years or decades" (NASA 2017). Although no model is perfect, the multitude of models used are compared, tested, and analyzed. Scientists compare their outcomes with historic data, including huge catastrophic events (e.g., large volcanic eruptions). Each model is built and calibrated independently. All models predict significant warming by increasing the amount of CO_2 added to the atmosphere. The predicted warming, with ever more complex models, has remained consistent over the past 30 years. Every model concludes that 100 percent of the earth's warming for the past 150 years is a result of human activity.
4. **We have heard premature claims from scientists before.** During the 1970s, some scientists warned that there might be global cooling, but they could not support their hypotheses with data, and this claim disappeared. Again, this process in science is normal.
5. **Climate change is a natural phenomenon.** It is natural for the earth's climate to change, but not in the relatively short time since the Industrial Revolution (last ±150 years). If you look at graphs from NASA scientists based on data from the 1880s until now, you will see that the surface temperature of the earth has increased incredibly quickly. A change of one or two degrees may not sound like much, but in such a short time, it is an extraordinary increase, and the increase continues. Every degree of warming on a global scale has outsized impacts as ice melt increases and sea level rises. With a continued increase at this rate, coastal areas will be flooded, and islands will disappear from the sea surface. Deniers like to use stories of the Little Ice Age (1300–1850) and the Medieval Warming (900–1300) to support the idea

that global climate change is a natural process. No one denies that these periods occurred, but they were *regional* events, affecting primarily northwestern Europe and eastern North America, not *global*. In roughly the past 2,000 years, the last 150 years is the only time the *global* climate has changed (NASA 2020).

6. **Sunspot activities or cosmic rays are heating the earth.** Sunspots (solar storms) may be associated with solar flaring. However, no recent evidence exists of an upward trend in the amount of the sun's energy reaching the earth's surface. Therefore, sunspots are not the culprit. Cosmic rays (CRs) are high-energy radiation that comes from outside of our solar system. There is a claim that cosmic rays seed the formation of clouds. If true, a lowering of the number of CRs hitting the earth would mean fewer clouds and increased warming by solar radiation. There are two problems with such reasoning—CRs are not efficient at generating clouds, and their levels have actually increased. So cosmic rays are not responsible for global warming (Brown 2013; NASA 2019).
7. **Carbon dioxide is a very small percentage of atmospheric gases, so it could not be responsible for the recent warming.** We have known for about the past 160 years that carbon dioxide traps heat (greenhouse effect). Nitrogen gas, which makes up almost 80 percent of the atmospheric gases, is essentially nonreactive. Other gases, produced by human activity, can surely contribute to the trapping of heat, but they tend to be very reactive and short-lived. The impact, not the proportion, of a gas in the atmosphere is the important factor in the trapping of heat.
8. **Scientists are manipulating the data.** With hundreds of the world's climate scientists studying climate change, such data manipulation would require a gigantic conspiracy. Peer review, as part of the scientific process, assures quality control. Scientists do make corrections or adjustments for historical data, because over time, instruments and techniques change and improve, and to account for these differences, scientists adjust some data slightly. If they did not, the differences would be even more stark and more worrying. In any event, what would the scientists' motives be? Getting another grant? On the other hand, who would benefit most from the continued effort to discredit scientists and their work? Fossil fuel companies and investors, perhaps?

There is no scientific support for *any* of the denial stories promulgated through media, social networks, and certain politicians.

QRQ 14. Select which of the following are similarities between the "tobacco strategy," denying a causal relationship between smoking and cancer, and the "climate change strategy," denying human impact on climate change: cast uncertainty and doubt about the evidence, maintain claims are premature, many of the same scientists (not experts in the field) deny the research, peer-reviewed data, proclaim no consensus among scientists, review articles, say computer models are not reliable, consult scientists that smoke, say science is not settled, support from industries that can benefit, town hall meetings

QRQ 15. Complete the following statement using a number to the nearest multiple of 10 (i.e., 10, 20, 30, etc.): Almost ______ percent of climate scientists today agree that warming of the earth is a reality and that the warming is primarily a result of the emissions of greenhouse gases.

QRQ 16. NASA defines ______ as "atmospheric conditions that occur locally over short periods of time—from minutes to hours or days" and defines ______ as the "long-term regional or even global average of temperature, humidity and rainfall patterns over seasons, years or decades."

1.4. Navigating in a Rich but Fallacious World of Information

It is disconcerting to live in a world of overwhelming amounts of information, knowing much of it is false or misleading. Some people term our times "post-truth," which is quite disheartening. How are we supposed to know what is true? Unfortunately, there are few shortcuts to help us identify reliable sources of information.

Most of us now access information primarily through the internet or social media, which are generally quite *unreliable*. A significant portion of information on social media seems to be advertising, gossip, spreading of rumors or propaganda, or bullying. Similarly, websites *may* contain useful and trustworthy information, but it is often very difficult to discern what is real and what is not. People on both social media and the internet have become very skilled at disguising their real purpose, making it very difficult to determine what is true and what is not.

Most of what we know, we learn from others, whether we read it or hear it from a teacher, parent, or other trustworthy source of information. Of course, we learn things through our own experiences too. You may not know how to fix a leak in your toilet, and you may not want to know. However, if you have a leaky toilet, you can call a plumber or a friend who knows how to repair toilets. On the other hand, if you learn for yourself, you can then repair a leaky toilet you encounter in the future. If you choose the last option, you could learn from your friend, you might find a video on the internet, or you could obtain a book that will instruct you. You were not born knowing how to fix a leaky toilet, and you have to learn from a source that can impart the knowledge gathered from experience.

1.4.1. Searching for the Elusive: Reliable Sources

We can learn from experts, who may have advanced degrees and who have extensive knowledge or skills in a particular area. In the process of sharing information, scientific experts typically describe central observations and questions, reveal alternative hypotheses, outline the experimental process utilized, and summarize the results. Then they describe findings and conclusions based on the evidence (experimental results). Wise decisions, proper diagnoses, or successful strategies in other fields (e.g., business, medical, financial, military) are also made methodically, based on evidence. Sometimes experts express opinions in interviews or editorials, and while the opinions are often interesting, they may contain considerable amounts of speculation and may not be on topics in which their expertise lies.

Here are prospective sources of information available to you, which you can rank for trustworthiness:

Articles from peer-reviewed journals

- Articles are reviewed by experts in the field; some are more rigorously reviewed; some have more impact on the scientific community.
- There are open-access journals that are good, along with some that aren't.

Talks, lectures, interviews, or articles authored by experts in the field of interest

- As long as the expert is speaking or writing within his or her area of specialization and has actually authored the material, the information is likely reliable.
- Opinions expressed by experts may be speculative, particularly if not within their area of expertise.

Research librarians at colleges, universities, and public libraries

- Generally, research librarians are very helpful, often underutilized resources.

Books by experts

- Probably not as reliable or as up-to-date as recent journal articles (peer-reviewed).
- Still, if a book is within an expert's area of expertise, the material may be quite reliable.

Newspapers

- Some newspapers (*New York Times, Washington Post, Wall Street Journal*) have a strong record of reliability.
- Although they do not always get stories exactly right, they usually try to get outside verification before they print.
- When in error, they print retractions.
- Some articles refer to a professional journal article; use to verify.
- Avoid using opinion pieces; concentrate on reporting that is verified.
- Some other newspapers print information that is often sensational and unverified; these rarely print retractions or acknowledge inaccuracies.

Magazines

- Vary in quality and reliability.
- Some magazines have excellent writing and may contain reliable information.
- Some magazines have articles that are biased; they still may contain reliable information but be wary.
- Be prepared to verify.

Web resources

- Faster publication but often *without* verification.
- Many websites are used for *propaganda* (information that is biased, misleading and meant to persuade or promote a particular point of view or political position); may be very subtle, stealthy. Be very wary and skeptical.
- Some websites are used to sell you a product, entertain you, or document an event.
- Be aware of the domain: .com and .net are not restricted like other domains, and the information is not as reliable.
- The domains .edu (educational institution), .gov (government), .org (nonprofit), and .mil (military) are more restrictive and more reliable; even so, verify.
- Many websites have names that make you think that they are nonprofit or government sites, but they aren't. For instance, at one time there was a website called "whitehouse.com" that turned out to be a pornography site.
- Regardless of domain, you should verify the information from another, trusted source.

Guidelines for websites

1. Does the information on the website cite sources? If so, are they credible? Primary or secondary?
2. How old is the information? For science articles, it is particularly important that the website is reviewed and updated. Many websites are only reviewed rarely, if at all. Some don't include a date, making it difficult to tell how old the information is.
3. Does the website's information seem to match reality or other sources you have read?
4. You may be able to gain some measure of the site's reliability by the quality of pages linked to it.
5. You may find reviews of the source.
6. Can you detect biases? Misleading information? Attempts to persuade you toward a particular viewpoint?
7. Is the terminology understandable?
8. Check the labeling of graphs and tables to see if they are meaningful and generated properly.
9. Wikipedia is often useful for an introduction to a particular topic, but there are some articles that are better researched, with better sources, than others. Some articles, in fact, are written at a more advanced level than just an introduction. Take a look at the resources they cite for preparing the article and see how credible they are.

Using information accessed through the internet is fraught with peril. It is a time-consuming and frustrating process, but you *can* obtain reliable information if you verify carefully.

PoI 17. Arthritis afflicts millions of people, particularly as they get older. You might have parents or grandparents that suffer from this disease. Look for the most effective treatments for osteoarthritis (degeneration of joint cartilage and underlying bone) using three website types with different levels of trustworthiness: one you would use yourself or recommend to a friend or family member; one that you think is a fraud, a waste of money, and perhaps dangerous; and one that might work but lacks evidence of merit. For each website, cite why you classified them as you did.

Image Credits

Fig. 1.1: Copyright © by Ratha Grimes (CC BY 2.0) at https://commons.wikimedia.org/wiki/File:Urchin.jpg.

Fig. 1.2: Adapted from Copyright © by JF Warner and DR McClay, (CC BY 3.0) at https://commons.wikimedia.org/wiki/File:Left-right_asymmetry_in_the_sea_urchin_-_journal.pbio.1001404.g001.png.

Fig. 1.3: Copyright © by Julius Senegal (CC BY-SA 4.0) at https://commons.wikimedia.org/wiki/File:Measles_US_1938-2019.png.

Fig. 1.4: Source: https://commons.wikimedia.org/wiki/File:Cancer_smoking_lung_cancer_correlation_from_NIH.svg.

LAB #1

Laboratory Exercises

1.5. Laboratory on Getting Started with NetLogo Fast

Each text module has a relevant laboratory employing a NetLogo simulation. In this laboratory, we learn how to use this software, which is free, extremely easy to use, and requires no programming experience. To start, download NetLogo (https://ccl.northwestern.edu/netlogo/) and view the short video "Getting Started with NetLogo Fast" on the text's website.

In all laboratories, *always* write your answers in the blanks. Some questions have boldface numbers, such as **1.** and **2.**, indicating places to complete a statement. When a list of choices, labeled with capital letters, appears, use a letter or letters in the corresponding answer blank(s). Sometimes these choices are in parentheses immediately after a boldface number.

1. By double-clicking the icon, start NetLogo. From the File menu select *Models Library*. Under Sample Models, select the *Biology* section, then open Rabbits Grass Weeds by clicking on the *open* button. This computer simulation is a predator-prey model, with rabbits as the predators and grass and weeds as the prey. It is an example of an *agent-based model* (*ABM*), or a model that incorporates autonomous, decision-making *agents*, each one of which has *states* and *behaviors*. Initially, we see the *Interface* level, with an empty black *world*, or a rectangular grid that agents will inhabit, and several controls and reporting features. If preferred, under the *Zoom* menu, select *Larger* for a larger display. In this question, we explore the action of several buttons.

   a. _______ In this simulation, *buttons*, which are in blue, set up, start, and stop the simulation. Click the *setup* button. Select everything that appears in the world: A. black squares, B. green squares (grass), C. purple squares (weeds), D. rabbits

      The basic agent in NetLogo is called a *turtle*. However, this model has designated a special breed of turtles called *rabbits*, with an individual turtle being a *rabbit*. (Who would have thought turtles could be rabbits!)

   b. _______ Initially, the world was covered in black squares, or *patches*. Pressing the *setup* button causes some squares to turn green, representing grass. Click *setup* several times. Select all you observe: A. different configurations of rabbits and grass, B. nothing, C. the same configuration of rabbits and grass

c. ________ This simulation is *stochastic*, or having an element of chance. Thus, we have different configurations with each click of *setup*. Click the *go* button to start the simulation. Click *go* again. After pressing *go* a second time, the simulation **1.** (A. continues at the same speed, B. runs faster, C. runs slower, D. stops). Click *go* several times, observing the simulation's action each time.

d. ________ At the top of the window, drag the SPEED slider about half-way to the right so that it registers "faster." Clicking *go*, run the simulation. By moving the SPEED slider to the right, the simulation **1.** (A. continues at the same speed, B. runs faster, C. runs slower, D. stops). Similarly, moving the slider to the left slows the simulation.

2. So far, we have been guessing how to work with the simulation, but the model's documentation can explain what the simulation does and how to use it and can suggest what to notice, things to try, and model extensions.

   a. ________ Click the *Info* tab at the top of the window to change to the *Info* level. Read "What Is It?" Select the kind(s) of competitive-advantage exploration(s) for which we can use the model: A. different growth rates for wolves, B. different growth rates of grass and weeds, C. differing amounts of rabbit energy

   b. ________ Read "How to Use It." The BIRTH-THRESHOLD slider sets the energy level at which **1.** (A. grass grows, B. the rabbits reproduce, C. weeds grow).

   c. ________________ Read "Things to Notice." Select all descriptions of the rabbit population over time: A. decreases, B. eventually stabilizes in a narrow range, C. goes through a dampened oscillation, D. increases, E. is out of phase with the total amount of grass, F. matches the total amount of grass

3. ________________ Click on the Interface tab. Inside the window, *sliders* and *switches*, which provide input values, are in green. This model has no switches. After stopping the simulation, change the NUMBER slider bar, which currently indicates 150, to 500. You can click on the slider to make incremental changes. Click setup. Select everything that happens: A. approximately the same amount of grass appears, B. approximately the same number of rabbits appear, C. less grass appears, D. fewer rabbits appear, E. lots more grass appears, F. lots more rabbits appear

4. ________________ Data output appears in plots and monitors, which are outlined in beige. For this simulation, Populations is a *plot*, and "count rabbits" is a *monitor*, showing the number of rabbits at the most recent tick. (A simulation time unit is a *tick*, which may represent any appropriate unit of time, such as a month or an hour.) Set the NUMBER slider to 150. Click setup. Run the simulation until Time on the plot indicates at least 100 ticks have occurred, and then stop the simulation by clicking go. Without clicking or holding down a mouse button, hover over the graph at the high point of the red plot of the rabbit population. The values that float on the left and at the bottom are "Pop" (rabbit population or count) and Time (number of ticks), respectively. Elsewhere, write down the maximum population. Set up and run the simulation at least four more times, each time recording the approximate maximum rabbit population. In the answer blank, give the average of these maximum populations.

Because the simulation is stochastic, we are likely to get different values with each execution. With labs, we frequently record the results of multiple runs in a spreadsheet and have the spreadsheet calculate the average of the values.

5. This question covers the *view updates* checkbox, which is toward the top of the interface, below the Interface, Info, and Code tabs.

   a. ______________ Click *setup*. Then, uncheck the box *view updates*. Click *go* twice. Select all that changes in the window over time: A. buttons, B. command center, C. count-rabbits monitor, D. populations graph, E. sliders, F. world

   b. _______ Without clicking *setup* or *go*, check *view updates*. Select what appears using the choices from part a. Sometimes, we want the simulation to run faster and only care about the final results. Unchecking the *view updates* box disables the animation and causes the simulation to execute faster.

Click on third tab, labeled *Code*, at the top of the page, but do not change anything. This level contains the NetLogo code for the simulation's program.

1.6. Laboratory on Getting Started with Spreadsheets

A laboratory covering spreadsheet features, useful for other laboratories, is available in various platforms on the text's website.

MODULE

2

Chemistry for Life—Essentials

2.1. Introduction

Living organisms are all built out of a set of important chemical compounds. The most obvious are *water* and a collection of *organic* (carbon-based) molecules. In this section, we'll consider the structures and general functions of the most important ones.

2.2. Immersing Yourself in Water

2.2.1. Essential Chemistry

Water is the naturally clear liquid forming springs, streams, rivers, estuaries, and oceans. Millions of organisms live in such aqueous environments, and water-based precipitation brings life to terrestrial habitats as well. Many forms of human recreation are conducted in or on water-based arenas. Among other uses, water can be used for drinking, to produce electricity, for irrigation, to extract minerals, and to carry away our wastes. And, critically, water is the *solvent* (dissolves other substances) for all living systems on earth. If you traveled to hot and dry places like the Sahara Desert, you would only find living creatures where they have access to water, at least periodically. Life would not be possible without water.

Water is comprised of two chemical elements—hydrogen and oxygen. Each water molecule is made up of one atom of oxygen and two atoms of hydrogen (H_2O). *Atoms* are the smallest units of any element that display all of the chemical properties of that element. If we examine the subatomic structures of an atom, we find three types of particles—*neutrons*, *protons*, and *electrons*. The core of an *oxygen* atom has eight neutrons, which have *no* charge, and eight protons, each of which has a *positive* charge. Around the core orbit eight *negatively* charged electrons, each of which resides in so-called energy shells, where the electrons increase in energy as you go from the innermost to the outermost shell. Each shell holds a specific number of electrons. For instance, the first shell holds only two electrons, whereas the second shell holds at most eight. So oxygen has two electrons in the first shell and six

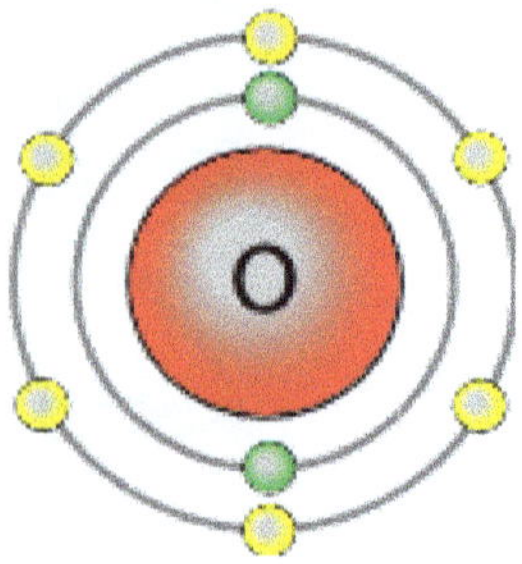

FIGURE 2.1a Oxygen with two shells, outermost with six electrons

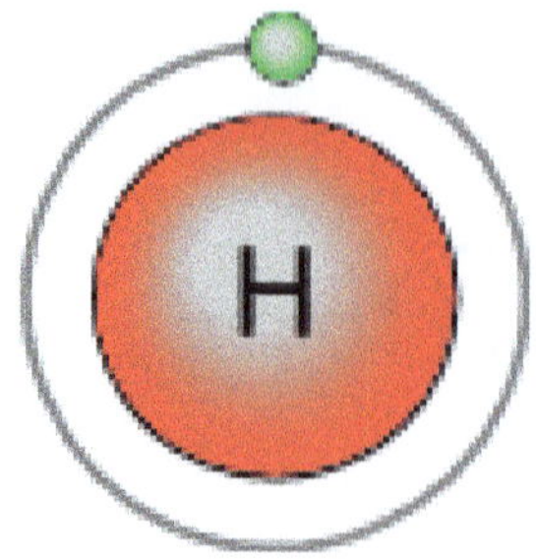

FIGURE 2.1b Hydrogen with one shell, one electron

in the second shell (see Figure 2.1a). The first shell is considered complete, but the outermost shell of oxygen would be most stable if it had two more electrons in that shell. Hydrogen does not have any neutrons, but it has one proton at its core and one electron in the nearest energy shell (see Figure 2.1b). Therefore, hydrogen would be more stable if it had one more electron in its outer shell.

Molecules are made up of two or more atoms that are held together by chemical bonds. Atoms with unfilled outer energy shells are likely to react with other atoms that can help them fill that shell. So if an oxygen atom shares one electron with each of two atoms of hydrogen, we get a molecule of water, and all of the outer shells are filled, at least part of the time (see Figure 2.2). The shared electrons between oxygen and hydrogen form strong, *covalent* bonds, which also bond the atomic members of other important molecules found in living organisms.

In the diagram of the water molecule, you will note that the hydrogen atoms are drawn near each other, toward one side. The reason for this arrangement is that by comparison, there is a much larger positive charge in the oxygen core, which has a stronger attraction for the electrons that the hydrogens are contributing. Therefore, the shared electrons spend more time orbiting the oxygen, giving oxygen a slightly negative charge. The hydrogens are left with only protons most of the time, and therefore they have a slightly positive charge. From this charge difference, water becomes a *polar* molecule, and this polarity means that water molecules tend to align themselves with their oxygen side attracted to the positive side of other water molecules. The attraction between the hydrogen of one water molecule with the oxygen of another water molecule is a weaker attraction called a *hydrogen bond* (see Figure 2.2 (a)). There is no sharing of electrons; the hydrogen bond is based on the slight attraction between the positive and negative charges.

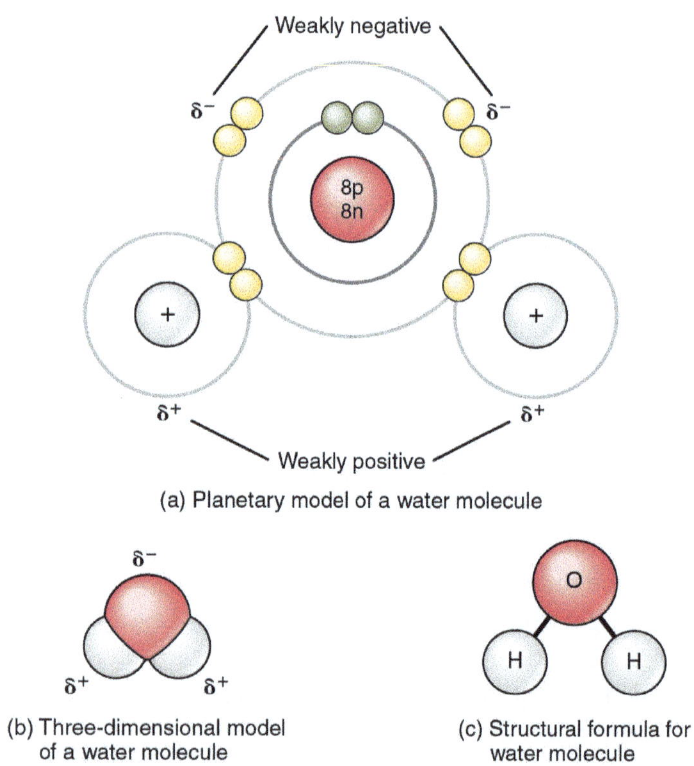

FIGURE 2.2 Water molecule

Sodium chloride (table salt) is a compound formed from one sodium (Na) and one chlorine (Cl) atom, NaCl. The two components are held together by a relatively weak attraction called an *electrostatic* or *ionic bond*. A sodium (Na) atom has 11 protons and 11 electrons, leaving only one electron in the outer shell. Chlorine has 17 protons and 17 electrons, with seven in the outermost shell. If sodium donates its outermost electron (its outer shell being now complete) to chlorine (filling its outer shell), sodium becomes a positively charged ion (11 protons and 10 electrons), and the chlorine becomes a negatively charged ion (17 protons and 18 electrons). Consequently, the two ions with opposite charges will be attracted to each other, forming a molecule of NaCl (see Figure 2.3). Crystallized table salt is made up of highly organized ions formed into a lattice of NaCl molecules.

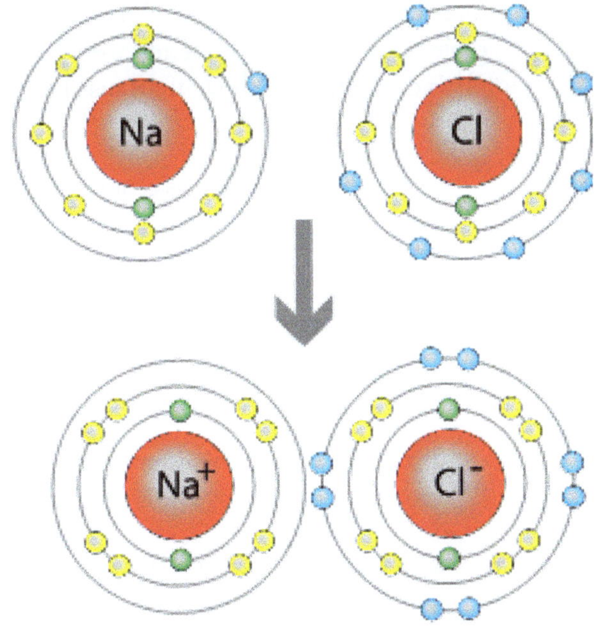

FIGURE 2.3 Ionic bonding of sodium and chloride ions to form salt

If we add NaCl crystals to water, the polar water molecules, with their positive and negative sides, will disrupt the weak ionic attraction in NaCl, forming two ions, Na^+ and Cl^-. Note in Figure 2.4 that the water molecules surround each ion. Water molecules (red and white) orient with their negative sides toward the sodium ions (yellow) and their positive sides toward the chloride ions (green). As we stir the mixture, more of the salt dissolves in the water, becoming evenly dispersed, and we term this mixture a *solution*. Salt is the *solute (substance dissolved)*, and water is the *solvent*.

Water is not only a polar molecule, but it also can form hydrogen bonds. These properties enable water to dissolve other polar molecules and ions. *Polar molecules* are molecules that are not charged but have an uneven charge distribution internally, so that they present somewhat negative and somewhat positive areas. Such molecules are termed *hydrophilic*, because they tend to dissolve in water and are therefore "water-loving." Examples include many organic molecules, like sugars and some amino acids. Molecules that do not dissolve in water tend to be *nonpolar* (no uneven charge

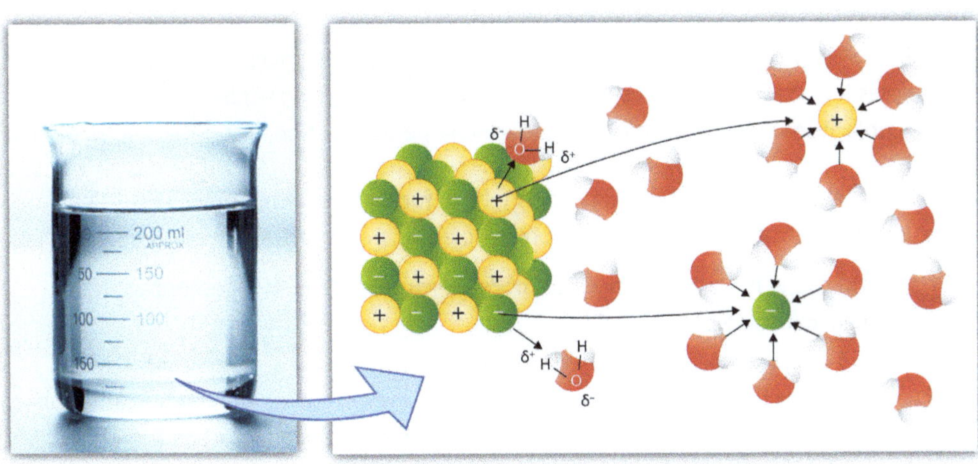

FIGURE 2.4 Salt crystal dissolved in water. Note the hydration spheres surrounding the sodium and chloride ions.

distribution) and are termed *hydrophobic* ("water-fearing"). Examples include fats and various molecules forming cell membranes. Some membrane proteins have both hydrophilic and hydrophobic regions, which we'll discuss later.

QRQ 1. Organic molecules contain the element ______.

QRQ 2. The smallest unit of any element that displays all of the chemical properties of that element is a(n) ______.

QRQ 3. In the subatomic structures of an atom, a(n) ______ has a positive charge; a(n) ______ has a negative charge; and a(n) ______ has no charge.

QRQ 4. Fill in the blanks with numbers: The first shell of an atom holds at most ______ electrons; whereas, once the first shell is filled, the second shell holds at most ______ electrons.

QRQ 5. Fill in the blanks with numbers: Hydrogen has ______ proton(s) and ______ neutron(s) in its core and ______ electron(s) in its outer shell.

QRQ 6. Fill in the blanks with numbers: Oxygen has ______ proton(s) and ______ neutron(s) in its core and has ______ electron(s) in its outer shell.

QRQ 7. Give the chemical symbol for each of the following:

a. chlorine
b. hydrogen
c. oxygen
d. sodium
e. sodium ion
f. table salt
g. water

QRQ 8. Select all that apply to hydrophilic molecules: dissolve substances, do not dissolve in water, examples include fats, examples include sugars and some amino acids, no uneven charge distribution, polar, tend to dissolve in water, uneven charge distribution, water-fearing, water-loving

QRQ 9. Select from the following the appropriate term for each statement below: covalent, electrostatic, hydrogen, hydrophilic, hydrophobic, molecule, nonpolar, polar, solute, solution, solvent

a. A(n) ______ is made up of two or more atoms that are held together by chemical bonds.
b. A(n) ______ dissolves other substances.
c. Molecules that are not charged but have an uneven charge distribution internally, so that they present somewhat negative and somewhat positive areas, are called ______.
d. The weak attraction between the hydrogen of one water molecule with the oxygen of another water molecule is called a(n) ______ bond.

e. The shared electrons between oxygen and hydrogen in water form strong, _____ bonds.

f. The two atoms of a molecule of table salt are held together by a relatively weak attraction, called a(n) _____ bond.

g. A substance that is dissolved in a solution is called a(n) _____.

2.2.2. Special Characteristics of Water

The large numbers of hydrogen bonds in water explain some of the unusual characteristics of water. For instance, water has a relatively *high heat capacity*. High heat capacity means that water doesn't change temperature without a large input of thermal energy, which will raise the temperature, or without a large release of thermal energy, which will lower the temperature. To raise the temperature of water, much of the energy input required doesn't speed up the movement of molecules, because so much energy is required first to break the hydrogen bonds. Likewise, much thermal energy is released as water molecules slow down with cooling temperatures. On a large scale, water's high heat capacity means that large bodies of water help to moderate climate conditions for coastal areas. As warming occurs, water absorbs large amounts of heat, but it releases the heat as the surroundings cool. In both cases, the water reduces the temperature range.

Water's polarity and hydrogen bonds promote *cohesion* (binding of water molecules to each other) and *adhesion* (binding of water to other types of molecules). Cohesive forces allow columns of water to be pulled upward in plant vessels from the roots, through the stem, into the leaves, and out through evaporation of water from the plant leaves. This movement is enhanced by adhesive forces that cause the water to cling to the inner surface of the plant vessels.

If you are fishing in quiet areas of a stream or lake, you may notice a host of insects (water striders, water boatmen, etc.) and a few spiders (fishing spiders) standing on the surface (see Figure 2.5). (Many of these insects, by the way, eat lots of mosquito larvae.) How can they stand on the water surface? Much of the reason is that cohesiveness gives the water a high *surface tension* (liquid's resistance to external forces). The animals tend to be lightweight and have thin, long, hairy legs that spread the weight out over a large area. The hairs tend to repel water and also trap air to increase flotation. These factors, plus the water's surface tension, resist the pull of gravity, which would normally sink the animals.

FIGURE 2.5 Water striders on water surface

Mammals are able to cool themselves by forming perspiration (sweat) on their skin surface. If you run a long distance, beads of sweat form on your skin. Eventually, the sweat evaporates, which helps to cool your body. These beads of sweat form because of the high surface tension of water. Evaporation of this sweat cools your body, because the fastest-moving water molecules leave your skin first, dissipating heat in the process. This evaporative cooling also takes place to cool bodies of water in hotter areas.

QRQ 10. Select all that are characteristics of water: good solvent, low surface tension, relatively low heat capacity

QRQ 11. Select from the following the appropriate term for each description below: adhesion, coherence, cohesion, high heat capacitance, high heat capacity, perspiration, surface tension

a. allows water to be pulled upward in plant vessels from the roots
b. binding of water molecules to each other
c. binding of water to other types of molecules
d. causes water to cling to the inner surface of plant vessels
e. helps to cool your body
f. liquid's resistance to external forces
g. causes large bodies of water to moderate climate in coastal areas

QRQ 12. (Possible answers in parentheses after each blank.) The temperature of water is increased by a ____ (large, moderate, small) ____ (input, release) of thermal energy.

2.3. Chemical Reactions

Water can dissolve many chemicals characteristically found in living things, and, therefore, it serves as a very able *solvent*. Accordingly, most chemical reactions in living systems occur in aqueous environments.

There are many chemical reactions in biological systems that we will refer to as *biochemical reactions*. Biochemical reactions will be described in more detail in another section of this text. Generally, a *chemical reaction* involves one or more *reactants* that are converted to *product(s)*. For this conversion, atoms in the reactants are rearranged, bonds are broken, and new bonds are made. Reactants must interact with the proper orientation and with sufficient energy for the reaction to proceed. For *reversible reactions*, products will also be converted back to reactants. In a cell, if the relative concentration of reactants and products does not change, the products and reactants reach an *equilibrium*.

Chemical reactions may require or release energy, and most chemical reactions, including those in cells, do not proceed spontaneously. Reactants in this system must reach an unstable *transition state*, which requires some sort of energy input to proceed. This energy is called the *activation energy* (E_A). Figure 2.6a shows the energy level for two reactants, A and B, that are converted to a product, AB. Note that the energy level is greater for the reactants than for the product, so this reaction is described as *exergonic*—energy is released. Nevertheless, to achieve the transition state, additional energy, E_A, is required. In Figure 2.6b, you see that the reactant energy level is lower than that of the product, so that the reaction is *endergonic*, with an $\mathbf{E}_A$ that is much greater than that required for the exergonic reaction. Again, both reactions require energy to initiate the reaction.

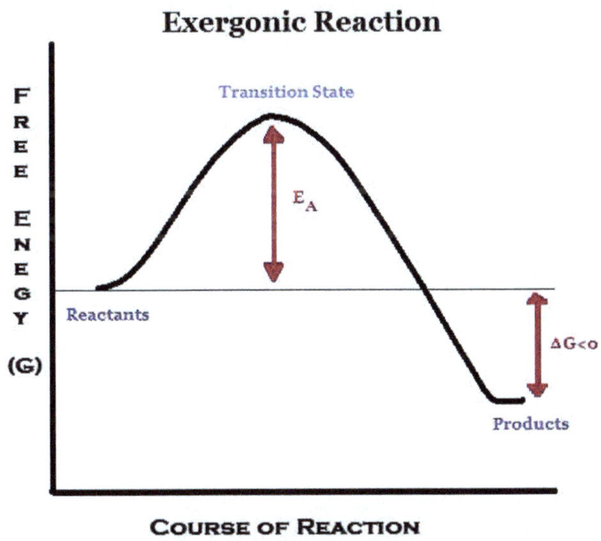

FIGURE 2.6a Plot of exergonic reaction

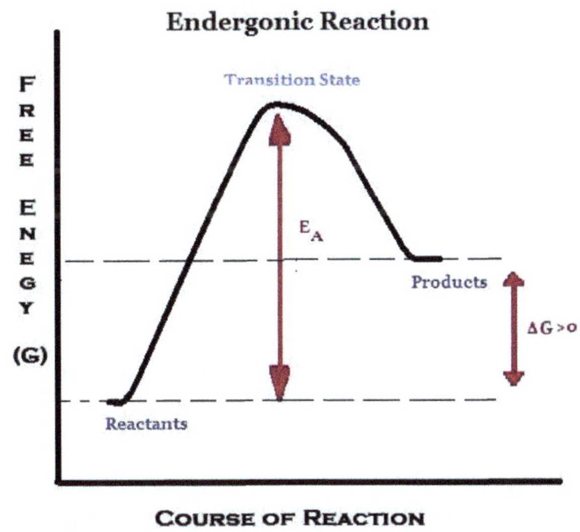

FIGURE 2.6b Plot of endergonic reaction

QRQ 13. A chemical reaction involves one or more ______ that are converted to one or more ______. In the cell, if their relative concentrations do not change, the reaction reaches a(n) ______.

QRQ 14. Select from the following the appropriate term for each description below: acting, activation, endergonic, endogenous, exergonic, transition, transitive

a. energy is released in this type of reaction
b. intermediate state of a reaction that is unstable, requiring energy input to proceed
c. reaction in which energy level is greater for the reactants than for the product
d. reaction in which the reactant energy level is lower than that of the product
e. type of input energy needed for a chemical reaction to proceed

QRQ 15. Select which of the following is the notation for activation energy: A, a, E, e, AE, A_E, E_A, A_e, E_a, a_E, e_A, a_e, e_a

2.4. Acids, Bases, and pH—Maintaining Balance

APPLYING SCIENCE TO LIFE 1 Alkaline Water

For years, a gentleman purporting to be a physician touted the health benefits of drinking alkaline water—pH of 9.5 or higher. He claimed it would help balance the alkaline-acid environment in the human body. He wrote books extolling this "cure" for various diseases, including cancer. Unfortunately, some innocent patients were taken in by his claims and then declined conventional medical treatment. At least one of them died from breast cancer, which the "doctor" had treated with alkaline water. The "doctor" held retreats, costing thousands of dollars, to sell his "curious" ideas. His and other proponents' thesis was that drinking the alkaline water and eating alkaline foods would alter the pH of your blood (and body) and that this change would defeat most diseases, reduce fat deposits around the belly, stop acid reflux, and so on. To most scientists, the recommendation for altering the pH of your blood is ridiculous, because human beings have elaborate homeostatic (balancing) mechanisms to keep your blood pH at or near 7.4.

As it turns out, the "doctor" had not completed an undergraduate education and had received his "PhD" from an online diploma mill. Finally, he was convicted for practicing medicine without a license and jailed. He lost one case recently with a considerable judgment, and other lawsuits are in the courts (Barrett 2017; CSPI 2018; Gorski 2018; Villines 2019).

2.4.1. pH Scale

What do *acidic* and *alkaline* (or *basic*) mean? Here we'll define *acids* as hydrogen ion (H^+) donors and *bases* as H^+ receptors. As it happens, some of the foods and beverages you consume are acidic, while others are basic (see Table 2.1).

TABLE 2.1. pH Measurement of Popular Beverages (Reddy et al. 2016)

Beverage	pH
Gatorade Lemon-Lime	2.97
Perrier Carbonated Mineral Water	5.25
Minute Maid Apple Juice	3.66
Country Time Lemonade	2.72
Pepsi	2.39
Sprite Zero	3.14
Monster Energy	3.48
Starbuck's Medium Roast	5.11

There are many routine sources of H^+ in your body. The most important source is the carbonic acid system. Metabolism produces CO_2, which combines with water to form carbonic acid, which may dissociate into $\mathbf{H^+}$ and HCO_3^-.

$$CO_2 + H_2O \leftrightarrow \underset{\text{carbonic acid}}{H_2CO_3} \leftrightarrow HCO_3^- + \mathbf{H+}$$

Other sources of H^+ are inorganic acids (sulfuric and phosphoric) produced during the digestion of meats. Meats tend to have a good supply of sulfur and phosphate, from which the acids are formed through metabolic processes. Furthermore, metabolism produces a host of organic acids, many of which are intermediates in the production of energy in the cell.

Most of the chemical reactions that occur in organisms do so in an aqueous environment. Acids in water tend to release or donate hydrogen ions (H^+), and bases tend to accept them (H^+). Water (H_2O) itself dissociates very little (about 1/10 million molecules) into H^+ and OH^-. Because the two ions are produced at equal proportion, water is considered to be neutral, neither acidic nor basic. The concentration of hydrogen ions ($[H^+]$) in pure water is equal to 10^{-7} moles/liter.[1] 10^{-7} is a very small number (0.0000001) and using it for calculations is cumbersome. But we can express that number using logarithms as $\log 10^{-7}$, which is equal to -7. If we use the negative log of 10^{-7} ($-\log 10^{-7}$), it equals $-(-7)$, which is 7. Thus, $-\log 10^{-5} = 5$, and so on. Using the negative log to express concentrations of H^+ or OH^- is convenient, and $-\log [H^+] = \text{pH}$ is now the standard (see Figure 2.7). Pure water has a pH of 7. Solutions that have more hydrogen ions than water have a lower pH, and those that have fewer hydrogen ions have a higher pH.

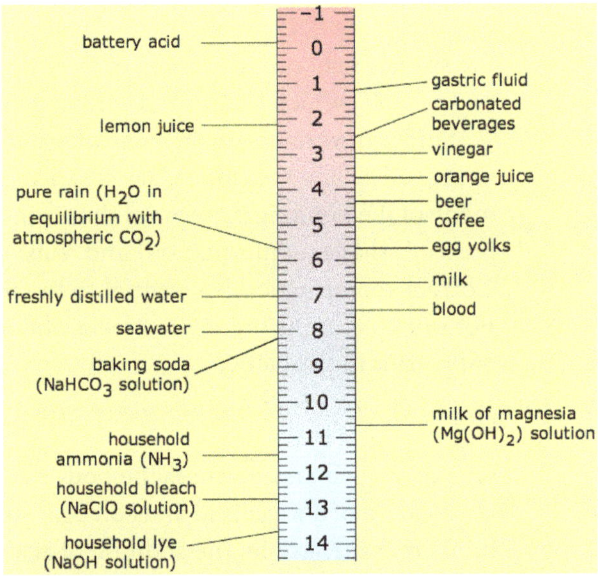

FIGURE 2.7 Scale of pH, with pH values given for familiar liquids

If a solution has a hydrogen ion concentration of 10^{-4} moles/liter, what would be the pH? Note that $10^{-4} > 10^{-7}$. $\text{pH} = -\log [H^+]$, and for this solution $\text{pH} = -\log [10^{-4}]$. Thus, $\text{pH} = -(-4) = 4$.

In this solution, how many more moles of hydrogen ions will be present than in pure water?

$$10^{-7}\text{ moles/liter} = 0.0000001\text{ moles/liter};\ 10^{-4}\text{ moles/liter} = 0.0001\text{ moles/liter}$$

The pH 4 solution has 1,000 times more hydrogen ions than pure water. Similarly, in a solution with a pH of 8, there would be tenfold fewer hydrogen ions than in pure water. From this exercise, you realize that small changes in pH can mean very large changes in hydrogen ion concentration.

1 [X] is shorthand for concentration of X. Therefore, $[H^+]$ is shorthand for concentration of H^+. A mole is a unit of measure for the atomic, molecular, or formula weight of the substance, expressed in grams; a mole is the mass of all atoms in a molecule.

2.4.2. Importance of Maintaining a pH Balance

The human body is complex, with about 37 trillion cells, each with thousands of metabolic reactions taking place. During our evolutionary history, we have developed quite a number of mechanisms to keep our bodies in a steady state. Human beings have evolved regulatory mechanisms to maintain proper temperatures, osmotic conditions, pH, O_2, nutrients, blood sugar, blood pressure, and so on. This balance, or steady state, is termed *homeostasis*. Your body needs your blood pH to be close to 7.4 for some very good reasons. Even a small deviation from this number would be dangerous. The alkaline water, which the bogus "doctor" touted, might have an effect on the pH of your urine, but not on the blood.

Fluctuations in the $[H^+]$ can affect several systems in your body. First, increases in $[H^+]$ can cause a depression of your central nervous system (CNS). You may become disoriented, lapse into a coma, and, perhaps, die. A decline in $[H^+]$ can lead to overexcitability of your peripheral nervous system (PNS), characterized by tingling and prickling in your extremities. Moreover, if the motor neurons of the PNS are overexcited, you may have muscle twitches and even spasms. More seriously, the spasms may occur in some of the muscles important for breathing. Even worse are the effects on the CNS, which can lead to convulsions.

Sufficient variation in $[H^+]$ can affect any of the enzymes in your body. Enzymes, which catalyze almost all cellular reactions, are sensitive to even small deviations in pH. Changes in pH can alter many of the bonding patterns of the amino acids in these proteins, which may potentially alter the protein's shape (conformation) and, subsequently, may depress or promote the enzyme's activity. Correct conformation is essential to the normal function of enzymes and other proteins, as well. For example, at the organismal level, changes in pH have strong ramifications for the functionality of proteins in the blood.

2.4.3. Buffers

One of the primary defenses the body has against erratic changes in pH is that of *buffers*. A buffer is a solution of a weak acid and conjugate (coupled) base, or vice versa, that will withstand a change in pH if challenged by the addition of acids and bases to the system. In our blood, pH needs to be kept to no more than a 0.05-pH variation from the desired pH of 7.4. The principal buffer system in blood is the acid-base pair *carbonic acid-bicarbonate system* (H_2CO_3 = acid and HCO_3^- = base).

$$CO_2 + H_2O \leftrightarrow H_2CO_3 \leftrightarrow HCO_3^- + H^+$$

If we add some acid to the system, increasing H^+, HCO_3^- will absorb the H^+, becoming H_2CO_3, reducing the H^+ in solution. If we add a base to the system, removing H^+, H_2CO_3 will dissociate to form HCO_3^- + H^+, increasing the H^+. Either way, the system works to maintain a stable pH.

The reactions in this buffer system are catalyzed by the enzyme *carbonic anhydrase*, which is located in red blood cells. As CO_2 is produced in the body's cells, it is released into the bloodstream (see Figure 2.8). A small portion dissolves in the blood plasma, but most of it moves into the red blood cell. A percentage attaches to hemoglobin, but almost three quarters of the amount released into the blood combines with water, which carbonic anhydrase converts to carbonic acid. Carbonic acid dissociates into a hydrogen ion (H^+) and a bicarbonate ion (HCO_3^-), which is transported into the plasma. Thus, most CO_2 is transported in the blood as bicarbonate. The hydrogen ions produced lower the pH inside the red blood cell, which decreases the hemoglobin's affinity for oxygen. Consequently, more oxygen is released into the tissues. When the blood returns to the lungs, many of these processes are reversed, allowing the release of CO_2 and the uptake of O_2.

Besides the carbonic acid-bicarbonate system, there are three other buffer systems in the human body: protein buffer, hemoglobin buffer, and phosphate buffer. The carbonic acid-bicarbonate system is the most important buffering system in the blood, but hemoglobin helps buffer the hydrogen ions generated by cellular metabolism. Protein buffering effects are mainly achieved in the intracellular fluid. Proteins have both acidic and basic groups and, thus, make effective buffers. Phosphate buffers are also important in intracellular fluids and are important as a urinary buffer.

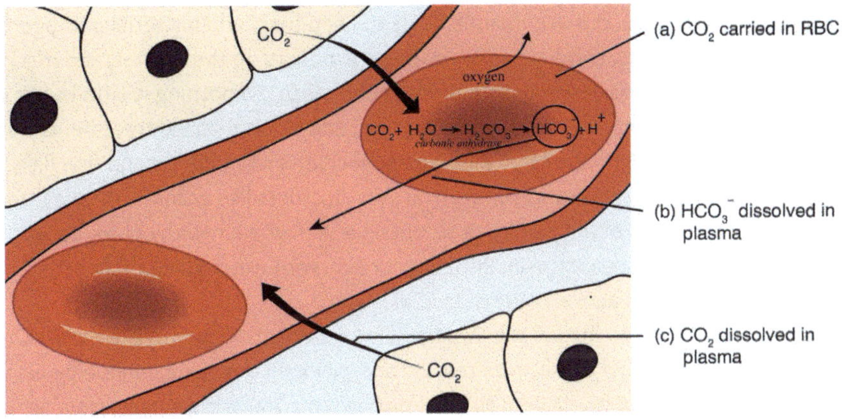

FIGURE 2.8 Carbon dioxide transport

2.4.4. Role of the Lungs and Kidneys

Chemical buffers are the first and quickest line of defense, but your body has two other, albeit slower, layers of defense against large pH fluctuations: the lungs and the kidneys. Some of the major arteries have receptors to detect changes in hydrogen ion concentrations. As $[H^+]$ increases, signals from these receptors send a message to the respiratory center in the brain, which sends signals that increase the rate and depth of breathing. This action increases the gas exchange in the lungs. There, HCO_3^- in the blood is combined with H^+ to form carbonic acid, which is converted into water, which is retained in the body, and CO_2, which is released into the air. Hence, with a decrease in hydrogen ion concentration, pH increases. Thus, increased respiratory activity results in increasing the pH of the blood as it picks up O_2 for transport back to the tissues. If $[H^+]$ decreases, ventilation would slow, removing less CO_2 and resulting in more bicarbonate and hydrogen ions remaining in the blood. The respiratory system influences the pH by *controlling the rate of carbon dioxide removal.*

The kidneys can vary the removal of H^+ in forming the urine. Furthermore, the kidneys can change the elimination and conservation rate of bicarbonates. If the kidneys eliminate an H^+, a bicarbonate will be able to take on another H^+. Unlike the lungs, which regulate CO_2 removal, the kidneys *regulate the rate of hydrogen ion removal.* Both systems are important, therefore, in the regulation of the body's pH.

APPLYING SCIENCE TO LIFE 2 Lifting a Protective Mantle

It seems that pH is important on the surface of your body as well as inside it. It may surprise you to know that most people have a skin pH of somewhere between 4 and 6. If you remember from our discussion of pH, you know that any pH below 7 is acidic. Of course, skin pH varies with age, region of the body, genetics, and so on, but generally it is acidic. Scientists from over 90 years ago described what they termed the "acid mantle" that covers your skin. Various acidic products are secreted onto the skin surface. For instance, your sweat glands secrete lactic acid and amino acids, and your sebaceous glands secret free fatty acids. Significantly, among other functions, this acid mantle helps to keep fungi and bacterial pathogens from taking hold on your skin's surface.

Today, there are many skincare products on the market, and marketers have persuaded people that they need to have skin that feels fresh and clean. So nothing will do but to scrub until your skin is "squeaky clean." Unfortunately, the abrasive, alkaline, and alcohol-based treatments help to break down the protective outer mantle. Symptoms of this damage include inflammation and dry, flaky skin. Evidence points to such a breakdown's contribution to various skin disorders like eczema, rosacea, and acne. People spend tremendous amounts of money trying to obtain a skin with a healthy glow, but many of the treatments work against maintaining a healthy skin. Rinsing your skin with water will temporarily raise the pH of your skin, and most soaps have pHs above 7 (Baranda et al. 2002).

Researchers in the Department of Dermatology at Wake Forest University Baptist Hospital recommend that care should be taken to restore and maintain the acidic mantle by using topical agents that are compatible with that protective layer (Ali and Yosipovitch 2013). Also, a dermatologist quoted in a recent *New York Times* article suggested using products that contain ingredients like petrolatum and glycerin, which may help to restore your barrier skin problems (Rubin 2019). It is your skin. You now know about pH, so you can choose to keep it healthy.

QRQ 16. Select from the following the appropriate term for each description below: acid, amount of hydrogen ions, base, buffer, carbolic anhydrase, carbonic anhydrase, concentration of helium ions, concentration of hydrogen ions, helium ion, hydrogen ion, neutral

a. bicarbonate
b. enzyme in red blood cells that catalyzes buffering reactions
c. H^+
d. $[H^+]$
e. hydrogen ion donor
f. hydrogen ion receptor
g. has a higher pH than water
h. has a pH less than 7
i. has more hydrogen ions than water
j. in water, tends to release or donate hydrogen ions
k. water

QRQ 17. Evaluate $-\log 10^2$.

QRQ 18. Calculate the pH of a liquid with a hydrogen concentration of 10^{-3} moles/liter.

QRQ 19. The hydrogen ion concentration of baking soda is about 10^{-9} moles per liter.

a. Give the pH of baking soda.
b. Is baking soda an acid or a base?
c. Pure water has how many times more hydrogen ions than baking soda?

QRQ 20. Stomach acid (hydrochloric acid, or HCl) is a strong acid that will ionize completely, meaning that a 10^{-1} solution in water will give a pH of about 1. Select from the following how many more hydrogen ions there are in this solution than in pure water: 10 times (10x), 100x, 1,000x, 10,000x, 100,000x, 1,000,000x, 10,000,000x

QRQ 21. The pH of water is ____, while blood has a pH of ____.

QRQ 22. Select which of the following are sources of hydrogen ions in the body: carbolic acid system, carbonic acid system, inorganic acids produced during the digestion of meats, inorganic acids produced during the digestion of vegetables, kidneys, liver

QRQ 23. The body's regulatory mechanisms tend to maintain ____, or a steady state for temperature, osmotic conditions, pH, O_2, nutrients, blood sugar, blood pressure, and so on.

QRQ 24. A(n) ____ is a solution of a weak acid and conjugate base, or vice versa, that will withstand a change in pH if challenged by the addition of acids and bases to the system.

QRQ 25. Select which of the following are the two major parts of the principal buffer system in blood: adenine, bicarbonate system, carbohydrates, carbon, carbonic acid, glycerol, hemoglobin, hydrolase, lipids, water

QRQ 26. (T/F) In the major buffering system in the blood, when acid is added to the system, HCO_3^- absorbs the additional H^+, becoming H_2CO_3.

QRQ 27. (T/F) In the major buffering system in the blood, when a base is added to the blood, H_2CO_3 will dissociate to $H_2O + CO_2$, decreasing H^+.

QRQ 28. (Possible answers in parentheses after blank.) The respiratory system helps regulate the body's pH by controlling the rate of removal of ____ (acid, bases, carbon dioxide, hydrogen ions, oxygen, water).

QRQ 29. (Possible answers in parentheses after blank.) The kidneys help regulate the body's pH by controlling the rate of removal of ____ (acid, bases, carbon dioxide, hydrogen ions, oxygen, water).

Image Credits

LAB #2

Laboratory Exercises

2.5. Laboratory on Strong Acids

2.5.1. Preliminary Work

In this laboratory, we consider the interactions of acids, bases, and buffers. From the text's website, view the introductory video and "Strong Acids Lab Notation and Formulas"; download and open *"Strong Acid Rev. nlogo."*

1. Click the Info tab at the top of the page and read "What Is It?" and "How It Works."
   a. Match a symbol from the following list to each substance: A^-, B, H^+, H-A, H-B^+, H_2O, H_3O^+, OH^-

      _______ hydrogen ion

      _______ hydronium ion

      _______ hydroxide ion

      _______ proton

      _______ strong acid

      _______ strong base

   b. _______ When an acid molecule donates a proton to a water molecule, a **1.** (A. hydrogen, B. hydronium, C. hydroxide) ion is formed.

   c. _______ When a base molecule takes a hydrogen atom from a water molecule, a **1.** (A. hydrogen, B. hydronium, C. hydroxide) ion is formed.

   d. **1.** _______ **2.** _______ A strong acid molecule **1.** (A. always, B. never, C. occasionally) dissociates in water, giving its hydrogen to a water molecule, converting the water molecule to **2.** (H^+, HA, H-B^+, H_3O^+, OH).

   e. **1.** _______ **2.** _______ A strong base molecule **1.** (A. always, B. never, C. occasionally) dissociates in water, taking a hydrogen from a water molecule so that the water molecule converts to **2.** (H^+, HA, H-B^+, H_3O^+, OH^-).

f. _______ pH is a measure of the number of **1.** (A. hydrogen ions or hydronium ions, B. hydrogen ions or hydroxide ions, C. hydronium ions, D. hydronium ions or hydroxide ions, E. water molecules) in a solution.

Consult other sources as necessary for the following parts of this question.

g. _______ A **1.** (A. liter, B. molarity, C. mole) is a unit of measure for the atomic, molecular, or formula weight of the substance, expressed in grams.

h. _______ A **1.** (A. liter, B. molarity, C. mole) is a unit of measure for the concentration of hydronium ions in a solution.

i. _______ Give the pH of water.

j. _______ The pH of an acid is **1.** (A. above, B. below, C. equal to) 7.

2. Read "How to Use It."

   a. Match each component below with its color in the model: A. black, B. blue, C. brown, D. burgundy, E. green, F. magenta, G. orange, H. red, I. rose, J. white, K. yellow

   _______ hydronium ion

   _______ hydroxide ion

   _______ water molecule

   b. _______ Give the number of water molecules initially in the solution.

2.5.2 Laboratory Work

The default values are VOL-ACID = 50 molecules and VOL-BASE = 5 molecules.

3. This question considers some of the interactions of acids, bases, and water.

   a. _______ Using the default settings, press SETUP and give the initial number of water molecules.

   b. _______ The initial amount of acid was 50 molecules (VOL-ACID), which is completely dissociated. Therefore, what is the number of hydronium ions after this dissociation?

   c. _______ Give the initial pH of the solution.

   d. _______ Press ADD BASE to add five molecules of base, which becomes quickly and completely dissociated. Therefore, what is the number of hydroxide ions after this dissociation?

   e. **1.** _______ **2.** _______ **3.** _______ Keep pressing STEP to have the simulation advance one step at a time until no more hydroxides remain. The pH increases, indicating the solution has become more **1.** (A. acidic, B. basic) and a hydroxide ion combines with a **2.** (C. hydronium ion, D. hydroxide ion, E. water molecule) in the same location to form **3.** (0, 1, 2, 3) additional water molecule(s).

   f. _______ The initial number of hydronium ions was 50. Give the number of hydronium ions after combination with the 5 hydroxide ions.

   g. _______ The initial number of water molecules was 100. Give the number of water molecules after the 5 hydroxide ions have been used.

h. ________ Suppose only 40 water molecules were available and VOL-ACID = 50 molecules, so that the initial number of strong acid molecules is 50. Give the number of hydronium ions that could be formed from dissociation in this situation.

i. ________ Let us consider a situation where VOL-ACID = 120 molecules and 1000 water molecules exist in solution. After pressing SETUP, give the number of hydronium ions.

j. ________ Continuing from part i, suppose VOL-BASE = 112. If we press ADD BASE, give the number of hydroxide ions that results.

k. Continuing from parts i and j, suppose we press GO and wait until the system stabilizes. Give the resulting number of each of the following:

________ *new* water molecules formed from hydronium and hydroxide reacting

________ hydronium ions

________ hydroxide ions

l. ________ Let us consider a situation where VOL-ACID = 115 molecules and 1000 water molecules exist in solution. After pressing SETUP, give the number of hydronium ions.

m. ________ Continuing from part l, suppose VOL-BASE = 140. If we press ADD BASE, give the number of hydroxide ions.

n. Continuing from parts l and m, suppose we press GO and wait until the system stabilizes. Give the resulting number of each of the following:

________ *new* water molecules formed from hydronium and hydroxide reacting

________ hydronium ions

________ hydroxide ions

An additional question on titration appears in the Active Learning version of this laboratory.

2.6. Laboratory on Buffers

2.6.1. Preliminary Work

We continue our studies of acids, bases, and buffers using a different model. From the text's website, view the introductory video and "Buffers Lab Notations and Formulas;" download and open "*Buffer Rev.nlogo.*"

1. Click the Info tab at the top of the page and read "What Is It?" and "How It Works."

   a. ________ (T/F) A buffer is a solution that resists change in pH when either an acid or a base is added into it, within limits.

   b. **1.** ________ **2.** ________ **3.** ________ **4.** ________ A weak acid molecule (H-A) **1.** (A. always, B. never, C. occasionally) dissociates in water, giving its hydrogen to a water molecule and converting the water molecule to **2.** (A^-, H^+, HA, H-B^+, H_2O, H_3O^+, OH^-), and converting the weak acid molecule to **3.** (A^-, H^+, HA, H-B^+, H_2O, H_3O^+, OH^-) called the weak acid's **4.** (D. conjugate base, E. dissociation constant, F. Ka, G. strong base).

c. ________ (T/F) A buffer is effective for every concentration of the weak acid and its conjugate base.

2. Read "How to Use It."

   a. Match each component below with its color in the model: A. black, B. blue, C. brown, D. burgundy, E. green, F. magenta, G. orange, H. red, I. rose, J. white, K. yellow

      ________ acid molecules

      ________ conjugate base molecules

      ________ hydronium ions

      ________ hydroxide ions

      ________ water molecule

   b. ________ percent Upon setup, what percentage of the weak acids dissociate?

   c. **1.** ________ **2.** ________ While the simulation is running, when H-A and H_2O are on the same patch, the result always becomes **1.** (A, A^-, A^+, H, H^-, H^+) and **2.** (H_3O^+, H-A, OH^-).

   d. **1.** ________ **2.** ________ Possible answers: A, A^-, A^+, H, H^-, H^+, H-A, H_2O, H_3O^+, OH^-. While the simulation is running, when H_3O^+ and OH^- are on the same patch, the result always becomes **1.** and **2.**

   e. **1.** ________ **2.** ________ While the simulation is running, when H-A and OH^- are on the same patch, the result usually, but not always, becomes **1.** (A, A^-, A^+, H, H^-, H^+, H_3O^+, OH^-) and **2.** (A, A^+, H, H^-, H^+, H_2O, H_3O^+, OH^-).

   f. **1.** ________ **2.** ________ While the simulation is running, when H_3O^+ and A^- are on the same patch, the result usually, but not always, becomes **1.** (A, A^+, H, H^-, H^+, H_2O, H_3O^+, OH^-) and **2.** (A, A^-, A^+, H, H^-, H^+, H-A, H_3O^+, OH^-).

   g. **1.** ________ **2.** ________ Pressing ADD ACID adds ADDED-ACID number of **1.**, while pressing the ADD BASE button adds ADDED-BASE number of **2.** Possible answers: A. conjugate base molecules, B. hydronium ions, C. hydroxide ions, D. weak acid molecules

   h. Suppose ADDED-BASE is 25 and ADDED-ACID is 15. After starting the simulation, suppose we click ADD BASE four times and ADD ACID three times. Give the following:

      ________ ADD-BASE-COUNT

      ________ ADD-ACID-COUNT

      ________ TOTAL-ADDED-BASE-MINUS-ADDED-ACID

2.6.2. Laboratory Work

The default values are STARTING-ACID = STARTING-CONJ-BASE 100 molecules, ADDED-ACID= ADDED-BASE = 15 molecules, and SEE-STARTING-WATER = OFF.

3. In this question, we observe the action of a buffer.

   a. ________ Using the default parameter values, press SETUP several times. Is the number of hydronium ions always the same, no (N) or yes (Y)?

b. _______ Repeat the following experiment several times: Press SETUP, observe the number of hydronium ions, press GO, let the pH stabilize (so that the number of hydronium ions becomes zero), and then observe the pH. Give the stabilized pH.

c. _______ The solution reaches the same equilibrium each time because of **1.** (A. buffering, B. the acidic nature of the solution, C. the basic nature of the solution, D. the initial number of water molecules).

d. _______ Press SETUP and GO and let the pH stabilize. Then, press ADD ACID and stop the simulation after pH stabilizes. Select all statements that are true about the pH graph after the addition of 15 molecules of acid (hydronium ions): A. after initial change goes to 7, B. initially drops below 4.5, C. initially rises above 9.5

e. When acid (or base) is added to the solution, the ions quickly disappear as they react. From the following list, select appropriate statements about each molecular count listed below after the addition of hydronium ions in part d: A. decreases by 15, B. decreases by 15 then increases to 0, C. does not change, D. increases by 15, E. increases by 15 then decreases to 0

_______ hydronium ions

_______ conjugate bases

_______ hydroxide ions

_______ acid molecules

_______ water molecules

f. _______ Without pressing SETUP, press ADD BASE and GO. Stop the simulation after pH stabilizes. Select all statements that are true about the pH graph after the addition of 15 molecules of base (hydroxide ions): A. after initial change goes to 7, B. initially drops below 4.5, C. initially rises above 9.5

g. Select an appropriate statement from the following list about each molecular count below after the addition of hydroxide ions in part f: A. decreases by 15, B. decreases by 15 then decreases to 0, C. does not change, D. increases by 15, E. increases by 15 then decreases to 0

_______ hydronium ions

_______ conjugate bases

_______ hydroxide ions

_______ acid molecules

_______ water molecules

h. _______ Does the buffered solution resist change in pH, no (N) or yes (Y)?

i. _______ Notice that after addition of an acid or a base, buffering with a weak acid and its conjugate base has returned pH to 7, indicating a neutral solution. What about the molecule counts? Select any molecule that has a different count than after setup and stabilization: A. hydronium ions, B. conjugate base, C. hydroxide ions, D. acid molecules, E. water molecules, F. none

j. _______ Press ADD ACID twice to add 30 hydronium ions, stopping when the pH has stabilized. Are we still in the buffering range so that the pH returns to 7, no (N) or yes (Y)?

k. We have added acid three times and base once, which is equivalent to adding acid twice (3 – 1 = 2). Thus, TOTAL-ADDED-BASE-MINUS-ADDED-ACID = –30. Uncheck *view updates* or maximize the speed of the simulation. Starting the simulation again, keep pressing ADD ACID and letting the pH stabilize until the solution *cannot* stabilize to a pH of 7. Give pH, TOTAL-ADDED-BASE-MINUS-ADDED-ACID, and count for each of the following:

_______ pH

_______ TOTAL-ADDED-BASE-MINUS-ADDED-ACID

_______ hydronium ions

_______ conjugate bases

_______ hydroxide ions

_______ acid molecules

l. _______ Select the two molecules with which hydronium can interact: A, A^-, A^+, H, H^-, H^+, H-A, H_2O, H_3O^+, OH

m. _______ Press ADD ACID again and give the pH after it stabilizes.

n. _______ Give the minimum value, which is a negative number, of TOTAL-ADDED-BASE-MINUS-ADDED-ACID for the buffering range.

o. _______ If TOTAL-ADDED-BASE-MINUS-ADDED-ACID is at this minimum and we click **1.** (A. ADD ACID, B. ADD BASE), pH of the solution will not return to 7.

p. _______ Give the maximum value, which is a positive number, of TOTAL-ADDED-BASE-MINUS-ADDED-ACID for the buffering range. Run tests if necessary to determine this value.

q. Summarize what you have observed.

MODULE

3 Chemistry for Life—Carbon-Based Molecules

3.1. Carbon

Carbon (C), with six protons and six electrons, turns out to be a very versatile element. With only four electrons in an outer shell that can accommodate eight, carbon can share up to four electrons with other elements (e.g., carbon, oxygen, hydrogen), thus forming up to four *covalent bonds*. If we examine methane (CH_4), a greenhouse gas, we find carbon bonded to four hydrogens (see Figure 3.1). Each hydrogen shares its single electron with one of the four outer-shell electrons of carbon.

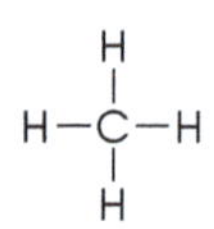

FIGURE 3.1 Diagram of methane

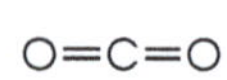

FIGURE 3.2 Diagram of carbon dioxide

Sometimes the carbon may share more than one electron with another element, most often two, forming what we call a *double bond*. *Carbon dioxide* (CO_2), a respiratory waste product in animals and absorbed by plants, has its one carbon double-bonded to two oxygens that each have two electrons in the outer shell (see Figure 3.2).

By connecting carbon atoms to one another, you can build larger, more complicated molecules called *biomolecules*. These molecules are the fundamental structural and functional components of all living things. Some small organic molecules, like sugars, which may have important functions of their own, can also be covalently linked together to form larger molecules (macromolecules).

The cells and bodies of living organisms contain four principal categories of organic molecules—*carbohydrates*, *proteins*, *nucleic acids*, and *lipids*. In the next few sections, we will examine these in greater detail.

QRQ 1. A(n) _____ bond is formed when an element shares more than one electron with another element.

QRQ 2. The chemical symbol for carbon is _____.

QRQ 3. By connecting carbon atoms to one another, you can build larger, more complicated molecules, called ____, which are the fundamental structural and functional components of all living things.

QRQ 4. Select from the following the four principal categories of organic molecules contained in cells and bodies of living organisms: biochemical, carbohydrates, carbon-based, endergonic, exergonic, lipids, nucleic acids, organic, proteins

3.2. Carbohydrates

Carbohydrates, made up of carbon, hydrogen, and oxygen, include simple sugars, short chains of sugars, or long chains of sugars. The carbohydrates get their name from the empirical formula of the sugars, where the C:H:O composition follows the ratio 1C:2H:1O, or CH_2O—literally, a hydrate (H_2O) of carbon (C). Most commonly, simple sugars (*monosaccharides*, meaning one sugar), which are characteristic of living things, contain five or six carbons. Hence, they are described as *pentoses* with five carbons or *hexoses* with six carbons. You may know many of the six-carbon sugars, because they are common in most people's diets: *glucose*, *fructose* (fruits), *galactose* (milk). So all of these sugars would have the following empirical formula: $\mathbf{C_6H_{12}O_6}$. Each one of them differs from the others only by the arrangement of their atoms. You do not need to know the structures, but we represent glucose as a six-sided chain and ring in Figure 3.3.

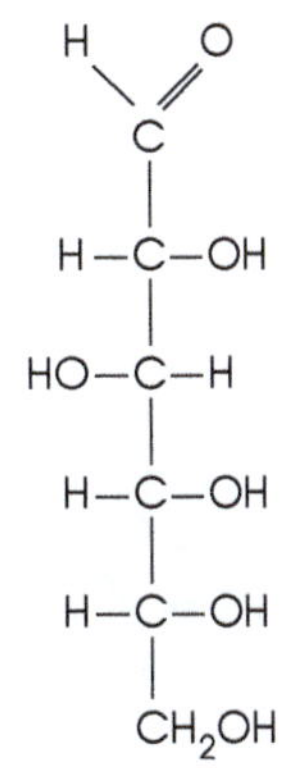

FIGURE 3.3a Glucose as a chain

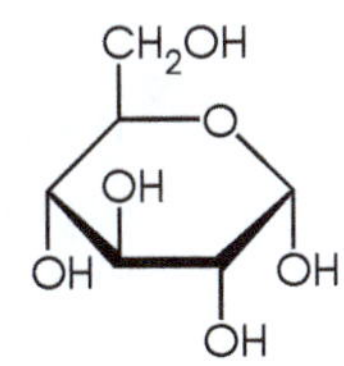

FIGURE 3.3b Glucose as a ring

We can combine two monosaccharides (sugars) to make a *disaccharide* sugar. A common disaccharide component of most diets is sucrose. In Figure 3.4, glucose and fructose (reactants) are combined to form sucrose, and water is released. Sucrose and water are products. This reaction is called a *condensation* or *dehydration* reaction, because in forming the covalent bond between the sugars, water is produced. Digestive enzymes (biological catalysts) in your small intestine use a *hydrolysis* reaction to break the bond between the two sugars by adding water back. In this case, the disaccharide and the water are the reactants, and the two monosaccharides are the products.

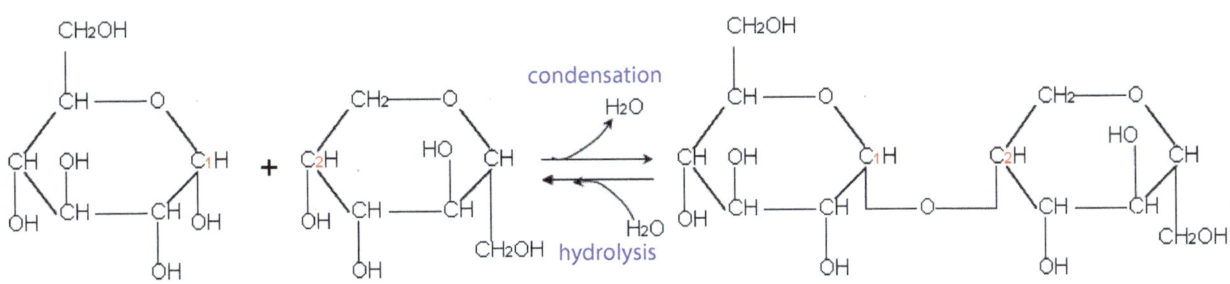

FIGURE 3.4 Condensation reaction, where the reactants glucose and fructose are connected by a covalent bond, resulting in production of sucrose and water

Using condensation reactions, we can make a disaccharide out of any two monosaccharides. We could use two glucoses and make maltose, or a glucose and a galactose and make lactose (milk sugar), and so on. (see Table 3.1). We could also make *oligosaccharides,* short chains of three to ten monosaccharides, using a series of condensation reactions. We could even make *polysaccharides,* like plant starch, which may contain hundreds or thousands of glucoses, using the same type of condensation reaction.

TABLE 3.1. Common Disaccharides and Their Component Monosaccharides

Disaccharide	Monosaccharide Components
sucrose	glucose + fructose
maltose	glucose + glucose
lactose	glucose + galactose

APPLYING SCIENCE TO LIFE 1 Not-So-Sweet Tooth

There are many controversies over the nutritional impact of dietary sugars. Some would like to credit sugars for most of the health ills of this century, including diabetes, cancer, obesity, and so on, and indeed there are some attractive links between sugars and these diseases. Undoubtedly, more sugars are consumed today through beverages and processed foods, which add sugars, as well as salt and fat. Examine a nutrition label on your next soft drink to see how many grams of sugar are present in one serving. Check out the amount of sugar and its calorie contribution in your next fast food meal. You can likely find that information on the company's website. Be sure to check the serving sizes when you tabulate the numbers. While you are there, check out the fat content and its calorie contribution. There seems to be an ongoing debate about which is worse for your diet—sugar or fat—but we won't enter that dispute right now. Eating too much of either will certainly increase the calories in your diet and will likely lead to weight gain.

We will consider the one health problem that we can lay on sugar—*dental caries* (cavities). Like almost all human health issues, there is rarely one underlying culprit. So dental caries are not caused by one factor but several. You need to consider the person's health, socioeconomic status, oral hygiene practice, immune state, and history of fluoride exposure. However, we do know that high amounts of sugar in any diet can promote the growth of bacteria, and the bacteria plus food debris can form a layer of *plaque* (an accumulating community of bacteria that adheres to the surface of your teeth). If high levels of sugar remain in the oral cavity, the bacteria metabolize the sugar and convert it to acidic products. The acid exposure to the enamel causes demineralization—caries. Without proper oral hygiene, accumulated plaque can inflame and damage your gums.

Most carbohydrates are potential sources of energy. Cells of living organisms metabolize sugars to produce energy-carrying molecules, like *ATP,* and heat.

Oligosaccharides are generally attached to other biomolecules, such as lipids to form *glycolipids* or proteins to form *glycoproteins.* The sugar sequences of oligosaccharides vary widely, which makes them quite useful. For instance, your red blood cells have membrane-associated lipids that possess oligosaccharides specifying your blood type. Your immune system recognizes your sequence but attacks blood cells with membrane glycolipids having different oligosaccharides attached. In other words, foreign blood cell glycolipids act as *antigens*, which cause an immune response to rid your body of these antigens. Consequently, if you were given a transfusion from a person with a different blood type, your immune system would create havoc in your body trying to rid you of such an alien invader. (There are a few exceptions).

Polysaccharides generally are storage forms of energy to be utilized in situations when you are out of sugars needed for energy. For instance, your brain, which demands enormous amounts of energy, requires sugar for energy. Other cells in your body can use another important source of energy—lipids. In plants, the main storage polysaccharide is *starch*. Animals can consume plant starch and break it down to sugars for energy. However, if an animal stores energy as a carbohydrate, it must produce an animal starch called *glycogen*. In your body, most glycogen is stored in your liver and in your muscles. We'll come back to glycogen later.

Some polysaccharides are used for structural support. A good example, *cellulose*, is found in plants. Another structural polysaccharide, *chitin*, is used to construct the exoskeleton, or external skeleton, of arthropods (insects, spiders, crustaceans, etc.) and the cell walls of fungi. Generally, to break them down, cellulose and chitin require specific enzymes (biological catalysts), identified by the *-ase* suffix attached to the enzyme name. In the digestive tract of ruminants, like cattle, are microorganisms that produce cellul*ase* (enzyme) to break down cellulose into sugar that they and the cattle can use. Otherwise, cattle would be unable to digest much of the grass they eat. Some bacteria, some fungi, and quite a variety of other organisms produce chitin*ase*, the enzyme that breaks down chitin. Interestingly, chitinases are now being studied for applications in various areas of medicine, such as for control of pathogenic fungi.

APPLYING SCIENCE TO LIFE 2 Putting Nature to Work

Polysaccharides, which are long chains of sugars, generally fall into two categories, storage (energy) and structure. Your digestive tract has enzymes that can break bonds in storage polysaccharides, like starch and glycogen, releasing monosaccharide (e.g., glucose) subunits, which cellular respiration can use for energy. Although your digestive system lacks the necessary enzymes to break down the polysaccharide cellulose, a structural component of plant cell walls, termites have these enzymes, allowing them to digest wood material.

Chitin is a very common polysaccharide, found in the exoskeletons of arthropods and in the cell walls of fungi. Your digestive system also lacks the enzyme *chitinase* to digest chitin. If you consumed chitin by eating an ant or an unshelled shrimp, you would not be able to take advantage of the energy stored in their exoskeletons. However, a diverse array of other organisms do have various forms of this enzyme, including types of bacteria that help to decompose dead organic matter, some plants as a way of discouraging pathogenic fungi, and of course, arthropods, because they must digest the chitin in their exoskeleton so they can shed it during growth (Kuddus 2014; Oyeleye and Normi 2018). As you can imagine, there is interest in using chitinase for human purposes. For instance, pharmaceutical, marine waste, and pest control companies are interested in harnessing chitinase as pesticides or antifungal medicines, for reducing marine wastes, and even for the production of biofuel.

QRQ 5. Select all of the following that are carbohydrates: simple sugars, short chains of sugars, long chains of sugars, simple fats, short chains of fats, long chains of fats

QRQ 6. Fill in the blanks with numbers to indicate the carbon-hydrogen-oxygen ratio in sugars: _____C: _____H: _____O.

QRQ 7. How many carbons are in a monosaccharide?

QRQ 8. Give the empirical formula for sugar.

QRQ 9. Select the following that are monosaccharide(s): clutose, fructose, fruitose, galactose, glactose, glucose, maltose, plantose, plant starch, sucrose, water

QRQ 10. Select from the following the appropriate term for each description below: antigen, APT, ATP, cellulase, cellulose, chitin, chitinase, condensation, disaccharide, drying, glycogen, glycolipid, glycoprotein, hydrolysis, monosaccharide, oligosaccharide, polysaccharides, progen, watering

a. animal starch
b. chain of two monosaccharides
c. chain of three to ten monosaccharides
d. energy-carrying molecule
e. enzyme that breaks down cellulose into sugar
f. enzyme that breaks down chitin into sugar
g. lipid with oligosaccharide attached
h. long chain of sugars
i. plant starch
j. polysaccharide used to construct the exoskeleton of arthropods
k. polysaccharide used for structural support in plants
l. protein with oligosaccharide attached
m. reaction in which water is added
n. reaction in which water is produced
o. storage form of energy

3.3. Lipids

Lipids are an assortment of hydrophobic molecules, usually formed from *hydrocarbons* (compounds composed of carbons and hydrogens) in chains or in rings. Probably the most familiar of these molecules are *triglycerides*, sometimes known as *fats*. If you receive a lipid profile from a blood test, the report will almost always include amounts of triglycerides and various forms of cholesterol. Triglycerides are another valuable source of energy for animals, actually more energy-rich than carbohydrates. Also, layers of fats can pad and protect internal organs, produce hormones, and insulate an organism, helping the animal survive in cold climates.

Structurally, fats are built on a backbone of *glycerol*, to which are bound three fatty acids. Glycerol is an alcohol made up of three carbons, five hydrogens, three alcohol (–OH) *functional groups* (see Figure 3.5). Functional groups are the small groups of atoms within a molecule that have their own characteristic properties and that confer specific chemical properties to the molecule.

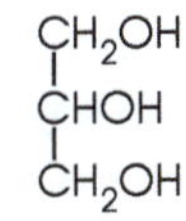

FIGURE 3.5 Diagrammatic representation of glycerol

Fatty acids are long hydrocarbon chains. At one end is a functional group called a *carboxyl group* (–COOH), circled in Figure 3.6. Commonly, there are even numbers of carbons in animal fatty acids, varying from 12 to 24. The top diagram in Figure 3.6 displays a fatty acid with 18 carbons forming a chain, with the carboxyl group (–COOH) on the left. The rest of the carbons are bonded not only to each other but to hydrogens. Note that each of the carbons is connected by a single covalent bond (drawn as a short line segment) to the greatest number of atoms possible (four). Fatty acids with this structure are called *saturated*, because each carbon is bound to the maximum number of hydrogens possible. The two lower diagrams in Figure 3.6 each have one double bond between two carbons. These fatty acids are called *unsaturated*, because the two double-bonded (drawn as two short, parallel line segments) carbons could potentially bind to two more hydrogens.

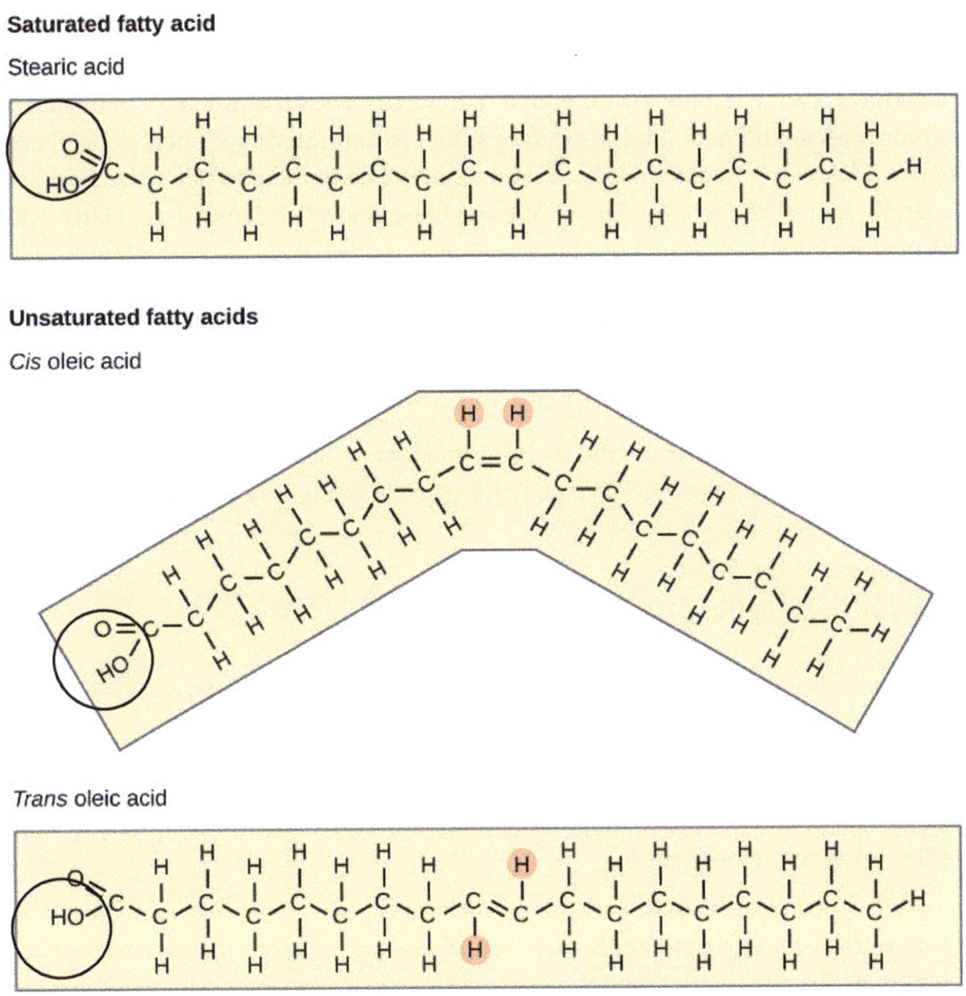

FIGURE 3.6 Diagrammatic representation of a saturated fatty acid (top) and two unsaturated fatty acids (middle, bottom); carboxyl (acid) groups circled

In Figure 3.7, we see three fatty acids forming covalent bonds with the three carbons of glycerol by a condensation reaction. The product is a triglyceride, or fat.

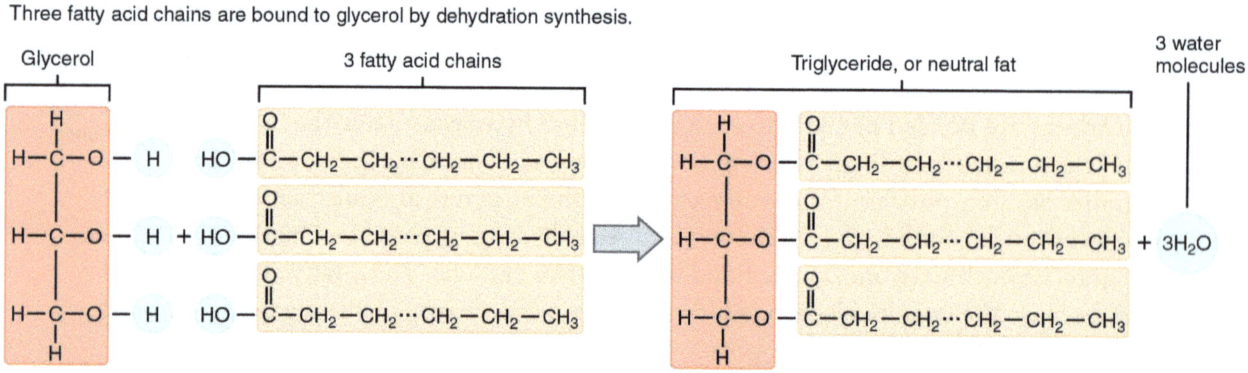

FIGURE 3.7 Triglyceride formed by adding three fatty acids to a glycerol molecule through condensation reactions

You might have read that it is recommended that we eat foods that are high in unsaturated fats, such as salmon, seeds, and nuts, and avoid foods high in saturated fats, such as beef, cream, and butter. Triglycerides, composed primarily of saturated fatty acids, tend to pack together more tightly and are generally solid at room temperature. We call these lipids *fats*. Unsaturated fatty acids, which have kinks in their fatty acid chains, do not pack together as tightly and are often liquid at room temperature. These lipids are called *oils*.

APPLYING SCIENCE TO LIFE 3 Fat of the Land

The Pew Foundation reviewed data from the US Department of Agriculture and found some interesting shifts that have taken place in the American diet. Comparisons of diets in 1970 and current diets show that we eat

- around 23 percent more calories
- more flours (mainly white, processed) and grains
- more fats and oils
- fewer eggs
- more chicken and less beef
- less milk but more cheese
- more high-fructose corn sweeteners

Food is generally abundant in the United States, and most Americans take full advantage. In 2010, the average American calorie intake was 2,481 calories per day, and about half of that came from two food groups, flours/grains (breads, pastries) and fats/oils (mostly vegetable sources, with significant amounts of butter). Breads and pastries are composed of a large percentage of carbohydrates, some protein, and some fat or oil. Carbohydrates and proteins, when metabolized, each yield about four calories per gram, but fats and oils yield about nine calories per gram. So you can see that fats and oils are very rich sources of energy. Americans consume about one-third of their daily calories in fat, or 827 calories.

Our digestive tract needs plenty of time to digest a meal containing fat. First, fats tend to form globs, or clumps, because they are hydrophobic. It takes some time for the stomach to break up the globs for passage into the small intestine, where most fat digestion takes place. In the small intestine, bile salts from the liver help to break up the fats into smaller clumps (*emulsification*) so that digestive lipases secreted by the pancreas can access the fat. The fat is converted into *fatty acids* and *monoacylglycerol* (glycerol plus one fatty acid), which are small enough to be absorbed by the cells that line the small intestine.

Once inside the intestinal cell, the fat components are reassembled with proteins and other lipids into lipoprotein packets (*chylomicrons*). Because of the hydrophobic nature of triglycerides, they must bind to hydrophilic proteins to improve solubility and transport in the blood. These packets are released into the vessels of the *lymph system* (a system of vessels, organs, and nodes that transport the lipoprotein packets into the bloodstream; also an important component of the immune system). These chylomicrons circulate through the body and pass along their contents to various tissues of your body. For example, in the capillaries of heart and skeletal muscles, an enzyme breaks up the fats into fatty acids and glycerol. Glycerol is taken up by the liver, where it may be used to produce sugar. The cells of these muscle tissues use the fatty acids to produce ATP. In fact, fatty acids provide 60 to 90 percent of the energy needs for the heart muscle. Adipose (fat storage) tissue can use fatty acids to make fats for energy reserves.

Many of our dietary fats come from meats, which are composed of saturated fatty acids. Besides the impressive number of calories contained in large portions of fats, saturated fats have another significant effect on your health. Saturated fats promote elevation of your *LDLs* (low density lipoproteins). LDLs are also lipoprotein packets that transport cholesterol and other lipids in your blood. High levels of LDLs can help form plaque in the walls of your arteries. Such plaque tends to stiffen and narrow your arteries. If a plaque fractures, blood platelets (which help form clots) will bind to the rupture, further obstructing your artery. Restricted blood supplies usually result in pain but also may lead to a heart attack or stroke.

After reading this information, you might consider limiting the amount of saturated fat in your diet. There are healthier fats, like those that are found in fish, olive oil, and nuts, but remember they are loaded with calories. However, eating them in moderation makes for a better diet.

There seems to be a debate about fats versus carbohydrates in your diet. You should know that carbohydrates make up about half of all the calories in most people's diets. Avoiding fats and oils and substituting sugar is really not such a good plan. Looking at the dietary information on many low- or no-fat foods, you may be surprised to see that the makers have substituted sugar for fat. Although sugar is lower in calories per gram than fat, excessive consumption of sugar can make you fatter too. The unpleasant truth is that sugar, in excess, can be converted into fat in a process promoted by insulin. Once you have the fat, you must metabolize it, because your body is unable to convert fat back to sugar.

Fats are important as energy sources and reserves, as sources of some hormones, as insulation, as padding for your soft organs, and for proper absorption and storage of fat-soluble vitamins. However, excess fat, like excess sugar, in your diet does not promote a healthy life (Desilver 2016; Pew Research Center 2016; USDA 2019).

APPLYING SCIENCE TO LIFE 4 Lost and Found

Body Mass Index (*BMI*), an indicator of body fat, is calculated by dividing your weight in kilograms by the square of your height in meters. For people whose bodies are lean and muscular, BMI may not be a very useful indicator, but BMI is an easily calculated indicator for most of the population. A BMI from 18.5 to < 25 is normal; 25 to < 30 is overweight; 30 to < 40 is obese; 40 or higher is extremely obese (NIH 2019, "Body Mass Index Table"). Most nutrition scientists and fitness experts would say that in the United States and many other countries there is an obesity epidemic. According to the CDC in 2016, 71.6 percent of those 20 years old and older and 20.6 percent of teens are overweight or obese (CDC 2018; CDC 2017; NIH 2019, "Body Mass Index").

Most of us have probably tried to lose weight. Movies, television, and magazines have displayed endlessly beautiful, handsome, slender, and fit examples for us to emulate. Comparing ourselves to these models of human perfection can be frustrating for the bulk of humanity. Added to that pressure is the body-shaming community on social media. It is little wonder that people in the United States undergo unnecessary, outlandishly expensive, elective alterations of their body parts. If not plastic surgery, we try diets, with and without exercise. If we succeed in keeping the weight we lose at bay, it is usually an exceptional success. Most people use some sort of fad or extreme dieting regimen, only to gain back any losses and then some.

Dieting in this country is not a new practice. Of course, there were times when a little extra flesh was a sign of success, an image to which many people aspired. Horace Fletcher during the early part of the 20th century persuaded many Americans to try chewing their food more thoroughly and spitting out fibrous bits as a way to lose weight. A more outlandish effort to induce weight loss was the taking of pills that supposedly contained tapeworm eggs. Once in the intestine, the tapeworm would hatch, attach, feed on partially digested food, and grow. This loss of nutrients available to the host would result in weight loss. Alas, it also might give the person diarrhea. Nevertheless, when the desired weight was obtained, the dieter would take a pill containing a chemical toxic to the worm and get rid of it. This diet was likely a scam, and strangely, it never really caught on.

For various reasons, many people go through diet cycling–diet, lose, gain, diet, lose, gain, and on and on. After many diet disappointments, people often give up. Many people also want a "quick fix" to their weight woes. Taking a pill is a much sought-after solution.

In 2003, the US Food and Drug Administration banned the sale of herbal weight-loss products containing *ephedra* because there were more than 150 deaths in the United States attributed to these products. Ephedra contains ephredrine and pseudoephedrine, considered stimulants, which may cause constriction of blood vessels. Low doses of the ingredients are useful for decongestants, but high doses raise blood pressure. Apparently, it does curb appetite but may also cause you to have a heart attack. If you purchase a decongestant like *Sudafed* at the pharmacy, you will likely have to show identification. Some amateur "chemists" use this over-the-counter medication to make the illegal and dangerous drug methamphetamine. So there was no magic bullet for weight loss with ephedra (USDA 2019; Harvard Health 2004). The battle for weight loss continues.

During the 2009 season of NBC's *The Biggest Loser*, 16 contestants, classified as extremely obese, spent seven months exercising and dieting intensively and lost an average of about 128 pounds. After their heroic efforts to lose weight, these results must have been particularly gratifying.

After six years, a team of scientists did a follow-up study of 14 contestants who agreed to participate. Overall, only one of the former contestants had not regained an appreciable amount of weight. The remainder had averaged a 90-pound weight gain. Sadly, the seven-month intervention was the most success many of the contestants had had in losing weight long-term (Fothergill et al. 2016).

What most of these participants had experienced was a degree of *metabolic adaptation*, a surprising response by the body to energy-intake restriction. It is surprising because during our fits of dieting (energy input restriction) and exercise, our *resting metabolic rate* (*RMR*) decreases. RMR, generally used interchangeably with *basal metabolic rate* (*BMR*), is the minimal amount of energy required to just keep your body maintained, wherein you are not eating and are not moving. Surprisingly, that makes up about two thirds of your entire energy expenditure each day. The remainder of your expenditure is from of feeding and physical activity (both exercise and non-exercise). So an RMR decline makes a big difference in your total metabolic expenditure. The average decline in RMR for the contestants was 660 calories per day. Some of that reduction was a result of the decrease in body mass. Taking size reduction into account, the contestants' RMRs still decreased by over 300 calories per day. And after six years, most of them still had lower RMRs, making it very difficult to maintain the weight loss. It is likely that most people could not have done much better (Dow 2018; Trexler, Smith-Ryan, and Norton 2014; Speakman and Selman 2003).

It is difficult to explain the complexities of why RMR declines as you diet to lose weight. There are at least six hormonal influences and other factors that come into play. As you reduce the intake of energy, your body prepares you for food scarcity. Your body slows down your metabolism so you don't use too much of your energy reserves. With oncoming food scarcity, you need to conserve your energy resources. This characteristic of metabolism was an advantage to our hunter-gatherer ancestors. There were no grocery markets for them to plunder. They had to be prepared for long periods of time between meals. Consequently, this physiological response is actually beneficial in times of famine. However, for the participants in the study, it must be very frustrating that their RMR stayed low even after they were off the extreme diet and exercise regimen. For the scientists trying to tease out an explanation, it must be intriguing to find one.

Finding some reasonable set of recommendations for managing weight is not an easy task. Understanding human nutrition and metabolism is complex, and we don't know all we need to know. Added to the outcomes and advances already outlined, we need to consider the complication of individual differences. It seems reasonable that we each choose a set of objectives for achieving and maintaining a weight that satisfies us as an individual. We must accept reasonable goals for weight loss and maintenance, and we need to adopt a set of habits that we can maintain. Even if we never reach that imagined "ideal" weight and shape, we will be healthier, and hopefully happier, with a lifestyle change we can live with. Almost every health scientist agrees that loss of any unhealthy amount of weight makes us less vulnerable to diseases. Obese and overweight conditions increase our chances of certain cancers, gallstones, high blood pressure, and cardiovascular diseases. The major threat, however, is diabetes. Over 45 percent of adults in the United States are prediabetic (34.5 percent) or have diabetes (13 percent; generally type II) (CDC 2020). Of the diabetics in the United States, 26 percent are overweight, and 61 percent are obese. Those who adopt a healthy lifestyle, based on a healthy diet and regular exercise, are going to be winners.

PoI 1. Calculate your own BMI. Use the following conversion formula if you are using height squared (in^2) and weight (lbs): 703 × weight (lbs) / [height² (in^2)]. Compare your answer to the chart in Figure 3.8.

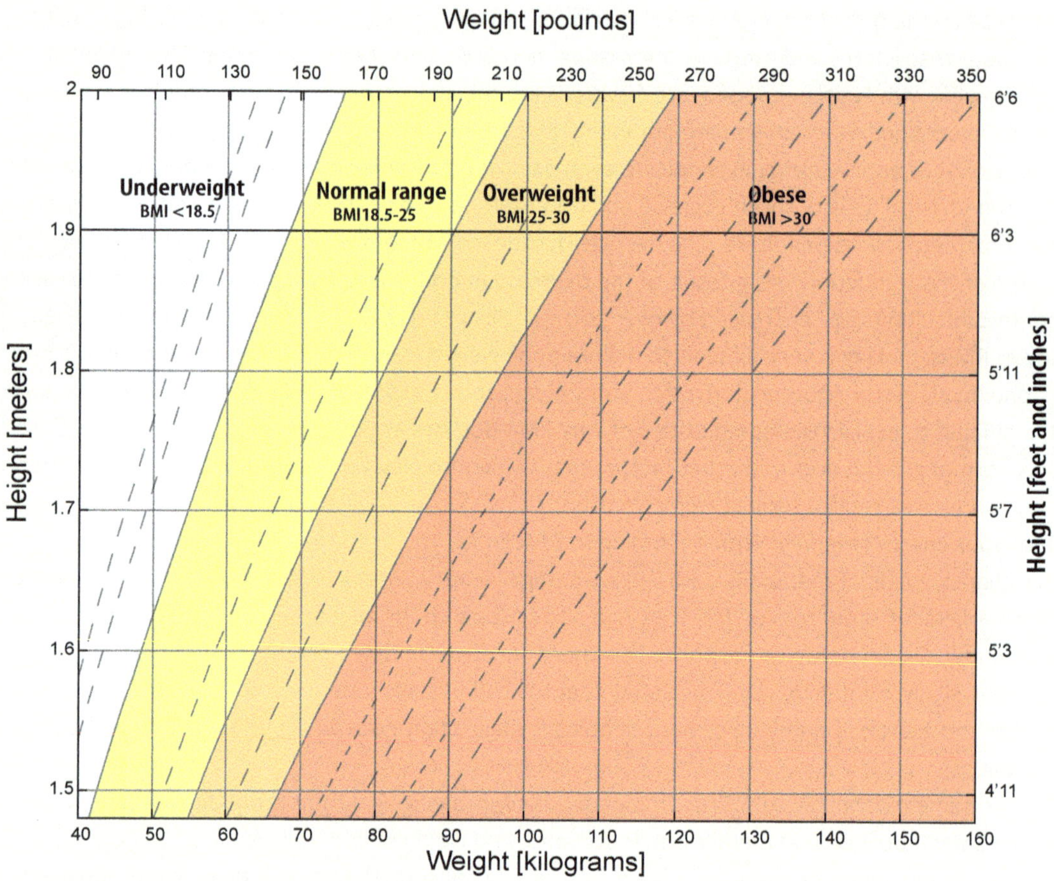

FIGURE 3.8 Body mass index (BMI) chart

PoI 2. Research two diets that are popular today.

a. What are the good and bad points of each?
b. Are these diets practical for most people?

PoI 3. In the follow-up studies of the show contestants, they found that the daily RMR, accounting for body mass lost, declined by about 300 calories. One pound of fat contains about 3,500 calories. Calculate the number of pounds each contestant would have gained in a month (30 days) if they did not adjust the caloric intake of their daily food intake by 300 calories each day.

PoI 4. How many pounds would the contestants gain in a year at the rate in question 3? After six years?

PoI 5. This question investigates the impact of small daily changes.

a. How many calories are in a 12-ounce can of a typical non-diet cola?
b. At a rate of drinking (or not drinking) such a cola every day, how many calories would be consumed (or not consumed) from this daily drink in a year?
c. Assuming no other changes to your diet, how many pounds would you gain (or lose) from drinking (or not drinking) such a cola every day?

PoI 6. Give impacts the following have on RMR: age, sex, ethnicity.

Other lipid molecules, called *phospholipids* (lipid molecules with a phosphate group), are important components of the cell membranes that surround any cell or membrane-bound structure within a cell. Chief among the phospholipids are *phosphoglycerides*. Like a triglyceride, a phosphoglyceride, such as in Figure 3.9a, has a backbone of glycerol (see Figure 3.5), with two fatty acids (see Figure 3.6) attached to the first two carbons. To the third carbon is attached a phosphate group, which in turn is bound to a chemical compound called an amino alcohol, such as ethanolamine in Figure 3.9a. If you substitute another amino alcohol for ethanolamine, such as choline in Figure 3.9b, the molecule becomes phosphatidylcholine. If you removed the phosphate and amino alcohol from the third carbon and replaced them with another fatty acid, the resulting molecule would be a fat or oil. What is important for you to remember is that the fatty acid/glycerol portion of this molecule is quite *hydrophobic*, but the phosphate/amino alcohol portion is *hydrophilic*. As we will see, these two portions make the molecule an ideal component for cellular membranes.

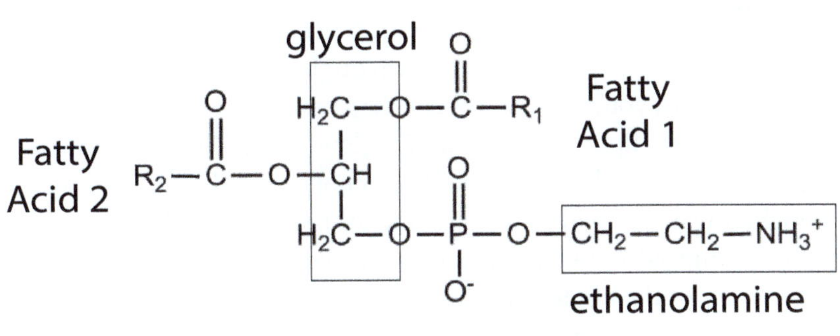

FIGURE 3.9a Phosphoglyceride with R = Ethanolamine

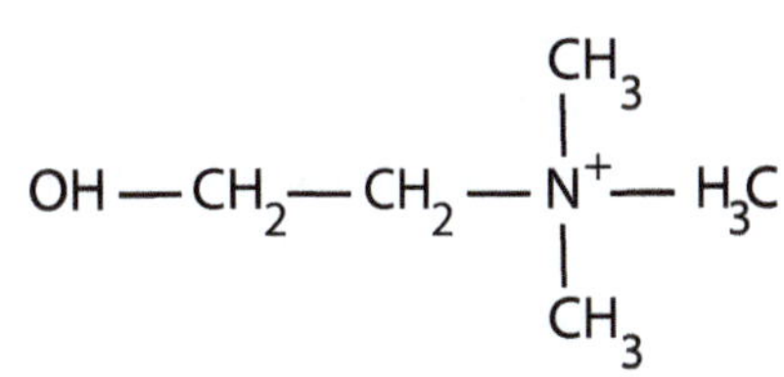

FIGURE 3.9b Choline

APPLYING SCIENCE TO LIFE 5 Phospholipid Failure

Patients diagnosed with Barth syndrome usually display poorly developed skeletal muscles, a weak but enlarged heart, and low white blood cell numbers, especially *neutrophils* (their functions include destruction of bacterial cells). This disease is an inherited disorder. Almost all Barth patients are males. For a female to inherit it, the father and mother would each need to contribute a copy of the defective gene. Males only need one copy of the gene. The mutated gene results in a defective enzyme that is responsible for the remodeling of phospholipids in the inner membrane of the *mitochondria*. One of these phospholipids is a very important one, *cardiolipin*. Cardiolipin is intimately associated with the structure and functions of the mitochondria. We will discuss mitochondrial function later, but it is essential for normal levels of energy production in your cells. Apparently, without remodeling, the cardiolipin is unable to carry out its normal function, disrupting the function of the mitochondria. The symptoms of Barth syndrome are the result. Most Barth individuals have a shortened life span, with many dying in infancy or childhood. Fortunately, the syndrome is quite rare (1/300,000 to 1/400,000 people in the world). Diagnosis can be determined by elevated organic acids in the urine and a genetic evaluation. Generally, there is no treatment but to address specific symptoms (NIH 2016, NIH 2019, "Barth Syndrome").

Other lipid molecules that have important functions in the body are *sterols*, which include cholesterol, vitamin D, bile acids, testosterone, and estrogen. *Cholesterol* has received much bad press but is a necessary component of all animal-cell membranes. Furthermore, cholesterol is the base molecule for the synthesis (production) of other sterol molecules, like fat-soluble vitamins, hormones, and bile acids. Produced primarily in the liver, cholesterol is made up of four fused hydrocarbon rings with a hydrocarbon tail. As Figure 3.10 shows, most of the molecule is hydrophobic (all carbons and hydrogens), with a hydrophilic –OH located at the third carbon in the first of the rings.

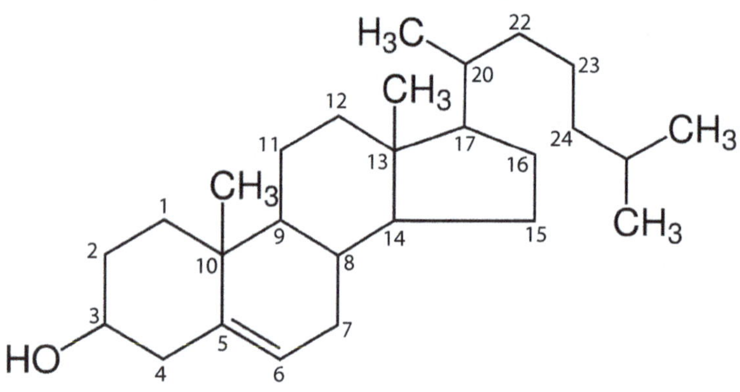

FIGURE 3.10 Diagrammatic representation of cholesterol

Being primarily hydrophobic, cholesterol doesn't dissolve in the blood, which is aqueous. So cholesterol must be transported through the blood attached to proteins, called *lipoproteins*. When you have your cholesterol checked, you obtain a value for *total cholesterol* plus values for *HDL* (high-density lipoprotein) and *LDL* (low-density lipoprotein). You have probably heard the terms "good" cholesterol (HDL) and "bad" cholesterol (LDL). HDL's pick up cholesterol from the tissues, including artery walls, taking some to the liver, which can convert the cholesterol to bile salts and acids. Bile salts emulsify fats in your intestine for better digestion and absorption. The LDLs deliver most of the cholesterol to the tissues, with certain cells of the immune system, called macrophages, taking up most of the cholesterol. These macrophages can accumulate in artery walls, where they contribute to lipid deposits that can promote inflammation and form *plaque* (arterial deposits containing fat, cholesterol, and calcium). The plaque can harden and narrow your arteries or even block them, which may lead to a heart attack.

APPLYING SCIENCE TO LIFE 6 Double-Edged Molecule

Rose Martin, age 32, recently lost her mother to a heart attack at the age of 53. Saddened and worried, Rose went to see her family physician. The physical examination indicated that Rose was 5'7" tall and weighed 165 pounds, which gave her a BMI (body mass index) of 26. Her blood pressure was 135/80. These numbers indicate that Rose was overweight with slightly elevated blood pressure. Rose's blood sugar was 116 mg/dL. The doctor had blood drawn and analyzed, and the significant results were as follows:

- Triglycerides: 180 mg/dL
- Total cholesterol: 320 mg/dL
- LDL cholesterol: 230 mg/dL
- HDL cholesterol: 35 mg/dL

PoI 7. Investigate normal ranges for each of these tests. What do these numbers tell you? Should Rose be worried? Do you think she is at risk for a heart attack? How should her doctor best advise and/or treat her?

PoI 8. What is atherosclerosis? What risks are associated with this disease? What is the underlying cause of this disease?

PoI 9. Two drugs commonly given to patients to lower their cholesterol are atorvastatin and ezetimibe. How do these two drugs work to reduce cholesterol and reduce plaque?

PoI 10. What is the normal range for blood sugar? Is this number of concern? Is there a connection between blood sugar and lipid metabolism?

QRQ 11. (T/F) Lipids are a mix of hydrophobic molecules.

QRQ 12. Select from the following the appropriate term for each description below: carbo-hydrogen, carbonic, fat, fatty acid, fatty base, functional group, glycerol, hydrocarbon, lipid, oil, saturated fatty acid, unsaturated fatty acid

a. a type of alcohol
b. built on a backbone of glycerol
c. compound with hydrogens and carbons
d. each carbon bound to the maximum number of hydrogens possible
e. generally solid at room temperature
f. has kinks when in a chain
g. long hydrocarbon chain with a carboxyl group at one end
h. small group of atoms within a molecule that has its own characteristic properties and that confers specific chemical properties to the molecule
i. tends to pack together tightly
j. triglyceride composed primarily of saturated fatty acids
k. triglyceride composed primarily of unsaturated fatty acids
l. usually formed from hydrocarbons in chains or in rings

QRQ 13. Select from the following the appropriate term for each description below: hydrophilic; hydrophobic, organelle, phosphorus group, phospho-oxygen group, phosphate group, phosphoglyceride, phosphoglycerine, phospholipid

a. fatty acid/glycerol portion of phosphoglyceride
b. has a backbone of glycerol
c. important component of the membranes
d. lipid that contains a phosphate group
e. membrane-bound structure in a cell
f. phosphate/amino alcohol portion of phosphoglyceride
g. phosphorus bound to four oxygens

QRQ 14. Select from the following all that apply to cholesterol: a fat; an oil; a sterol; dangerous, with no known uses; dissolves in the blood; most of the molecule is primarily hydrophilic; necessary component of all animal cell membranes; produced primarily in the liver; produced primarily in the spleen; transported through the blood attached to lipoproteins; used in the production of fat-soluble vitamins, hormones, and bile acids

3.4. Proteins

3.4.1. Protein Composition

Proteins are large organic molecules containing one or more chains of amino acids, called *polypeptides*. An *amino acid* is characterized by an amino group ($-H_3N^+$) and an acid carboxyl group ($-COO^-$) attached to a central carbon (alpha [α]-carbon). The central carbon is also attached to a hydrogen and a variable (R)-group (see Figure 3.11) It is the *R-group* that determines the chemical behavior of an amino acid in proteins. The carboxyl group of one amino acid interacts with the amino group of the next amino acid to form a *peptide bond* (see Figure 3.12) in a chain of amino acids (polypeptide) (see Figure 3.13). Note that the formation of this bond is another example of a condensation reaction—water is removed. With each having a different R-group, there are 20 amino acids common to animals. The chart in Figure 3.14 groups the amino acids according to the chemical nature of their side chains. You don't need to know the details

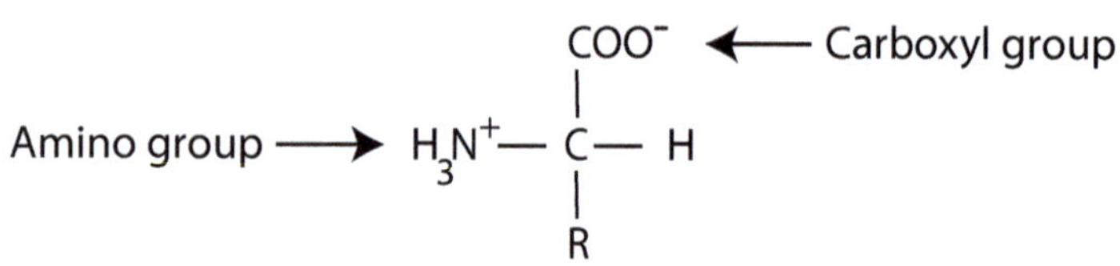

FIGURE 3.11 Generalized structure of an amino acid

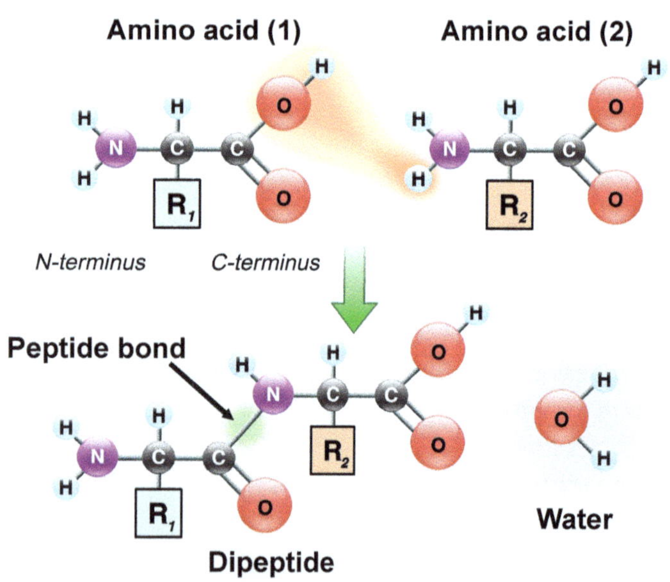

FIGURE 3.12 Formation of peptide bond

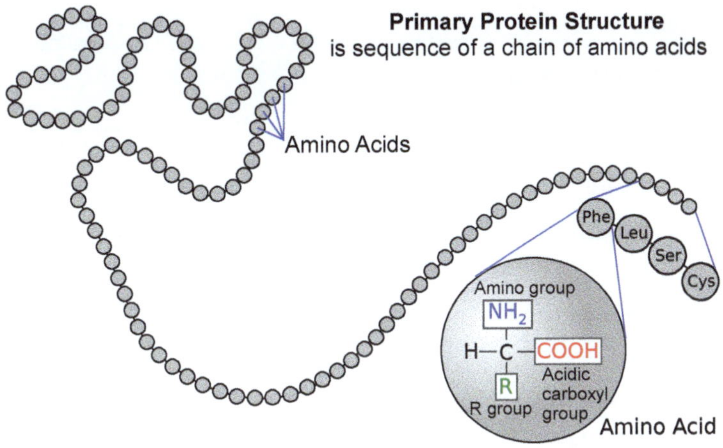

FIGURE 3.13 Primary protein structure

of the chart, but be aware of the major classes of R groups, which are very important in determining the final structure of the protein.

Although the amino acid sequence is essential for determining the final protein shape (*conformation*), protein structure is much more complex. Individual polypeptides are folded and twisted into specific three-dimensional shapes that are crucial for the proper functioning of the protein (see Figure 3.15). The shapes are maintained by bonds that form within the same polypeptide and between polypeptides. In Table 3.2 is a review of common bond types that are important for protein shape. In addition to these bonds are *hydrophobic interactions*, where amino acids bearing nonpolar (water-repelling) R-groups cluster in an attempt to avoid the polar, aqueous environment of the cell.

TABLE 3.2. Basic Bond Types Important to Forming the Functional 3-D Protein Shape

Bond Type	Description
disulfide	covalent bond formed between SH-containing cysteines
hydrogen	attraction between a somewhat positively charged hydrogen atom attached to a strongly electronegative atom (e.g., nitrogen, oxygen) and another electronegative atom
hydrophilic	amino acids orienting outward to be close to water
hydrophobic	amino acids orienting towards the polypeptide center to avoid water
ionic	bond formed between oppositely charged R-groups (acidic and basic)

Examine Figure 3.15 and note the types of bonding that help shape the final protein. You see examples of hydrogen, ionic, and covalent (disulfide bridge) bonds, plus *hydrophobic interaction* between two hydrophobic R-groups (hydrophobic molecules tend to cluster with others and away

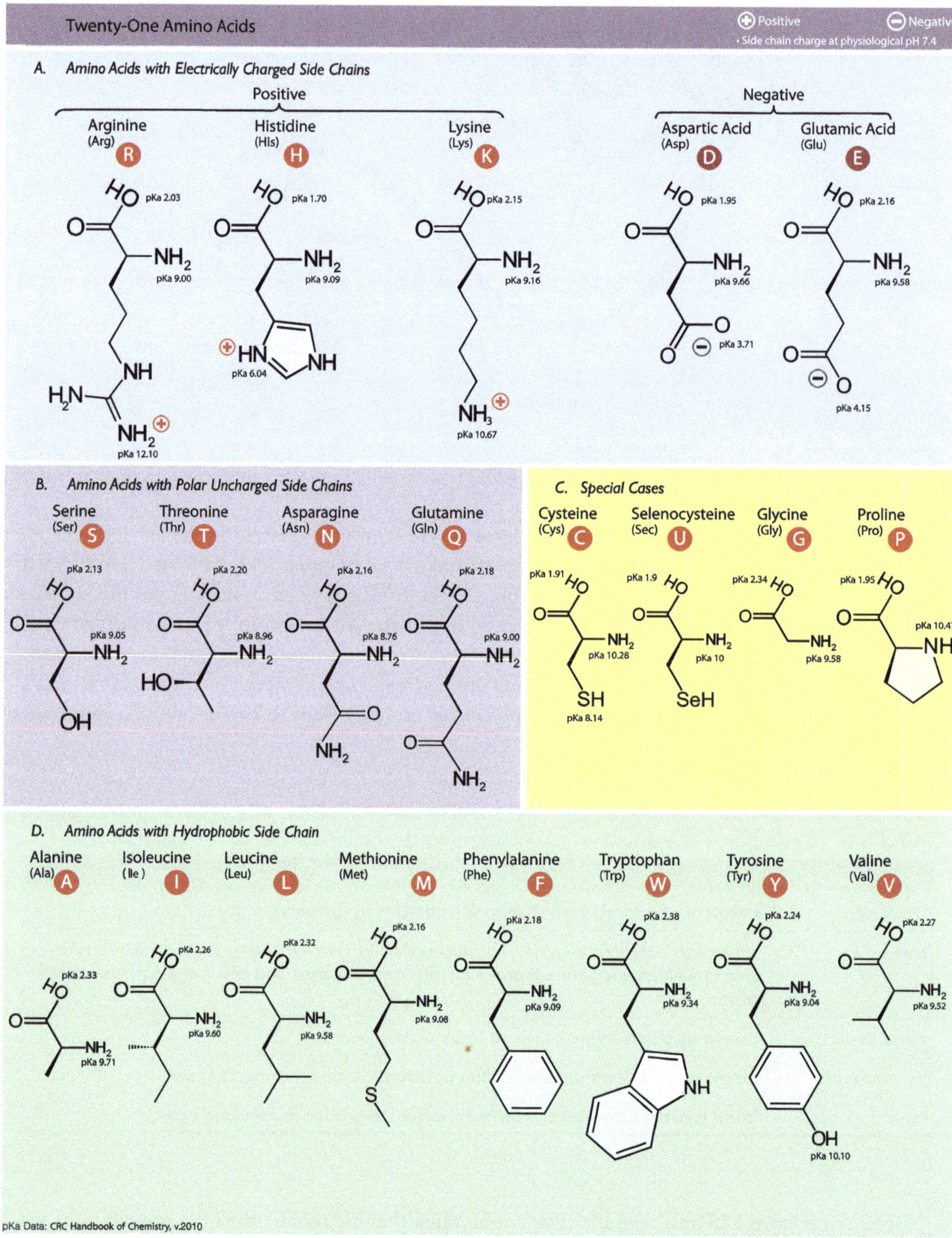

FIGURE 3.14 Amino acids classified by chemical nature of R-group

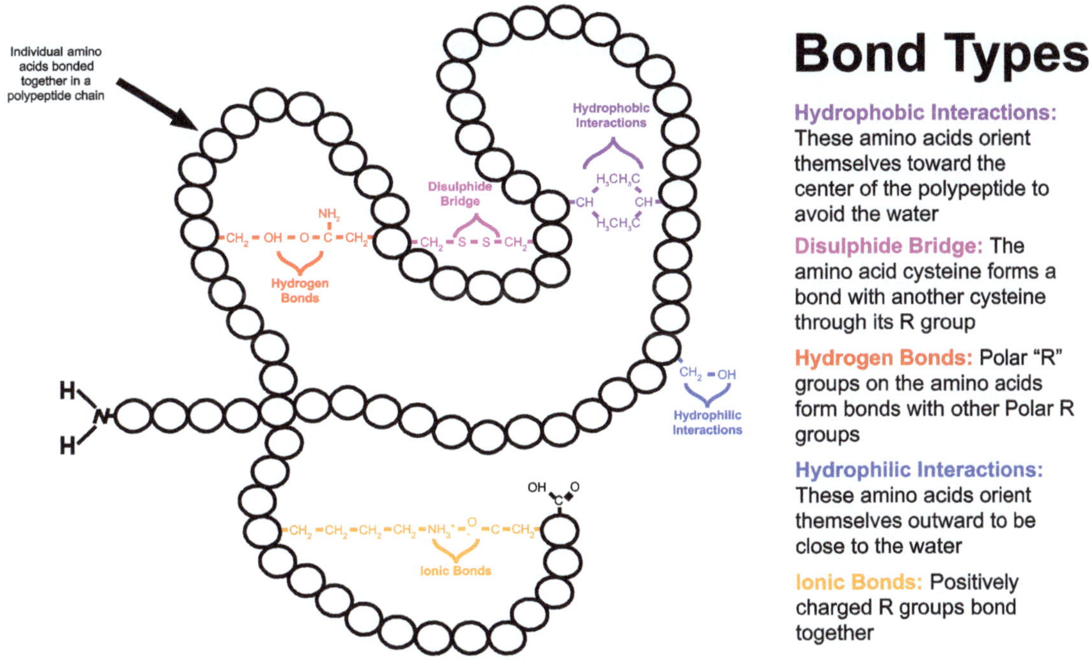

FIGURE 3.15 Folding of polypeptide chain. 3-D shape is maintained by bonding of R-groups between amino acids in the chain.

from polar or hydrophilic molecules). If you heat proteins enough, you will destroy some of the bonds that hold the protein in its functional shape. This process is called *denaturation*. Other ways of disrupting protein shape and function are to alter the pH or to add lots of salt (ionic concentration). Can you think of any reason why pH or salt concentration may lead to denaturation?

Some proteins are made up of only one polypeptide, while others may be combined with other polypeptides to form a functional protein. For instance, *myoglobin* is a heme-containing (iron-containing) protein with only one polypeptide that is found in striated and cardiac muscle. Myoglobin primarily serves to store O_2 for quick transfer to the muscles. However, *hemoglobin*, which transports O_2 in the blood, is also a heme-containing protein but has four polypeptides (see Figure 3.16).

FIGURE 3.16 Hemoglobin, made up of four polypeptides

APPLYING SCIENCE TO LIFE 7 Thalassemia: Sea Sickness

Hemoglobin (Hb) is a protein crucial to supplying sufficient oxygen to all the cells in our bodies. As we discussed earlier, hemoglobin is made up of four polypeptide (globin) subunits. Two of the subunits are made up of *α-globin*, and the other two subunits are made up of *β-globin*. Sometimes the genes that code for these two globins are altered. These alterations are called *mutations*, and they are *autosomal* (not on sex chromosomes) *recessives* (takes more than one copy of the defective gene for expression). If the mutation is critical enough, it will lead to a decrease or failure of synthesis of that globin type. The mutation can occur in the alpha or the beta genes, leading to an imbalance of alpha- and beta-globins, which will form defective hemoglobin molecules. Defective hemoglobin proteins can lead to the destruction of red blood cells (RBC). Low RBC counts result in a range of anemias and, in some severe cases, death. Some forms are mild enough to escape detection.

There are two pairs of α-globin genes, and one of each pair is located on chromosome 16. You get one chromosome 16 from your mother and one from your father. The most commonly seen mutations in the α-globin genes are deletions of portions of the gene. On each of the two chromosome 11s is a copy of the β-globin gene. There are about 400 mutations that occur in this gene, most involving a change in only one subunit.

The resulting disease is called thalassemia, which comes from the Greek word for sea (*thalassa*). Alpha-thalassemia is most common in ethnic groups from African and Southeast Asian descent, and beta-thalassemia is most common in those groups, also, as well as in those from Mediterranean descent.

Thalassemia represents a good example of how a defect in one protein may have calamitous consequences. In most cases, treatment involves frequent transfusions. Eventually, with our advances in molecular biology and genetics, we may be able to repair some of these mutations (NIH 2019, "HBA1 Gene").

APPLYING SCIENCE TO LIFE 8 Mutation in β-Globin: Misshaped Future

Beta-thalassemia is not the only inherited disease that results from a mutation in the β-globin gene. Another inherited defect, caused by the change in one amino acid in the β-globin gene, is sickle cell anemia, which is a well-known variant of sickle cell disease. If the amino acid in position six in the beta-globin is changed from glutamic acid to valine, the globin molecule forms *Hemoglobin S* (*HbS*). Sickle cell manifests in a patient if both genes (genes from each parent) are mutated. Both β-globins in the red blood cells will be HbS. These abnormal molecules will adhere to each other, forming a rather stiff and stringy combination, which causes the red blood cells (RBCs) to assume an unusual, sickled shape. These abnormal red blood cells tend to have a shorter life span (lowering RBC count), causing the characteristic anemia. Moreover, their rigid form can block small blood vessels, leading to damaged organs and considerable pain. One organ that can be damaged is the spleen, which recycles worn out RBCs, stores a reserve of blood, and participates in the immune system (producing antibodies and removing bacteria). This disease is also characterized by swelling of hands and feet, delayed growth and development, and vulnerability to infections. Moreover, sickle cell anemia may lead to stroke or blindness. It occurs most commonly in persons of African ancestry.

Little was understood about this disease until the 1970s. Since then, we have learned much about effective treatments. Bone marrow transplant is effective, but it is a risky procedure, and matching donors are difficult to obtain. Such a transplant also must be done for patients under the age of 16. Other treatments are given to relieve discomfort.

In areas like Africa, where malaria is prevalent, a person with the mutated β-globin gene has an advantage. Malaria is caused by a protozoal parasite (*Plasmodium* species) that completes its life cycle in blood cells of the host. When the parasite enters a red blood cell of a person who only has one copy of the sickle Hb gene, and thus lacks sickle cell characteristics, the parasite induces normal-looking RBCs to sickle. Then the deformed red blood cells are picked up by the immune system and destroyed. The consequent malaria is rarely fatal and may be milder than in people without the mutation (Luzzatto 2012; Mayo Clinic 2019; Vichinsky 2008; CDC 2019, "HBB Gene").

3.4.2. Protein Functions

Proteins have a number of vital functions in living organisms, both at the cellular and organismal levels. Some of the major functions include enzymes (catalysts), structural components, transport, antibodies, and messengers. We will consider enzymes here and the other functions in subsequent pages.

PoI 11. What may have delayed research on sickle cell anemia?

3.4.3. Enzymes

Enzymes as Catalysts

Enzymes are mostly proteins that *catalyze* (speed up) chemical reactions. Living organisms need enzymes because, at the relatively moderate temperatures harmonious with life, chemical reactions do not proceed fast enough to sustain living creatures. Without enzymes, the only way to speed up these enzymatic reactions would be to add heat energy, raising the temperature. Think of the tremendous heat required to bring water to a boil. Living organisms could not withstand the temperatures required.

Cells contain hundreds or thousands of enzymes that function to support the chemical activities required for normal function. Enzymes speed up reactions and can be reused many times. Depending on cellular needs and conditions, they can speed up reactions in either direction. Furthermore, enzymes only speed up the reaction but do not change the reactant-product equilibrium.

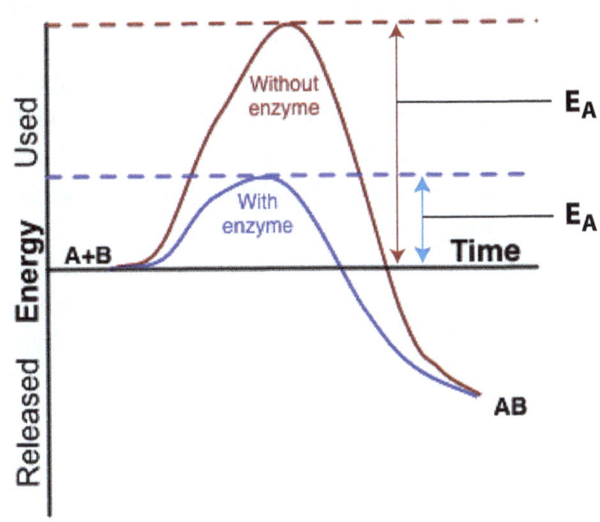

FIGURE 3.17 Reaction sequence for an exergonic reaction, with and without an enzyme

An enzyme catalyzes a reaction by reducing the *activation energy* (E_A) for the reaction, lowering the energy input required for the reaction to take place. Note the reduction of E_A for the exergonic reaction in Figure 3.17. The reaction is exergonic, so as the reactants interact, some energy is released. With the enzyme, the transition state occurs with less energy input, and the reaction proceeds more easily.

There are many different enzymes, and each interacts with a certain reactant, called a *substrate*, by binding it to a specific region, called the *active site*. Assuming the enzyme is in a functional conformation, the recognized substrate can bind noncovalently (using weaker bond types) to form a loose fit with the enzyme, which induces shape changes in the active site so that there is a better, closer fit. This change helps to orient the reactants properly for catalysis to take place. Following the catalytic action, products are released, no longer fitting tightly to the active site (see Figure 3.18).

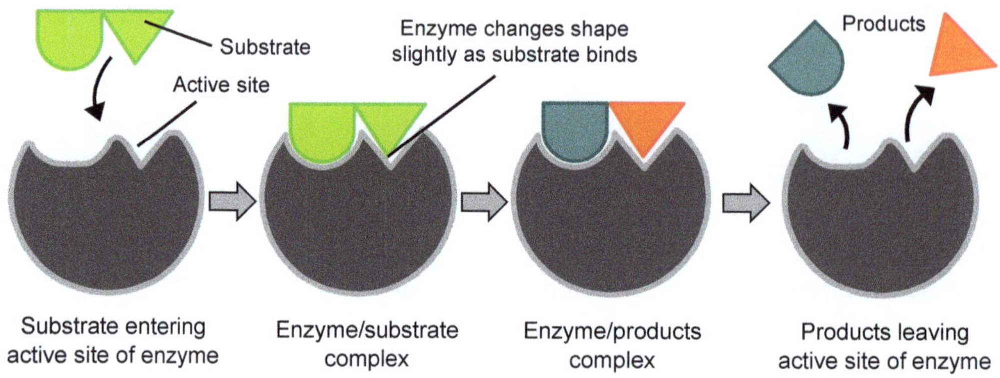

FIGURE 3.18 Diagram showing the binding (loose fit), induced, tight binding, and release of products

Enzyme Reaction Types

There are at least six major types of enzymatic reactions, but we are only going to describe two significant examples: oxidoreductases and hydrolases. One of the most important type of chemical reaction in the cell is the *oxidation-reduction* (*redox*) reaction. In this reaction, one molecule loses or donates electrons, a process known as *oxidation*, and one molecule gains electrons, a process known as *reduction*. Thus, the participant that donates the electrons (i.e., is oxidized) is called the *donor*, and the participant that gains electrons (i.e., is reduced) is called the *acceptor*. In these types of reactions, catalyzed by an *oxidoreductase*, there is always an acceptor and a donor. Oxidoreductases are usually called *dehydrogenases* or *reductases*. If an oxygen molecule is the acceptor, we call the enzyme an *oxidase*.

Some enzymes use water to break covalent bonds in a molecule. Reactions of this type are catalyzed by a class of enzymes called *hydrolases*. Specific examples of hydrolases include those that break down lipids (*lipases*), carbohydrates (*glycosidases*), proteins (*peptidases* or *proteases*), and nucleic acids (*nucleases*).

APPLYING SCIENCE TO LIFE 9 Out, Damned Spot!

It is too bad that William Shakespeare didn't know about enzymes, or he could have helped Lady Macbeth get the blood off her hands. You have undoubtedly spilled food or drink on a favorite shirt or blouse.

Or perhaps you have cut your finger and some of the blood found its way to the leg of a pair of slacks. Removal of many stains used to require harsh and/or elaborate treatments, sometimes performed by a professional launderer. Now, many of these same stains can be removed by the presence of enzymes in your liquid laundry detergent. As you remember, enzymes are biological catalysts that work quickly and in small amounts on specific target substrates. Most of the enzymes in detergents are *hydrolases*, which act by using water to split covalent bonds. As the enzyme digests the stain molecules, smaller, more easily removed molecules are produced that are flushed out during the wash.

Not all laundry detergents contain enzymes, but if you examine the ingredients to find out, many of the major brands contain a minimum of three types of enzymes: amylase, mannanase, and protease. In addition to these enzymes, some brands also have one or more of the following enzymes: lipase, pectin lyase, and cellulase. For most of these enzymes, their names tell you what their potential substrates are. Proteases digest (break up) proteins; lipases digest fats or oils; cellulase digests cellulose. Less obvious to you may be amylases (digest starches), mannanase (digests polysaccharides containing the sugar mannose), and pectin lyase (a non-hydrolase enzyme that digests pectin, a polysaccharide found in plant [fruits] cell walls). Table 3.3 summarizes information on enzyme stain removal.

TABLE 3.3. Stain-Removing Enzymes Commonly Found in Commercial Laundry Detergents

Enzyme	Substrate	Stain Type Removal
amylase	starches	starch-based gravies, pasta dishes
mannanase	plant polysaccharides containing mannose (called mannans)	thickeners found in barbecue sauce, salad dressing, shampoos, etc.
protease	proteins	blood, sweat, grass, chocolate
lipase	fats and related lipids	grease
pectin lyase	pectins in fruits	wine, tomato

You may note that cellulase is not found within the table. Cellulase is an enzyme that digests the plant polysaccharide cellulose and is particularly useful for laundering cotton garments. During normal wear, cotton fabrics produce small fibers (microfibers) on the surface that not only change the texture but also tend to dull the color. Cellulase in the detergent acts on these small fibers but does not seem to affect the main fibers of the garment. Removal of these microfibers also removes stains clinging to them, returning the fabric to its former, brighter color.

Of course, laundry detergents have many other ingredients that are important for cleaning clothes. However, enzymes make it possible to remove stains at lower temperatures (saving energy) and without the use of more stringent treatments.

PoI 12. Which enzymes would be most useful for removing stains from

a. spaghetti with tomato sauce spilled on your clothing?
b. giblet gravy on your grandmother's tablecloth?
c. your soccer uniform after a big match?
d. barbeque chicken stains on your t-shirt after a picnic?
e. an abattoir (slaughterhouse) worker's white cotton uniform after a day at work?

QRQ 15. A large organic molecule containing one or more chains of amino acids is a(n) ______.

QRQ 16. A chain of amino acids is a(n) ______.

QRQ 17. In a chain of amino acids, the carboxyl group of one amino acid interacts with the amino group of the next amino acid to form a ______.

QRQ 18. Fill in the blank with a number: There are ______ amino acids common to animals.

QRQ 19. Select from the following the appropriate term for each description below: alpha helix, amino acid, C-group, conformation, denaturation, hemoglobin, ironglobin, myoglobin, R-group

a. characterized by an amino-group, a carboxyl group attached to a central carbon that is also attached to a hydrogen and a variable R-group
b. determines the chemical behavior of an amino acid in proteins
c. final 3-D shape of a protein
d. iron containing protein with four polypeptides; transports O_2 in blood
e. iron containing protein with one polypeptide; found in striated and cardiac muscle
f. primarily serves to store O_2 for quick transfer to the muscles
g. process of destroying, perhaps by heating, some of the bonds that hold the protein in its functional shape

QRQ 20. Proteins that speed up chemical reactions are called ______.

QRQ 21. (T/F) Enzymes alter the reactant-product equilibrium.

QRQ 22. (T/F) An enzyme reduces the activation energy for a reaction.

QRQ 23. (T/F) An enzyme raises the energy input required for a reaction.

QRQ 24. In a redox reaction, one molecule, known as a(n) ______, loses or donates electrons in a process known as ______.

QRQ 25. In a redox reaction, one molecule, known as a(n) ______, gains electrons in a process known as ______.

QRQ 26. Select from the following the appropriate term for each description below: active region, active site, binding region, binding site, catalyze, catatonic, hydrolase, oxidation-reduction, substrate

a. binding region of an enzyme
b. reactant that interacts with a certain enzyme
c. redox
d. speed up
e. uses water to break covalent bonds in a molecule

3.5. Nucleic Acids

Consider how vital the *nucleic acids*, double-stranded DNA (*deoxyribonucleic acid*) and single stranded RNA (*ribonucleic acid*) must be, conveying coded information necessary for the synthesis of *all* proteins. Proteins play crucial roles in almost all cellular and many organismal activities. Examples of protein functions include biological catalysts (enzymes), transporters, signals and receptors, immune molecules, structural molecules, and locomotory molecules. To make a protein, RNA receives the code from the DNA by the process of *transcription*, and the RNA code is then *translated* into an amino acid sequence for a new polypeptide. Of course, the translated product will undergo further modification on its road to becoming a functional protein, but the essential information has come from the DNA. DNA is also a molecule that is capable of replication, which places it at the foundation for all reproduction, both cellular and organismal.

Nucleic acids are *polymers* (macromolecules comprised of repeated subunits that are structurally similar), made up of a succession of subunits called *nucleotides*. Each nucleotide is a *monomer* (one subunit of a chain of subunits in a polymer). Each nucleotide monomer is made up of a 5-carbon *sugar* (pentose), a *phosphate* (PO_4), and a *nitrogen base* (nitrogen-containing, with the chemical properties of a base). In DNA the sugar is *deoxyribose*, and in RNA the sugar is *ribose*. As Figure 3.19 shows, each phosphate binds covalently to two sugars. This chain of sugars and phosphates forms the backbone of the DNA (also in RNA), and each nitrogen base links to a sugar but not to other nitrogen bases in the same chain. In each type of nucleic acid, DNA or RNA, there are four different nitrogen bases. In both DNA and RNA are the nitrogen bases *adenine* (A), *guanine* (G), and *cytosine* (C). In DNA, the fourth nitrogen base is *thymine* (T), and in RNA, the fourth nitrogen base is *uracil* (U).

Nucleotides also serve as the foundation for a number of important molecules that participate in cellular metabolism. *ATP* (adenosine triphosphate) is considered the universal coupling agent that connects energy-producing reactions and energy-demanding reactions. *FAD* (flavin adenine dinucleotide) and *NAD* (nicotinamide dinucleotide) are classified as *coenzymes*, which help to transfer hydrogens (electron and proton) for oxidation-reduction reactions in the cell. Stay tuned—we'll return to these molecules later.

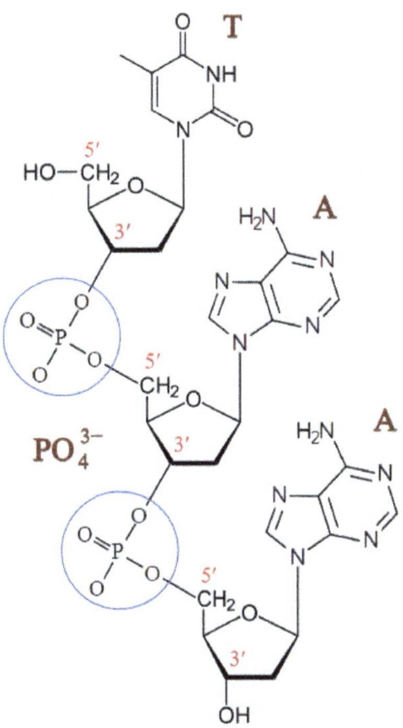

FIGURE 3.19 Short segment of a single strand of DNA with phosphate bonds circled

QRQ 27. To make a protein, ______ receives a code from ______ by the process of ______, and this code is then ______ into a(n) ______ acid sequence for a new polypeptide.

QRQ 28. A subunit of a nucleic acid is a(n) ______.

QRQ 29. Select the three that are components of a nucleotide: five-carbon sugar, three-carbon sugar, amino acid, ATP, nitrogen base, phosphate

QRQ 30. Select from the following the appropriate term for each description below: coenzyme, deoxyribonase, deoxyribose, DNA, monomer, polymer, ribonase, ribose, RNA, single

a. deoxyribonucleic acid
b. helps transfer hydrogen for an oxidation-reduction reaction in the cell
c. macromolecule with repeated, structurally similar subunits
d. molecule capable of replication
e. one subunit of a chain of subunits in a polymer
f. ribonucleic acid
g. sugar in DNA
h. sugar in RNA

QRQ 31. Select all that are bases in DNA and/or RNA: adecil, adenine, adesine, cytocil, cytonine, cytosine, guancil, guanine, guasine, thycil, thymine, thytosine, uracil, uramine, urasine

Image Credits

Fig. 3.1: Source: https://commons.wikimedia.org/wiki/File:Methane-2D-square.png.
Fig. 3.2: Source: https://commons.wikimedia.org/wiki/File:Carbon-dioxide.png.
Fig. 3.3a: Copyright © by Yvn (CC BY-SA 3.0) at https://commons.wikimedia.org/wiki/File:GlucoseTeX.PNG.
Fig. 3.3b: Source: https://commons.wikimedia.org/wiki/File:Alpha-D-Glucopyranose.svg.
Fig. 3.4: Copyright © by Belgarath007 (CC BY-SA 3.0) at https://commons.wikimedia.org/wiki/File:Formation_du_saccharose.PNG.
Fig. 3.5: Source: https://commons.wikimedia.org/wiki/File:Glycerin.svg.
Fig. 3.6: Copyright © by OpenStax College (CC BY 4.0) at https://commons.wikimedia.org/wiki/File:Figure_03_03_05.jpg.
Fig. 3.7: Copyright © by OpenStax College (CC BY 4.0) at https://commons.wikimedia.org/wiki/File:220_Triglycerides-01.jpg.
Fig. 3.8: Source: https://commons.wikimedia.org/wiki/File:Body_mass_index_chart.svg.
Fig. 3.9a: Source: https://commons.wikimedia.org/wiki/File:Phosphatidylethanolamine.png.
Fig. 3.10: Source: https://commons.wikimedia.org/wiki/File:Cholesterol_(chemical_structure).svg.
Fig. 3.11: Copyright © by Ppfk (CC BY-SA 3.0) at https://commons.wikimedia.org/wiki/File:Aminoacid_general_structure.png.
Fig. 3.12: Source: https://commons.wikimedia.org/wiki/File:Peptidformationball.svg.
Fig. 3.13: Source: https://commons.wikimedia.org/wiki/File:Protein_primary_structure.svg.
Fig. 3.14: Copyright © by Dan Cojocari (CC BY-SA 3.0) at https://commons.wikimedia.org/wiki/File:Molecular_structures_of_the_21_proteinogenic_amino_acids.svg.
Fig. 3.15: Copyright © by WikiComTD (CC BY-SA 4.0) at https://commons.wikimedia.org/wiki/File:Tertiary_Structure_of_a_Protein.svg.
Fig. 3.16: Copyright © by OpenStax College (CC BY 3.0) at https://commons.wikimedia.org/wiki/File:1904_Hemoglobin.jpg.
Fig. 3.17: Copyright © by IMeowbot (CC BY-SA 3.0) at https://commons.wikimedia.org/wiki/File:Enzyme_activation_energy.png.
Fig. 3.18: Source: https://commons.wikimedia.org/wiki/File:Induced_fit_diagram.svg.
Fig. 3.19: Copyright © by Kes47 (CC BY-SA 3.0) at https://commons.wikimedia.org/wiki/File:Phosphodiester_Bond_Diagram.svg.

LAB #3

Laboratory Exercises

3.6. Laboratory on Enzyme Kinetics

3.6.1. Preliminary Work

In this lab, we examine the kinetics of an enzyme catalyzing a single substrate. From the text's website, view the introductory video; download and open "*Enzyme Kinetics Multi.nlogo.*"

1. This question refers to some prerequisite definitions. If necessary, look up requested information in other sources.

   a. **1.** ________ **2.** ________ Complete the following statement to define a "catalyst" of a chemical reaction: A catalyst of a chemical reaction is a substance that **1.** (A. decreases, B. does not change, C. increases) the rate of the reaction but **2.** (D. does, E. does not) undergo permanent chemical change during the reaction.

   b. ________ Select all that apply to the term *covalent bond*: A. does not involve the sharing of pairs of electrons by atoms, B. imbalance of attractive and repulsive forces between atoms, C. involves atoms sharing an electron pair, D. involves atoms sharing exactly one electron, E. stable balance of attractive and repulsive forces between atoms, F. unstable balance of attractive and repulsive forces between atoms.

2. Click the Info tab at the top of the page and read "What Is It?"

   a. Select the appropriate notation (Kc, Kd, Kr, E-S, E, S, P, E+S, E+P, Kc / Kd, V_{max}) for each of the following:

      ________ rate constant for complex formation

      ________ rate constant for complex dissociation

      ________ rate constant for catalysis

      ________ maximum rate at which the reaction can proceed

   b. ________ Select the first step in catalysis: A. a molecule of E and a molecule of P combine to create a molecule of ES, B. enzyme catalyzes conversion of substrate to product, C. E-S complex dissociates into a molecule each of E and S, D. E-S complex formation

c. _______ Select the reason(s) why the rate of complex formation is very fast: A. determined by collision of the molecules, B. determined by dissociation of the molecules, C. involves four molecules, D. requires high energy

d. **1.** _______ **2.** _______ Complete the following statement: The rate for an enzyme to catalyze the conversion of substrate to product is much slower than complex formation/dissociation because the required **1.** (A. electromagnetic force, B. energy, C. enzyme, D. substrate) is **2.** (E. much higher, F. much lower, G. somewhat higher, H. somewhat lower) than that required for collision or separation.

e. **1.** _______ **2.** _______ Complete the following statement: The Michaelis-Menten Curve can be plotted by measuring the rate of complex **1.** (A. dissociation, B. formation,) at different **2.** (C. enzyme, D. E-S complex, E. product, F. substrate) concentrations.

f. **1.** _______ **2.** _______ **3.** _______ Complete the following statement: Inhibitors are molecules that are structurally similar to **1.** (A. enzyme, B. E-S complex, C. product, D. substrate) molecules that can complex with the **2.** (A. enzyme, B. E-S complex, C. product, D. substrate) and interfere with the **3.** (E. E-S complex formation, F. E-S dissociation, G. product formation).

3. Read "How to Use It."

a. Match each component with its color in the model below: A. black, B. blue, C. brown, D. cyan, E. gray, F. green, G. magenta, H. orange, I. red, J. white, K. yellow

_______ enzyme

_______ product

_______ substrate

b. _______ Select the slider that establishes the initial number of substrate molecules: A. CONCENTRATIONS, B. Kc, C. Kd, D. Kr, E. MICHAELIS-MENTEN, F. VOLUME

c. **1.** _______ **2.** _______ Complete the following statement: If NUMBER-EXPERIMENTS is 5, MICHAELIS-MENTEN is OFF, VOLUME is 50, Kc = 20, Kd = 25, and Kr = 50, the simulation runs the experiment **1.** (1, 2, 5, 10) time(s), displaying each **2.** (A. estimated final velocity of the reaction, B. estimated initial velocity of the reaction, C. time for complete conversion of complexes to substrates, D. time for complete conversion of substrates to complexes).

d. _______ Give the default value of Kr.

e. _______ Will product be produced with the default value of Kr, no (N) or yes (Y)?

f. _______ For the simulation, give the molecule that does not move: A. enzyme, B. E-S complex, C. product, D. substrate

3.6.2. Laboratory Work

*For each indicated experiment below, using NUMBER-EXPERIMENTS = 10, run each experiment at least **10** times unless otherwise instructed. Use a **spreadsheet** to record results, calculate means, and produce graphs not included with the model.*

4. This question examines the impact of Kc, the rate of complex formation, on the speed of conversion of substrate to product.

   a. _______ Complete the following hypothesis: With the other rate constants held constant, as Kc increases, the time for conversion of substrate to product **1.** (D. decreases, I. increases).

   b. _______ Set NUMBER-EXPERIMENTS to 1, VOLUME to 50 molecules, MICHAELIS-MENTEN to OFF, and Kc = 20, Kd = 25, and Kr = 50. Set up and run the simulation until all substrate has been converted to product. Give the final value of product concentration.

   c. _______ Start a spreadsheet named *EnzymeKinetics,* as in Table 3.4. Set NUMBER-EXPERIMENTS to 10, and click SETUP-MULTI. Uncheck *view updates* or increase the simulation speed. Run the simulation, which executes the experiment ten times; and copy and paste the ten output times for complete conversion into the "Kc = 20" row of your spreadsheet. Because tabs separate times in the output, when you paste the line of times into the spreadsheet, each value will appear in a different column. In the last column of the spreadsheet, calculate the mean time to convert all the substrate to product. Give this mean, rounded to a whole number.

TABLE 3.4. Initial Format of Spreadsheet for Laboratory

Name(s)											
Question 4											
Substrate volume = 50, Kd = 25, Kr = 50											
For each Kc value (10 experiments each), time to convert all substrate to product											
Kc	**Time**										**Mean Time**
20											

   d. _______ Repeat part c with Kc = 40, recording the times (ticks) in the next row of the spreadsheet. Be sure to click SETUP-MULTI before GO-MULTI.

   e. _______ Repeat part c with Kc = 60.

   f. _______ Repeat part c with Kc = 80.

   g. _______ Repeat part c with Kc = 100.

   h. _______ In the spreadsheet, plot the mean time versus Kc. Select all statements that are true about the graph: A. decreasing, B. increasing, C. horizontal, D. concave up, E. concave down, F. linear

   i. Discuss how your results support or refute your hypothesis from part a.

5. Michaelis and Menten devised a mathematical equation that expresses the relationship between reaction rate and substrate concentration. In this question, for a particular situation we develop a Michaelis-Menten Curve and determine V_{max}, the maximum reaction rate or maximum velocity of the reaction. Michaelis and Menten made several simplifying assumptions, including the following:

   - The rate of product formation is slow in comparison to the rate of ES formation and the rate of ES dissociation to E + S. Thus, we should have Kr be much smaller than Kc and Kd.

- The concentration of substrate is much greater than the concentration of enzyme. Thus, when the MICHAELIS-MENTEN switch is ON, we have 50 enzyme molecules instead of 150, as is the case when OFF, and we use values of substrate volume greater than 50.

a. _______ Because we will be generating a Michaelis-Menten Curve, turn ON the MICHAELIS-MENTEN switch. Satisfying the first assumption of Michaelis and Menten above, set Kc to 80, Kd to 62, and Kr to 10. Set NUMBER-EXPERIMENTS to 10 and VOLUME to 75. Thus, setup will place 75 substrate molecules and 50 enzyme molecules in the world at the start of each of the ten experiments and satisfy the second Michaelis-Menten assumption. We will be running the simulation for several different substrate concentrations and recording the values in the spreadsheet. Because the initial number of enzyme molecule is always 50, as a simplifying assumption, we use VOLUME to represent substrate concentration [S]. Start a second sheet in your spreadsheet document similar to Table 3.4. Click SETUP-MULTI and then GO-MULTI. Copy and paste the ten output velocities into the first data row of the table. Give the mean velocity to two decimal places.

For each indicated volume in the next eight parts, set up and run the simulation, enter the volume and the output values into another row of the spreadsheet, have the spreadsheet calculate the mean velocity, and give the mean velocity to two decimal places.

b. _______ *Volume* = 100

c. _______ *Volume* = 150

d. _______ *Volume* = 200

e. _______ *Volume* = 300

f. _______ *Volume* = 400

g. _______ *Volume* = 500

h. _______ *Volume* = 600

i. _______ *Volume* = 800

j. _______ Plot this mean *v* versus *Volume*, which represents [S]. Select all statements about the graph: A. concave down, B. concave up, C. decreasing, D. horizontal, E. increasing, F. linear

k. _______ Estimate the eventual maximum reaction rate, V_{max}, from this graph.

l. **1.** _______ **2.** _______ **3.** _______ **4.** _______ **5.** _______ Complete the following statement explaining why the Michaelis-Menten curve increases steeply initially and then becomes almost horizontal: Initially, there is a lot of **1.** that can interact with the **2.**, and there is no **3.** Eventually, little **4.** remains to be converted to the desired **5.** Answer choices: A. enzyme, B. product, C. substrate

Two additional parts to question 5 related to the Michaelis-Menten constant and another question exploring the effects of inhibitors on catalysis are available in Active Learning.

UNIT II

Cells: Basic to Life

MODULE

4 Understanding the Importance of Cells

4.1. Introduction

Consider the huge numbers of different types of plants and animals on the earth. How many different types do you think there are? Scientists estimate there are 8.7 million (± 1.3 million) plant and animal species, and only about one million of the animal species have been described so far (Census of Marine Life 2011). Beetles alone make up 1.5 million of the 7.7 million species of animals (Stork 2018). And there may be as many as one trillion (1×10^{13}) bacterial species (Locey and Lennon 2016).

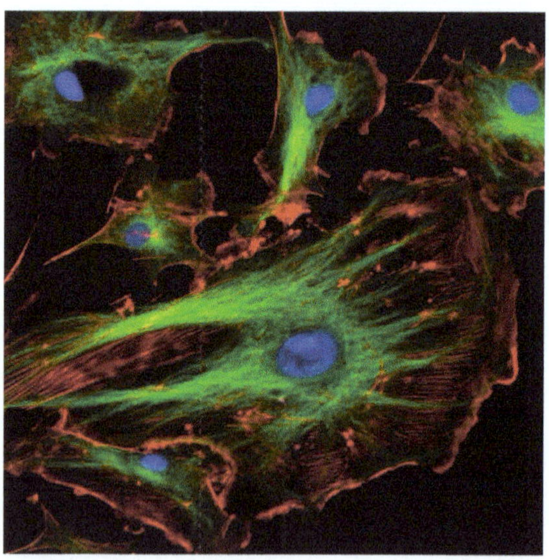

FIGURE 4.1 Bovine pulmonary artery endothelial cells, stained to reveal nucleus (blue), microtubules (green), and actin filaments (red)

All living things are composed of one or more cells. One-celled organisms, like *Paramecium*, are able to carry out all necessary life functions. As organisms became multicellular, they grew larger, cell numbers increased, and cells became modified to perform specific functions. For instance, over 37 trillion cells make up the average human body, and almost all of these cells are specialized in structure and function (Bianconi et al. 2013). Human cells are categorized into about 200 specific types (AAAS 2019).

All cells, regardless of specialty, are surrounded by a membranous envelope called the *plasma membrane*. Inside this membrane is the *cytoplasm* (fluid plus membranous and particulate components), and the fluid part is the *cytosol*. Although some features are common to all kinds of cells, the cells of plants, animals, fungi, and protozoans contain structures formed from internal (intracellular) membranes, as well as nonmembranous particles, tubules, and fibers (see Figure 4.1). These intracellular membranes compartmentalize the cytoplasm, establishing proper environments and platforms for various cellular functions. Thus, membranes are integral to the structure and function of these cells.

4.2. Life Support—Cell Membranes: Structure Underlying Function

Whether the membrane surrounds the cytoplasm or a nucleus or forms a cytoplasmic compartment, the basic organization is the same—a double layer of *phospholipids* associated with a variable set of *proteins*, depending on the actual function of the membrane. Let's take a look at the components of membrane structure, but as we examine these components, focus on their chemical nature. Avoid getting bogged down in the finer details. It is the chemical nature of each component that enables membranes to carry out their functions for cells.

4.2.1. Membrane Components

All types of membranes are composed of similar components—mostly lipids and proteins. *Lipids* are carbon-based molecules that are not soluble in water, but they are soluble in *hydrophobic* solvents, which repel water, such as gasoline. *Proteins* are large molecules made up of one or more chains of *amino acids*, which we discuss in a later section. The proportions of and specific types of each component vary with the type and function of the membrane involved.

As predicted by extensive experimental work, a double layer of phospholipids establishes the basic structure of any membrane. The exact composition of phospholipids varies, but the primary players are *phosphoglycerides* (as described in module 3). As you learned in that module on organic molecules, phosphoglycerides, such as that pictured in Figure 3.9a, are a type of phospholipid formed from a backbone of a three-carbon molecule, called *glycerol* (Figure 3.5), to which a *fatty acid* (Figure 3.6) is attached to each of the first two carbons. To the third carbon is attached a phosphate group, which in turn is bound to a chemical compound called an *amino alcohol*, such as ethanolamine, shown in Figure 3.9a.

Recall that fatty acids bound to glycerol provide a *hydrophobic* (repels water), or what is called *nonpolar*, character to the phosphoglyceride. The *hydrophilic* (attracted to water), or *polar*, phosphate and amino alcohols groups, together with the hydrophobic hydrocarbon chains, give the phosphoglycerides two distinctly different chemical regions. One region, composed of fatty acids, is hydrophobic, and the other, composed of the amino alcohol, is hydrophilic. These two contrasting chemical characteristics are essential to phosphoglycerides' roles in membrane structure and functions.

Figure 4.2 diagrams sections of two phosphoglyceride *bilayers*, such as that found in cell membranes. The upper bilayer diagram is composed of phosphoglycerides that have only *saturated* fatty acids, and the lower diagram contains both *saturated* and *unsaturated* fatty acids. In both diagrams, the two layers are oriented so that the hydrophilic phosphates + amino alcohols face either the exterior or interior, aqueous compartments, with the fatty acids (hydrophobic) of each layer facing toward the fatty acids of the other layer. Observe in the lower diagram in Figure 4.2, the double bonds of the unsaturated fatty acids cause the hydrocarbon tails to kink or bend, preventing tight packing of the bilayer.

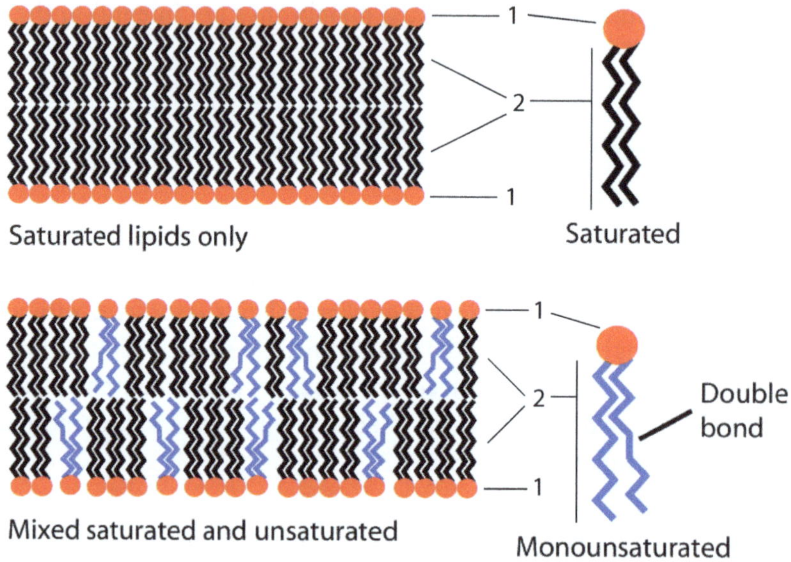

FIGURE 4.2 Upper diagram: bilayer of phosphoglycerides with only saturated fatty acids; lower diagram: bilayer of phosphoglycerides with both saturated and unsaturated fatty acids; 1 = hydrophilic and 2 = hydrophobic

Many cell membranes contain other lipid components, including *sterols*. For example, in human plasma membranes, that sterol is *cholesterol* (Figure 3.10), which is a largely hydrophobic molecule, but which also has a polar (hydrophilic) group at one end. You are probably more familiar with cholesterol as a risk factor for cardiovascular disease. However, cholesterol has a number of essential roles in the body, and one of these is in the structure of cell membranes. In the membrane, cholesterol is oriented with the polar portion of the molecule toward the polar groups of the phospholipids (see Figure 4.3). Here, the cholesterol decreases the permeability of the membrane. Cholesterol's

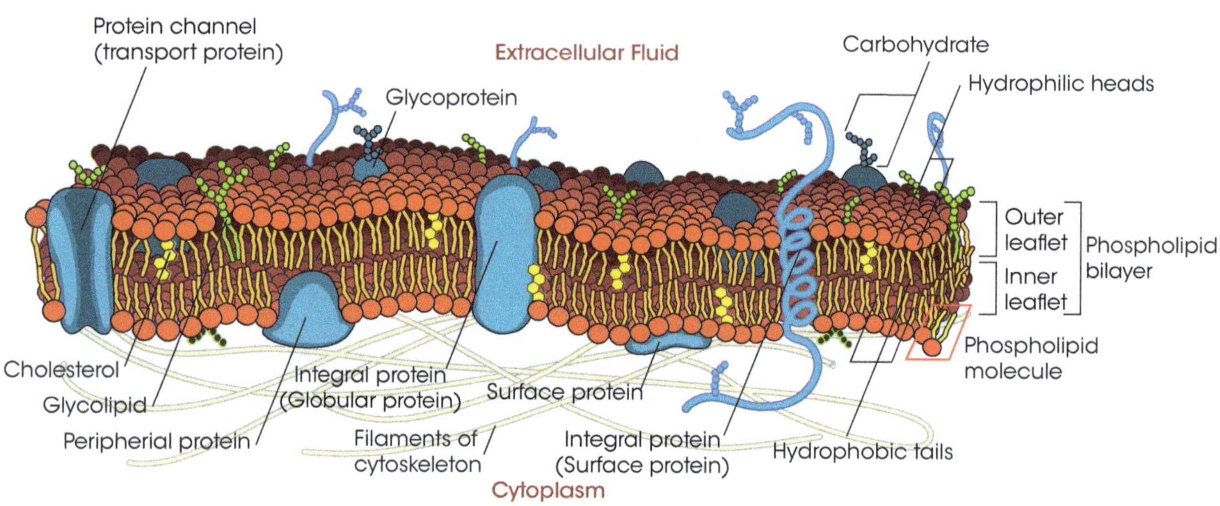

FIGURE 4.3 Diagram of plasma membrane

nonpolar (hydrophobic) portions dangle among the fatty acid chains and keep them from packing tightly and solidifying.

The other common components of cell membranes are *proteins*. Previously, we discussed the structural organization of proteins. Recall that proteins are composed of one or more chains of amino acids held together by peptide bonds. Technically, each chain of amino acids is called a *polypeptide*. An *amino acid* is characterized by an amino group ($-H_3N^+$) and a carboxyl group ($-COO^-$) attached to a central (alpha [α])-carbon. The central carbon is also attached to a hydrogen and a variable (R)-group (see Figure 3.11). It is the *R-group* that determines the chemical behavior of an amino acid and in a protein contributes to its final conformation.

We can separate two leaflets, the outer and inner, of the cell membrane. Membrane proteins may associate primarily with one leaflet of the cell membrane, and others may pass through both leaflets, with exposure to the cytoplasm and the extracellular fluid (see Figure 4.3).

QRQ 1. The outer boundary of a cell is called the plasma ______.

QRQ 2. The fluid plus membranous and particulate components inside the cell is called the ______, while the fluid part is called the ______.

QRQ 3. Select the appropriate term for each blank: amino acid, amino alcohol, fatty acid, glycerol, molecule of water, phosphate group, phospholipid. Phosphoglycerides are a type of ______ formed from a backbone of a 3-carbon molecule, called ______, to which a(n) ______ is attached to each of the first two carbons. To the third carbon is attached a(n) ______, which in turn is bound to a chemical compound, called a(n) ______.

QRQ 4. Characterize each of the following as hydrophilic or hydrophobic:

a. amino alcohols
b. fatty-acid part of a phosphoglyceride
c. interior of a membrane
d. lipids
e. outer surfaces of a bilayer
f. something attracted to water
g. something polar
h. something that repels water

QRQ 5. The major sterol of animal membranes is ______.

QRQ 6. Select which are the characteristics of cholesterol: decreases the permeability of a membrane, has a polar group at one end, largely a hydrophobic molecule, nonpolar portion dangles among the fatty acid chains, keeps fatty acid chains from packing tightly and solidifying, oriented with the polar portion of the molecule toward the nonpolar groups of the phospholipids

QRQ 7. A protein is a large molecule made up of one or more chains of ______ held together by ______. Each chain is called a(n) ______.

QRQ 8. An amino acid is characterized by a(n) ______ ($-H_3N^+$) and a(n) ______ ($-COO^-$) attached to a(n) central ______. The latter is also attached to a(n) ______ and a variable ______.

QRQ 9. (T/F) The R-group determines the chemical behavior of an amino acid and in a protein contributes to its final conformation.

QRQ 10. (T/F) Membrane proteins can only associate with one leaflet of a cell membrane.

4.3. Cell Membranes: Major Functions

We have discussed the structure of membranes, but why are they so important? In this section, we discuss the following major functions of membranes:

- To protect, organize cells, and serve as physical barriers
- To regulate transport
- To facilitate communication
- To serve as reaction platforms

4.3.1. To Serve as Physical Barriers

Cells are essentially living compartments separated from an external aqueous environment but containing an internal aqueous environment that is quite different chemically from the external milieu. For proper function, it is important to maintain those chemical differences. Remember that the lipids of membranes are hydrophobic. If you were designing a membrane as a barrier, where water-based solutions are on both sides, wouldn't it make sense for the barrier to be hydrophobic? Additionally, within the cells of plants and animals are other compartments (*organelles*). Moreover, the surrounding membranes help to control what molecules and other chemicals enter the organelle and the exit of reaction products from these organelles. The *nucleus*, surrounded by a double layer of membrane, contains genetic information necessary for replication, growth, and maintenance of the cell. Other membrane-bound compartments found in the cytosol include *mitochondria*, *endoplasmic reticulum*, *Golgi*, *lysosomes*, *peroxisomes*. Additionally, there are other membranous compartments, such as vacuoles and vesicles.

4.3.2. To Regulate Transport

Membranes are formed from two layers of phospholipids that form a hydrophobic, physical barrier. This barrier, with its hydrophobic interior, limits the movement of nutrients and wastes into or out of the cell or cellular compartment. Some things can move in or out by diffusion, but for others, it is more difficult or impossible without a channel or transporter. Consequently, we say the membrane is *selectively permeable*. Some entities, like water and small hydrophobic molecules, may pass easily, with a concentration gradient, through such a barrier, but larger and/or highly charged, hydrophilic molecules might not pass through a pure phospholipid bilayer at all. However, the cell membrane is not composed of phospholipids only. There are proteins that may form *channels* or *transporters* for large molecules, polar molecules, and *ions* (electrically charged atoms or molecules). Chemical gradients are established by differences in solute concentration on either side of a membrane. If the solute can move through the membrane, either by diffusion or using a protein channel or transporter, the solute will move from an area of high concentration to an area of low concentration. Ions, exhibiting both a chemical and electrical gradient, will still move from a high to low concentration through ion channels. Whatever the mechanism used, membranes, external and internal, help to *regulate* what moves into or out of the cell or the internal compartments.

Molecules that can move through a membrane will generally move in the direction of the lowest concentration of that molecule (*diffusion*). For instance, water, which is often in higher concentration outside the cell, will move into the cell through the plasma membrane, which has a lower water concentration. Some water can diffuse, but scientists have also found water channels that allow more water to pass through the membrane. This special type of *diffusion* is called *osmosis*. Most bacteria have sturdy cell walls outside the plasma membrane. Often their environment has a higher concentration of water molecules than in their cytoplasm. Thus, water has a tendency to flood into the bacterial cell, which might cause the cell to burst. The cell wall is able to prevent the cell from rupturing.

Other small hydrophobic molecules will diffuse through the membrane from the area of highest concentration of that molecule. As described above, for molecules that are too large, polar, or charged to pass through the membrane, there are special protein channels or carriers to transport them with the gradient, and this transport requires no energy. Biologists call this type of transport *facilitated diffusion*. A portion of the transporter recognizes the molecule to be moved. For a *protein channel*, the recognized molecule diffuses through with the gradient (labeled in Figure 4.4). When the molecule binds to a *carrier protein*, it induces a shape change (conformational change) in the transporter molecule, and this change is such that the molecule moves across the membrane (labeled in Figure 4.4). For this system to work, there must be a specific receptor site for recognizing the molecule, and the movement must be with the concentration gradient (high concentration toward low concentration) (see Figure 4.4).

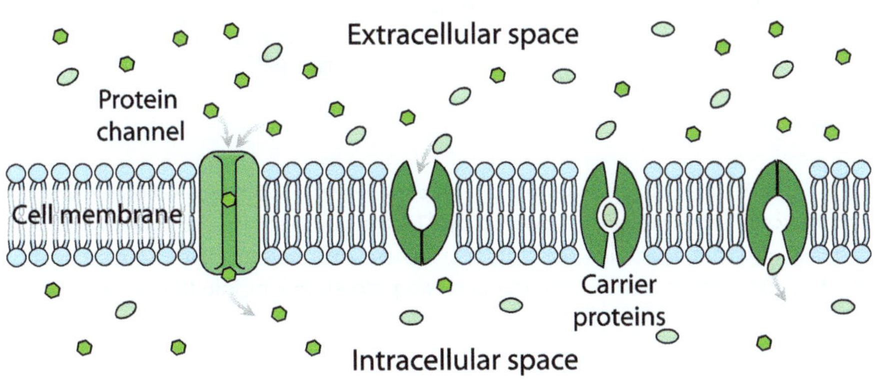

FIGURE 4.4 Facilitated diffusion, using protein channel or carrier protein

Sometimes the cell or organelle requires import or export of molecules against a concentration gradient (low concentration to high concentration). For this movement the cell needs *active transport* (see Figure 4.5), which requires the membrane to have a transporter that recognizes and can bind to the molecule, and there must also be an input of energy. The energy can be garnered from *ATP* or some other energy-rich molecule. It happens that there are many molecules where the cell needs to move ions or molecules against a concentration gradient.

One good example is the *sodium–potassium pump*, found in the plasma membrane of most cells. This pump helps to maintain concentration gradients across cell membranes, which are important for maintaining cellular volumes, membrane electrical potentials, transmission of cell signals, filtration function of kidneys, and so on. The pump, utilizing energy from hydrolysis of ATP, moves sodium ions (Na^+) out of the cell and potassium ions (K^+) into the cell, both against concentration gradients (see Figure 4.5). Three sodium ions bind to the pump protein's intracellular sites, followed by the removal of a phosphate from ATP, which is transferred to a binding sight on the pump. The bindings cause the pump protein to change shape, so that the sodium ions no longer bind, and they are released outside. The shape change exposes potassium ion binding sites, and two potassium ions from outside bind. This binding promotes the release of the phosphate, and the protein returns to its previous shape. In this way, the potassium ions are released into the interior of the cell.

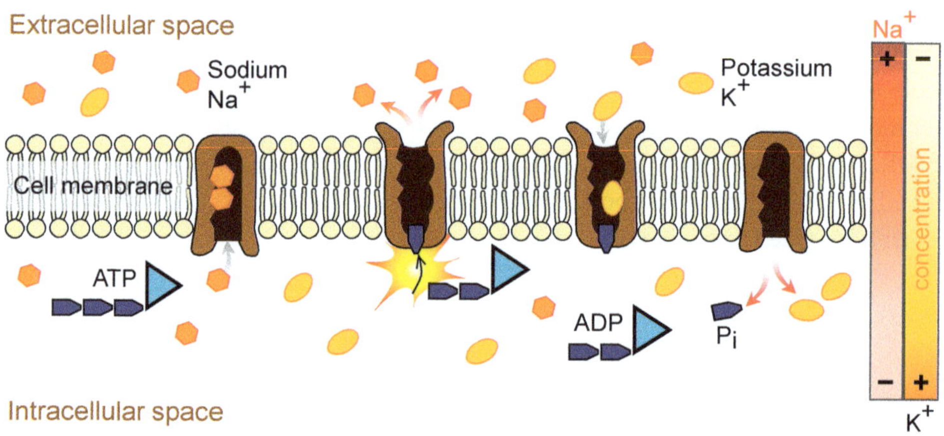

FIGURE 4.5 Sodium–potassium pump, a type of active transport, located on plasma and many other cellular membranes. Powered by ATP, the pump transports sodium and potassium ions in opposite directions, each against its concentration gradient. One cycle of pump activity transports three sodium ions out and two potassium ions into the cell or organelle.

Membranes also participate in physically transporting much larger materials into or out of cells. If the material is being exported, membrane-bound vesicles surround such materials and fuse with the plasma membrane. Releasing the unbound contents, the vesicle membrane becomes part of the plasma membrane. This process is called *exocytosis* (see Figure 4.6a). In contrast, material being imported is captured by an *invagination* (folding inward) of the plasma membrane, which closes off into an internal vesicle. This ingress is referred to as *endocytosis* (see Figure 4.6b). Certain vesicles

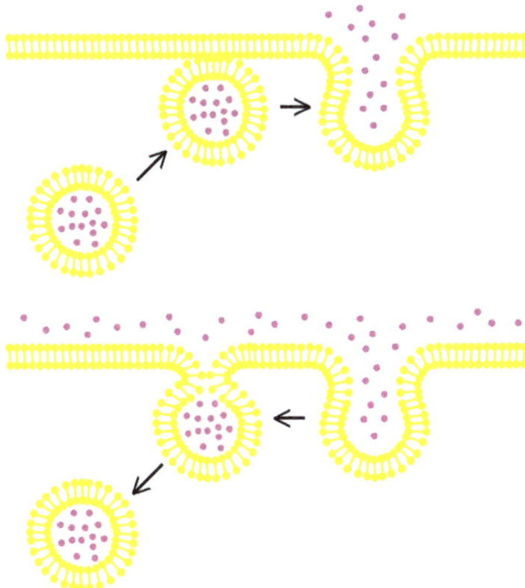

FIGURE 4.6a (top) Exocytosis b (bottom) Endocytosis

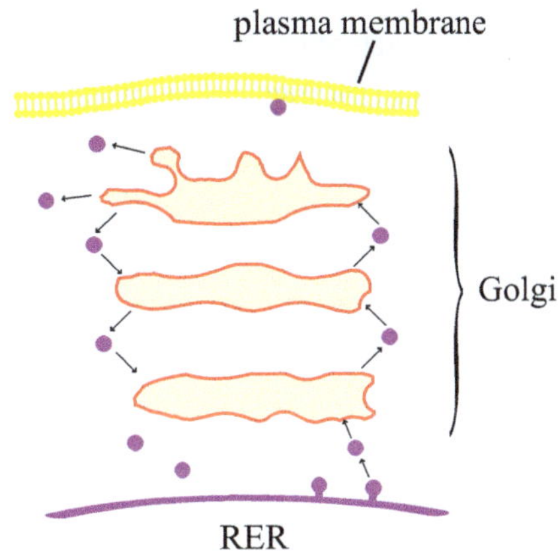

FIGURE 4.7 Transport of materials via vesicles to and from the rough ER to and from the Golgi

may fuse with other membrane-bound organelles within the cytosol. For instance, an incoming vesicle may fuse with a *lysosome,* an organelle that contains enzymes to break down imported material. Vesicles can also be formed from the membranes of some internal compartments for transport to other compartments (see Figure 4.7) or to the outside of the cell.

APPLYING SCIENCE TO LIFE 1 Extreme Adaptations

In some of the coldest and hottest places on earth, there are microbes that grow and reproduce. The microbes that live in the coldest places (polar, deep ocean) are called *psychrophiles,* and those that live in the hottest places (hot springs, geothermal vents) are called *thermophiles.* Psychrophiles have optimal temperatures of 15 °C (59 °F) but may live at temperatures lower than 0 °C (32 °F). Thermophiles have optimal temperatures of 60 °C (140 °F), but some may live at temperatures as high as 120 °C (248 °F). Obviously, they are highly adapted to tolerate these extremes in temperature. Some of the most important adaptations involve the molecules that make up their membranes. We have described two types (saturated/unsaturated) of phospholipids that form cell membranes.

PoI 1. Hypothesize which type of phospholipid you expect to find in psychrophiles and explain your answer.

PoI 2. Hypothesize which type of phospholipid you expect to find in thermophiles and explain your answer.

GtM 1. For each of the following, select the transport mechanism that would likely be used for transportation through the membrane. Give the reason you made your choice for each.

a. Water
b. Sodium ions (Na+) against concentration gradient
c. Glucose into muscle cells
d. Bacterial cell into white blood cell

QRQ 11. The ______, surrounded by a double layer of membrane, contains genetic information necessary for replication, growth, and maintenance of the cell.

QRQ 12. Select from the following the appropriate term for each description below: active transport, calcium ion pump, carrier protein, chemical gradient, conformational change, diffusion, electrical gradient, endocytosis, exocytosis, facilitated diffusion, invagination, ion, osmosis, protein channel, selectively permeable

a. ability of water to pass through the membrane with the aid of water channels, based on differences in concentration on either side of the membrane
b. aided by special protein channels or carriers but requiring no energy, the import or export of molecules that are too large, polar, or charged to pass through a membrane
c. allows a specific recognized molecule to diffuse through the membrane with the gradient
d. binds to a specific recognized molecule, inducing a shape change in the transporter molecule, thus enabling the recognized molecule to move across the membrane with the concentration gradient
e. characterization of the membrane indicating that some things can move in or out of the cell easily, but for others it is more difficult or impossible without help
f. electrically charged atom or molecule
g. established by differences in charge across a membrane
h. established by differences in solute concentration across a membrane
i. movement of molecules through a membrane, aided only by the differences in concentration of that molecule
j. example of active transport
k. folding inward
l. import or export of molecules against a concentration gradient in a cell or organelle
m. physically exporting material by having a membrane-bound vesicle surround such material, fuse with the plasma membrane, and release the unbound contents, with the vesicle's membrane becoming part of the plasma membrane
n. physically importing material captured by an inward folding of the plasma membrane, which closes off into an internal vesicle
o. requiring a transporter that can bind to a specific molecule and with an input of energy, the import or export of that molecule through a membrane

p. shape change

q. when a muscle relaxes, using energy from ATP, it moves released Ca^{++} against a concentration gradient across the sarcoplasmic reticulum (SR) membrane back into the internal space

QRQ 13. (Possible answers in parentheses after each blank.) Movement across a gradient is from an area of a. ______ (high, low) concentration to an area of b. ______ (high, low) concentration.

4.3.3. To Facilitate Communication

Cells are bombarded with signals from their environment and other cells. Most of these signals are chemical, but cells can control the movement of ions at the membrane level, making cells capable of controlling electrical current flow and cross-membrane electrical potentials. Nerve cells (*neurons*) utilize this ability to transmit nerve impulses in only one direction from one neuron to another. In animals, neurons join other neurons indirectly at a synapse. To bridge the synaptic space (*cleft*), a neuron produces *neurotransmitters*. These chemicals travel a short distance to a receptor on a nearby neuron (see Figure 4.8). In the figure, neurotransmitters are produced in a presynaptic neuron (labeled A) and packaged into vesicles. When this neuron is stimulated to release the neurotransmitter, the vesicles are moved to and fuse with the neuronal membrane, releasing a chemical message into the *synaptic cleft*, or space at the synapse (B). The neurotransmitter molecules diffuse across the cleft to bind to receptors of the postsynaptic neuron (C), stimulating an electrical signal along the neuronal membrane of neuron (C).

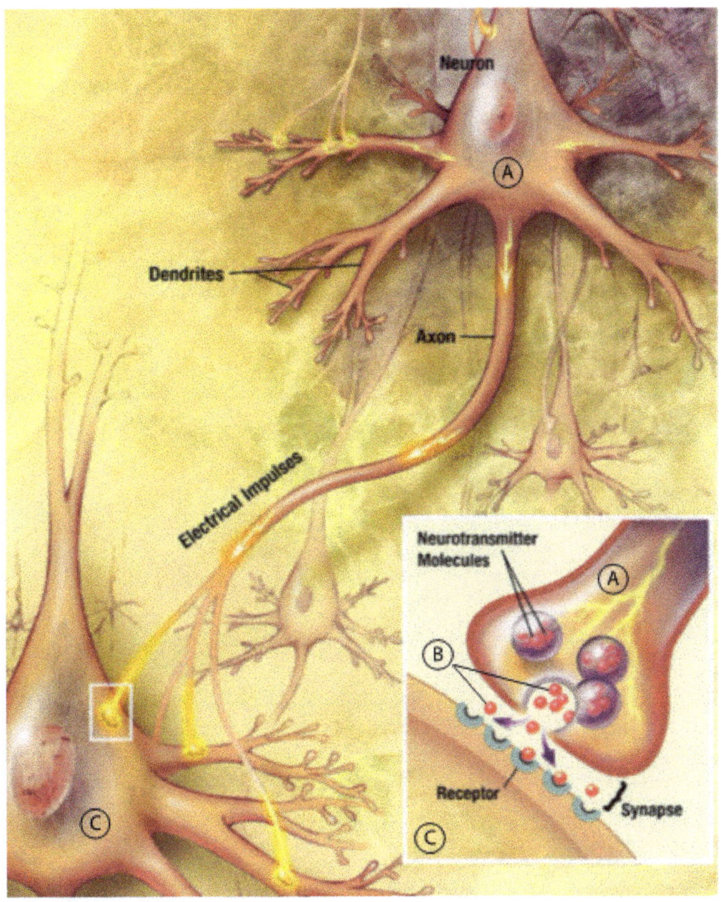

FIGURE 4.8 Synapse between presynaptic and postsynaptic neurons

APPLYING SCIENCE TO LIFE 2 Making a Difference in Many Lives

According to estimates by the National Institutes of Mental Health for 2017, 1.3 million or 7.1 percent of adults in the United States had one or more major episodes of depression in that year. Prevalence of such episodes was higher among females (8.7 percent) versus males (5.3 percent), with the highest prevalence being in the 18–25 age group (13.1 percent). Frequency of depressive episodes was also higher in adults whose racial backgrounds were from two or more racial origins (11.3 percent) than for any single race (NIMH 2018, NIMH 2019).

One of the most commonly prescribed classes of drugs prescribed to treat depression is the *selective serotonin reuptake inhibitors (SSRIs)*, like Paxil, Lexapro, Prozac, and Celexa. *Serotonin* (5-hydroxytryptamine [*5-HT*]) is a neurotransmitter that presynaptic neurons release into the synaptic cleft. The compound then diffuses toward the postsynaptic neuron and binds to the receptor sites on the membrane, stimulating depolarization of the postsynaptic cell (see Figure 4.9). Normally, the unused and released 5-HT molecules are taken up (reuptake) by transporters in the membrane of the presynaptic neuron, where the serotonin may be recycled. The SSRI drugs work by blocking the reuptake of the 5-HT molecules by the presynaptic neuron, prolonging high levels of serotonin outside that cell (Guzman 2019).

Serotonin (5-HT) is produced by many types of neurons, including those in the brain stem. This neurotransmitter is associated with many body functions, including a sense of happiness and well-being. We still have much to learn about how serotonin might accomplish these and other functions, but the SSRIs do seem to help improve the lives of people suffering with depression. Of course, all drugs have side effects, and physicians should fully disclose those before prescribing them for their patients.

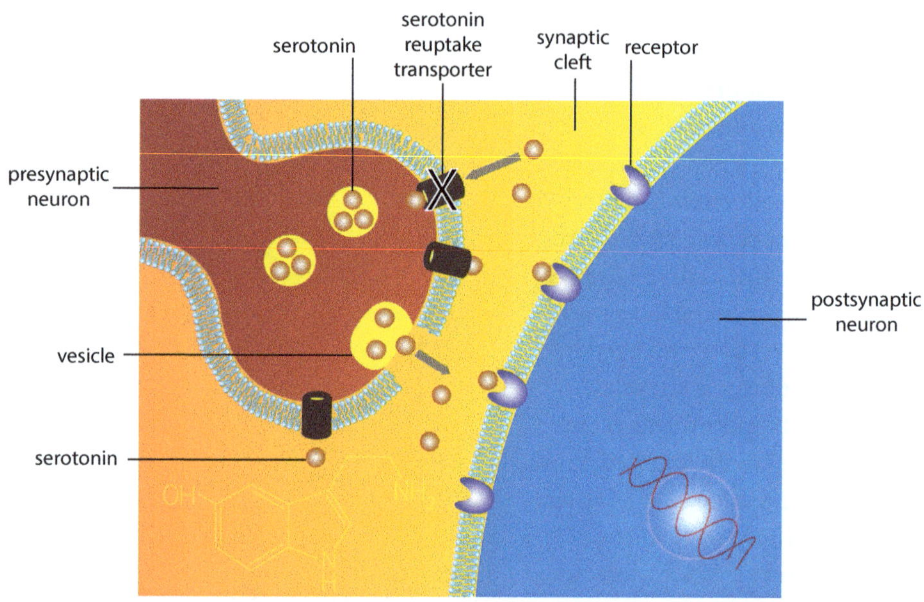

FIGURE 4.9 Action of serotonin

In animals, glands release chemical messages called *hormones*, which are transported by the blood throughout the body. Thus, a hormone's effects may be at some distance from the source of the signal (see example below). Some cells produce signals that affect the source of the signal. For example, during an immune response, certain *T-lymphocytes* produce signals that cause these same cells to proliferate.

In the plasma membrane, there are proteins that act as *receptors* for chemical signals. Each receptor is generally a transmembrane protein, specialized to interact with one or several chemically similar signals. Because many of these signals are too big or too highly charged to pass through the membrane into the cytosol, signal interaction with the receptor transduces the message through the membrane. Some signal-receptor interactions initiate the production of a "second signal," which may set off a series of cellular reactions. In other cases, the interaction activates an *enzyme* (organic catalyst in a chemical reaction for a biological system), which will subsequently activate or inactivate other enzymes. In still other cases, the interaction will open or close an ion channel.

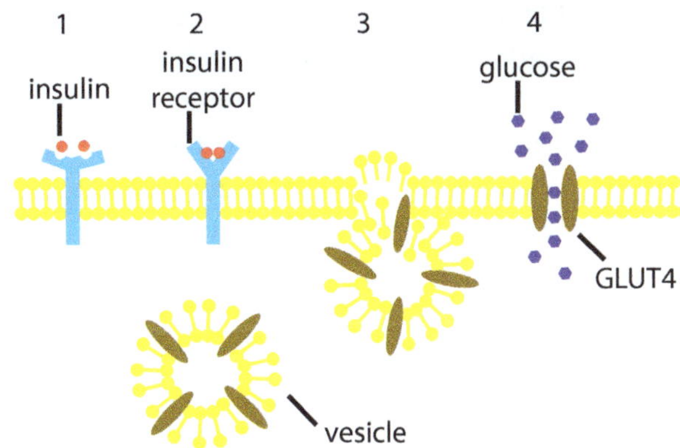

FIGURE 4.10 Insulin receptor activation, promoting glucose uptake

One interesting example is the *insulin receptor* interaction on the plasma membrane of striated muscle (see Figure 4.10). These receptors, after binding to insulin, transduce the signal from the binding subunit of the receptor to the enzymatic subunit, activating an enzyme. This enzyme activity will induce a cascade of internal changes in the cell. One of the major effects from this cascade is to promote the movement of vesicles, whose membranes contain glucose transporters. The vesicles are moved to the plasma membrane, where the vesicular membrane fuses with the plasma membrane, resulting in more glucose transporters (GLUT4) onto the surface of the muscle cell. Glucose in the blood will bind to the transporter on the outside and induce a shape change, which moves the glucose into the cytoplasm. Once inside the muscle cell, the glucose can be metabolized (used to supply energy for muscle contraction) or used to synthesize glycogen, which can later release glucose for metabolism.

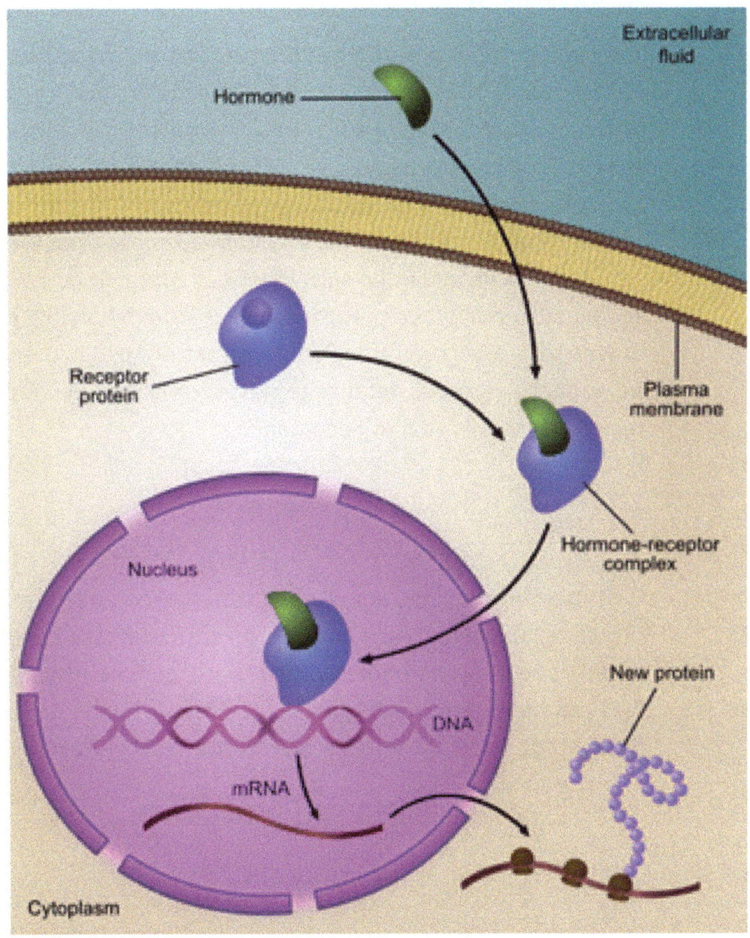

FIGURE 4.11 Hormone binds to nuclear receptors, activating transcription and subsequent protein synthesis

Signals that can pass through the plasma membrane have internal receptors either in the cytosol or in the nucleus (see Figure 4.11). For instance, steroid hormones, like thyroid hormone (TH), are hydrophobic molecules and can pass through the membrane. If responsive to

TH, the cell will have receptor molecules for the hormone in the nucleus of the cell. The hormone binds to the receptor and changes its conformation (shape), activating it. The hormone-receptor complex then recognizes and binds to specific, regulatory DNA sequences and promotes *transcription*, creating an RNA (messenger RNA, mRNA) (copy of a complementary strand of DNA) of select DNA sequences into specific RNAs. These selected RNAs are translated into protein sequences that may affect metabolism, growth, and development of the organism. In other cases, the hormone binding may inhibit transcription. Either way, the effects of the hormone are greatly expanded.

Cells not only have signal receptors, they also produce chemical signals. Many cells may send out a plethora of chemical signals for local or remote reception. In other cases, the signals remain attached to the outer part of the plasma membrane and interact with receptors on other cells. These interactions are important for recognition during an immune response or in the establishment of tissues.

APPLYING SCIENCE TO LIFE 3 Signaling Transplant Success

The United Network for Organ Sharing (UNOS) happily revealed a record number of organ donations during 2018 (UNOS 2019). Transferring cells, tissues, or organs from one individual to a genetically different recipient poses a number of problems. Each cell in your body, except nonnucleated cells such as red blood cells, displays a set of surface proteins (major histocompatibility complex [MHC]). One of the functions of this complex is to help your immune system to recognize your own cells. Because of this recognition, the immune system does not attack and damage your cells. So it is the interaction between *receptors* on the surface of our white blood cells and the MHC *signals* on the surface of almost all of our other cells that save us from a horrible internal battle. However, if cells, tissues, or organs from a person having a foreign set of MHC protein signals are introduced, the immune system will mount a vigorous defense to destroy the intruder. Ensuring that new tissue is compatible with the recipient's cells is vital to acceptance.

4.3.4. To Serve as Reaction Platforms

Cell membranes also serve as platforms for enzymatic pathways and systematic transport. For instance, an assemblage of electron transporters for the electron transport system in the plasma membrane of some bacteria and the inner membrane of mitochondria facilitates production of *ATP*, the most common source of energy in the cell (see Figure 4.12). Another electron transfer system, used for energy generation and reduction, is found in the inner membrane of plant chloroplasts.

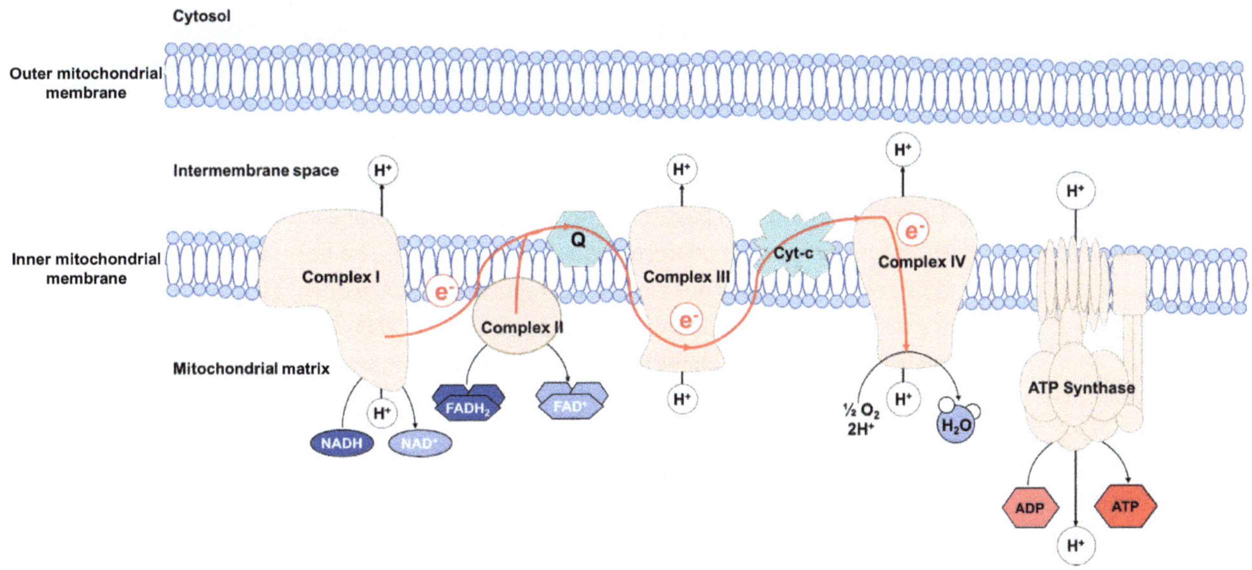

FIGURE 4.12 Electron transport chains shown within inner membrane of mitochondrion

QRQ 14. A nerve cell is called a(n) ______. ______ are chemicals secreted by one nerve cell that travel a short distance to a receptor on a nearby nerve cell.

QRQ 15. ______ Arrange in order the following steps in transmission of nerve impulses:

a. chemical message released into the synaptic cleft
b. neurotransmitters are produced in a neuron and packaged into vesicles
c. neurotransmitters bind to receptors on a neuron
d. neurotransmitters diffuse across the cleft
e. neurotransmitters stimulate an electrical signal along the neuronal membrane
f. when neuron is stimulated to release the neurotransmitter, vesicles are moved to and fuse with the neuronal membrane

QRQ 16. The most common source of energy in the cell is ______.

QRQ 17. Characterize each application below as to which of the following cell membrane functions it best illustrates: to facilitate communication, to protect and organize cells and serve as physical barriers, to regulate transport, to serve as reaction platforms

a. After binding to insulin, the insulin receptor on the plasma membrane of striated muscle transduces the signal from the binding subunit of the receptor to the enzymatic subunit, activating the enzyme.
b. An assemblage of electron transporters for the electron transport system in the plasma membrane of some bacteria and the inner membrane of mitochondria facilitates production of ATP.

c. Interaction between receptors on the surface of your white blood cells and a set of surface proteins (MHC) on the surface of almost all of your cells helps your immune system to recognize your own cells.
d. Membranes separate interior regions of an organelle from the rest of the cell's interior, allowing each compartment to serve specific functions.
e. Thyroid hormone can pass through the plasma membrane and bind to a certain receptor, changing its conformation and activating it. The hormone-receptor complex then recognizes and binds to specific, regulatory DNA sequences and promotes transcription of select DNA sequences into specific RNAs.
f. Water channels allow water to pass through the membrane.

Image Credits

Fig. 4.1: Source: https://imagej.nih.gov/ij/images/FluorescentCells.jpg.

Fig. 4.2: Source: https://commons.wikimedia.org/wiki/File:Lipid_unsaturation_effect.svg.

Fig. 4.3: Adapted from Source: https://commons.wikimedia.org/wiki/File:Cell_membrane_detailed_diagram_en.svg.

Fig. 4.4: Source: https://commons.wikimedia.org/wiki/File:Scheme_facilitated_diffusion_in_cell_membrane-en.svg.

Fig. 4.5: Source: https://commons.wikimedia.org/wiki/File:Scheme_sodium-potassium_pump-en.svg.

Fig. 4.8: Adapted from Source: https://commons.wikimedia.org/wiki/File:Chemical_synapse_schema_cropped.jpg.

Fig. 4.9: Copyright © 2011 Depositphotos/meletver.

Fig. 4.11: Copyright © by Ali Zifan (CC BY-SA 4.0) at https://commons.wikimedia.org/wiki/File:Regulation_of_gene_expression_by_steroid_hormone_receptor.svg.

Fig. 4.12: Copyright © by Dw001 (CC BY-SA 4.0) at https://commons.wikimedia.org/wiki/File:ElectronTransportChainDw001.png.

LAB #4

Laboratory Exercises

4.4. Laboratory on Membrane Formation

4.4.1. Preliminary Work

In this lab, we consider the formation of membranes. From the text's website, view the introductory video; download and open "*Membrane Formation Rev.nlogo.*"

1. Click the Info tab at the top of the page and read sections "What Is It?" and "How It Works."

   a. _______ The simulation involves the formation of membranes in **1.** (A. air, B. blood, C. intestines, D. saline solutions, E. skin, F. tissues, G. water).

   b. Give the color (A. black, B. blue, C. brown, D. cyan, E. gray, F. green, G. magenta, H. orange, I. purple, J. red, K. white, L. yellow) for each of the following:

      _______ circle representing water molecule

      _______ circle representing hydrophilic end

      _______ circle representing hydrophobic end

   c. _______ What structure represents a lipid? (A. bar, B. blue circle, C. orange circle, D. purple circle, E. purple circle connected to orange circle)

   d. _______ Select all that apply to the term *hydrophilic*: A. attracted to water, B. mixes readily with water, C. molecule repels water, D. not attracted to water

   e. _______ Using the same possible answers as in part d, list all that apply to the term *hydrophobic*.

   f. _______ At every tick, each molecule selects a random molecule within INTERACTION-DISTANCE. Select all circumstances where the current, or acting, molecule then moves away from the randomly selected molecule: A. both are hydrophilic, B. both are hydrophobic, C. both are water, D. one is hydrophobic and one is hydrophilic, E. one is water and the other is hydrophilic, F. one molecule is hydrophobic and the other is water

   g. **1.** _______ **2.** _______ **3.** _______ Thus, lipid molecules tend to align with **1.** (L. like, U. unlike) ends being adjacent to each other, water molecules tend to **2.** (A. attract, R. repel) each other, and water molecules and **3.** (I. hydrophilic, O. hydrophobic) ends of lipids tend to be adjacent to each other.

h. _______ Suppose a hydrophobic molecule selects a random molecule that is water within INTERACTION-DISTANCE. By how much (as a positive number) do they move away from each other? A. LIPID-LENGTH, B. RANDOM-FORCE, C. TOO-CLOSE-FORCE, D. –TOO-CLOSE-FORCE, E. WATER-OIL-FORCE, F. –WATER-OIL-FORCE, G. WATER-WATER-FORCE

i. _______ In all other situations than the answer for part f, the acting molecule **1.** (A. moves away from, B. moves toward, C. stays in the same relative position to) the randomly selected molecule.

j. _______ By how much? Use possible answers from part h.

k. _______ When are molecules connected by a link? A. orange circles forming a lipid, B. orange circles forming water, C. purple and orange circles forming a lipid, D. purple and orange circles forming water, E. purple circles forming a lipid, F. purple circles forming water

l. _______ If it is a member of such a linked pair, the acting molecule moves what distance from its partner? Use possible answers from part h.

m. _______ The hydrophobic isolation plot shows the average percentage of each **1.** (A. hydrophilic molecule's neighbors that are hydrophilic, B. hydrophilic molecule's neighbors that are hydrophobic, C. hydrophobic molecule's neighbors that are hydrophobic, D. hydrophobic molecule's neighbors that are hydrophilic).

n. _______ Higher values in the hydrophobic isolation plot indicate that a greater number of **1.** (A. hydrophilic molecules are associated with water and hydrophobic molecules, B. hydrophilic molecules are isolated from water and hydrophobic molecules, C. hydrophobic molecules are associated with water and hydrophilic molecules, D. hydrophobic molecules are isolated from water and hydrophilic molecules).

2. Read "How to Use It" and "Things to Notice."

a. _______ TOO-CLOSE-FORCE is **1.** (A. an indicator that a molecule is "too close" to another molecule, B. how much a molecule should move when it is "too close" to another molecule, C. how much force a molecule should exert on any other molecule, D. how much is "too much" force to exert on a close molecule).

b. _______ Each molecule will move in a random direction **1.** (A. LIPID-LENGTH, B. RANDOM-FORCE, C. TOO-CLOSE-FORCE, D. WATER-OIL-FORCE, E. WATER-WATER-FORCE) amount each tick.

c. _______ To increase the temperature in the system, we should move the **1.** (A. LIPID-LENGTH, B. RANDOM-FORCE, C. TOO-CLOSE-FORCE, D. WATER-OIL-FORCE, E. WATER-WATER-FORCE) slider to the right.

d. **1.** _______ **2.** _______ **3.** _______ **4.** _______ The Info section defines a micelle as a circular structure of lipids where their **1.** (A. hydrophilic, B. hydrophobic) ends all point outward toward surrounding **2.** (C. lipid, D. oil, E. random, F. water) molecules. However, a reverse micelle is a circular structure of lipids in which their 3. (A. hydrophilic, B. hydrophobic) ends all point inward toward a collection of 4. (C. lipid, D. oil, E. random, F. water) molecules.

e. Find in other sources at least three applications of micelles. Place in quotation marks quoted material and list references in parentheses.

4.4.2. Laboratory Work

In the laboratory, we will be examining the impact of the forces one at a time. Use a spreadsheet to record data and calculate averages. *Pressing DEFAULT-VALUES sets default parameters as follows: NUM-WATER = 750, NUM-LIPIDS = 250, WATER-WATER-FORCE = 0.2, WATER-OIL-FORCE = −0.7, TOO-CLOSE-FORCE = −0.3, RANDOM-FORCE = 0.10.*

3. This question explores the impact of WATER-WATER-FORCE (the amount an acting molecule should move in the direction of a randomly selected molecule that is within INTERACTION-DISTANCE) in several situations.

   a. _______ Press DEFAULT-VALUES and turn the STOP-AT-300 switch ON to stop the simulation at tick 300. Set up and run the simulation at least *five* times, recording the hydrophobic isolation and noting the structures that form. View simulation updates for at least one simulation. Afterward, speed the process or uncheck *view updates* before pressing GO, and after completion, check *view updates* to observe the formations. Select all formations that you observe at the end of 300 ticks: A. bilayers, B. micelles, C. reverse micelles, D. single layers, E. stars, F. tri-layers

   b. _______ Give the average hydrophobic isolation for this experiment.

   c. _______ Two water molecules move closer together with WATER-WATER-FORCE value (A. 0.1, B. 0.2, C. 2.0).

   d. **1.** _______ **2.** _______ **3.** _______ Complete the following hypothesis: As WATER-WATER-FORCE increases, reverse micelles are **1.** (L. less, M. more) likely to form, and hydrophobic isolation **2.** (D. decreases, I. increases), indicating that hydrophobic molecules are becoming **3.** (L. less, M. more) isolated from water and hydrophilic molecules.

   e. _______ Change WATER-WATER-FORCE to 0.3. Set up and run the simulation at least *five* times, recording the hydrophobic isolation and noting the structures that form. To speed the experiment, uncheck *view updates* before pressing GO, and after completion, check *view updates* to observe the formations. In comparison to part a, we are *less* likely to observe **1.** (A. bilayers, B. micelles, C. reverse micelles, D. single layers, E. stars, F. tri-layers).

   f. _______ Give the average hydrophobic isolation for this experiment.

   g. _______ Change WATER-WATER-FORCE to 0.1. Set up and run the simulation at least *five* times, recording the hydrophobic isolation and noting the structures that form. In comparison to part a, we are *less* likely to observe **1.** (A. bilayers, B. micelles, C. reverse micelles, D. single layers, E. stars, F. tri-layers).

   h. _______ Give the average hydrophobic isolation for this experiment.

   i. **1.** _______ **2.** _______ **3.** _______ **4.** _______ **5.** _______ **6.** _______ For blank 2 use choices D. decreases, I. increases, and for all other blanks use L. less, M. more. Based on these results, as WATER-WATER-FORCE increases, like ends of lipids (i.e., both hydrophilic or both hydrophobic) have **1.** of an affinity for each other, and hydrophobic isolation **2.** When WATER-WATER-FORCE is too small, bilayers are **3.** likely to form and reverse micelles are **4.** likely to form. When WATER-WATER-FORCE is too large, bilayers are **5.** likely to form and reverse micelles are **6.** likely to form.

   j. Discuss the results, possible reasons for these outcomes, and whether the results support or refute your hypothesis.

4. This question explores the impact of WATER-OIL-FORCE, the amount an acting molecule should move in the direction of a randomly selected molecule that is within INTERACTION-DISTANCE. Because WATER-OIL-FORCE is negative, the molecules repel each other.

a. _______ A water molecule and the hydrophobic molecule of a lipid move farther apart with WATER-OIL-FORCE value **1.** (A. -2.0, B. -0.7, C. -0.1).

b. **1.** _______ **2.** _______ Complete the following hypothesis: As WATER-OIL-FORCE becomes more negative, reverse micelles are **1.** (L. less, M. more) likely to form, and hydrophobic isolation **2.** (D. decreases, I. increases).

c. **1.** _______ **2.** _______ **3.** _______ Press DEFAULT VALUES and set WATER-OIL-FORCE to -0.1. Turn the STOP-AT-300 switch OFF. Set up and run the simulation until about tick 500, viewing the world at that point. (For comparisons, you may choose to take a screenshot.) Decrease WATER-OIL-FORCE to -0.7. Do *not* press *setup*. Run the simulation until about tick 1,000 and view the world. After the WATER-OIL-FORCE decreases, water molecules are **1.** (A. closer together, B. farther apart), bilayers become **2.** (C. less, D. more) defined, and the hydrophobic isolation graph initially **3.** (E. decreases gradually, F. decreases sharply, G. increases gradually, H. increases sharply, I. is horizontal).

d. **1.** _______ **2.** _______ **3.** _______ Without pressing *setup*, decrease WATER-OIL-FORCE to -1.0 and run the simulation until about tick 1,500, observing this world. (For comparisons, you may choose to take a screenshot.) Water molecules are **1.** (A. closer together, B. farther apart), **2.** (C. bilayers, D. reverse micelles) form, and the hydrophobic isolation graph **3.** (E. decreases gradually, F. decreases sharply, G. increases gradually, H. increases sharply, I. is horizontal).

e. **1.** _______ **2.** _______ **3.** _______ **4.** _______ **5.** _______ **6.** _______ Without pressing *setup*, decrease WATER-OIL-FORCE to -2.0 and run the simulation until about tick 2,000, viewing the final results. (For comparisons, you may choose to take a screenshot.) Water molecules are **1.** (A. closer together, B. farther apart), reverse micelles **2.** (C. are, D. are not) clearly formed, bilayers **3.** (C. are, D. are not) clearly formed, and the hydrophobic isolation graph **4.** (E. decreases gradually, F. decreases sharply, G. increases gradually, H. increases sharply, I. is horizontal) before reaching an equilibrium. This equilibrium is **5.** (J. greater, K. less) than the hydrophobic isolation percent for a WATER-OIL-FORCE value of -1.0 and **6.** (J. greater, K. less) than the hydrophobic isolation percent for a WATER-OIL-FORCE value of -0.1.

f. **1.** _______ **2.** _______ **3.** _______ **4.** _______ **5.** _______ **6.** _______ **7.** _______ **8.** _______ With WATER-OIL-FORCE decreasing from -0.1 to -0.7, as hydrophobic molecules and water molecules become **1.** (A. less, B. more) repellent, bilayers **2.** (C. collapse, D. form), and hydrophobic isolation **3.** (E. decreases, F. increases). As this repelling force continues to become stronger, with WATER-OIL-FORCE going from -0.7 to -1.0, bilayers **4.** (C. collapse, D. form), reverse micelles **5.** (C. collapse, D. form), and hydrophobic isolation **6.** (E. decreases, F. increases). As this repelling force becomes even stronger, with WATER-OIL-FORCE going from -1.0 to -2.0, the lipids around groupings of water molecules move **7.** (G. faster, H. slower), and hydrophobic isolation **8.** (E. decreases, F. increases) before reaching equilibrium.

g. Discuss the results, possible reasons for these outcomes, and whether the results support or refute your hypothesis.

h. **1.** _______ **2.** _______ **3.** _______ **4.** _______ Change both WATER-WATER-FORCE and WATER-OIL-FORCE to 0. Set up and run the simulation. The hydrophobic isolation graph **1.** (A. decreases gradually, B decreases sharply, C. increases gradually, D. increases sharply, E. is horizontal) so that we have neutral hydrophobic isolation, water molecules **2.** (F. are, G. are not) evenly distributed, reverse micelles **3.** (F. are, G. are not) formed, and bilayers **4.** (F. are, G. are not) formed.

5. This question explores the impact of TOO-CLOSE-FORCE, the amount an acting molecule should move in the direction of a randomly selected molecule that is within TOO-CLOSE-DISTANCE. Because TOO-CLOSE-FORCE is negative, molecules that are "too close" repel each other.

   a. _______ With TOO-CLOSE-FORCE value **1.** (A. –2.0, B. –0.7, C. –0.1), molecules that are "too close" move farther apart.

   b. **1.** _______ **2.** _______ **3.** _______ Complete the following hypothesis: As TOO-CLOSE-FORCE becomes more negative, reverse micelles are **1.** (L. less, M. more) likely to form, bilayers are **2.** (L. less, M. more) likely to form, and hydrophobic isolation **3.** (D. decreases, I. increases).

   c. **1.** _______ **2.** _______ **3.** _______ Press DEFAULT VALUES and set TOO-CLOSE-FORCE to –0.2. Turn the STOP-AT-300 switch OFF. Set up and run the simulation until about tick 500, viewing the world at that point. (For comparisons, you may choose to take a screenshot.) Decrease TOO-CLOSE-FORCE to –0.3. Do *not* press *setup*. Run the simulation until about tick 1,000 and view the world. As the TOO-CLOSE-FORCE decreases, water molecules are **1.** (A. closer together, B. farther apart), bilayers become **2.** (C. less, D. more) defined, and the hydrophobic isolation graph initially **3.** (E. decreases gradually, F. decreases sharply, G. increases gradually, H. increases sharply, I. is horizontal).

   d. **1.** _______ **2.** _______ **3.** _______ **4.** _______ Without pressing *setup*, decrease TOO-CLOSE-FORCE to –0.4 and run the simulation until about tick 1,500, observing this world. Water molecules are **1.** (A. closer together, B. farther apart), **2.** (C. bilayers, D. reverse micelles) form, and the hydrophobic isolation graph **3.** (E. decreases gradually, F. decreases sharply, G. increases gradually, H. increases sharply, I. is horizontal) before **4.** (J. decreasing gradually, K. decreasing sharply, L. increasing gradually, M. increasing sharply).

   e. **1.** _______ **2.** _______ **3.** _______ **4.** _______ **5.** _______ Without pressing *setup*, decrease TOO-CLOSE-FORCE to –0.7. Molecules move **1.** (A. faster, B. slower), water molecules are **2.** (C. closer together, D. further apart), reverse micelles **3.** (E. are, F. are not) clearly formed, bilayers **4.** (E. are, F. are not) clearly formed, and the hydrophobic isolation graph **5.** (G. decreases gradually, H. decreases sharply, I. increases gradually, J. increases sharply, K. is horizontal) before reaching equilibrium.

   f. Discuss the results, possible reasons for these outcomes, and whether the results support or refute your hypothesis.

6. This question explores the impact of RANDOM-FORCE, the amount each molecule moves in a random direction each tick.

   a. _______ A water molecules move farther apart with RANDOM-FORCE value **1.** (A. 0.0, B. 0.1, C. 1.0).

   b. **1.** _______ **2.** _______ **3.** _______ Complete the following hypothesis: As RANDOM-FORCE increases, reverse micelles are **1.** (L. less, M. more) likely to form, bilayers are **2.** (L. less, M. more) likely to form, and hydrophobic isolation **3.** (D. decreases, I. increases).

   c. **1.** _______ **2.** _______ Press DEFAULT VALUES and set RANDOM-FORCE to 0.0. Turn the STOP-AT-300 switch OFF. Set up and run the simulation until about tick 500, viewing the world at that point. (For comparisons, you may choose to take a screenshot.) Increase RANDOM-FORCE to 0.1. (Be sure to select 0.10, not 0.01.) Do *not* press *setup*. Run the simulation until about tick 1,000 and view the world. As the RANDOM-FORCE increases from 0.0 to 0.1, bilayers become **1.** (A. faster, B. slower) defined, and the hydrophobic isolation graph **2.** (C. decreases gradually, D. decreases sharply, E. increases gradually, F. increases sharply).

d. **1.** _______ **2.** _______ Without pressing *setup*, increase TOO-CLOSE-FORCE to 0.30 and run the simulation until about tick 1,500, observing the final world. Molecules move **1.** (A. faster, B. slower), and structures **2.** (C. cannot, D. can still) hold together, while hydrophobic isolation decreases slightly.

e. Discuss the results, possible reasons for these outcomes, and whether the results support or refute your hypothesis.

An additional question on the impact of water concentration is available in Active Learning.

MODULE

5

Inside the Plasma Membrane

5.1. Prokaryotic Cell Structure and Function

We have not previously described the cells of bacteria, which are described as *prokaryotic* (lacking a membrane-bound nucleus) (see Figure 5.1). Essentially, a prokaryotic cell lacks an organized, membrane-bound nucleus, which characterizes all *eukaryotic* cells, such as those of humans. In the cytoplasm, bacteria do have a circular chromosome composed mostly of double-stranded *DNA* (deoxyribonucleic acid), plus some associated *RNAs* (ribonucleic acids) and proteins. Most of these RNAs are *messenger RNAs* (*mRNA*), which transcribe the DNA code that is then transferred to *ribosomes* for synthesis of proteins. The proteins that are present include enzymes, *transcription factors* (promoting or inhibiting gene expression), and proteins that help to organize the DNA. There may be more than one copy of the DNA, and smaller satellite DNAs also. This concentration of the bacterial genome and associated molecules forms what is called the *nucleoid*. This structure has a less definite shape than the nucleus of eukaryotes, but it serves the same general function.

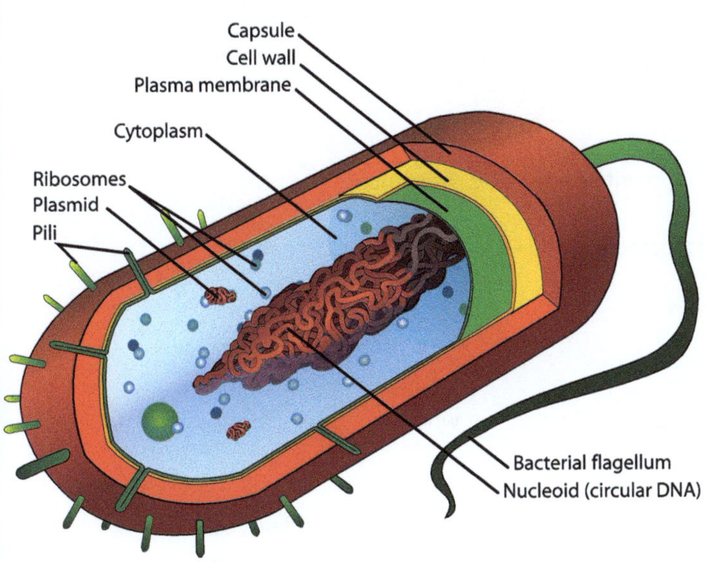

FIGURE 5.1 Diagram showing basic structure of a bacterium (prokaryote)

Prokaryotes have a plasma membrane, surrounded by a sturdy *cell wall*, whose structure is based on polymers made up of nitrogen-containing sugars and amino acids. Its primary purpose is to prevent the cell from bursting from water absorption. (Some bacteria also have an outer covering called the capsule). Prokaryotes also lack the internal compartmentalization by organelles. There are ribosomes and storage granules in the jelly-like cytosol. In some bacteria, the plasma membrane may be folded internally to increase surface area for various metabolic processes.

Despite the comparatively simplified cell structure, bacteria have been and are extremely successful life forms. Although we tend to associate bacteria with illness, the majority of bacterial species are benign or quite beneficial. For instance, in nature, many bacteria are decomposers of organic matter, making the soil richer and more productive. Others help to fix nitrogen, which in the atmospheric form (N_2), plants can absorb as nutrients for their growth and reproduction. In your own digestive tract, we are learning how important the bacterial flora is to your health. Bacteria are also useful in the production of foods (like yogurt and cheese) and antibiotics. Bacteria have been genetically manipulated to produce specific proteins (like insulin) in large quantities and more cheaply.

PoI 1. Human beings manufacture and use many antibacterial compounds, generally called *antibiotics* or, more specifically, *antibacterial drugs*. Some are taken internally, while others are used topically (including *antiseptics*). Some kill bacteria outright, whereas others merely inhibit their growth. Antibiotics might also target other potentially harmful organisms, such as some fungi, protozoans, flatworms, and a few viruses. To decontaminate surfaces, we use antimicrobial compounds called *disinfectants*, which are too toxic for use in or on the body. The commonly used antibacterial agents below are termed antibiotics, although principally we use them only against bacteria. For each, determine the way it works (very generally) and the type of organism that produced it.

a. Penicillin
b. Tetracycline

PoI 2. One of the most vexing problems in medicine is antibiotic resistance, because when resistance by a bacterial species develops, that particular antibiotic is no longer useful. We now have some bacteria that are resistant to most of our currently available drugs. Antibiotic resistance has developed by a majority of pathogenic (disease-causing) bacteria. For the following antibiotics, how has resistance developed?

a. Tetracycline
b. Streptomycin (aminoglycoside)

PoI 3. What factors are important in the development of antibiotic resistance?

PoI 4. How can you help to prevent acquiring a pathogen and reduce the development of antibiotic resistance?

QRQ 1. Cells that lack a membrane-bound nucleus are called ______, while ______ cells contain a membrane-bound nucleus.

QRQ 2. Select all statements that apply to bacteria: cell has a membrane-bound nucleus, cell has a nucleoid, cell has a plasma membrane, surrounded by a sturdy cell wall, cell lacks internal compartmentalization by organelles, cytoplasm contains a circular chromosome, plasma membrane's main purpose is preventing cell bursting from water absorption, vast majority of bacterial species cause disease

5.2. Eukaryotic Cell Structure and Function

APPLYING SCIENCE TO LIFE 1 Baby Talia's Medical Mystery

Baby Talia was brought to the emergency room by her mother. Over the six months she has lived, she has been to the ER several times with recurring respiratory infections. Talia is small for her age, never having been a good eater. Her distraught mother says that her baby seems to be breathing too fast and is wheezing. The doctor also notices that the infant shows very poor muscle tone, medically termed hypotonia, and has an enlarged tongue. An X-ray reveals an enlarged heart. Given the evidence, the doctor orders blood tests, but before the test results are available, Talia goes into cardiac arrest, and the medical team is unable to resuscitate her.

Talia's catastrophic death can be explained on the basis of a cellular-level defect. But before we develop a hypothesis about where the defect might be, we need to learn more about the structure and function of cells.

5.2.1. The Nucleus

Cells of animals, plants, fungi, and protozoa have a *eukaryotic* (with a membrane-bound nucleus) cell structure, composed of cells that have a membrane-bound nucleus and internal, membrane-bound compartments, usually referred to as *organelles*. In Figure 5.2 are diagrams of cells from each of the

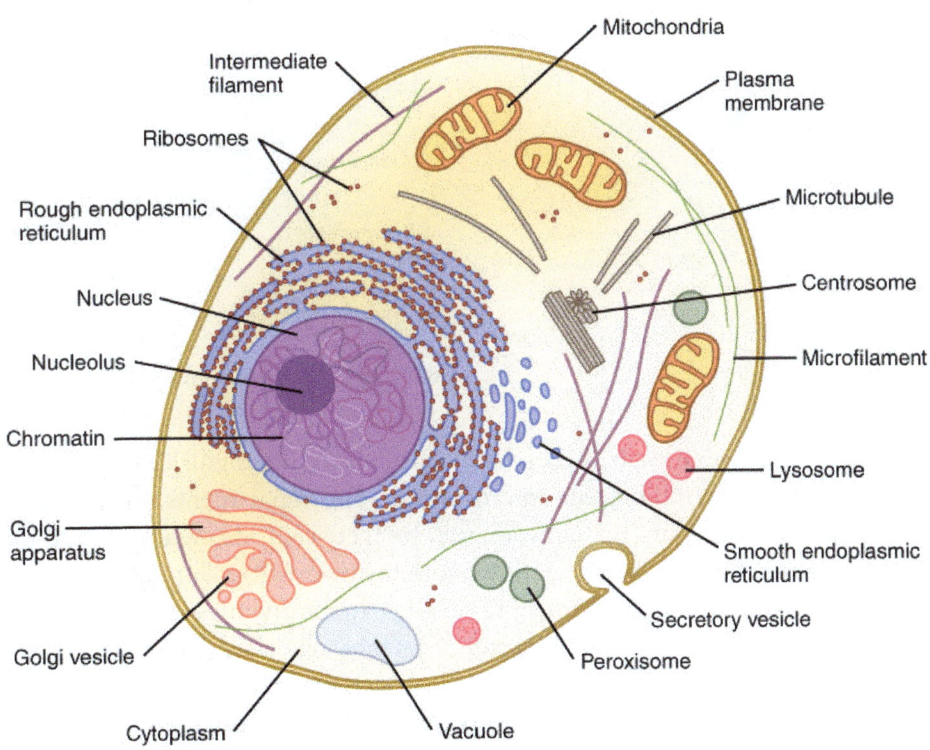

FIGURE 5.2a Animal cell

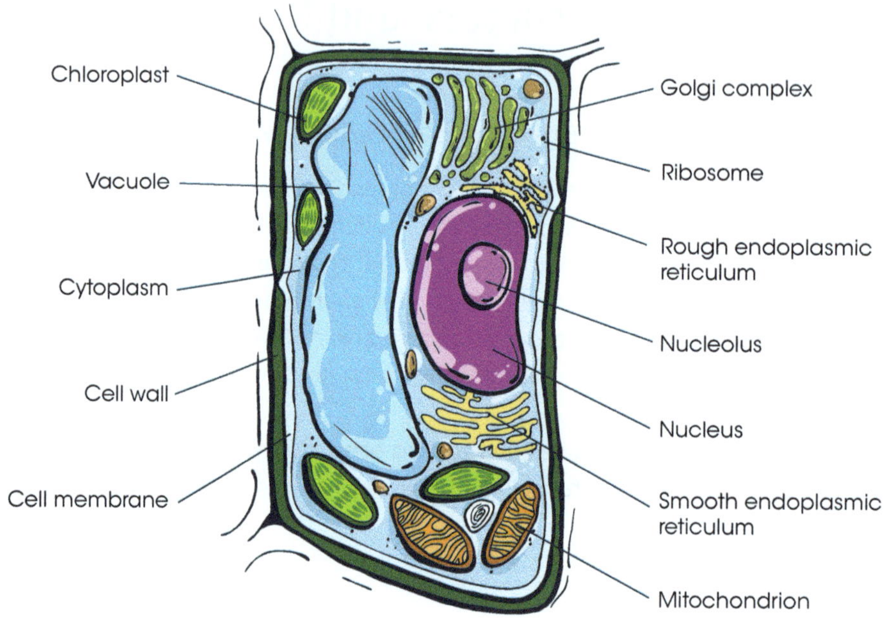

FIGURE 5.2b Plant cell

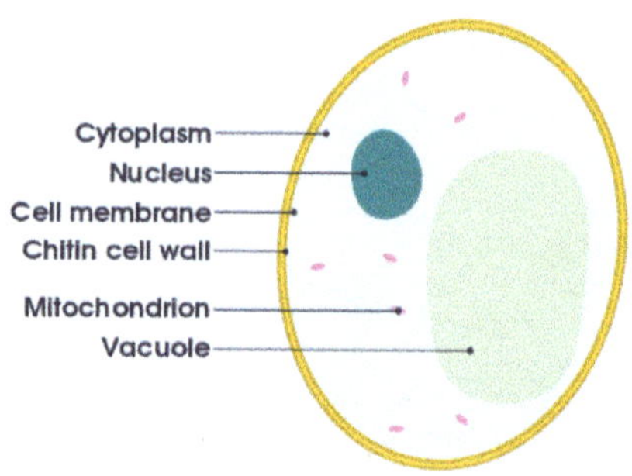

FIGURE 5.2c Yeast cell

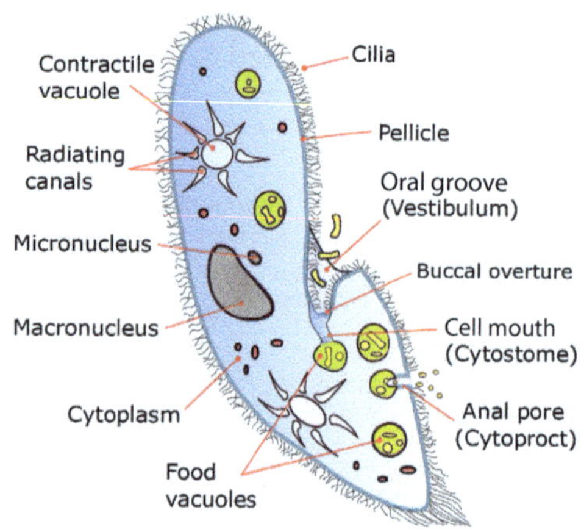

FIGURE 5.2d Paramecium (Protista)

eukaryotic groups. Notice that each of these cells has its own distinguishing characteristics, but they all have a membrane-bound nucleus and membranous compartments, such as mitochondria, endoplasmic reticulum, and so on. For convenience, we'll concentrate on animal and plant cells.

To understand better the life of a cell, let's explore the cytoplasm of an animal cell (see Figure 5.3). After passing through the plasma membrane, you can hardly miss the *nucleus*, a prominent, consolidated body surrounded by a double layer of membrane called the *nuclear envelope*. If you could watch activity around the nucleus, you would see various molecules entering and exiting the nucleus through *pores* in the nuclear envelope. Water-soluble molecules, particularly small ones, can move through the pores by diffusion. All of the larger molecules, such as proteins, are transported through the pores, bearing distinctive signals that bind to pore-complex proteins. These signals earmark the larger molecules for transport through the pore into or out of the nucleus.

Inside the nucleus is most of the cell's DNA, arranged into the chromosomes. The genetic material appears to be jumbled, but it is highly organized around protein particles. Although the condensed, identifiable chromosomes are only visible during cell division, much of the DNA is quite busy in the meantime making various types of RNA—*messenger RNA (mRNA), transfer RNA (tRNA), ribosomal RNA (rRNA),* and *small nuclear RNA (snRNA).* Each of these RNA types has an important role to play so that the DNA code is properly transferred to direct the synthesis of specific protein sequences. Ribosomal RNA is actually produced in an identifiable region of the nucleus called the *nucleolus.*

Because the nucleus plays such an essential informational role in the cell, it is not surprising that the contents are separated by the surrounding double membrane. Aside from protecting the genetic instructions for cell structure and function, the membrane and nuclear pores help to regulate the molecules entering and exiting.

The nuclear envelope is continuous with the membranes of another compartment, the endoplasmic reticulum (ER) (see Figures 5.2a and 5.3). You can see that there are two interconnected but structurally and functionally distinct types of ER. That nearest to the nucleus is the *rough endoplasmic reticulum* (*RER*), which generally appears as a series of flattened, membranous sacs whose outer surfaces are dotted with small, protein-RNA complexes called *ribosomes.* Extending from the RER is the *smooth endoplasmic reticulum* (*SER*), which is typically made up of a set of somewhat more tubular sacs that lack ribosomes. Beyond the SER is an unconnected set of membranous sacs that appear stacked one upon another, called the Golgi. Outside all of these organelles are various

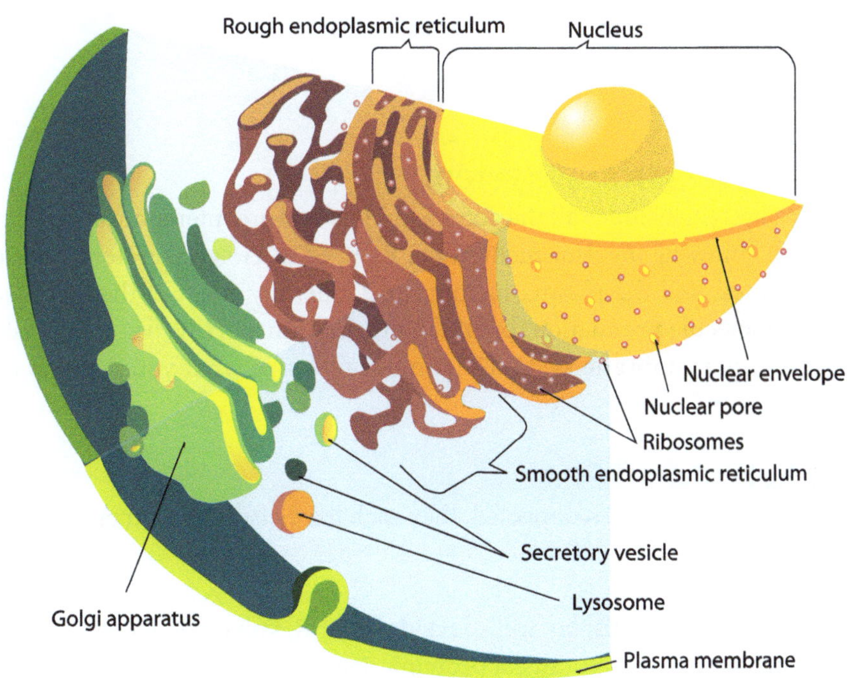

FIGURE 5.3 Endomembrane system

types of vesicles, some of which are transport vesicles, moving materials from one membranous compartment to another or to the plasma membrane. Other vesicles, such as lysosomes and peroxisomes, have other, specific functions. Together, biologists consider them a complex of interrelated membranous compartments, which they call the *endomembrane system*.

QRQ 3. The double layer of membrane surrounding the nucleus is called the ______ ______.

QRQ 4. The general term for a membrane-bound, organized or specialized compartment inside a cell is a(n) ______.

QRQ 5. Water soluble molecules, particularly small ones, can move by ______ through ______ in the nucleus's membrane. However, larger molecules, such as proteins, must be transported.

QRQ 6. The double membrane of the nucleus is continuous with the membranes of another compartment, the ______ ______, which is abbreviated ______.

QRQ 7. Select from the following the appropriate term for each description below: DNA, endomembrane system, eukaryotic, Golgi apparatus, lysosome, nucleolus, nucleus, peroxisome, prokaryotic, RER, ribosome, SER, vesicle

a. animal and plant cells
b. complex of interrelated membranous compartments in the cell
c. fungi and protozoa cells
d. outer surfaces are dotted with ribosomes
e. region of the nucleus producing ribosomal RNA
f. small protein-RNA complex

5.2.2. The Endomembrane System

The endomembrane system includes all the compartments that are interconnected, directly or indirectly, to include the following: nuclear membrane, RER, SER, Golgi, secretory vesicles, vacuoles, lysosomes, peroxisomes, and the plasma membrane. Generally, most of the included organelles are associated with molecule production, breakdown, and transport. What are possible functions of members of this endomembrane system? Think in terms of what you think is necessary for supporting life.

Endoplasmic Reticulum (ER)

Although the RER and SER are interconnected, they each have distinctive functions (see Figure 5.3).

RER

If you wanted to study cells that display well-developed RER, you could look for them in the liver. Cells of this organ produce large amounts of proteins, some of which are secreted (transported out of the cell). You can identify the RER because it has an external membrane surface that is dotted with *ribosomes* (see Figure 5.3). As Figure 5.4 illustrates, ribosomes are rRNA and protein particles that convert the mRNA code, produced from the DNA sequences, into amino acid sequences (*translation*) during protein synthesis. Translation is the second of two major processes required to make proteins. The first is the synthesis of mRNA from a code on DNA (*transcription*). The mRNA is then transported into the cytoplasm and attaches to a ribosome for translation. It follows that one of the primary functions of the RER is protein synthesis.

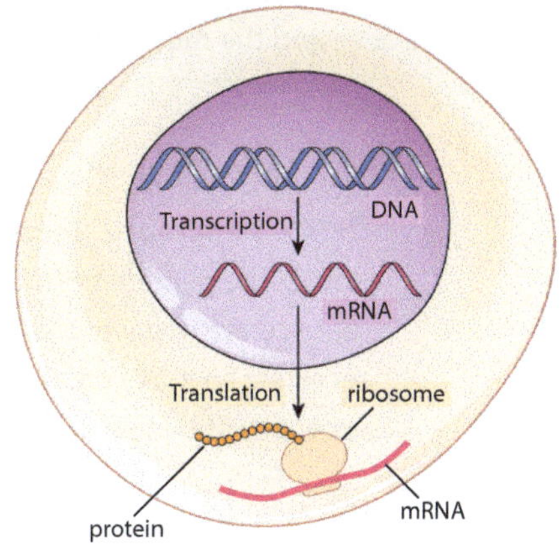

FIGURE 5.4 Transcription and translation

Translation of a new polypeptide addressed for the RER begins on an unattached ribosome (free) in the cytosol. The new polypeptide contains a set of amino acids, called the *signal sequence*. A polypeptide with such a sequence earmarks the free ribosome for the RER membrane. The ribosome then attaches to the outer membrane of the RER, and translation continues. The polypeptide is threaded through a channel in the membrane into the *lumen* (internal space) of the RER. Sometimes, the complete polypeptide is transported into the lumen, and in other cases, the polypeptide becomes part of the RER membrane.

Once in the RER lumen, *chaperone* proteins (to assist with proper folding or unfolding of polypeptides) and enzymes are present that can fold the polypeptide into a functional *conformation* (shape), assemble polypeptide subunits into more complex proteins, and add sugars to proteins (*glycosylation*). Such glycosylated proteins are called *glycoproteins*. These sugars help to influence folding, and hence the shape of the protein. Many proteins synthesized on the RER may, by way of vesicles, be exported to the *Golgi*.

PoI 5. The complexity of the processing of proteins is interesting, impressive, and essential. Consider that each of the enzymes involved in the glycosylation and other processing is coded for by the cell's DNA. So slight changes in the genetic information can alter the function of those enzymes. What do you think would be the impact of a mutation that causes a glycosylation enzyme to fail or do a poor job?

SER

The smooth endoplasmic reticulum synthesizes a variety of lipids, including cholesterol, the phospholipids for membranes, and steroids, on the outside leaflet of the SER membrane. You would expect and find a generous amount of SER in cells that secrete steroid hormones, like those of male and female gonads. The SER also contains enzymes that help detoxify metabolic wastes, drugs, and alcohol. Thus, liver cells also contain extensive amounts of SER.

Muscle cells have a specialized type of SER, called *sarcoplasmic reticulum*, which plays a special role in storage of calcium ions (Ca^{++}). When a nerve impulse moves over the muscle membranes, it causes a sudden release of some of the Ca^{++}, which indirectly prompts the muscle to contract. Once the stimulation has passed, the calcium ions are transported back into the sarcoplasmic reticulum, against a concentration gradient, and the muscle relaxes.

APPLYING SCIENCE TO LIFE 2 Life or Death

It might be of interest that after death, ATP production ceases, resulting from lack of oxygen and lost glycogen reserves. Without ATP, which supplies energy to the calcium pumps on the sarcoplasmic reticulum (and on the plasma membrane), these pumps are unable to remove calcium ions from the muscle cytoplasm, and the muscle remains constricted (rigor). The resulting condition is termed *rigor mortis.*

GtM 1. Pancreatic Secret(e) The pancreas is best known for its role in producing and secreting *insulin,* the peptide hormone that regulates blood glucose levels. The pancreas has other cells that secrete many of the digestive enzymes for the small intestine. With its role in producing and secreting so many peptide and protein products, what membrane-bound organelle would you expect to find in abundance in pancreatic cells?

QRQ 8. The first major process required to make proteins, which is the synthesis of mRNA from a code on DNA, is called ______.

QRQ 9. The second major process of protein synthesis, which is the conversion of mRNA code into amino acid sequences, is called ______.

QRQ 10. An rRNA and protein particle in the cytosol that converts mRNA code into amino acid sequences (second major process of protein synthesis) is a(n) ______.

QRQ 11. Select from the following the appropriate term for each description below: chaperone protein, conformation, glucoprotein, glucosylation, glycoprotein, glycosylation, gonad, liver, lumen, membrane, pancreas, RER, ribosome, sarcoplasmic reticulum, SER, signal sequence

a. addition of sugars to proteins
b. area of RER where polypeptide subunits are assembled into more complex proteins
c. arrangement of amino acids prompting newly translated polypeptide to move to RER
d. assists in the proper folding or unfolding of polypeptides
e. cells in this area display well-developed RER
f. contains enzymes that help detoxify metabolic wastes, drugs, and alcohol
g. in muscle cells, a specialized type of SER that stores calcium ions
h. internal space
i. part of the RER where folding of a polypeptide into a functional shape occurs
j. protein with added sugars
k. proteins synthesized on the RER may become part of the RER
l. shape
m. synthesizes a variety of lipids

QRQ 12. Give the sequence of events for muscle contraction and subsequent muscle relaxation from the following: A. indirect prompting of the muscle to contract, B. movement of nerve impulse over the muscle membranes, C. muscle relaxation, D. stimulation passes, E. sudden release of Ca++, F. transportation of calcium ions against a concentration gradient

Golgi

The Golgi body, or Golgi, is usually associated with modifying proteins arriving from the ER (Figure 5.3). Further sugar additions (glycosylation) or phosphate additions (phosphorylations) help to direct the protein to the correct location in the cell. For instance, digestive enzymes, meant for lysosomes, arrive in the Golgi with sugars attached in the RER, and Golgi enzymes add phosphates to them. Then the enzymes are gathered into a vesicle and moved to forming lysosomes. On the other hand, if a protein is intended for the plasma membrane or secretion, the Golgi will modify the sugar composition on the protein so that it is targeted by vesicles for the plasma membrane. Therefore, proteins heading for the plasma membrane may become part of that membrane, or the glycoproteins may be released as *secretory proteins*. These modifications include signals that address the protein for its proper destination.

The Golgi also contain enzymes that synthesize lipids, other than phosphoglycerides, important for plasma membrane structure and function. In plant cells, the Golgi is also is the site for the synthesis of complex carbohydrates that combine with cellulose to form the cell's wall.

Vesicles

Produced from the ER or Golgi or budded from the plasma membrane, vesicles move from one part of the endomembrane system to another, fusing with other compartments. Vesicles provide a way to share or secrete materials, and their membranes may provide a way to renew, expand, or rebuild the membrane of compartments and the plasma membrane.

Vacuoles

Generally, vacuoles are fairly large vesicles generated by the ER and Golgi. In cells that phagocytize (ingest), such as some protozoans and human macrophages, vacuoles are compartments formed during the phagocytic process. In some protozoa, we find specialized *contractile vacuoles* that help to remove excess water from the cytoplasm.

In plants, vacuoles, formed from the coalescence of many smaller vacuoles as more and more water accumulates, may be the largest of the internal structures. Plant vacuoles can swell to allow a cell to expand. They also store vital salts, pigments, food reserves, and certain metabolic wastes. In fact, some retain toxic materials to inhibit herbivores from eating the plant. Further, plant vacuoles can remove excess hydrogen ions to help maintain a stable pH (measure of hydrogen ion concentration).

Lysosomes

Lysosomes are membrane-bound packets of digestive enzymes and are found in many types of eukaryotic cells. The enzymes are activated by low pH (acidic conditions). As lysosomes are produced, protein pumps move hydrogen ions into the lumen of the developing lysosome, which lowers the pH sufficiently to activate the digestive enzymes inside. Mature lysosomes are responsible for breaking down incoming organic materials. They also recycle materials from old organelles or defective or unneeded internal macromolecules.

As an example, you would find that protozoa, like *Amoeba*, engulf food by surrounding it with pseudopodia, forming a food vacuole. The food vacuole fuses with a lysosome, and the imported materials are broken down into smaller molecules, like sugars and amino acids, which can then be transported out of the vacuole into the cytoplasm for use.

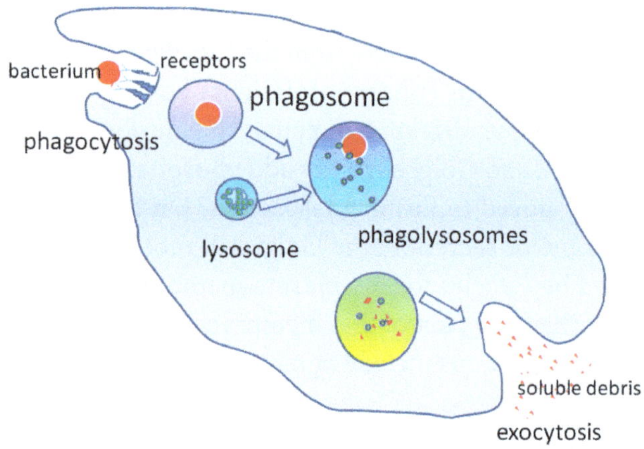

FIGURE 5.5 Diagram of cell phagocytizing and processing a bacterium

In your body, there are white blood cells of your immune system, such as macrophages, neutrophils, and dendritic cells, which can engulf bacteria or organic debris in a manner similar to protozoa (see Figure 5.5). After fusion of the phagocytic vesicle with a lysosome, the material is broken down into some useful smaller molecules for assimilation by the cell or waste materials that are expelled. A good example of macrophages may be found associated with the air sacs (alveoli) of your lungs. These phagocytes (cells that ingest) help to clean up the alveoli by phagocytosing toxic allergens and even infectious bodies that did not get eliminated by the cilia/mucous removal system operating in the lungs. After engulfing the target within a vesicle, the phagocyte can digest it or, if the target is a pathogen (e.g., bacterium), kill it and digest it. Lysosomal fusion with the vesicle provides digestive enzymes to break down the target's organic molecules into smaller, useful subunits for the cell. Any wastes are disposed of by exocytosis.

Peroxisomes

Like lysosomes, peroxisomes are specialized vesicles, surrounded by one layer of membrane. After budding off from the SER, they import needed enzymes and other proteins. Peroxisomes can divide but lack any genetic instructions. Cells that produce, store, or break down lipids are good sources of peroxisomes.

Peroxisomes have quite a variety of functions, including the oxidation of very long chains of fatty acids. Enzymes in this organelle remove hydrogens from target molecules and add them to oxygen, producing hydrogen peroxide (H_2O_2), which is toxic to cells. However, the peroxisomes use the peroxide to detoxify other chemicals, like alcohol. In the process, peroxide is converted to water and oxygen. This process is particularly significant in the liver and kidneys. Peroxisomes also synthesize phospholipids for the insulating cover on nerve cells. Abnormal products in phospholipids often result in cardiac, neurological, and muscular problems.

In seeds, stored fats are converted to sugars in specialized peroxisomes (glyoxysomes). Sugars are then available for energy or synthesis. Interestingly, human beings are unable to convert fats directly into sugars.

QRQ 13. The _____ are usually associated with modifying proteins arriving from the ER by adding sugars or phosphates to direct the proteins to the correct locations in the cell.

QRQ 14. Produced from the ER or Golgi or budded from the plasma membrane, _____ move from one part of the endomembrane system to another, fusing with other compartments.

QRQ 15. Generally, a(n) _____ is a fairly large compartment generated by the ER and Golgi during a phagocytic process.

QRQ 16. In many types of eukaryotic cells, a membrane-bound packet of digestive enzymes called a(n) ______ is responsible for breaking down incoming organic materials and recycling materials from old organelles or defective or unneeded internal macromolecules.

QRQ 17. Specialized vesicles, called ______, have a variety of functions, including the oxidation of very long chains of fatty acids.

QRQ 18. Select from the following the compartment important to each function/application below: Golgi, lysosomes, peroxisomes, RER, SER, vacuole, vesicles

a. after bacteria or organic debris are engulfed by white blood cells of your immune system, the material is broken down into some useful, smaller molecules for assimilation by the cell or waste materials that are expelled
b. can remove excess hydrogen ions to help maintain a stable hydrogen ion concentration in plant cells
c. contain enzymes that synthesize lipids, other than phosphoglycerides, important for plasma membrane structure and function
d. in a specialized form of this type of compartment, stored fats in seeds are converted to sugars, which are then available for energy or synthesis
e. in plant cells, these structures retain toxic materials, inhibiting consumption by herbivores
f. in some protozoa, specialized forms of this type of compartment help to remove excess water from the cytoplasm
g. protozoa surround food, fusing the packet with this type of compartment, where digestive enzymes break the food down into smaller molecules
h. provide a way to share or secrete materials
i. site for the synthesis of complex carbohydrates that combine with cellulose to form the walls of plant cells
j. store vital salts, pigments, food reserves, and certain metabolic wastes in plant cells
k. synthesize phospholipids for the insulating cover on nerve cells
l. where enzymes remove hydrogen from target molecules and add it to oxygen, producing hydrogen peroxide, which is used to detoxify alcohol

5.2.3. Energy-Providing Organelles

All eukaryotic cells have mitochondria, which provide most of the needed energy for the cell. Photosynthetic eukaryotes also have chloroplasts, which are also involved in energy production.

Mitochondria

Mitochondria, which convert the energy contained in food materials into a readily usable form of energy in ATP, are in all types of eukaryotic cells, even photosynthetic ones. Photosynthetic eukaryotes can convert light energy into food, like glucose, but to get energy for work, they must convert the energy stored in food into ATP or related high-energy compounds.

The mitochondrion is formed from a double layer of membrane—*outer* and *inner*. The outer membrane is very permeable and is not much of a barrier. The inner membrane expands surface area for reactions. Two compartments are generated by this arrangement—the *intermembranous space*, between the outer and inner membranes, and the *matrix*, bordered by the inner membrane. The mitochondrion has some of its own DNA, circular and double-stranded. Human mitochondrial DNA contains a total of 37 genes, which code for some proteins, some tRNA, and rRNA. Almost all of the proteins needed for the mitochondrion structure and function are produced outside the mitochondrion, and the proteins are imported. The proteins that are translated on the mitochondrial ribosomes are used in the production of ATP through the electron transport system (Sylvester et al. 2004). Inside the mitochondrion are the key pathways for most of the energy production from carbohydrates and fatty acids. In the mitochondrial matrix are the enzymes for the Krebs cycle (to generate energy during aerobic respiration), and within the folded inner membrane is the apparatus for the electron transport system and ATP synthesis. The mechanism for obtaining energy for cellular functions will be discussed later.

APPLYING SCIENCE TO LIFE 3 A House Dividing

Mitochondria are self-replicating organelles. They contain their own DNA (mtDNA) and ribosomes (where proteins are made), but most of the proteins required for mitochondrial activity and replication are coded for by nuclear genes. The replication of both nuclear and mitochondrial DNA seem to be coordinated, although they are not necessarily simultaneous. Overall, the ratio of mitochondria numbers per nucleus for any cell remains relatively constant.

The nuclear-directed proteins are translated in the cytoplasm and imported by the mitochondria. However, the mitochondrial chromosomes contain vital genes for normal function. Of course, there is much less DNA on each circular chromosome, and consequently there are only 37 genes. Most of the mtDNA is devoted to coding for RNAs needed for protein production, but 13 code for proteins. These proteins code for key subunits of enzymes responsible for most ATP production in aerobic cells.

Mitochondrial DNA is very prone to mutation, some of which occurs during replication. Furthermore, the apparatus for repairing mutations is not as reliable as that for nuclear DNA. These mutations, if numerous and debilitating enough, can result in poorly functioning mitochondria and cellular death.

Mitochondria, during the course of normal function, will generate a number of *reactive oxygen species* (*ROS*). ROS are chemicals containing oxygen that are very reactive, and these reactive species can damage proteins and nucleic acids. (Your cells do have countermeasures to detoxify some of these compounds, but they are not perfect.) Mitochondrial DNA is quite vulnerable to damage from these ROS, and the mutations, either unrepaired or imperfectly repaired, can lead to malfunctions in the mitochondria. One effect of these "weakened" mitochondria is a decline in ATP production. Any cells that have a high energy demand will then be sensitive to this deterioration. One set of such cells are cells in your inner ear, which is responsible for hearing. Malfunctioning of these cells can be one of the factors in the onset of deafness as people age. Of course, there are other genetic and environmental influences as well (Chial and Craig 2008; Falkenberg 2018; NIH 2019; Sylvester et al. 2004).

Chloroplasts

Chloroplasts, like mitochondria, are surrounded by an outer and inner layer of membrane. Within these membranes is another set of membranous sacs. Within the membranes of these sacs are the pigment and protein complexes that form the platform for the *light reactions* of photosynthesis. Also, in that membrane is the ATP synthesis complex. Using light energy and electrons from water, the light reactions produce ATP and NADPH. The light reaction also produces oxygen as a byproduct. The reactions that effectively use energy from ATP and NADPH to assimilate carbon dioxide into sugar, sometimes referred to as the *dark reaction* of photosynthesis, are located in the stroma.

Chloroplasts' primary function is to produce glucose and other organic products that furnish food and building materials for the plant cell. Like the mitochondria, chloroplasts also contain their own DNA and ribosomes. Chloroplasts import most of the proteins required for their structure and function.

QRQ 19. _____ convert the energy contained in food materials into a readily usable form of energy in ATP.

QRQ 20. (T/F) The organelle for converting energy in food to energy in ATP occurs in animal cells but not photosynthetic cells.

QRQ 21. Select from the following the appropriate term related to mitochondria for each description below: DNA, inner membrane, intermembranous space, matrix, mitochondrion, outer membrane, RNA

a. between the outer and inner membranes
b. bordered by the inner membrane but surrounded by both membranes
c. circular and double-stranded in mitochondrion
d. formed from a double layer of membrane
e. imports most proteins required for its structure and function
f. quite selective, not very permeable
g. self-replicating organelle
h. very permeable

QRQ 22. The primary function of a(n) _____ is to produce glucose and other organic products that provide food and building materials for a plant cell.

QRQ 23. In chloroplasts, the _____ reaction uses light energy and electrons from water to produce ATP and NADPH, with oxygen as a byproduct, while the _____ reaction uses energy from ATP and NADPH to assimilate carbon dioxide into sugar.

5.3. Cytoskeleton—Keeping the Cell in Shape

What are some important functions of your skeleton? Support, muscle attachment and movement, protection of soft tissues like your brain. But why would a cell need a skeleton? Although the cellular functions might not be as obvious, the functions relate to those of your skeleton. The cellular skeleton, known as the cytoskeleton, is made up of proteins and, although it may appear static in a diagram, is instead a very dynamic structure that may be quickly rearranged depending on the needs of the cell.

There are three key players for the cytoskeleton: *microfilaments, microtubules,* and *intermediate filaments* (see Figure 5.6). Each of these plays an active supportive role in the cell, but each may make its own distinct contribution to other cellular functions as well. Moreover, the observed function of each can be altered by interactions with other proteins.

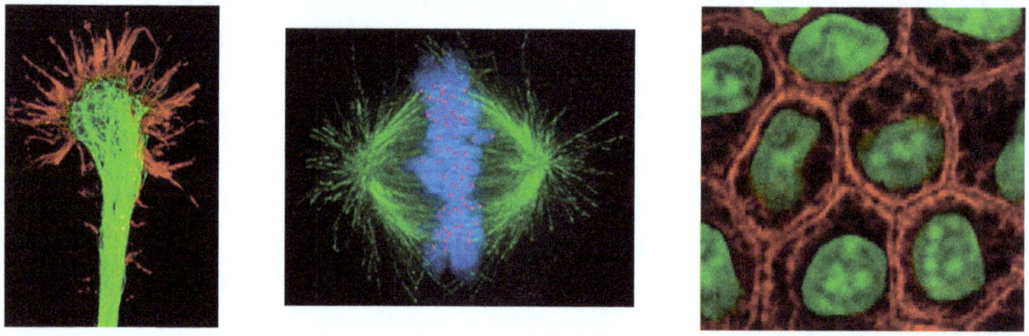

FIGURE 5.6 Elements of the cytoskeleton: **a.** Microfilaments (in red) at tips of neuronal growth cone; **b.** Microtubules of the mitotic spindle apparatus (in green); **c.** Intermediate filaments surrounding epithelial cells (in red)

Individual *microfilaments* are made up of two chains of individual actin protein molecules entwined into a helical filament. Microfilaments in muscle cells are important contributors to muscle contraction during interactions with the protein bundles containing myosin. Interestingly, each filament can be rapidly lengthened or shortened at either end, so it is not surprising that they also are important support for other types of cellular movements. For instance, some cells extend or grow, probing new territory at the forward edge, or growth cone, as in the growth cone of a neuronal axon, shown in Figure 5.6a. The extension is supported by the microfilaments, which can rapidly assemble and grow with the advancing growth cone. Similarly, the actin microfilaments are important players in the production of advancing extensions of phagocytic (engulfing) cells, like macrophages or protozoa. Moreover, they are involved in cell division.

Microtubules are hollow tubes composed of filaments of the protein tubulin (see Figure 5.7). Each filament is made of chains, which contain dimers (protein with two polypeptide chains) of α and β tubulin. Then 13 of these filaments are assembled into a cylinder to form the microtubule. Like the microfilaments, microtubules are also dynamic, growing or shortening as needed. Usually, however, one end tends to grow or shrink more readily, and this is the end that generally is pointed toward the periphery (see Figure 5.6b).

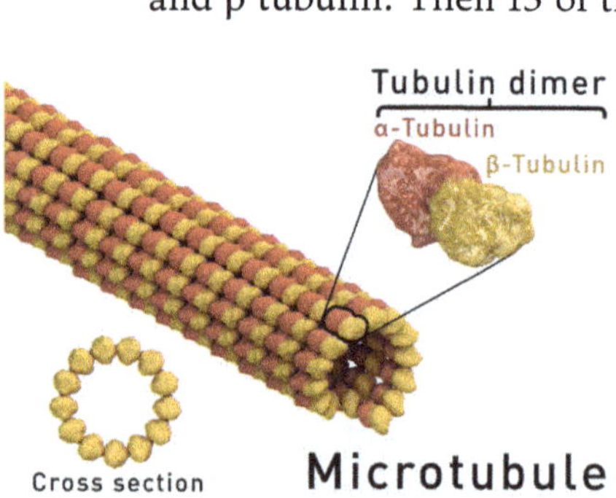

FIGURE 5.7 Structure of the microtubule, showing the tubulin heterodimer that composes each of the 13 filaments

The main functions of microtubules include support and movement. For instance, they are the major components in cilia and flagella. Cilia may move protozoa through their environment, and ciliated cells lining your respiratory tract may remove debris and microbes in mucus streams (see Figure 5.8).

Microtubules help to organize the cytoplasm and form the spindle fibers during cell division (see Figure 5.6b). Importantly, with molecular motor molecules, they also guide secretory vesicles (vesicles that

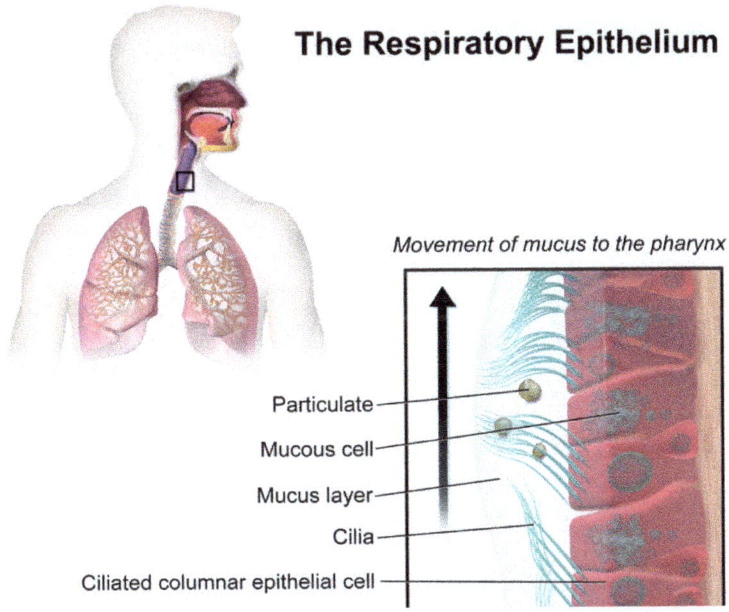

FIGURE 5.8 Respiratory epithelium that is lined with ciliated and mucus-secreting cells

fuse with the plasma membrane and release their contents) to the plasma membrane. For example, neurotransmitters, contained in secretory vesicles, are moved to the plasma membrane at the synaptic cleft for release (see Figure 5.9). Mitochondria can also be relocated in the cytoplasm along tracks of microtubules.

Scientists have studied *intermediate fibers* much less intensively, perhaps because they do not seem to be involved in any other type of cellular function. Several major types of intermediate

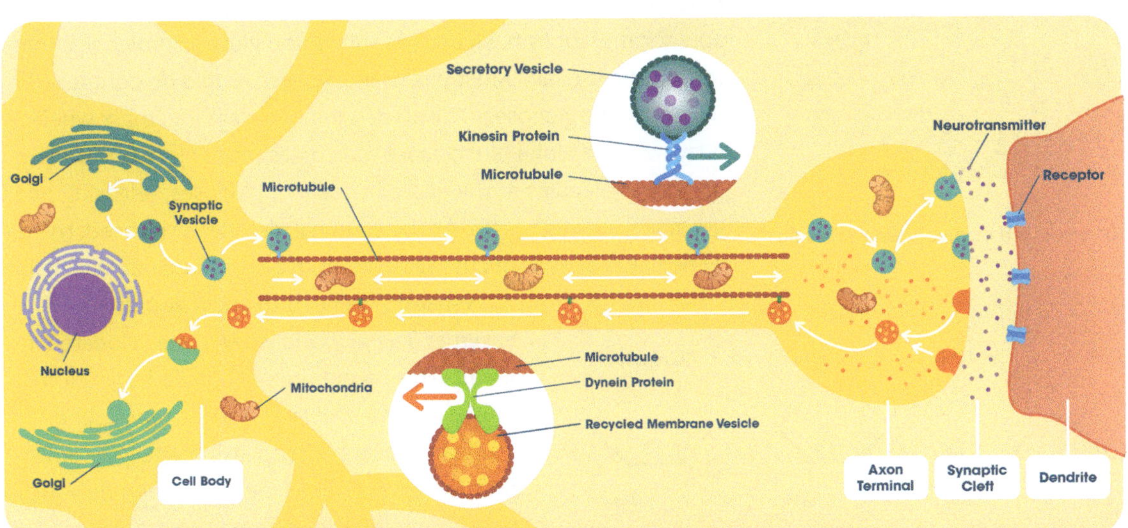

FIGURE 5.9 Transport of vesicles along microtubular tracks in the axon of a nerve cell

GtM 2. Think of the functional organs and organ systems of your body. Give examples of cellular organelles that perform functions similar to the following of your body's systems:

a. Digestive system
b. Excretory system
c. Circulatory system
d. Skeletal system
e. Nervous system

filaments are known. They seem to be more stable than microfilaments and microtubules, and their principal function seems limited to support, shape, and organization of the cell, particularly in its interactions with other cells in tissues. They are quite prevalent in epidermal (skin) cells. With cell division, they must reorganize to permit cytoplasmic division, but less is known about those mechanisms.

APPLYING SCIENCE TO LIFE 4 The Medical Mystery of Baby Talia Resolved

Now that you know more about the structure and function of the cell, let's take another look at the case of Baby Talia. The test results were available after Talia's death. Based on her signs and symptoms, the doctor ordered an assay for the enzyme *acid alpha-glucosidase* (*GAA*). GAA is an enzyme that breaks down glycogen in the lysosome. There are two organs that store most of the body's glycogen—the liver and muscles. The diagnosis is *Pompe disease*.

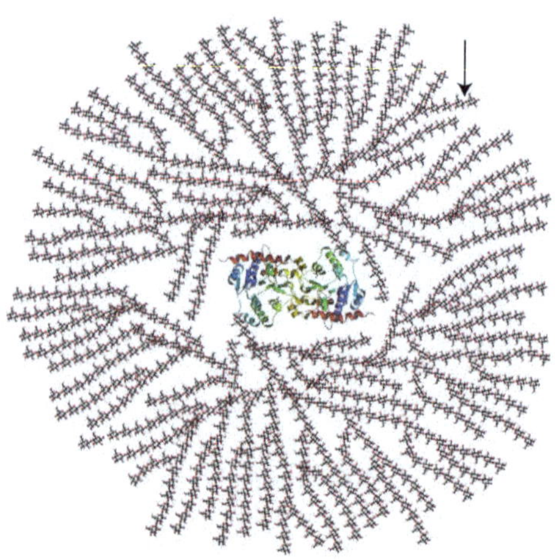

FIGURE 5.10 Structure of glycogen molecule

So, you know the location of the defect and the defective component. Now we need to understand why the defective lysosomal enzyme would cause death in an infant. In Figure 5.10, you can see a drawing of the glycogen molecule. Essentially, the glycogen molecule structure is composed of multiple glucose molecules linked together to form chains that are in turn linked to other chains, arranged around a protein (enzymatic) core. You can see that in this diagram of a glycogen molecule, there are many sugar molecules that can be enzymatically severed from glycogen, becoming available for metabolism. Figure 5.11 shows you glucose molecules that form a short chain. Most chains in glycogen would be composed of 8–12 of these glucoses. The arrows in both diagrams mark where an enzyme (acid alpha-glucosidase) can digest the links between two glucose molecules at the end of a chain of glucoses, releasing one glucose for metabolism. As the terminal glucose is released, the enzyme can make more cuts sequentially from the ends of the chains.

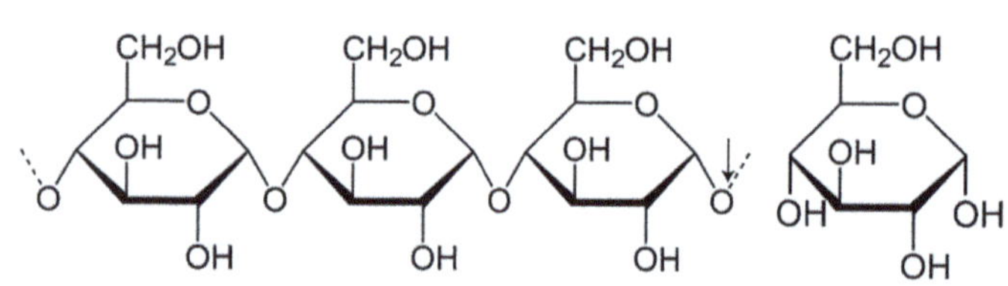

FIGURE 5.11 Three glucose molecules. Arrow indicates bond broken by GAA.

What do we need to know about this enzyme? The glucosidase enzyme is located in a membrane-bound organelle called a *lysosome*, one of several specialized structures found inside cells. Generally speaking, the lysosome's primary function is to digest or degrade various types of molecules, including any glycogen brought into the lysosome. In Pompe disease, which is inherited as an autosomal (not on a sex chromosome) recessive gene, the absence of a functional glucosidase enzyme results in the accumulation of glycogen in the lysosomes, malfunctioning lysosomes, and leakage of glycogen into the cytoplasm. The accumulation of these large molecules can overwhelm and disrupt the normal functioning of a cell. In this disease, the most obvious cells affected are found in muscle, including skeletal and cardiac muscles. In the heart, the cardiac muscles become much thicker (*hypertrophy*), particularly those in the larger chambers (*ventricles*), and the thickening may partially obstruct the outflow of blood (see Figure 5.12). Insufficient blood flow from the heart may reduce the volume of oxygenated blood for transport to the body tissues, starving these tissues of needed oxygen.

Presently, physicians can use *enzyme replacement therapy* (*ERT*), which reduces the effects that result from an absence of a key enzyme in one organelle. Each patient may respond differently to this therapy, and some patients are unable to tolerate it. However, when the disease is diagnosed early and treatment begun early, many of the pathologies of this disease are diminished. If Talia had been diagnosed earlier, perhaps ERT might have helped Talia.

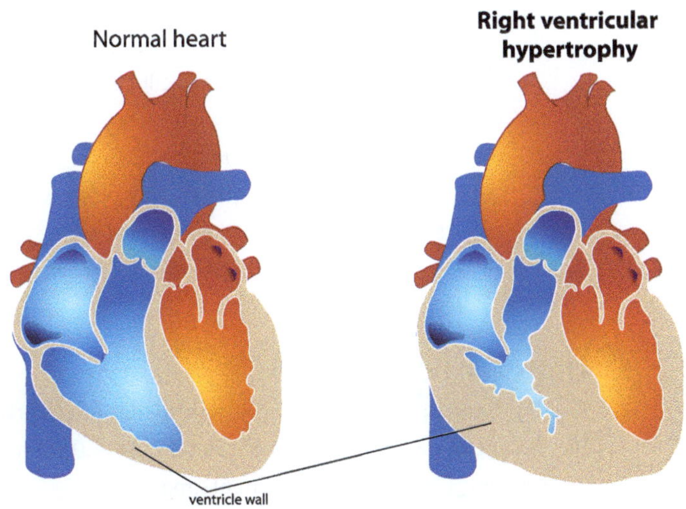

FIGURE 5.12 Comparison of ventricular walls in normal vs. heart with ventricular hypertrophy in a Pompe patient's heart

Hopefully, you now see that seemingly small changes, even at the subcellular level, can affect the organism in some impressive ways. Pompe disease is classified both as a type of glycogen storage disease (GSD) and a lysosomal storage disease (LSD). There are at least 13 known types of the former and more than 50 known types of the latter. GSDs can manifest in other organs, but they all involve some sort of defect for glycogen metabolism. LSDs involve defects in the metabolism in a variety of organic compounds that lysosomes encounter. Also, there are defects that affect other organelles, like mitochondria, with varying degrees of severity. To understand many health issues, it helps to understand about the structure and function of the units basic to life—cells.

QRQ 24. The cellular skeleton is known as the ______.

QRQ 25. (T/F) Composed of proteins, the cellular skeleton may be quickly rearranged depending on the needs of the cell.

QRQ 26. Classify each statement below by the following most relevant player in the cellular skeleton: intermediate filaments, microfilaments, microtubules

a. each filament can be rapidly lengthened or shortened at either end
b. filaments are assembled into a cylinder
c. form the spindle fibers during cell division
d. grows or shortens as needed, but the end pointed toward the periphery tends to grow or shrink more readily

e. hollow tubes composed of filaments of the protein tubulin
f. important contributors to muscle contraction
g. important players in the production of advancing extensions of phagocytic (ingulfing) cells
h. less studied of the three
i. made up of two chains of individual actin protein molecules, entwined into a helical filament
j. major components in cilia that move protozoa through their environment
k. major components in ciliated cells lining your respiratory tract that remove debris and microbes
l. most stable of the three
m. principal function seems limited to support, shape, and organization of the cell
n. quite prevalent in epidermal-type cells
o. support cells that extend or grow, probing new territory at the forward edge

Image Credits

Fig. 5.1: Source: https://commons.wikimedia.org/wiki/File:Average_prokaryote_cell-_en.svg.
Fig. 5.2a: Copyright © by OpenStax (CC BY 4.0) at https://commons.wikimedia.org/wiki/File:0312_Animal_Cell_and_Components.jpg.
Fig. 5.2b: Adapted from Copyright © 2017 Depositphotos/aglia83.
Fig. 5.2c: Copyright © by domdomegg (CC BY 4.0) at https://commons.wikimedia.org/wiki/File:Simple_diagram_of_yeast_cell_(en).svg.
Fig. 5.2d: Copyright © by Deuterostome (CC BY-SA 4.0) at https://commons.wikimedia.org/wiki/File:Paramecium_diagram.png.
Fig. 5.3: Source: https://commons.wikimedia.org/wiki/File:Endomembrane_system_diagram_en.svg.
Fig. 5.4: Copyright © by OpenStax (CC BY 4.0) at https://commons.wikimedia.org/wiki/File:0328_Transcription-translation_Summary.jpg.
Fig. 5.5: Copyright © by Graham Colm (CC BY-SA 3.0) at https://commons.wikimedia.org/wiki/File:Phagocytosis2.png.
Fig. 5.6a: Source: https://commons.wikimedia.org/wiki/File:Growthcone.jpg.
Fig. 5.6b: Source: https://commons.wikimedia.org/wiki/File:Kinetochore.jpg.
Fig. 5.6c: Copyright © by John Schmidt (CC BY-SA 3.0) at https://commons.wikimedia.org/wiki/File:Epithelial-cells.jpg.
Fig. 5.7: Copyright © by Thomas Splettstoesser (CC BY-SA 4.0) at https://commons.wikimedia.org/wiki/File:Microtubule_structure.png.
Fig. 5.8: Adapted from Copyright © by BruceBlaus, (CC BY 3.0) at https://commons.wikimedia.org/wiki/File:Blausen_0766_RespiratoryEpithelium.png.
Fig. 5.9: Adapted from Copyright © 2018 Depositphotos/VectorMine.
Fig. 5.10: Source: https://commons.wikimedia.org/wiki/File:Glycogen_structure.jpg.
Fig. 5.11: Source: https://commons.wikimedia.org/wiki/File:Amylose3.svg.
Fig. 5.12: Source: https://commons.wikimedia.org/wiki/File:Right_Ventricular_hypertrophy.svg.

LAB #5

Laboratory Exercises

5.4. Laboratory on Blood Sugar Regulation

5.4.1. Preliminary Work

In this lab, we examine the liver's and pancreas's regulation of blood glucose through the production of the hormones insulin and glucagon. Glucose is a prime source of energy for your cells, especially those of your brain and muscles. Much dietary glucose is absorbed by liver cells and stored as glycogen (starch). This absorption is stimulated by high levels of the hormone insulin, which is produced by cells of the pancreas. When blood glucose levels fall, other pancreatic cells will produce the hormone glucagon, which stimulates the liver cells to convert glycogen into sugar. The sugar enters the blood for use by other cells for energy. From the text's website, view the introductory video; download and open "*Blood Sugar Regulation Rev.nlogo.*"

1. Click the Info tab at the top of the page and read the section "What Is It?"
   a. **1.** ______ **2.** ______ **3.** ______ **4.** ______ **5.** ______ Consult other sources as needed to complete the following statement: The **1.** (A. blood, B. liver, C. pancreas) produces the **2.** (D. hormone, E. enzyme, F. lipid, G. protein) glucagon, which stimulates the breakdown of a large carbohydrate molecule, **3.** (H. glucose, I. glycogen, J. insulin), to the sugar **4.** (H. glucose, I. glycogen, J. insulin) in the **5.** (A. blood, B. liver, C. pancreas).
   b. **1.** ______ **2.** ______ **3.** ______ **4.** ______ Consult other sources as needed to complete the following statement: The **1.** (A. blood, B. liver, C. pancreas) produces the **2.** (D. enzyme, E. hormone, F. lipid, G. protein) insulin, which regulates **3.** (A. blood, B. liver, C. pancreas) **4.** (H. glucose, I. glycogen, J. insulin) levels.
   c. Match each component below with its color and shape in the model: A. black dots, B. blue dots, C. brown patches, D. burgundy dots, E. gray patches, F. magenta dots, G. orange dots, H. red dots, I. rose dots, J. white dots, K. yellow patches

      ________ glucagon signal molecule

      ________ glucose signal molecule

      ________ insulin signal molecule

      ________ liver cells

      ________ pancreatic cells

d. ____________ A balanced, or steady, state of glucose is termed glucose **1.**

e. ____________ The process in which a product of an activity causes more or less of that activity to occur is called a **1.** loop.

f. **1.** ______ **2.** ______ **3.** ______ Possible answers for each blank: A. a central regulator, B. feedback loops, C. glucose homeostasis. The model demonstrates how **1.** is/are maintained in the body through **2.** even without direction from **3**.

g. ____________ Select the conditions that the model demonstrates have an effect on glucose homeostasis: A. breathing, B. eating, C. flu, D. mental illness, E. metabolic disorders, F. physical activity, G. sleeping

h. ______ (T/F) In the model, interactions between cells and signal molecules result in changes to the blood sugar level.

2. Read "How It Works" and "How to Use It."

a. **1.** ______ **2.** ______ Possible answers for each blank: A. glucagon, B. glucose, C. insulin. If glucose levels are low, pancreatic cells produce **1.**, while if glucose levels are high, pancreatic cells produce **2.**

b. **1.** ______ **2.** ______ Possible answers for each blank: A. release stored glucose, B. sequester glucose from the bloodstream. If liver cells detect a higher glucagon signal than insulin signal, the liver cells **1**. However, if liver cells detect a higher insulin signal than glucagon signal, the liver cells **2.**

c. ____________ For simplicity, the model does not consider glycogen but starts with the liver having a store of **1.**

d. ____________ To regulate the level of glucose in the blood, the system relies on feedback loops between which two things? A. blood, B. glucose, C. glycogen, D. insulin, E. liver, F. pancreas

e. **1.** ______ **2.** ______ When the cells are not very sensitive to a particular signal, the cells **1.** (A. are, B. are not) very good at detecting the presence of those signal molecules, so that it will typically take **2.** (C. less, D. more, E. the same number) of those signal molecules to produce a particular response from the cell.

f. ____________ The signal molecules produced by the cells move through the world by a random walk that simulates their **1.** through the body in the bloodstream.

g. ______ (T/F) There exist unlimited supplies of glucose in the blood and glucose (formed from glycogen, not considered in the model) stored in the liver cells.

h. ____________ Pressing the **1.** button will add small amounts of glucose over time, simulating the absorption of nutrients by the digestive system after a meal.

i. ____________ Select all indicated differences between the model and the real world: A. Adjustments in the model take much longer than they would in a real body. B. Blood sugar fluctuation in this model is mainly because it has a higher grain size for time and energy than in a real body. C. Differences of only one or two molecules can have a much larger impact in the model than they would in a body. D. There is more glucose in the model than in a real body.

j. ______ The approximate number of seconds in a model tick is **1.** sec.

k. Indicate which button(s), switch(es), and/or slider(s) we should employ to examine the effect of the factors below on glucose homeostasis: A. EAT, B. GLUCAGON-SENSITIVITY, C. GLUCOSE-SENSITIVITY, D. INSULIN-SENSITIVITY, E. METABOLIC-RATE, F. THREE-MEALS

_______ activity level or exercise

_______ eating

_______ metabolic disorders

l. Match the following button, slider, monitor, or plot to the descriptions below: A. BLOOD GLUCOSE, B. EAT, C. FOLLOW-GLUCOSE, D. FOLLOW-INSULIN, E. GLUCOSE-SENSITIVITY, F. INSULIN-SENSITIVITY, G. INSULIN, H. METABOLIC-RATE, I. SIGNAL MOLECULES, J. STORED GLUCOSE, K. THREE-MEALS

_______ Plots the counts of glucose, insulin, and glucagon in the bloodstream

_______ Shows the number of glucose molecules in the bloodstream

_______ Shows the number of glucose molecules stored in the liver

_______ Shows the number of insulin molecules in the bloodstream

_______ The number of glucose molecules consumed by the body on each tick

_______ The probability that liver cells will detect an insulin molecule that is present

_______ The probability that pancreatic cells will detect a glucose molecule that is present

5.4.2. Laboratory Work

Use a spreadsheet to record results and perform calculations. The default slider values are as follows: METABOLIC-RATE = 100, INSULIN-SENSITIVITY = 0.75, GLUCOSE-SENSITIVITY = 0.75, GLUCAGON-SENSITIVITY = 0.75. Before each experiment, verify that parameters are correct, as the slider bars can change inadvertently.

3. This question involves modeling the baseline levels of insulin, glucagon, and glucose for someone with a healthy metabolism. Use the default slider values and turn off THREE-MEALS.

a. ______ Run the simulation for a while, observing the graphs of the signal molecules along with the monitors (above the graphs) with the specific levels of the signal molecules in the bloodstream and the number of glucose molecules stored in the liver. After an initial period of about 20 or 30 ticks, do the levels of the signal molecules in the bloodstream change very much over time, no (N) or yes (Y)?

b. Give an approximate equilibrium value for the amount of each in the bloodstream:

_______ glucose (black plot)

_______ glucagon (red plot)

_______ insulin (blue plot)

c. ______________________________ Have the simulation run faster or uncheck *view updates*, and let the model run until it stops. Give the user message that appears.

d. ____________ In a spreadsheet, type your name(s) at the top along with "Blood Sugar Regulation Spreadsheet." Label an area "Question 3d" and have a column header of "Ticks to no stored glucose." Observing the monitor below the graph, record the number of ticks it took for the stored glucose to deplete in the absence of food. Run the experiment two more times, recording the ticks. On the spreadsheet, calculate the mean.

e. ______ There are 400 sec/tick. Indicate how to calculate the total number of seconds until depletion of the stored glucose: A. (400 sec/tick) / (number of ticks until depletion), B. (number of ticks until depletion) / (400 sec/tick), C. (400 sec/tick) × (number of ticks until depletion). Using your answer to part d, complete the computation on your spreadsheet. Suggestion using *units cancellation*: On a piece of paper, write the numbers along with the units, such as ticks or sec/tick. Perform the calculation, such as multiplication or division, that results in the desired unit, such as sec.

f. ______ How many seconds are in an hour?

g. ______ Indicate which of the following will calculate the total number of hours until depletion of the stored glucose, and complete the computation on your spreadsheet: A. (number of seconds until depletion) / (number of seconds in an hour), B. (number of seconds in an hour) / (number of seconds until depletion), C. (number of seconds until depletion) × (number of seconds in an hour)

h. ______ Indicate which of the following will calculate the total number of days until depletion of the stored glucose, and complete the computation on your spreadsheet: A. (24 hr/dy) / (number of seconds until depletion), B. (number of seconds until depletion) / (24 hr/dy), C. (number of seconds until depletion) × (24 hr/dy)

i. Discuss possible reasons why the number of days to death by starvation (total depletion of stored glucose) that you calculated for part h and actual time for starvation when someone has water (typically a month or two) are so different.

4. In this question, we consider the impact of eating on levels of insulin, glucagon, and glucose in a healthy person. Thus, turn on the switch THREE-MEALS, so the simulated person has three meals a day, at 6:00, 12:00, and 18:00, using a 24-hour clock, and use the default slider values.

a. ______ Run the simulation for about 300 ticks. Other than during nutrient absorption by the digestive system after a meal, do the levels of the signal molecules in the bloodstream change very much over time, no (N) or yes (Y)?

b. ______ Are any of the equilibrium levels (other than during nutrient absorption) of the signal molecules in the bloodstream appreciably different than those in question 3b, no (N) or yes (Y)?

c. Discuss possible reasons for your answers to parts a and b.

d. ______ When a meal occurs, about how much does blood glucose increase over the equilibrium level in question 3b?

e. ______ When a meal occurs, about how much does glucagon change from the equilibrium level in question 3b? Use a negative number for a value below the equilibrium and a positive number for a value above the equilibrium.

f. ______ When a meal occurs, about how much does insulin change from the equilibrium level in question 3b? Use a negative number for a value below the equilibrium and a positive number for a value above the equilibrium.

g. ______ When eating occurs, what happens to stored glucose? A. continues to decrease, B. predominantly unchanged, C. temporarily increases

h. Discuss the reasoning behind what you observed, particularly increases and decreases in the numbers of molecules, for parts d–g.

i. ______ Estimate the number of ticks it takes for blood glucose, glucagon, and insulin to return to their baseline levels after a meal begins. On your spreadsheet, with the area marked as "Question 4i," calculate the time in hours, and give your answer here.

5. In this question, we consider the impact of exercise on levels of insulin, glucagon, and glucose in someone with a healthy metabolism. We can simulate the body during periods of exercise by increasing METABOLIC-RATE to 200.

   a. **1.** ______ **2.** ______ **3.** ______ **4.** ______ Complete the following hypothesis related to the impact of exercise on levels of insulin, glucagon, and glucose for someone with a healthy metabolism: If the metabolic rate of a healthy person is raised, then in the bloodstream, the amount of glucose will be **1.** (A. about the same, H. higher, L. lower), the amount of glucagon will be **2.** (A. about the same, H. higher, L. lower), and the amount of insulin will be **3.** (A. about the same, H. higher, L. lower), while the amount of stored glucose will **4.** (F. diminish faster, N. not diminish, R. diminish at about the same rate, S. diminish slower) than when the person has a lower metabolic rate.

   b. ______ Run the simulation for a while. Give an approximate equilibrium value for the amount of glucose in the bloodstream.

   c. ______ With an increased metabolic rate, the equilibrium amount of glucose in the bloodstream was **1.** (A. about the same, H. higher, L. lower) than the equilibrium when METABOLIC-RATE was 100 (question 3b). (Additionally, compare your experimental results to the relevant part of your hypothesis.)

   d. ______ Give an approximate equilibrium value for the amount of glucagon in the bloodstream.

   e. ______ With an increased metabolic rate, the equilibrium amount of glucagon in the bloodstream was **1.** (A. about the same, H. higher, L. lower) than the equilibrium when METABOLIC-RATE was 100 (question 3b). (Additionally, compare your experimental results to the relevant part of your hypothesis.)

   f. ______ Give an approximate equilibrium value for the amount of insulin in the bloodstream.

   g. ______ With an increased metabolic rate, the equilibrium amount of insulin in the bloodstream was **1.** (A. about the same, H. higher, L. lower) than the equilibrium when METABOLIC-RATE was 100 (question 3b). (Additionally, compare your experimental results to the relevant part of your hypothesis.)

   h. ______ In your spreadsheet, label an area "Question 5h" and include the metabolic rate. Have a column header of "Ticks to no stored glucose." Observing the monitor below the graph, record the number of ticks it took for the stored glucose to deplete with exercises and in the absence of food. Run the experiment two more times, recording the ticks. On the spreadsheet, calculate the mean and enter the number here.

   i. ______ With an increased metabolic rate, the amount of glucose stored in the liver **1.** (F. diminished faster, N. did not diminish, R. diminished at the same rate, S. diminished slower) than when METABOLIC-RATE was 100 (question 3d). (Additionally, compare your experimental results to the relevant part of your hypothesis.)

   j. Discuss why increasing the metabolic rate impacts the amounts of glucose, glucagon, and insulin in the bloodstream and stored glucose as observed in your experiments.

6. This question involves modeling levels of insulin, glucagon, and glucose for someone with type 2 diabetes, where the body detects or responds to insulin very poorly. In the NetLogo program, we can model type 2 diabetes by decreasing the slider INSULIN-SENSITIVITY to 0.25. Otherwise, use the default slider values, making sure to change METABOLIC-RATE back to 100.

   a. **1.** ______ **2.** ______ **3.** ______ **4.** ______ Complete the following hypothesis: If cells do not detect insulin well, then in the bloodstream, the amount of glucose will be **1.** (A. about the same, H. higher, L. lower), the amount of glucagon will be **2.** (A. about the same, H. higher, L. lower), and the amount of insulin will be **3.** (A. about the same, H. higher, L. lower), while the amount of stored glucose will **4.** (R. diminish at about the same rate, F. diminish faster, S. diminish slower, N. not diminish) than in a person with a healthy metabolism.

   b. ______ Run the simulation for a while. Give an approximate equilibrium value for the amount of glucose in the bloodstream.

   c. ______ With a decreased ability for cells to detect insulin, the equilibrium amount of glucose in the bloodstream was **1.** (A. about the same, H. higher, L. lower) than the equilibrium when METABOLIC-RATE was 100 (question 3b). (Additionally, compare your experimental results to the relevant part of your hypothesis.)

   d. ______ Based on your experiments, should someone with type 2 diabetes restrict the amount of sugar in their diet, no (N) or yes (Y).

   e. ______ Give an approximate equilibrium value for the amount of glucagon in the bloodstream.

   f. ______ With a decreased ability for cells to detect insulin, the equilibrium amount of glucagon in the bloodstream was **1.** (A. about the same, H. higher, L. lower) than the equilibrium when METABOLIC-RATE was 100 (question 3b). (Additionally, compare your experimental results to the relevant part of your hypothesis.)

   g. ______ Give an approximate equilibrium value for the amount of insulin in the bloodstream.

   h. ______ With a decreased ability for cells to detect insulin, the equilibrium amount of insulin in the bloodstream was **1.** (A. about the same, H. higher, L. lower) than the equilibrium when METABOLIC-RATE was 100 (question 3b). (Additionally, compare your experimental results to the relevant part of your hypothesis.)

   i. ______ In your spreadsheet, label an area "Question 6i" and include the insulin sensitivity. Have a column header, "Ticks to no stored glucose." Observing the monitor below the graph, record the number of ticks it took for the stored glucose to deplete in the absence of food. Run the experiment two more times, recording the ticks. On the spreadsheet, calculate the mean.

   j. ______ With a decreased ability for cells to detect insulin, the amount of glucose stored in the liver **1.** (F. diminished faster, N. did not diminish, R. diminished at about the same rate (say, within about 20 ticks until complete use of stored glucose), S. slower) than the equilibrium when METABOLIC-RATE was 100 (question 3d). (Additionally, compare your experimental results to the relevant part of your hypothesis.)

   k. Discuss why increasing the ability for cells to detect insulin impacts the amounts of glucose, glucagon, and insulin in the bloodstream and stored glucose as observed in your experiments.

MODULE 6

Channeling Energy

PHOTOSYNTHESIS

6.1. Introduction

Whatever living organisms do, they need energy to maintain an organized structure and to metabolize for maintenance, growth, and reproduction. However, being alive comes with a cost. Every organism must be able to supply itself with enough energy to support essential life functions. Biologists once thought organisms that convert the energy from sunlight into chemical, potential energy (food and such) directly or indirectly supported *all* life on earth. In other words, every living organism depended on *photosynthesis*. For the most part, that idea still is valid, but technology for exploring extreme depths of our oceans has yielded some surprising and transforming information.

In the deepest parts of the ocean, scientists found life; in fact, life was thriving. Despite the cold, the dark, and the extreme pressure, interesting and remarkable communities of organisms were found clustered around deep-water vents. These somewhat short-lived flows of hot, acidic water containing a potpourri of potentially harmful chemicals, including heavy metals, provide an environment that sustains a remarkable array of microorganisms. Unlike microbes that grow in shallower water or on land, these microorganisms can obtain the energy needed to operate quite successfully without sunlight. They sustain themselves by a process called *chemosynthesis*, where energy is obtained by oxidizing some of the chemicals in the bubbling soup from the vent. Furthermore, there are animals, like snails, that graze on this carpet of microorganisms, which then may be eaten by shrimp or crabs, which in turn may be eaten by fish or octopi. Some worms and clams incorporate the microbes into their own flesh, where they protect and nourish the bacteria, and the bacteria provide them with home-delivered food.

APPLYING SCIENCE TO LIFE 1 Partners in Brine

In 1977, using the deep submergence vehicle DSV *Alvin* two marine geologists were exploring the Galapagos Rift (see Figure 6.1a), submerged about 2,500 m in the dark, cold waters of the Pacific

Ocean. Here, off the west coast of South America, is a region where the seafloor is spreading on either side of a ridge that marks the junction between two *tectonic plates* (Cocos and Nazca, see Figure 6.1b). What they found was a field of *hydrothermal vents*, which are interesting enough, particularly to geologists. However, imagine how surprised and excited they must have been to discover that associated with

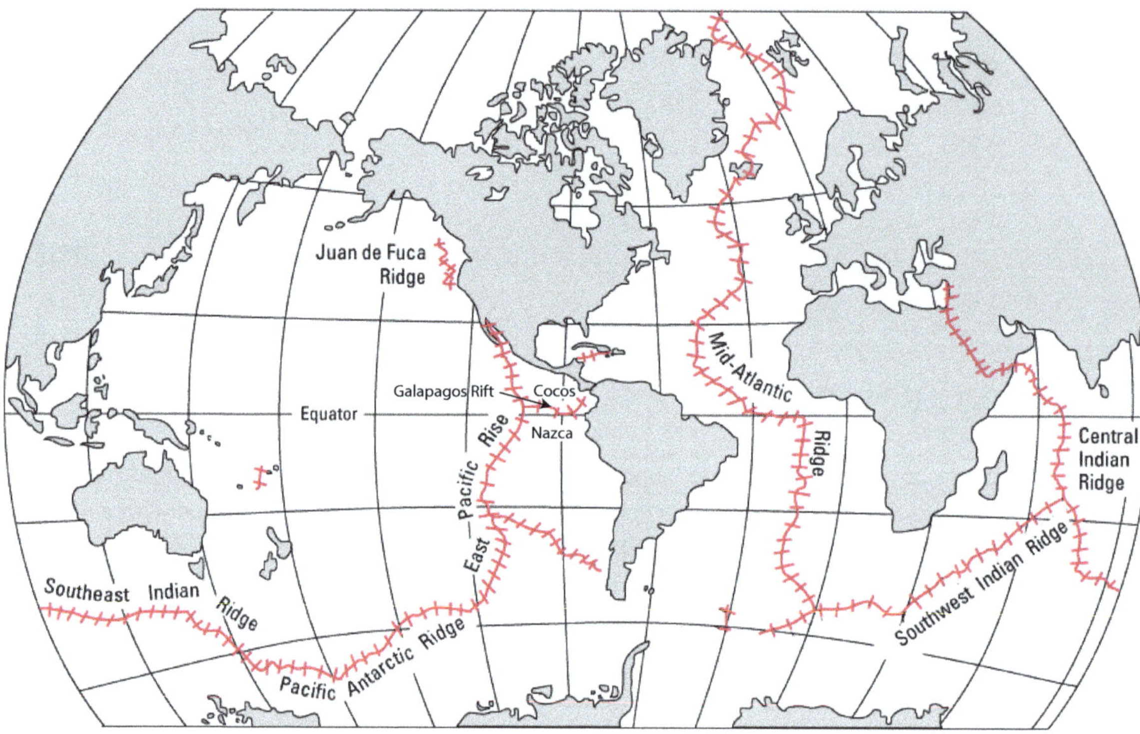

FIGURE 6.1a Mid-oceanic ridges and location of Galapagos Rift

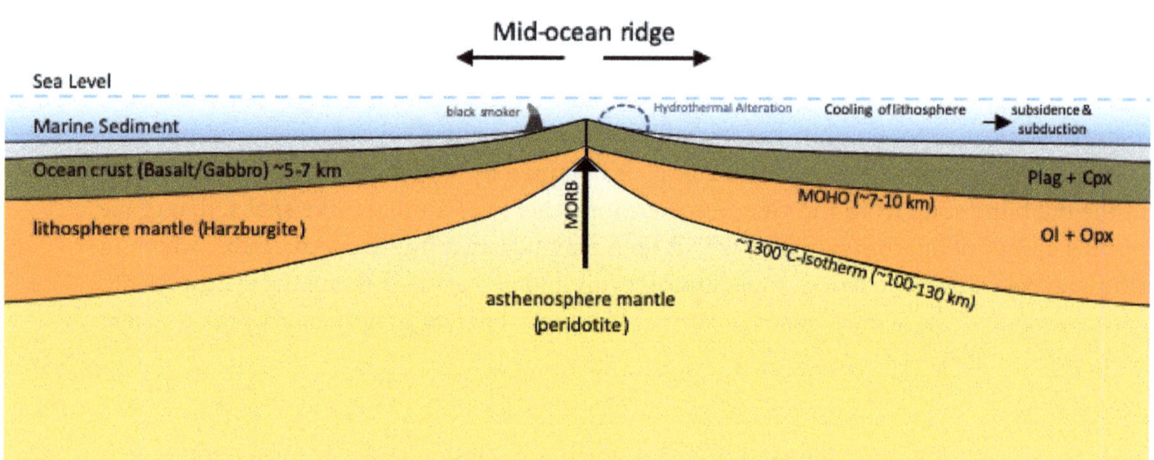

FIGURE 6.1b Mid-ocean ridge, where a submerged volcano produces molten lava between two spreading tectonic plates

those vents was a somewhat bizarre biological world that no one expected to be there. Their team did not include a biologist, but what these explorers catalyzed was an enormous scientific effort to explore and understand this unfamiliar world. Scientists might have anticipated microbial life under the extreme environment of the deep ocean, but remarkably there were crustaceans, fish, worms—a plethora of unfamiliar animals flourishing there. How could they live at those depths and harsh conditions without photosynthetic producer organisms? After years of hard work, careful observations, and exciting experiments, we are beginning to understand.

One of the interesting critters biologists have discovered around these vents are some very large worms—giant tube worms (*Riftia pachyptila*). Arranged in rather sizeable groups, these worms grow to about two meters in length. In Figure 6.2, you can get a good idea of what they look like. *Riftia* are the larger white worms with the red structures at the tip end. The white, tube-shaped body is covered in a protective coat made of chitin. In this photo are the unusual and beautiful adult worms, which also have a fascinating and highly modified anatomy. The red structure is called the *plume* (a respiratory organ that takes up oxygen for respiration and other chemicals needed by the symbiotic microbial community living within the worm). Over much time, these worms have drastically reduced their digestive tract, replacing it with an extensive internal organ called the *trophosome*, a specialized internal structure housing symbiotic microbes.

Without a photosynthetic source for the organic molecules needed to support the lives of most other animals, a very interesting symbiotic relationship has evolved between these worms and microorganisms that live in cells within the internal tissues of the worm. These internalized microbes are able to use a type of chemosynthesis, where they oxidize hydrogen sulfide molecules (H_2S), discharged from the vents, to generate ATP and NADPH. These oxidation products (ATP and NADPH) are used to reduce carbon dioxide, which is accomplished much the same way as in photosynthesis. The organic derivatives from carbon dioxide (e.g., sugars) provide the worm with energy sources and materials for synthetic purposes.

It would be a good question to ask, "How do these microbes get into the worm's flesh?" To answer that question, let us first consider briefly the life cycle of the worm. Gametes produced by the worms are released, and after fertilization, larval stages settle onto the available substrate (hard surface). During

FIGURE 6.2 Giant tube worms (white, with red tips) from the Galapagos Rift vents

metamorphosis (transformation from immature to adult form), the immature worm (larva) is "infected" by the microbes. The microbes invade the skin and make their way into the internal region of the worm, where they make themselves at home in the special cells of the future trophosome, called *bacteriocytes* (specialized cells where symbiotic bacteria reside) (Bright, Klose, and Nussbaumer 2013). Now, the worm provides a place to live, but how else does it benefit the microbes?

To answer that question, let's go back to the plume, the red, bushy tip at the end of the worm. The plume is red because of a very special type of hemoglobin in circulation. As you would expect, this hemoglobin picks up any available oxygen from the seawater and transports it throughout the body of the worm. Just like in the cells of your body, oxygen supports respiration and the production of energy. However, this hemoglobin also picks up sulfides for transport to the microbes in the bacteriocytes. Additionally, the plume absorbs carbon dioxide from seawater, which can be reduced, producing organic molecules that can be used by the worm.

These partners have coevolved to ensure thriving populations in a very hazardous environment. In this partnership, the worm provides the microbes with a relatively protected place to live, and its respiratory and circulatory systems provide the necessary chemicals for metabolism. The microorganisms can oxidize sulfides, supplied by the worm, to obtain energy and reducing power to convert carbon dioxide into organic products for themselves and their benevolent landlords (Bright, Klose, and Nussbaumer 2013).

6.2. Light into Life: The Wonders of Photosynthesis

Even if we could survive the conditions, we don't live near a thermal vent in the ocean floor. Thus, humans are very dependent on photosynthetic organisms (e.g., most plants and algae and some bacteria) to do their job, providing us with two things upon which we and most other heterotrophs are absolutely dependent: oxygen and food. Photosynthetic organisms are considered *autotrophs* because they are capable of making sugars and meeting their other organic needs using the energy from sunlight. Because we have to acquire pre-prepared food for our nutrition, humans are considered *heterotrophs*. Heterotrophs include a variety of organisms including animals and decomposers, like fungi. For certain heterotrophs, plants play other important roles, but let's concentrate on them now as the primary food sources, oxygen suppliers, and carbon dioxide consumers. Photosynthetic organisms are considered *producers*. You aren't required to eat plants directly (although it is better for your health and happiness), but even if you eat meat, you are eating to gain needed energy, and that energy has been channeled through plants. You are a *consumer*.

Photosynthetic organisms utilize light energy, absorbed from the sun, and carbon dioxide (CO_2) to make sugars and other organics for themselves and that consumers can use. The generalized equation usually given for photosynthesis is

$$6CO_2 + 6H_2O \xrightarrow{\text{light energy}} C_6H_{12}O_6 + 6O_2$$

The equation does not reveal the mechanics of photosynthesis but is meant to show two basic outcomes of the process: carbon dioxide is chemically reduced (gains electrons, i.e., H^+) to form sugar, and water is split to yield oxygen and hydrogens. Although it might not be obvious from the equation, the hydrogens yield electrons to reduce CO_2, and hydrogen ions help to form an electrochemical gradient. This hydrogen ion gradient helps to power the formation of ATP, which can provide energy for various chemical reactions in the cell. To better understand how photosynthesis works, we will consider the process in more detail after this short case study.

APPLYING SCIENCE TO LIFE 2 Long-Term Complications

For some time, there has been much discussion about rising levels of CO_2 in the atmosphere. Because CO_2 is a greenhouse gas, which can promote warming temperatures, scientists want to understand what the effects of rising atmospheric CO_2 on living things might be. For instance, what might be the effects of increased carbon dioxide on the productivity of trees? One hypothesis is that green plants might benefit from more CO_2 and use the excess chemical energy from photosynthetic activity to increase growth and productivity (energy converted to organic products). In one study, Oak Ridge National Laboratory (ORNL) and Duke University collaborated in long-term field studies to test the effects on productivity of increased carbon dioxide concentrations. Previous experiments tested this hypothesis using plants in growth chambers and greenhouses, where conditions are more easily controlled. However, field studies are much more complicated and difficult to control. ORNL's experimental plant was the sweetgum tree, and Duke's plant was the loblolly pine. Both tree species grow widely in the southeastern United States. Each study was able to increase the $[CO_2]$ in plots of trees artificially, monitoring the various environmental conditions and measuring the trees' productivity. During the first few years, both studies found that increasing carbon dioxide concentrations boosted productivity. However, as the study proceeded, they found that the relationship was more complicated, as growth started to slow.

Scientists always look to see if the data support their hypothesis. In this example, there was support for the hypothesis, but there is not solid support in the longer term. In this situation, scientists at ORNL and Duke would ordinarily propose alternative hypotheses. That does not mean that a study has failed or that their hypothesis was bad. Even if the scientists analyze the results and accept a hypothesis, the experiments still must be replicated by other scientists (Schlesinger et al. 2006; Warren et al. 2015).

PoI 1. What other factors might be important for determining the relationship between CO_2 and tree productivity?

PoI 2. How would you modify the hypothesis for future experiments?

6.3. Shining the Light on Photosynthesis

Plants, algae, and some protozoa and bacteria have the ability to photosynthesize. In all these organisms, except the bacteria, photosynthesis occurs within cytoplasmic organelles called *chloroplasts*. The process of photosynthesis in bacteria is similar, but to understand the process, we will focus on flowering green plants. In the diagram of a potted plant in Figure 6.3, you see the main functional systems of a green plant. The root system functions to anchor the plant and to absorb water and nutrients from the soil. The other major system is the shoot system, which includes the stem, the leaves, and the flowers. The stem grows toward the source of light, bearing the light-absorbing leaves, where most of the photosynthesis takes place. From the stem, a short branch bears the flower. Naturally, there are a large number of variations and specializations of all these primary parts, but you now know enough to appreciate the material in this section.

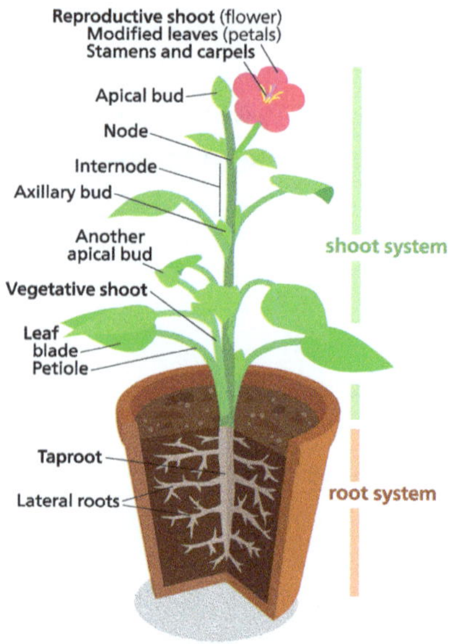

FIGURE 6.3 Diagram of potted plant, showing roots, stems, leaves, and flowers

6.3.1. Leaf Anatomy

To orient ourselves for our discussion of photosynthesis, examine the cross section of a plant leaf in Figure 6.4. This section is taken across the center of a leaf, showing the central vein, which contains transport vessels, some carrying water and minerals and some carrying sugars. Below the epidermal layer on the leaf's upper side is a densely packed set of elongated cells that make up the *palisade mesophyll*. Below that layer is another layer, called the *spongy mesophyll*, of loosely packed cells and lots of air spaces. Most of the chloroplasts are packed into the palisade cells, and therefore these cells are centers for photosynthesis. In the photograph in Figure 6.4, the *stomata* (pl. of *stoma*) are located primarily on the lower surface of the leaf. A stoma (labeled) is a tiny pore, specialized to allow CO_2 into and O_2 out of the leaf. Figure 6.5a is a photograph of an open stoma, made up of two *guard cells*. Inside those cells are some chloroplasts (red). Figure 6.5b contains diagrams of an open and closed stoma. When the stomata are open, they allow for gas exchange (CO_2 in and O_2 out). When closed, gas exchange is blocked.

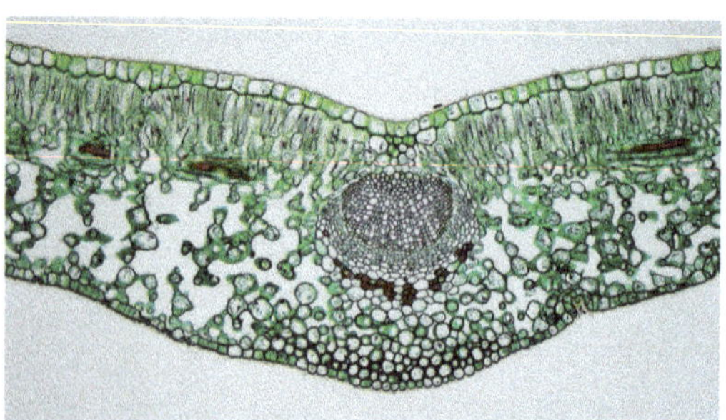

FIGURE 6.4 Cross section through a leaf, showing central vein surrounded by mesophyll

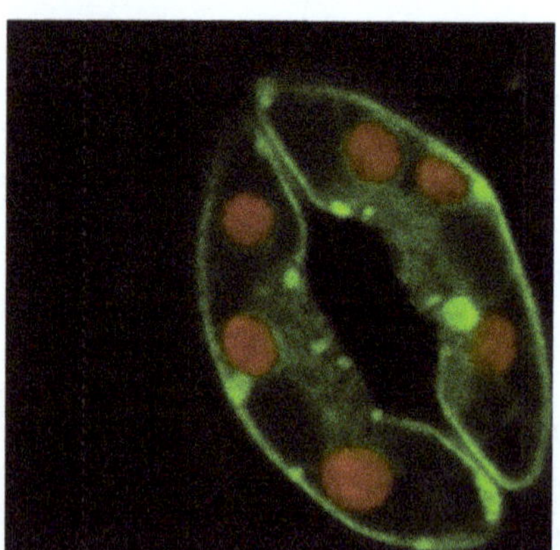

FIGURE 6.5a Photograph of open stoma

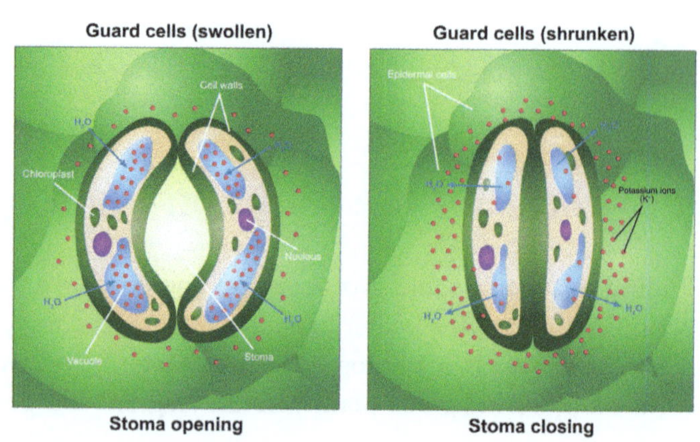

FIGURE 6.5b Diagrams of stoma, open and closed

APPLYING SCIENCE TO LIFE 3 All That Photosynthesizes May Not Be Leaves

In our consideration of photosynthesis, we looked at a typical leaf structure, but leaves may not always appear like they do on the plants in our front yard. As plants established themselves, particularly in a new climate, various specializations were often required to better adapt the plant to that climate. For instance, in the southwest deserts of the United States, we find a group of unusual plants that have evolved for life in the hot sun and limited water. Cactuses have shallow roots to maximize absorption of water from infrequent rainfall. More obviously, they also have highly modified stems and leaves (see Figure 6.6). The leaves have developed into spines, which reduce moisture loss and may protect the plant against herbaceous animals. The normal function of the leaves has been taken on by modified stems, which serve to store water, reduce evaporation (sunken stomata), produce flowers, and, importantly, photosynthesize.

FIGURE 6.6a Young Saguaro Cactus

FIGURE 6.6b Close-Up of Spines

As you see, leaves perform other functions than being platforms for photosynthesis. Spines are just one modification of leaves. In aloe plants, the leaves are thickened and succulent, like the cactus stems, serving for water storage. Other plants, like grapes, sweet peas, and squash, produce threadlike outgrowths (tendrils), which help by caressing supportive plants or other structures, allowing upward growth.

Even more highly specialized leaves can be found in carnivorous plants, like the pitcher plant and the Venus flytrap (see Figure 6.7). These plants often grow on soils that are nitrogen-deficient, so insects they catch help fill that nutritional gap. The trapping apparatus in both plants represent modifications of leaves. The flytrap has two opposing, hinged lobes covered with hairs. When an insect is lured by nectar to the lobes and touches one hair followed by touching another hair within 20 seconds, the lobes snap shut. As the victim squirms, the trap closes tighter, sealing the poor devil's fate. Digestive glands on the surface secrete enzymes that degrade the insect for nutrients. Full digestion takes place over about a week, after which the leaves reopen. These leaves have a limited life, digesting only four or five insects. They eventually fall off, but there are replacement traps, and the plant also photosynthesizes.

FIGURE 6.7 Modified leaf of Venus flytrap, showing trigger hairs

The pitcher plant is a passive trapper (see Figure 6.8). It lures the insect by the seductive aroma of its nectar. The naïve animal falls through a pitcher-shaped tube into a pool of nectar and drowns. There are backward-turned hairs or waxy scales to prevent the insect from crawling out. Digestion takes place in the fluid at the bottom of the tube. Obviously, most leaves have photosynthetic functions, but variation is not only important, it makes life more interesting.

FIGURE 6.8a Pitcher plants in flower

FIGURE 6.8b Isolated tubular leaf

6.3.2. Chloroplast Structure

In terrestrial green plants, photosynthesis takes place in chloroplasts (see Figure 6.9). One leaf mesophyll cell might contain 20 or more chloroplasts. During cell division, parental cell chloroplasts are distributed to each of the progeny cells. Chloroplasts are self-replicating and can pinch in two to increase the number of chloroplasts in the new cells. Chloroplasts, like mitochondria, are surrounded by an outer and inner layer of membrane (see Figure 6.10). The area surrounded by these membranes is called the *stroma*, and within the stroma are ribosomes, starch granules, numerous copies of circular DNA, and another set of interconnected, membranous sacs called *thylakoids*. The outer membrane is permeable to all but large molecules, but the inner membrane, like the inner membrane of mitochondria, is much less permeable. Within the membrane of the thylakoids are the pigment and protein complexes, where the *light reactions* of photosynthesis take place. Also in that membrane is the ATP synthesis complex. Thankfully, the light reaction also produces oxygen as a byproduct. The reactions that effectively use energy from ATP and NADPH to assimilate carbon dioxide into sugar, sometimes referred to as the *dark reaction* of photosynthesis, are located in the stroma.

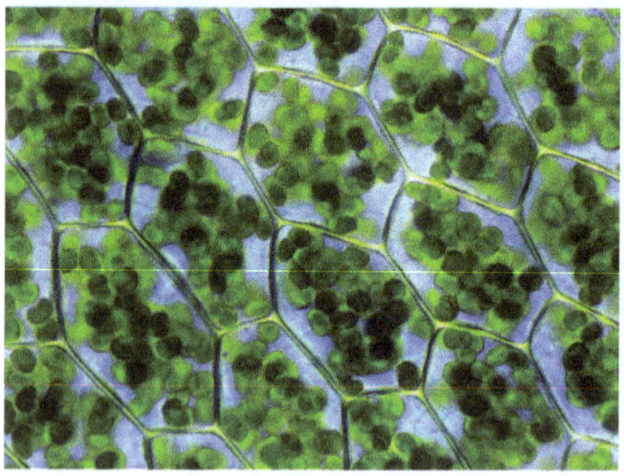

FIGURE 6.9 Chloroplast found in moss cells

Keeping in mind that most of the living things on earth are dependent on photosynthesis, it is time to explore that

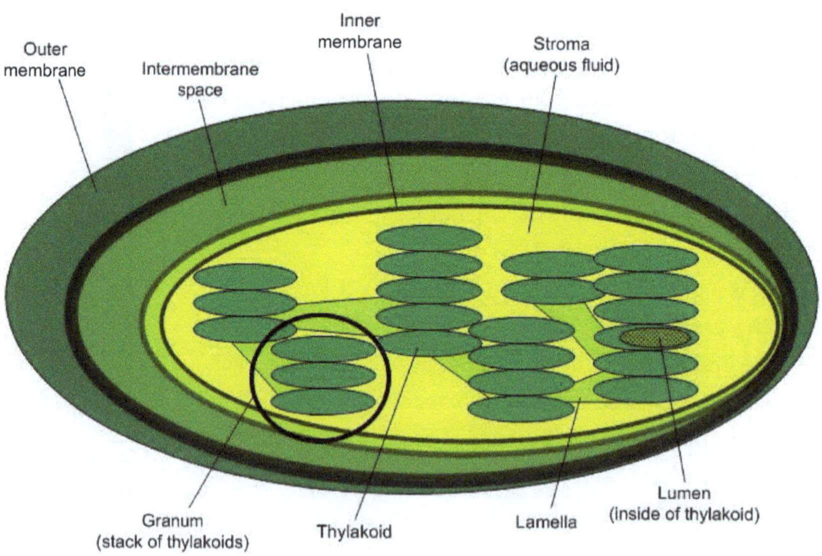

FIGURE 6.10 Chloroplast structure

remarkable process. Photosynthesis is basically two major, interconnected sets of reactions—the *light reaction* and the *dark reaction*. The light reaction captures light energy and converts it into *ATP* and the reduced coenzyme *NADPH*. Fortunately, as a byproduct, the light reaction produces O_2 as well. The dark reaction uses ATP for energy and NADPH for reducing power (ability to donate electrons [and H^+] to *reduce* another molecule) to produce sugar. The sugar is then available for energy, storage, or conversion into some other needed organic molecules.

QRQ 1. Select the following terms that apply to a cow: autotroph, consumer, heterotroph, producer

QRQ 2. Select the following terms that apply to a blade of grass: autotroph, consumer, heterotroph, producer

QRQ 3. In the generalized chemical equation for photosynthesis, a plant begins with carbon dioxide and ______ and, with the capture of light energy, yields ______ and oxygen.

QRQ 4. The number that should appear in each blank for the generalized chemical equation for photosynthesis is ______.

$$CO_2 + H_2O \xrightarrow{\text{light energy}} C_6H_{12}O_6 + O_2$$

QRQ 5. In plants, photosynthesis occurs within cytoplasmic organelles called ______.

QRQ 6. Select from the following the appropriate term for each description below: A. illumin, B. guard cell, C. guard duct, D. lumen, E. spongy mesophyll, F. spongy thylakoid, G. stoma, H. stroma, I. thylakoid, J. thylightoid

a. aqueous fluid in the interior of a chloroplast
b. its membrane contains pigment-protein complexes that capture light energy and convert it to chemical energy
c. membranous sack in the stroma that interconnects with others in a stack
d. chloroplasts are packed into this kind of leaf cell
e. space inside the thylakoid
f. tiny pore specialized to allow CO_2 into and O_2 out of the leaf
g. two of these surround the stoma

QRQ 7. The light reaction, which is one of two major reactions of ____, captures ____ energy and converts it to ____ and the coenzyme ____ with ____ as a byproduct.

QRQ 8. The dark reaction, which is one of two major reactions of ____, utilizes ____ for energy and ____ to produce ____.

6.3.3. Light Reaction

Within the membranes of the thylakoids of the chloroplast are complexes of pigments and proteins. In each plant leaf, there are huge numbers of chloroplasts with pigments that absorb certain wavelengths of light from the sun. The principal pigment found in most plants is *chlorophyll a*. Chlorophyll a has a complicated structure that includes a hydrocarbon ring with nitrogen atoms surrounding a central magnesium atom and a hydrophobic tail of carbons and hydrogens. Naturally, you do not need to learn this structure, but do admire it in Figure 6.11. Interestingly, hemoglobin has a similar ring that surrounds iron instead of magnesium.

The light reaction depends on two types of pigment-protein clusters, called *photosystems*, embedded in the thylakoid membranes—*PSII* and *PSI*. These photosystems are composed of hundreds of molecules of chlorophyll a, along with accessory pigments. The pigments are associated with proteins that form a framework for the pigments. The chief role of such a complex is to harvest light, and the accessory pigments allow the plant to capture a wider range of light wavelengths. Furthermore, some accessory pigments help to protect chlorophyll from sunlight's damage. At the center of these complexes is a reaction center, containing a pair of special chlorophylls.

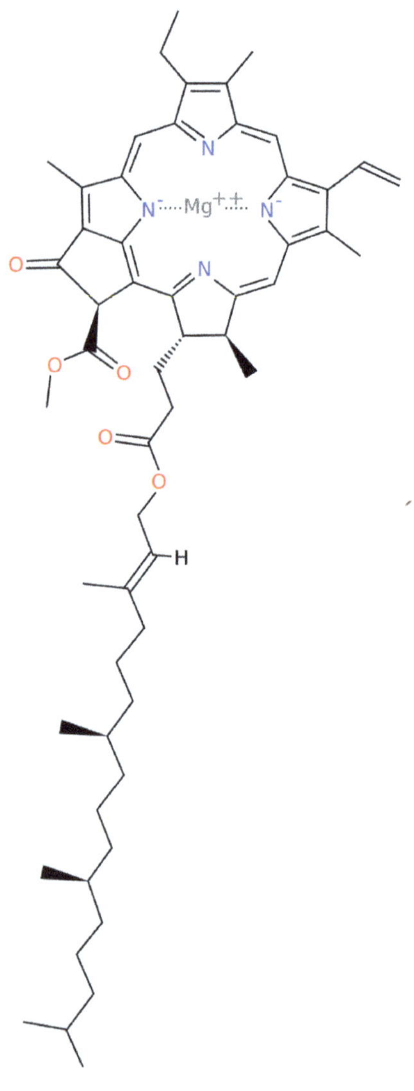

FIGURE 6.11 Chlorophyll structure

We begin with PSII. The pigment molecules in both photosystems are arranged to transfer the light energy onward, focusing on the centrally located *reaction center*. As absorbed light energy excites a pigment molecule (raises an electron to a higher energy level), that molecule passes this energy on to an adjacent molecule, which passes it to an adjacent molecule, and on and on. Finally, the energy is focused onto the reaction center, where the energy excites electrons of the special chlorophylls, P_{680} (absorb maximally at a light wavelength of 680 nm). The excited electrons are passed onto an initial electron acceptor, which passes the electrons on through an *electron transport chain* (*ETC*). The P_{680} chlorophylls in PSII are now missing electrons, and these chlorophylls attract electrons away from water, splitting it. That splitting leaves O_2 and H^+. Small amounts of O_2 are used by the plant's *mitochondria* (organelles where respiration, using O_2, produces energy), and the rest of it exits through the stomata for non-photosynthetic organisms, such as humans, to use (see Figure 6.12).

The electrons from PSII pass from one acceptor to another (redox reactions) along the ETC, releasing small amounts of energy along the way, and finally arriving at the reaction center of PSI (see Figure 6.12). Some of the energy released from the ETC of PSII powers the movement of hydrogen ions from the stroma into the thylakoid lumen. These H^+, added to the hydrogen ions from the splitting of water, form an electrochemical gradient of H^+. The gradient represents a reservoir of stored energy. This gradient interacts with a large protein complex embedded in the thylakoid membrane—*ATP synthase*. Remember that ions will move from an area of high concentration (thylakoid lumen) to an area of lower concentration (stroma). Thus, the ATP synthase

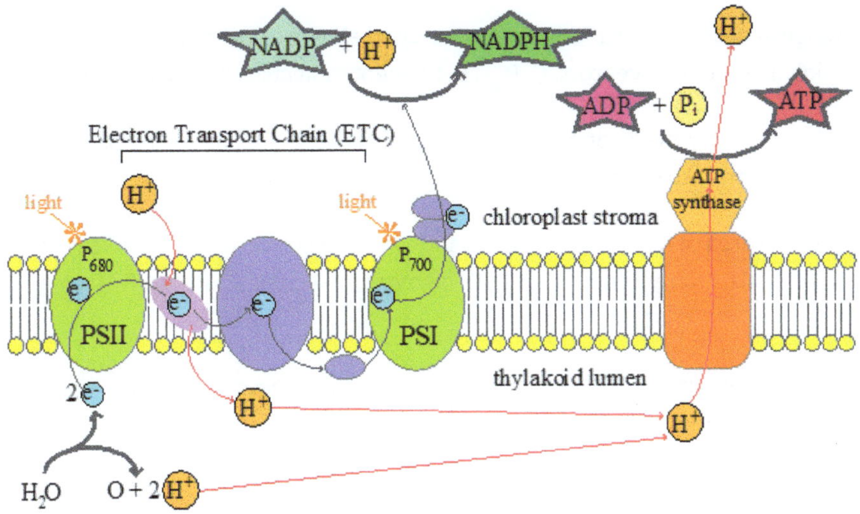

FIGURE 6.12 Light reaction of photosynthesis

(see Figure 6.12) can harness the energy of the gradient to combine the organic molecule *ADP* with a *phosphorus* (*Pi*) to make *ATP*. In this case, the ATP is generated in the stroma, where it can be used to power the dark reaction.

We left some electrons in the reaction center of PSI, and those electrons need to become energized to be of benefit to the plant (see Figure 6.12). Thus, light absorbed by the pigments of PSI focus on exciting those electrons of the special chlorophylls in that reaction center. These special chlorophylls absorb maximally at wavelengths of light of 700 nm, so these chlorophylls are called P_{700}. The electrons from P_{700} are then energized and are passed to a short ETC with the final electron acceptor, an enzyme that transfers the electrons to the coenzyme $NADP^+$, which attracts a hydrogen ion, making *NADPH*.

QRQ 9. The principal pigment found in most plants absorbing certain wavelengths of light is ________.

QRQ 10. Select from the following the appropriate term for each description below: A. electron thylakoid center, B. electron transport chain, C. excite, D. P_{680}, E. P_{700}, F. P_{860}, G. photosystem, H. pigment, I. reaction center of PSII

a. centrally located area that receives electrons from pigment molecules
b. ETC
c. molecule that transfers the light energy onward
d. molecule which, after losing electrons in PSII, attracts electrons away from water, splitting it into oxygen and hydrogen ions
e. pigment-protein mechanism of light reaction, centered in thylakoid membranes
f. special chlorophyll in reaction center of PSII, excited by energy from pigments
g. to raise an electron to a higher energy level

QRQ 11. Arrange in order the locations to which excited electrons are passed in PSII: A. ETC acceptors, B. initial electron acceptors, C. P_{680}, D. pigment molecules, E. reaction center of PSI

QRQ 12. Arrange in order the locations to which hydrogen ions are passed in PSI: A. ATP synthase, B. P_{700}, C. stroma, D. thylakoid lumen

QRQ 13. Select the area of highest H^+ concentration in the electrochemical gradient of H^+: A. ETC acceptors, B. initial electron acceptor, C. P_{680}, D. pigment molecules, E. stoma, F. stroma, G. thylakoid lumen

QRQ 14. Use chemical symbols, abbreviations, or acronyms for each of the following blanks: The ATP synthase can harness the energy of the electrochemical gradient of H^+ of PSI to combine the organic molecule _____ with a high-energy _____ to make _____.

QRQ 15. Arrange in order the locations to which excited electrons are passed in PSI: A. NADP-reductase, B. $NADP^+$, C. P_{700}, D. pigment molecules, E. reaction center of PSI, F. short ETC

QRQ 16. In PSI, $NADP^+$ attracts a hydrogen ion to make _____.

6.3.4. Dark Reaction

Overall, the dark reaction takes an inorganic form of carbon (CO_2) and reduces it to form an organic molecule (initially a sugar). To accomplish this conversion, the dark reaction utilizes ATP and NADPH, produced by the light reaction. The dark reaction, called the *Calvin cycle*, is a series of reactions that must begin with two molecules: *CO_2* and a five-carbon *ribulose bisphosphate*. CO_2, produced by respiration, enters the leaf through the stomata. At the end of a turn of the Calvin cycle, we have the molecule of *ribulose 1,5 bisphosphate* (*RuBDP*), which then can be used for another turn of the cycle (see Figure 6.13).

A special enzyme, originally called ribulose bisphosphate carboxylase but now called *rubisco*, adds a carbon dioxide molecule to the existing ribulose bisphosphate by a process called *carbon fixation*. This reaction generates an unstable six-carbon product that splits into two 3-C molecules. After energy and electrons are added, the intermediate product is a triose sugar (three-carbon monosaccharide) phosphate, some of which can be used to make larger sugars (like glucose) and other organic molecules. You need more than one of these triose sugars to make glucose and lots more to regenerate RuBP. It actually takes six turns of this cycle to have enough triose phosphates make glucose and to regenerate RuBP. Six turns requires 18 ATP and 12 NADPH to make one molecule of glucose.

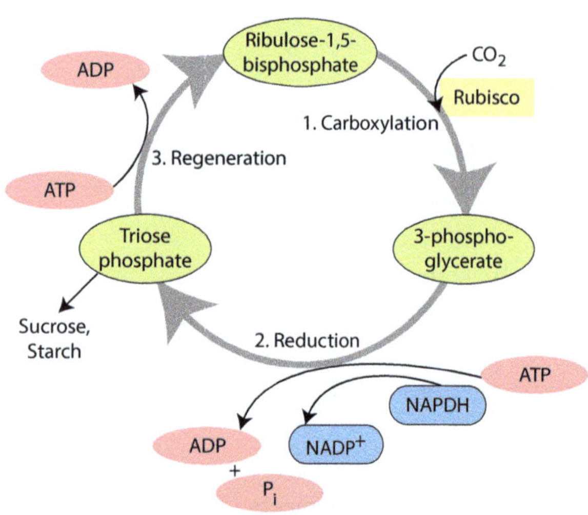

FIGURE 6.13 Calvin cycle

APPLYING SCIENCE TO LIFE 4 Complications: More Data to Consider

Scientists set up a three-year study at the University of Minnesota to examine the effects of warming temperatures on photosynthesis. They used both local broadleaf and conifer tree seedlings in outdoor plots, where they elevated the ambient temperature of the above and below ground of between 2.9° and +3.4° C between April and November. They recorded plant and soil temperatures and soil moisture (0–20 cm depth) continuously. What the researchers found is that increased temperature, as expected, increased the rates of photosynthesis, provided there was sufficient soil moisture. When the rain decreased during dry periods, the rate of photosynthesis declined across all species tested.

These results indicate that soil moisture is an important limiting factor on plant photosynthetic rate in seedlings of tree species in more northerly, boreal forests. (Such forests include broadleaf [maple, poplar, birch, oak] and conifer species [pines, firs and spruces]).) Additionally, higher temperatures may increase loss of moisture from the soil. Thus, any gains in productivity achieved with rising temperatures are likely to be offset or negated by reduced soil moisture (Reich et al. 2018).

QRQ 17. The dark reaction takes an inorganic form of carbon, _____ and, using the energy of _____ and the _____ (select oxidizing or reducing for this blank) power of the coenzyme _____, forms a(n) _____.

QRQ 18. Another name for the dark reaction is the _____ _____.

QRQ 19. The process of converting inorganic carbon dioxide to an organic carbon molecule is called carbon _____.

Image Credits

Fig. 6.1a: Source: https://commons.wikimedia.org/wiki/File:World_Distribution_of_Mid-Oceanic_Ridges.gif.

Fig. 6.1b: Copyright © by BrucePL (CC BY-SA 4.0) at https://commons.wikimedia.org/wiki/File:Mid-ocean_ridge_cut_away_view.png.

Fig. 6.2: Source: https://commons.wikimedia.org/wiki/File:Riftia_tube_worm_colony_Galapagos_2011.jpg.

Fig. 6.3: Copyright © by Kelvinsong (CC BY-SA 3.0) at https://commons.wikimedia.org/wiki/File:Plant.svg.

Fig. 6.4: Source: https://commons.wikimedia.org/wiki/File:Angiosperm_Morphology_The_Mesophytic_Dicotyledonous_Leaf_of_Ligustrum_(36845190976).jpg.

Fig. 6.5a: Copyright © by Alex Costa (CC BY 4.0) at https://commons.wikimedia.org/wiki/File:Plant_stoma_guard_cells.png.

Fig. 6.5b: Copyright © by Ali Zifan (CC BY-SA 4.0) at https://commons.wikimedia.org/wiki/File:Opening_and_Closing_of_Stoma.svg.

Fig. 6.7: Copyright © by Noah Elhardt (CC BY-SA 2.5) at https://commons.wikimedia.org/wiki/File:Venus_Flytrap_showing_trigger_hairs.jpg.

Fig. 6.9: Copyright © by Kristian Peters -- Fabelfroh (CC BY-SA 3.0) at https://commons.wikimedia.org/wiki/File:Plagiomnium_affine_laminazellen.jpeg.

Fig. 6.10: Source: https://commons.wikimedia.org/wiki/File:Chloroplast-new.jpg.

Fig. 6.11: Copyright © by Wc010394 (CC BY-SA 4.0) at https://commons.wikimedia.org/wiki/File:Chla.png.

LAB #6

Laboratory Exercises

6.4. Laboratory on Algae

6.4.1. Preliminary Work

In this laboratory, we investigate a simplified aquatic ecosystem in which algae move upward in a column of water to gather sunlight for photosynthesis and downward in that column to obtain nutrients. From the text's website, view the introductory video; download and open "*Algae Rev.nlogo.*"

1. Click the Info tab at the top of the page and read sections "What Is It?" and "How It Works."
   a. ____________ Select which nutrients the description lists for algae: A. ammonia, B. calcium, C. carbon, D. chlorine, E. iron, F. magnesium, G. nitrogen, H. oxygen, I. phosphate , J. potassium, K. sulfur
   b. ______ Where are most of the nutrients in the lake? A. deeper parts, B. middle, C. surface, D. throughout
   c. ______ Where is most of the light in the lake? Use the same answer choices as in part b.
   d. ______ What causes algae to sink? A. deflation of balloon-like structures, B. dense substances they burn, C. dense substances they make from light, D. fins, E. flagella, F. inflation of balloon-like structures, G. oxygen
   e. ______ What causes algae to rise? Use the same answer choices as in part d.
   f. ____________ At every tick, algae use dense substances to produce what kind of energy?
   g. ______ How many patches are in the column, including the air, water, and bottom patches?
   h. Give the color (A. black, B. blue, C. brown, D. light yellow, E. lime, F. pink, G. red, H. sky blue, I. yellow) of each of the following:

      ________ bottom patch

      ________ air patch during daylight

      ________ air patch during night

i. Give the light level of the following:

_______ bottom patch.

_______ air patch during daylight.

_______ air patch during night.

j. ___________ Select all possible algae colors: A. black, B. blue, C. brown, D. light yellow, E. lime, F. pink, G. red, H. sky blue, I. yellow

2. Read "How to Use It."

a. ______ Suppose the LIGHT-SPREADINESS slider has value 0.8 and the light level of the air patch is 1. Give the light value of the patch at location 0 9 to one decimal place.

b. ______ Give the light value of the patch at location 0 8 to two decimal places.

c. ______ Give the light value of the patch at location 0 7 to three decimal places.

d. ______ hour(s) A tick represents how many hours?

e. ______ tick(s) Plots are not displayed until after how many ticks?

f. Select the appropriate model component for each description below: A. ALGAE-DISTRIBUTION, B. AVERAGE-24-HR-HEIGHT, C. CURRENT-AVERAGE-24-HR-HEIGHT, D. CURRENT-AVERAGE-HEIGHT, E. DAYLIGHT-LENGTH, F. LIGHT-SPREADINESS, G. VARIABLE-DAYLIGHT-LENGTH?

_______ Changes how much the light penetrates the water

_______ Current average height of all algae

_______ Histogram of average algae heights at each tick

_______ Number of hours of sunlight in a day

_______ Over the previous 24 ticks, the average height of all algae

_______ Plot or monitor of the history of the average height of algae over the past 24-hour period

_______ When off, the day length can only be changed manually by the user

_______ When on, during execution, the length of each day changes, simulating the changes of the seasons

6.4.2. Laboratory Work

The default parameter values are as follows: LIGHT-SPREADINESS = 0.75, STOP-AT-DAY-7? is off, VARIABLE-DAYLIGHT-LENGTH? is off, and DAYLIGHT-LENGTH = 12.

3. This question examines the behavior of the algae over several days with fixed light penetration and daylight length. Use default parameter values, except turn on STOP-AT-DAY-7?.

a. **1.** ______ **2.** ______ Complete the following hypothesis: During daylight, algae tend to **1.** (A. move down, B. move up, C. stay at the same level), while during the night, algae tend to **2.** (A. move down, B. move up, C. stay at the same level).

b. ______ Press SETUP several times and observe the distribution of the algae in the water column, as well as ALGAE-DISTRIBUTION and CURRENT-AVERAGE-HEIGHT. True (T) or false (F): Initially, the algae are randomly distributed.

c. **1.** ______ **2.** ______ Change the speed to be slower and run the simulation. During daylight, algae tend to **1.** (A. move down, B. move up, C. stay at the same level), while during the night, algae tend to **2.** (A. move down, B. move up, C. stay at the same level). Compare your observations to your hypothesis in part a.

4. In this question, we consider the influence of light penetration on algae's average 24-hour height in the column of water. Use default parameter values, except have STOP-AT-DAY-7? turned on.

   a. **1.** ______ **2.** ______ **3.** ______ Complete the following hypothesis: For a daylight length of 12 hours, on average, algae are **1.** (A. about the same depth, B. deeper, C. higher, D. vary in depth) at the end of daylight as light penetration increases and **2.** (A. about the same depth, B. deeper, C. higher, D. vary in depth) at the end of night as light penetration increases. Moreover, their average 24-hour height **3.** (E. decreases, F. increases, G. remains the same, H. varies) as light penetration increases.

   b. Set LIGHT-SPREADINESS to 0.5, so that light does not penetrate the water well. To record your data, start a spreadsheet (*Algae*) as in Table 6.1. Setup and run the simulation at least five times with *view updates* unchecked or with a fast speed, and record the output values. Because tabs are separators, we can copy the three output values at once and paste them into the spreadsheet so that the values are in different columns. Calculate the average of each column and record the averages below.

   ______ Mean height at end of daylight on Day 6

   ______ Mean height at end of night on Day 6

   ______ Current average 24-hour height at end of Day 6

TABLE 6.1. Start of Algae Lab Spreadsheet

Your Names			
Algae Lab			
	Question 4b		
	LIGHT-SPREADINESS		**0.5**
	DAYLIGHT-LENGTH		**8**
	Mean height on Day 6		
	At end of daylight	**At end of night**	**Current average 24-hr height**

   c. Set LIGHT-SPREADINESS to 0.55, so that the light penetrates the water somewhat better than in part b. Repeat part b, labeling your spreadsheet appropriately.

   ______ Mean height at end of daylight on Day 6

   ______ Mean height at end of night on Day 6

   ______ Current average 24-hour height at end of Day 6

d. Repeat part b with LIGHT-SPREADINESS = 0.65.

_______ Mean height at end of daylight on Day 6

_______ Mean height at end of night on Day 6

_______ Current average 24-hour height at end of Day 6

e. Repeat part b with LIGHT-SPREADINESS = 0.75.

_______ Mean height at end of daylight on Day 6

_______ Mean height at end of night on Day 6

_______ Current average 24-hour height at end of Day 6

f. Repeat part b with LIGHT-SPREADINESS = 0.85, so that light penetrates the water well.

_______ Mean height at end of daylight on Day 6

_______ Mean height at end of night on Day 6

_______ Current average 24-hour height at end of Day 6

g. ______ Using your observations, complete the following statement: For a daylight length of 12 hours, on average, algae are **1.** (A. about the same depth, B. deeper, C. higher, D. vary in depth) at the end of daylight as light penetration increases. (Compare your conclusions with your hypothesis in part a.)

h. ______ Considering only your observations from parts d, e, and f, complete the following statement: For a daylight length of 12 hours, as light penetration increases from 0.65 to 0.85, on average, algae are **1.** (A. about the same depth, B. deeper, C. higher, D. vary in depth) at the end of night. (Compare your conclusions with your hypothesis in part a.)

i. Discuss the answers for mean height at the end of night in parts b, c, and d. Why do these answers occur?

j. ______ Using your observations, complete the following statement: For a daylight length of 12 hours, on average, the algae's average 24-hour height **1.** (A. decreases, B. increases, C. remains the same, D. varies) as light penetration increases. (Compare your conclusions with your hypothesis in part a.)

k. **1.** ______ **2.** ______ Using data from the averages of the third column for parts b–f, have your spreadsheet generate a graph of the algae's average 24-hour height at the end of Day 6 versus LIGHT-SPREADINESS. The graph is **1.** (A. decreasing, B. horizontal, C. increasing) and **2.** (D. concave down, E. concave up, F. linear).

l. ______ We can fit a smoother graph that captures the trend of data points of part k. Using the fitted graph, we can estimate the algae's average 24-hour height at the end of Day 6 for LIGHT-SPREADINESS values where we do not have data. However, we should be careful not to use the graph to estimate values outside the range of data. Thus, we should not estimate the average 24-hour height for LIGHT-SPREADINESS less than 0.5 or greater than 0.85. Estimate the algae's average 24-hour height at the end of Day 6 for LIGHT-SPREADINESS = 0.7.

5. This question studies the impact of daylight length on the behavior of the algae. Use default parameter values, except have STOP-AT-DAY-7? turned on.

   a. **1.** ______ **2.** ______ **3.** ______ Complete the following: For LIGHT-SPREADINESS = 0.75, as the number of daylight hours increases, on average, algae are **1.** (A. about the same depth, B. deeper, C. higher, D. vary in depth) at the end of daylight and **2.** (A. about the same depth, B. deeper, C. higher, D. vary in depth) at the end of night. Moreover, their average 24-hour height **3.** (E. decreases, F. increases, G. remains the same, H. varies).

   b. Set DAYLIGHT-LENGTH to be 8 hours. In another part of your *Algae* spreadsheet, indicate Question 5b and list the values for LIGHT-SPREADINESS and DAYLIGHT-LENGTH and have appropriate headers for the output values. Set up and run the simulation at least five times with *view updates* unchecked or with a fast speed and record the output values under the appropriate headers. Calculate the average of each column and record the averages below.

      ______ Mean height at end of daylight on Day 6

      ______ Mean height at end of night on Day 6

      ______ Current average 24-hour height at end of Day 6

   c. Repeat part b for DAYLIGHT-LENGTH = 10.

      ______ Mean height at end of daylight on Day 6

      ______ Mean height at end of night on Day 6

      ______ Current average 24-hour height at end of Day 6

   d. Repeat part b for DAYLIGHT-LENGTH = 12.

      ______ Mean height at end of daylight on Day 6

      ______ Mean height at end of night on Day 6

      ______ Current average 24-hour height at end of Day 6

   e. Repeat part b for DAYLIGHT-LENGTH = 14.

      ______ Mean height at end of daylight on Day 6

      ______ Mean height at end of night on Day 6

      ______ Current average 24-hour height at end of Day 6

   f. Repeat part b for DAYLIGHT-LENGTH = 16 hours.

      ______ Mean height at end of daylight on Day 6

      ______ Mean height at end of night on Day 6

      ______ Current average 24-hour height at end of Day 6

   g. ______ Considering only your observations, complete the following statement: For LIGHT-SPREADINESS = 0.75, as the number of daylight hours increases, on average, algae are **1.** (A. about the same depth, B. deeper, C. higher, D. vary in depth) at the end of the *night*.

   h. ______ Considering only your observations from parts b, c, and d, complete the following statement: For LIGHT-SPREADINESS = 0.75, as the number of daylight hours increases from 8 to 12 hours, the algae's average 24-hour height **1.** (A. decreases, B. increases, C. remains the same, D. varies).

i. **1.** ______ **2.** ______ **3.** ______ Using data from the averages of the third column for parts b–f, have your spreadsheet generate a graph of the algae's average 24-hour height at the end of Day 6 versus DAYLIGHT-LENGTH. The graph is mostly **1.** (A. decreasing, B. horizontal, C. increasing) and **2.** (D. concave down, E. concave up, F. linear). However, for longer daylight lengths, the graph **3.** (G. becomes fairly horizontal, H. decreases fairly steeply, I. increases fairly steeply).

j. ______ Although we generally should *not* make estimates out of the range of the data, use the graph to estimate the algae's average 24-hour height when DAYLIGHT-LENGTH = 20 hours.

k. Now, let us see the danger of making such extended estimates by repeating part b for DAYLIGHT-LENGTH = 20 hours. Notice how different the estimate in part j is from the experimental data.

   ________ Mean height at end of daylight on Day 6

   ________ Mean height at end of night on Day 6

   ________ Current average 24-hour height at end of Day 6

l. ______ Run the simulation again for DAYLIGHT-LENGTH = 20 hours at a much slower speed to observe the heights of the algae, particularly before the end of daylight. True (T) or false (F): Before the end of daylight, algae reverse directions and start rising.

m. ______ Repeat part l for DAYLIGHT-LENGTH = 12 hours.

n. Discuss why you think the experimental data, particularly for the mean height at end of daylight on Day 6 and the average 24-hour height at end of Day 6, behave as they do.

o. What additional experiments might you run to clarify the situation? You can consider adjusting the model to obtain additional data.

6. This question examines the results when the simulation runs for several years, with the daylight changing according to the seasons. Use default parameter values, except turn on VARIABLE-DAYLIGHT-LENGTH?.

   a. **1.** ______ **2.** ______ Complete the following hypothesis: During the summer months, when daylight is longer, algae are **1.** (A. higher, B. lower) in the water, while in the winter months, when daylight is shorter, algae are **2.** (A. higher, B. lower) in the water.

   b. ______ Run the simulation at a fast speed, noticing the behavior of daylight length and average 24-hour height. In this model, the shortest day of the year is the first day of the simulation and every January 1 thereafter. Assuming 365.25 days in a year, which accounts for leap year, how many ticks (hours) are in a year?

   c. ______ Is the algae's average 24-hour height maximum on January 1, no (N) or yes (Y)?

   d. ______ Does the algae's average 24-hour height have a significant dip at the longest day of the year, no (N) or yes (Y)?

   e. Discuss these results, considering your experiments in question 5.

MODULE

7 Metabolism of the Cell—Using Energy Constructively

APPLYING SCIENCE TO LIFE 1 Thermostat Wars

Office workers or families living in the same residence often have the "battle of the thermostat." Generally, males tend to like the thermostat at a lower temperature. Females often prefer the thermostat set to make it warmer. Because energy consumption of residences and offices account for about one third of CO_2 emissions and because the behavior of the occupants (including setting the thermostat) is responsible for the overwhelming majority of the energy consumption for each building, it is important to understand completely the energy required of all occupants of a building. Unfortunately, the indoor climate regulations for buildings have been set based on recommendations made by ASHRAE (the American Society of Heating, Refrigerating, and Air-Conditioning Engineers), and their calculations are based on the resting metabolic rate (RMR) for adult males. According to two Dutch scientists at Maastricht University, the calculations may overestimate the female metabolic rate by as much as 35 percent (Kingma and van Marken Lichtenbelt 2015). Consequently, women may feel more comfortable at a warmer ambient temperature than men.

Naturally, there is individual variation, but men generally like the ambient temperature cooler, and it is in part due to their proportionately greater amount of muscle mass. Skeletal muscles demand a larger energy supply than some other tissues. The resting metabolic rate (RMR, in kcal/min) measures the energy required at rest for basic body functions, like heartbeat, respiration, and blood pressure, and accounts for about two-thirds of the energy used by the body. A much smaller percentage is needed to process and absorb nutrients. The remaining 4 percent is used for physical activities. As we will see later, a higher rate of metabolic activity also produces higher amounts of heat.

Actually, there may be no way to put a halt to "thermostat wars." Women and men differ structurally and physiologically. However, when setting standards for internal climate in buildings, we can take into account these differences.

7.1. Metabolic Pathways

The metabolism of any organism, from a bacterium to a person, is really based on the metabolism of their cells. We define cellular *metabolism* as all of the chemical reactions that provide energy and supply needed organic molecules for growth, maintenance, reproduction, and response to signals in their external and internal environments. Chemical reactions are arranged into a series of interconnected, enzyme-catalyzed *metabolic pathways*. Enzymes provide a crucial mechanism for organizing and regulating all cellular chemical activity. Each reaction in a pathway utilizes a specific enzyme to convert a molecule or molecules, called the *substrate(s)*, into metabolic *intermediate(s)*. These intermediates then become the substrates for the next enzyme in the pathway, and so on, until production of the final molecule(s), or *product(s)*. The metabolic pathways devoted to producing energy for the cell and organism are termed *catabolic*. Those pathways, where basic materials or subunits are used to construct functional biomolecules for the cells, are termed *anabolic*. These pathways require energy input. Thus, the two types of metabolic pathways that make up metabolism are catabolism and anabolism.

APPLYING SCIENCE TO LIFE 2 Anabolic Steroid Use: Getting to Your Heart

Thomas was 28 years old when he collapsed and died, apparently of cardiac arrest. Although his roommate called for emergency services, resuscitation failed. Thomas was a bodybuilder who worked out daily and apparently had used various types of anabolic steroids to improve his build and competitive performance for over a year. An autopsy revealed an enlarged liver and some atrophy of his testicles. Although physical examination of his heart showed no outward anomaly, microscopic examination revealed areas of stiffened cardiac muscle (fibrotic), which physicians usually see in individuals with progressive forms of heart damage, often leading to heart failure.

Whether you are an athlete or not, you probably know someone who has used *anabolic steroids* or performance-enhancing drugs. These are synthetic substances that resemble the human male sex hormone. Physicians prescribe such steroids occasionally for patients with specific hormonal issues, like loss of muscle mass (disease-related), insufficient testosterone production, and certain types of anemia. Unfortunately, these drugs are often misused by athletes or bodybuilders to increase muscle mass or improve athletic performance. These uses, often unsupervised, are dangerous and illegal.

Using these drugs for the wrong reasons can cause a variety of medical problems. It is estimated that up to four million people in the United States have used anabolic steroids to build muscle (Baggish et al. 2017). The National Institute of Drug Abuse (NIH) has enumerated a number of possible effects from long-term use, including the following (NIH 2019):

- breast growth and shrinking of the testicles in males
- deepening of voice and facial hair growth in females
- high blood pressure
- cardiovascular disease, including heart attack
- liver disease (e.g., liver cancer)
- kidney damage

In a study published in 2017, physicians studied the cardiovascular function of 140 male weightlifters, aged 34 to 54, which included 86 users and 54 nonusers. Of the users, 28 had not taken the drugs during the previous 15 months. They found serious evidence of cardiovascular damage, including weakened diastolic function (left ventricle relaxes and fills with blood), for both current and past users. Furthermore, the users had higher blood pressure, higher levels of "bad" cholesterol (LDLs), and higher levels of plaque buildup in their arteries (Baggish et al. 2017).

QRQ 1. We define ______ as all of the chemical reactions that provide energy and supply needed organic molecules for growth, maintenance, reproduction, and response to signals in their external and internal environments.

QRQ 2. Metabolic pathways devoted to producing energy for the cell and organism are termed ______.

QRQ 3. Metabolic pathways that use energy input and basic materials or subunits to construct functional biomolecules for the cells are ______ pathways.

QRQ 4. In a metabolic pathway, a chemical reaction utilizes a specific enzyme to convert a molecule or molecules, called the ______(s), into metabolic ______(s). The latter is used with the next enzyme in the pathway, and so on, until production of the final molecule(s), or ______(s).

7.2. Breaking Down Cellular Respiration

APPLYING SCIENCE TO LIFE 3 The Costs of Hunger

Malik's father had been killed last winter. He got caught in crossfire between the police and a robbery suspect. The family missed him, but they also missed his steady income. Malik's mother, Patrice, was working two jobs trying to keep her children fed, clothed, and sheltered. Income from her regular work as a hotel housekeeper and weekend work as a waitress usually takes care of the rent, utilities, and bus/subway fares. Consequently, she doesn't have much left over for food. Fortunately, the children's schools provide inexpensive and some free food for breakfast and lunch for the children, but only during the week. Also, Patrice is able to obtain some food from a food bank that a group of churches run in the neighborhood. Of course, she never knows what that food will be until she picks it up. Otherwise, she often runs out of money before the month's end. When Malik is not looking after his little brothers, he can sometimes earn a bit of money helping neighbors carry groceries home from the market.

Malik misses his dad, but the worst thing is that now there is never enough food. Malik, 13, is starting to grow more rapidly, and he always feels hungry. His brothers keep him busy most of the time, but he really misses the fun he used to have with his friends and the chance to play sports after school. He takes his brothers to elementary school on his way to middle school and picks them up on his way home. Everyone is hungry when they return home, but actually, the three boys feel hungry all of the time. Sometimes they will split a small pack of chips, but Malik usually tries to get his little brothers to watch TV while he attempts to study. When his mother returns from work in the early evening, she hurries to prepare dinner. Malik helps clean up afterward while Patrice does some laundry. Three boys dirty lots of clothes.

In frustration, Malik does something he would normally not do. While he is in the local market, he stuffs some candy into his pocket. He doesn't realize that the store owner, Mr. Fischer, sees him. Mr. Fischer has known the family for years, and he believes Malik is a good, well-behaved boy. The market owner feels bad that the family is having a hard time, so Mr. Fischer talks with Malik on his next visit. In the back office, Malik begins to sob, and he promises to pay Mr. Fischer back for the candy. He said he and his brothers were so hungry, and what he did was wrong but desperate. Mr. Fischer tells Malik that he won't report this incident to his mother, but he will let him help him in the market at night. In exchange for Malik's work, he will pay him in food, and the pay will be nutritious. The candy he took is not the right antidote for hunger.

Millions of people go through their days hungry, and many of them are children who suffer various consequences of growing up malnourished. Of course, most people in the United States think that hunger only happens in developing countries. The United States is blessed with abundant food. Few understand that many people right here at home are also going hungry. Furthermore, here there is so much processed and generally unhealthy food available. Much of it is high in sugar and fat, but it is sometimes the cheapest and most accessible.

Food insecurity may be defined as the lack of *dependable* access to a sufficient quantity of affordable and nutritious food. Malik's family is one of millions of families in the United States that are faced with food insecurity. In 2017, the US Department of Agriculture estimated that in the United States, food insecurity affected 15.7 million households, about 40 million people, including 12 million children.

Malik has an older sister with a newborn baby. His sister is only 16, and her boyfriend just dumped her. With nowhere else to go, she has returned home. She has an unfinished high school education, and she has no job. Besides, she needs to take care of her daughter. Now Patrice has two more mouths to feed, and she may need Malik to find a job.

PoI 1. What are the underlying factors that bring about food insecurity in the United States?

PoI 2. Teenagers, like parents, in a food insecure household will often sacrifice their own nutrition or health to help younger siblings. Consequently, they may eat less to provide more for others. Under more severe circumstances, teens may try other ways to obtain food.

a. What might they do?
b. Why might some of the actions taken by desperate teens be termed "self-sabotaging"?

PoI 3. Food banks are available in many areas of the United States.

a. What is a food bank?
b. What kinds of food do food pantries typically have?
c. Where does a food bank get its food?

PoI 4. Why might parents or teens not take advantage of food from such a source?

PoI 5. Support for households that experience food insecurity is available from various governmental and charitable sources. What are some of the support services available in your area?

PoI 6. Which of these services have been most successful in helping people to escape poverty and/or food insecurity?

PoI 7. What are some remedies for reducing food insecurity in the United States?

PoI 8. What detrimental effects might Malik's taking a job have on his own life?

PoI 9. Children are considered vulnerable to all types of problems when they don't get the quantity or quality of food they need. What are some of these problems?

PoI 10. What are the consequences of childhood nutritional deprivation?

In the module on photosynthesis, we learned that photosynthetic organisms (primary producers) are able to absorb light energy and convert it into chemical potential energy (ATP), which then can be used to reduce CO_2 to synthesize sugars and other organic compounds. Primary producers, as well as all consumers, must be able to obtain energy needed for maintaining structure, growth, reproduction, and other activities necessary for life. Consumers take in food, which they metabolize for energy. As we discuss the important aspects of cellular metabolism, bear in mind the interconnectedness of pathways for photosynthesis with the pathways of catabolism (producing energy) and anabolism (constructing functional biomolecules).

7.2.1. Aerobic Cellular Respiration

Cellular respiration in humans and other consumers includes a sequence of reactions that generates energy from the chemical bonds of organic compounds (food) and uses that energy to synthesize ATP, which can be used subsequently to power other cellular reactions that require energy. Don't confuse cellular respiration with bodily respiration, the gas exchange that takes place in your lungs. Later you will see that both respirations are connected, however. We begin with one molecule of glucose (sugar), which could have come from starch in plants or in animals may have come from a food source or stored glycogen. Following is the general equation for cellular respiration:

$$C_6H_{12}O_6 + 6O_2 \rightarrow 6CO_2 + 6H_2O + {\sim}28ATP + heat$$

The cell captures about one third of the energy (ATP) available in a molecule of glucose ($C_6H_{12}O_6$), and the rest is released as heat. However, the general equation hides the steps and redox reactions of cell respiration.

After the glucose is transported into the cell, it will enter the first major part of cell respiration, a series of reactions that will convert it from one six-carbon sugar into two three-carbon sugars—two molecules of *pyruvate*. If you ignited and burned a sugar molecule, all of the energy would be released at once as heat. In cellular respiration, the breakdown of sugar is gradual, done in steps.

Some of these steps involve oxidation (removal of electrons and accompanying H^+ = hydrogens). If you look back at the equation above, you see that the products are CO_2, water, ATP, and heat. This equation reflects complete breakdown (oxidation) of glucose without the details. What you can see is that the CO_2 lacks the hydrogens associated with carbons in glucose. Each hydrogen is made up of one electron (e^-) and one proton (H^+). Thus, the oxidation involved the removal of the electrons as part of the hydrogen. The hydrogens seem to be part of water molecules now. This equation has skipped way ahead to the complete breakdown of glucose. We are breaking glucose down in enzyme-catalyzed steps, and the hydrogens are actually now attached to a coenzyme, *NAD*, and another coenzyme, *FAD*. *Coenzymes* are nonprotein organic molecules that assist an enzyme with its catalytic function. Table 7.1 describes the coenzymes we will encounter in this module. Note that all of those described are derivatives of vitamins we obtain from our diet.

TABLE 7.1. Coenzymes for Photosynthesis and Respiration

Coenzyme		Functions	Vitamin	Source	Deficiency
Coenzyme A		Transfer of two C groups Krebs cycle Fat catabolism	Vitamin B_5 Pantothenate	meat, vegetables, beans, grains, eggs	Inability to break down some fats, vomiting, low blood sugar, lack of energy
FAD	Flavin Adenine Dinucleotide	Redox reactions Krebs cycle	Vitamin B_2 Riboflavin	plants, grains, dairy, eggs, fish, meat	Mouth sores, inflammation of mouth lining, dry skin, sore throat
NAD	Nicotinamide Adenine Dinucleotide	Redox reactions Glycolysis Krebs cycle	Vitamin B_3 nicotinamide niacin	dairy, beans, meat, eggs	Pellagra (alcoholics), lack of appetite, fatigue, abdominal pain
NADP	Nicotinamide Adenine Dinucleotide Phosphate	Redox reactions Dark reaction of photosynthesis			

As we *oxidize* (remove electrons from) glucose, we *reduce* (add electrons to) NAD or FAD. Note that NAD and FAD are oxidized forms of the coenzymes. Technically, NAD can only accept two electrons and one proton, so we usually symbolize the reduced NAD as NADH + H^+. For simplicity, we will symbolize the reduced form of NAD as *NADH*. Similarly, FAD is reduced by taking on two hydrogens (two protons and two electrons), giving us *FADH2*. NAD and FAD have roles in various pathways in the cell, but we will focus only on their roles in oxidation-reduction (redox) reactions of cell respiration.

Glycolysis

We now consider this first major part of cell respiration, glycolysis, in greater detail. Located in the cytoplasm, *glycolysis* (*glycol* = sugar; *lysis* = split), or the breaking down of glucose to produce energy and pyruvic acid, is shown in an abbreviated form in Figure 7.1. Even though there are multiple steps, for our purposes, this figure is sufficient. Starting at the top, with glucose (6-C sugar), two important things occur: sugar is *phosphorylated* (phosphate is added) twice, and it is then split into two 3-C sugars, each with one phosphate (molecule containing PO_4). This requires the input of two high-energy phosphates from two molecules of ATP. After this preparatory step, the three-carbon sugars are oxidized, donating electrons to NAD and yielding two NADH, and then the sugars each receive another phosphate, giving them two phosphates each. In the succeeding steps, both phosphates are removed from each and added to a total of four ADPs to form four ATPs. Because the phosphates are removed from enzymatic substrates to form ATP, we call this process *substrate phosphorylation*. We are left with two 3-C molecules of pyruvate, two NADH + H^+, and a net of two ATPs. Remember, we used two ATPs to phosphorylate glucose at the beginning of glycolysis.

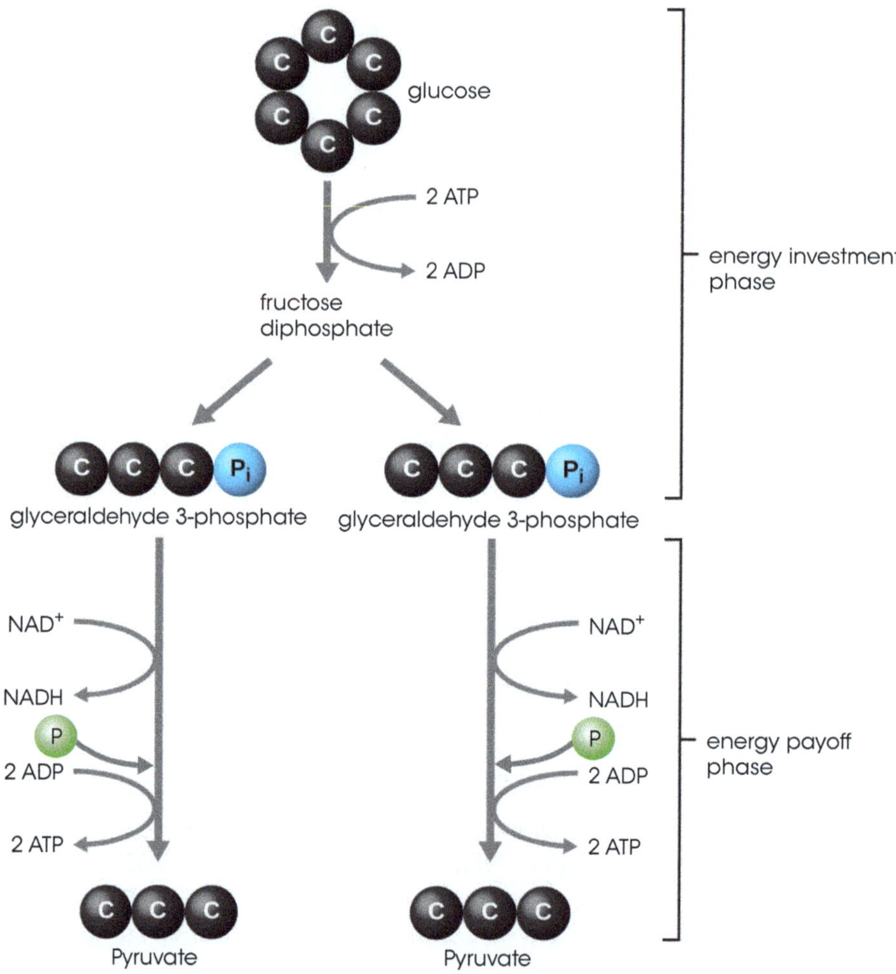

FIGURE 7.1 Glycolysis

The results of glycolysis seem somewhat disappointing. We end up with *only* two molecules of ATP, and we have only partially oxidized the glucose. On the other hand, we still have two molecules of sugar (pyruvate), and these reduced molecules can still be oxidized further. And what does the cell do with the NADH?

QRQ 5. Complete the general equation for cellular respiration, filling in numbers for four blanks and selecting items for **1.**, **2.**, **3.**, and **4.** from the following: ADP, ATP, C_6, CO_2, $C_6H_{12}O_6$, H, H^+, O_2, heat, $NADP^+$, NADPH, reductase

1. + ______ O_2 → ______ **2.** + ______ H_2O + ~______ **3.** + **4.**

QRQ 6. In the first major part of cell respiration, through a series of reactions, one six-carbon sugar, ______, is converted to two three-carbon sugars—two molecules of ______. The reactions also yield ______ NADH(s) and a net of ______ ATP(s).

QRQ 7. A nonprotein organic molecule that assists an enzyme with its catalytic function is called a ______.

QRQ 8. Complete the statement with answers selected from the following: ADP, ATP, FADH, $FADH_2$, NADH, $NADH_2$, oxidized, reduced, respired. In redox reactions of cell respiration, the coenzyme FAD is ______ to ______, while NAD is ______ to ______.

QRQ 9. When phosphate is added to a molecule, we say the molecule is ______.

QRQ 10. The first major part of cell respiration, which starts with one six-carbon sugar, is called ______.

The Krebs Cycle

The mitochondrion is formed from a double layer of membrane—outer and inner. The outer membrane is very permeable and is not considered much of a barrier. The inner membrane, which is folded to form the cristae (see Figure 7.2), is much less permeable and is quite selective. Two compartments are generated by this arrangement—the intermembranous space, between the outer and inner membranes, and the matrix, bordered by the inner membrane. Inside the mitochondrion are the key pathways for most of the energy production from carbohydrates and

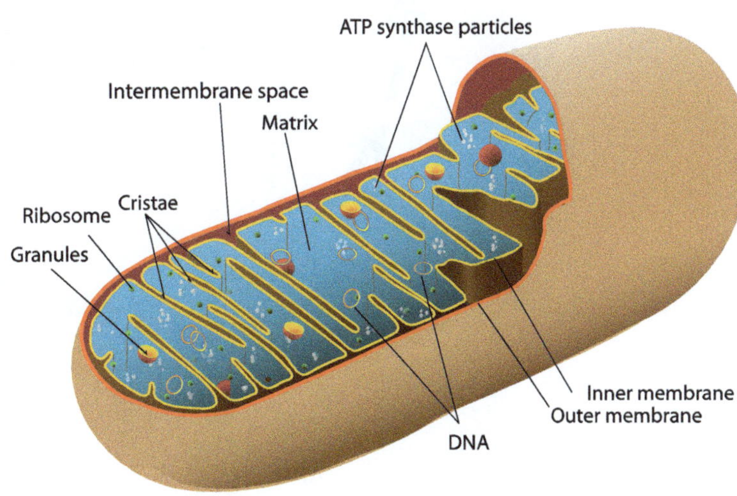

FIGURE 7.2 Diagram showing major regions and structures of the mitochondrion

fatty acids. In the mitochondrial matrix are the enzymes for the *Krebs cycle*, and within the folded inner membrane is the apparatus for the *electron transport system* and *ATP synthesis*.

We begin the next steps of glucose oxidation with the *two* pyruvate molecules produced in glycolysis. Pyruvates produced in the cytoplasm by glycolysis are transported into the *mitochondrial matrix* (fluid interior of a mitochondrion), where the two molecules will be converted into two 2-C molecules of *acetyl CoA (AcCoA)* (see Figure 7.3a). This step introduces our second coenzyme,

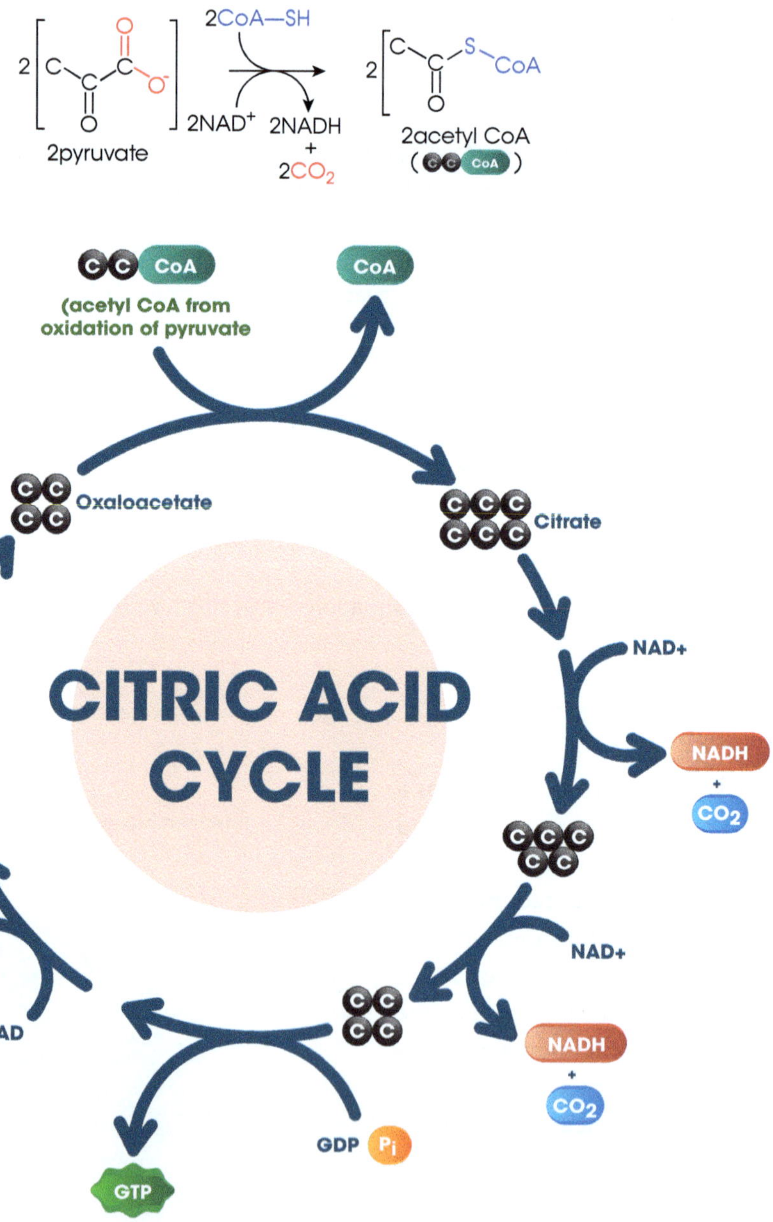

FIGURE 7.3 a (top) Diagram of the oxidation of pyruvate b (bottom) Diagram of the Krebs (citric acid) cycle

coenzyme A, which is described in Table 7.1. This conversion releases one molecule of CO_2 from each pyruvate. After this conversion, cell respiration has formed two of six potential CO_2 that could be produced during the complete oxidation of glucose. Furthermore, each pyruvate is oxidized to transfer electrons to the coenzyme NAD, yielding a total of *two* NADH. Therefore, we still have little to brag about in our efforts to obtain energy from oxidizing glucose. However, let's not forget that we have two AcCoA molecules. AcCoA molecules then enter a cyclic pathway called the *citric acid cycle* or the *Krebs cycle*, which is also in the mitochondrial matrix and is a sequence of reactions to generate energy during aerobic respiration (see Figure 7.3b). The steps of the cycle were elucidated by Sir Hans Krebs, and the first product in the cycle is the 6-C molecule *citric acid*—hence the alternate cycle names.

Each AcCoA is enzymatically joined to a 4-C molecule (oxaloacetic acid), forming the 6-C citric acid. Oxaloacetic acid, as we shall see, is generated at the end of one turn of the Krebs cycle. Citric acid then is processed through a series of enzymatic reactions that finish the oxidation of the glucose, which began back in glycolysis. During one turn of the Krebs cycle, we have two decarboxylation reactions that produce two CO_2. The cycle also produces one molecule of *ATP* by substrate phosphorylation, and four oxidation reactions produce three molecules of *NADH* plus one molecule of *FADH2*. Because we have two AcCoAs to oxidize, the cell runs the cycle twice. Products from *two* turns of the cycle total to the following:

$$2\ ATP,\ 6\ NADH,\ 2\ FADH_2,\ 4\ CO_2$$

Adding to the cycle totals, which includes the two NADH and two CO_2 (from the two pyruvates converted to two AcCoA), the cell produces six CO_2, two ATP, eight NADH, and two $FADH_2$ in all. If you add to this total the two ATP and the two NADH from glycolysis, you still have only four ATP and ten NADH (see Table 7.2). Doesn't this seem like too much work to produce only four ATP?

TABLE 7.2. Products of Glycolysis and the Krebs Cycle

Source	ATP	NADH	$FADH_2$	CO_2
glycolysis	2	2		
pyruvate → AcCoA		1		1
pyruvate → AcCoA		1		1
Krebs cycle (one turn)	1	3	1	2
Krebs cycle (one turn)	1	3	1	2
Total	4	10	2	6

Obviously, we have not yet revealed the entire story. Although we have only produced four net molecules of ATP, we do have 12 molecules of reduced coenzymes. If those molecules can be reduced, we can also oxidize them to get at the electrons and hydrogen ions. To do that we need to use another system—the *electron transport system* (*ETS*).

QRQ 11. The two three-carbon sugars that are products of glycolysis are transported into the ______ matrix, where each will be converted into one 2-C molecule of ______. In this reaction, the substrate is oxidized to transfer electrons to coenzyme ______, yielding a total of two ______. This oxidation-reduction reaction also releases a total of ______ molecules of ______ ______. The two 2-C molecule products of this redox reaction enter a cyclic pathway, called the ______ ______ cycle or the ______ cycle.

QRQ 12. Fill in a term for the first blank and numbers for the remaining blanks: Products from two turns of the Krebs cycle include two 6-C ______ acid molecules, ______ ATP, ______ NADH, ______ $FADH_2$, and ______ CO_2. Combined with earlier values, to this point, the original one glucose has given rise to a total of ______ ATP, ______ NADH, ______ $FADH_2$, and ______ CO_2.

The Electron Transport System

We move into another part of the mitochondrion for the ETS—the *inner membrane*. If you remember from our study of cellular organelles, the mitochondrion is composed of an inner and an outer membrane, as shown in Figure 7.4. Embedded in the inner membrane are multiple sets of the complexes that make up an *electron transport chain* (*ETC*), as well as the ATP synthase apparatus we encountered in the light reaction of photosynthesis.

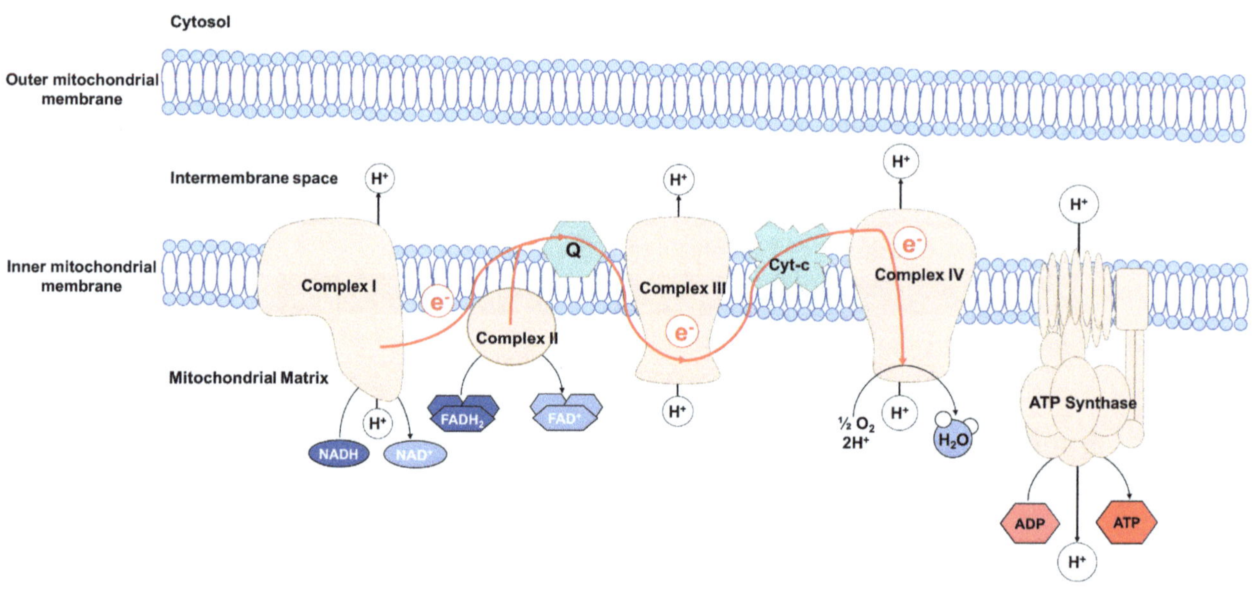

FIGURE 7.4 Electron transport system, including the electron transport chain and the ATP synthase complex

The electron transport chain is composed of four complexes, which receive electrons from the reduced coenzymes NADH and $FADH_2$. The diagram in Figure 7.4 shows the organization of this system. The electrons are then passed through and between the complexes from electron donors to electron acceptors, with the final electron acceptor being molecular oxygen. Each donor is temporarily oxidized, and each acceptor is temporarily reduced. No sooner has the donor received the electrons than it is, in turn, oxidized, and it is then ready to receive more electrons. NADH brings and transfers its electrons and protons to Complex I, leaving NAD for reuse. The electrons are then passed along by a mobile carrier molecule in the membrane (Q) to Complex III, and then the electrons are picked up by another mobile carrier (cyt-c) and transported to Complex IV, where the electrons with some protons finally reduce O_2 to form water.

$FADH_2$ transfers its electrons and protons to Complex II, which also passes them onto the mobile carrier Q, and so on, as with the electrons from NADH (see Figure 7.4). The final acceptor for these electrons is also O_2.

As the electrons are passed along, they are releasing some energy. So they are essentially running downhill. By various mechanisms, this release powers protons from the mitochondrial matrix across the inner membrane into the intermembranous space (see Figure 7.4). As these protons accumulate, they become concentrated, and an electrochemical gradient is formed. Just as in the chloroplasts, the protons, moving toward a lower concentration, pass through the ATP synthase complex, driving the production of ATP from ADP and phosphate.

Hopefully, you will anticipate where we get more ATP from glucose. According to experimental evidence, the protons from the NADH transfer, which enter at Complex I, can provide enough gradient power to yield 2.5 ATP/electron pair transported by NADH. $FADH_2$, which entered the ETS at a lower level, provides less gradient power and yields 1.5 ATP/electron pair transported by $FADH_2$.

If glucose requires two turns of the Krebs cycle, then one glucose gives rise to ten molecules of NADH and two molecules of $FADH_2$. Plus, we have four ATP from glycolysis and the Krebs, produced by substrate phosphorylation. Ten pairs of electrons from NADHs yield four ATP (2.5×10), and two pairs of electrons from $FADH_2$ yields three ATP (1.5×2). If we add the additional four ATP, we get *28 ATP*. The overwhelming majority of the ATP produced in respiration is dependent on oxygen as the final electron acceptor. This form of respiration is termed *aerobic* (with oxygen). If oxygen is not present in sufficient amounts, then the cells must generate ATP by another path . . . an *anaerobic* path (without oxygen).

APPLYING SCIENCE TO LIFE 4 The Expense of Free Diving

In 2016, William Trubridge from New Zealand set a new free immersion record while free diving at Dean's Blue Hole in the Bahamas—407 feet. At that depth, he was under more than 12 atmospheres of pressure. He survived, but not many people would have. It is beyond any reasonably comfortable depth for a human. Recreational SCUBA divers are warned not to venture below 128 feet.

Marine mammals (e.g., seals, sea lions, whales, dolphins, etc.) often have incentives and adaptations to dive much deeper. For instance, elephant seals can dive more than 1,500 m (~5,000 feet), and they can remain there for about two hours. Beaked whales, which tend to inhabit deeper waters, take the prize with dives to 2,792 m (~9,800 feet), and they can stay there for over two hours. Incredibly, they can repeat such deep dives several times a day.

Great depths in the ocean are cold, dark, and under tremendous pressure. Why would animals go to such extreme environments? Apparently, they do these dives for the same reason you make a trip to the grocery store. They find food there. The beaked whales, for instance, love the delicious deepwater squid they find.

Many marine mammals have adaptations that help them make such dives. For instance, the greatly increased hydrostatic pressure at great depths can compress any air pockets, including the lungs. Think of the pressure you feel in your sinus spaces when you dive to the bottom of a pool. The pressure is much greater at hundreds of feet. Collapsing their lungs isn't as hard on these marine divers as it would be for you. Most of your oxygen is stored in your lungs, with some in your hemoglobin (blood) and your myoglobin (muscles). Diving marine mammals may have collapsible ribs cages and lungs, which minimizes the air pockets and decreases buoyancy. Most of their oxygen is stored in the blood, where they have a much higher percentage of red blood cells and hemoglobin, and in their muscle, where they have much more myoglobin for oxygen storage. Many of these mammals actually exhale up to 90 percent of the oxygen in their lungs before they dive. Even if you are a fit, athletic person, you likely can't exhale more than about 10 percent of your lungs' volume of oxygen. In fact, if you were about to dive down 20 feet, you would probably inhale and not exhale.

In addition to storing the maximum amount of available oxygen for their dives, they also try to use as little of the oxygen as possible. So they decrease their buoyancy by deflating their lungs, which makes the dive easier. Thus, many of them just glide toward the bottom, and they don't need to use any energy swimming. Many of them can selectively slow down their heart and metabolic rates to reduce the need for oxygen. They also shuttle blood toward their brains, hearts, and muscles, shunting it away from other organs and the skin. With little blood flowing toward the skin, they reduce heat loss. Under extreme circumstances, the animals' tissue can turn to anaerobic respiration, but some of the products are toxic at high levels.

It is cold in the depths, so most of these diving animals have sausage-shaped bodies, which gives them a low surface-to-body ratio, reducing loss of heat across the skin. Water withdraws heat from your body much faster than air, but many of these animals have insulating layers of fat (Costa 2001; Costa and Sinervo 2004; Hindell et al. 1992; Hooker et al. 2011; Mirceta et al. 2013; Pavid 2017; Starrs 2016).

Although we don't know everything about the amazing abilities these animals have for these deep dives, we are doing more research to better understand them. We now have better remote sensing devices and other instruments, but some diving mammals, like the beaked whales, are pretty difficult to locate. Studying animals in the wild is challenging but also immensely interesting.

QRQ 13. Omitting unused choices, arrange in order the locations to which excited electrons are passed in the electron transport chain: A. combines with molecular oxygen and H+ to form water in mitochondrial matrix, B. Complex I, C. Complex II, D. Complex III, E. Complex IV, F. NADH

QRQ 14. Omitting unused choices, arrange in order the locations to which excited electrons are passed in the electron transport chain: A. combines with molecular oxygen and H+ to form water in mitochondrial matrix, B. Complex I, C. Complex II, D. Complex III, E. Complex IV, F. $FADH_2$

QRQ 15. Omitting unused choices, arrange in order the locations to which hydrogen ions are passed in the electron transport system: A. ATP synthase complex, B. inner membrane, C. intermembranous space, D. mitochondrial matrix, E. powers synthesis of ATP from ADP and inorganic phosphate in mitochondrial matrix

QRQ 16. In the electron transport system, an electrochemical gradient of ______ ions is formed.

QRQ 17. Cell respiration produces a total of ______ ATP from one six-carbon sugar.

QRQ 18. Fill in each blank with one or more words and no abbreviation or acronym: The major steps of cell respiration discussed, in order, are ______, ______, and ______.

QRQ 19. Cell respiration that uses oxygen is called ______, while respiration that generates ATP without oxygen is called ______.

7.2.2. Anaerobic Alternatives to Aerobic Respiration

During vigorous physical activity (running, swimming laps, singles tennis, bicycling, etc.), your muscles need all the energy they can muster. Your skeletal muscles have a small amount of ATP available immediately, but that doesn't last for long. Your muscles also store a compound called *creatine phosphate (CP)* (see Figure 7.5). When needed, an enzyme can remove the phosphate from CP and add it to an ADP molecule to make ATP. Your muscles can also break down sugar in aerobic respiration to yield more ATP, but that is a slower process that requires sufficient O_2.

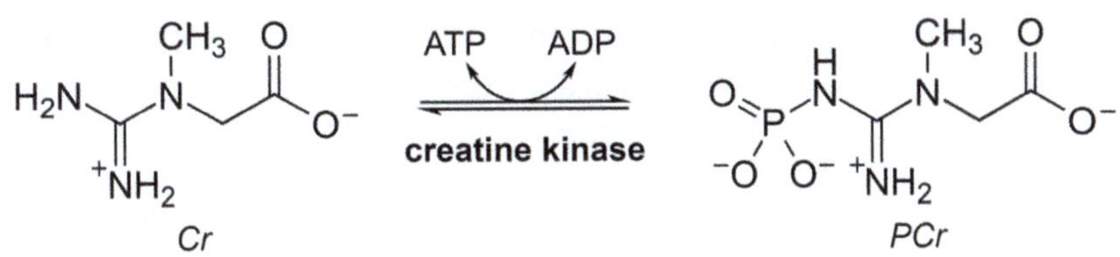

FIGURE 7.5 Creatine kinase can generate ATP from a phosphate donation from creatine phosphate.

If the oxygen levels drop enough, your muscles turn to an anaerobic source of ATP—*fermentation*, but not the alcoholic version. Your cells use a type of fermentation that produces *lactic acid* instead. You might be surprised to know that no extra ATP is produced by this process. So why would the muscle cells utilize this pathway?

Examine Figure 7.6, which shows the pathway from pyruvate, in the absence of oxygen, to lactic acid. What is the advantage of this pathway? If you look at the top part of this drawing, you see glucose converted to two pyruvates, and in the process, a net of two ATP and two NADH are generated. However, the cell can run short of NAD to pick up the electrons for the initial oxidation of glucose. If there is no oxygen, the NADH just accumulate, with no place to take their electrons. Converting pyruvate to lactic acid doesn't produce ATP directly, but it does oxidize NADH to NAD so the latter can return to glycolysis to pick up more electrons. This regeneration of NAD

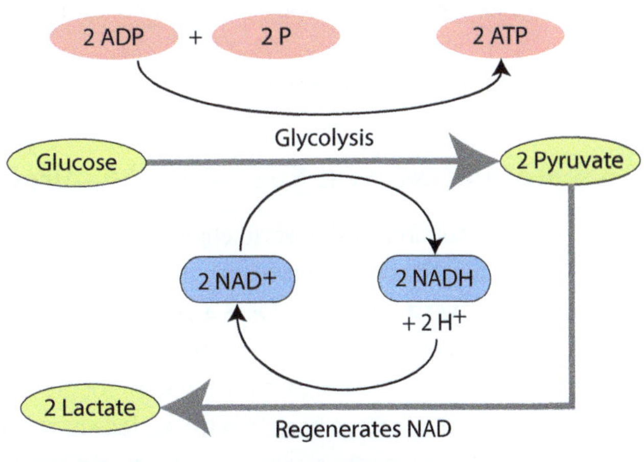

FIGURE 7.6 Lactic acid fermentation

keeps glycolysis going, and the small amount of ATP produced provides the cell with some energy, although much less than through aerobic respiration. With rest, oxygen levels again rise in your muscles, and lactic acid can be converted back to pyruvate, which can be oxidized by the Krebs cycle.

APPLYING SCIENCE TO LIFE 5 Muscle Fatigue

For many years physiologists and fitness experts have hypothesized that it is the accumulation of lactate (lactic acid) that causes muscle fatigue during heavy physical activity. More recently, scientists have found little evidence to support this hypothesis. Currently, much evidence supports a new hypothesis attributing fatigue to changes in levels of *inorganic phosphate* (P_i) and *calcium* (Ca++) in the muscle cell cytoplasm. Muscle contractions depend heavily on energy (ATP) and increased [Ca^{++}], which under nerve impulse stimulation, is released by the sarcoplasmic reticulum (smooth endoplasmic reticulum of muscle) into the cytoplasm. As available ATP is used up, you are left with ADP and P_i. So [P_i] increases, which may directly interfere with contractile fibers activity, but it also seems to reduce the availability of calcium ions needed for contraction. One hypothesis is that P_i binds to and causes calcium to precipitate. There are other hypotheses, and there is likely more than one mechanism. The jury is still out, and that is part of the joy of science. We are curious, and we constantly work to improve our hypotheses and experiments so that we can advance our understanding of nature. Scientists tend to embrace uncertainty but work to reduce it (Allen and Trajanovska 2012; Westerblad, Allen, and Lannergren 2002).

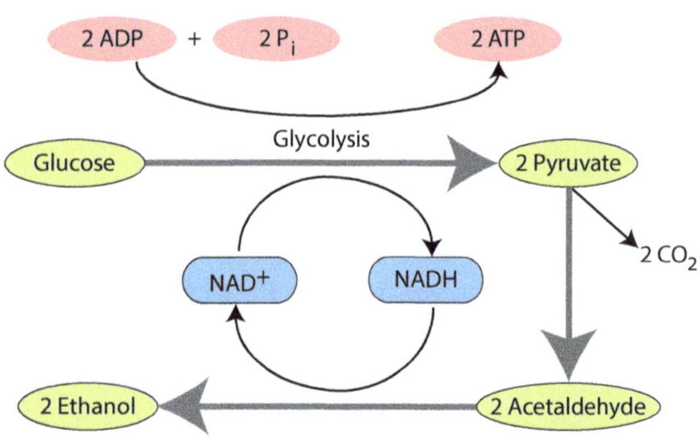

FIGURE 7.7 Alcoholic fermentation

There are actually various fermentation pathways in prokaryotic organisms, as well as in yeasts. In yeasts, there is alcoholic fermentation. That pathway exists for the same reason that the lactic acid fermentation pathway exists—to regenerate NAD for glycolysis and ATP production. Examine Figure 7.7, where you see that pyruvate is converted initially to acetaldehyde and CO_2. An enzyme reduces acetaldehyde to ethanol and regenerates NAD.

APPLYING SCIENCE TO LIFE 6 Brewing Up a Profit

Many cultures of the world enjoy the alcoholic products provided by the humble single-cell eukaryotic yeasts, and cultures have been making alcoholic drinks for millennia. The US beer-brewing industry presides over a 114 billion-dollar-a-year market, according to the Brewers Association Data.

Besides yeasts, there are some bacteria used to produce fermented products. For instance, the agave plant, native to Mexico, has been used for an extremely long time for the production of *pulque*, a type of fermented "cactus beer." On the surface of the agave is a bacterium, *Zymomonas*, which helps to ferment the plant pulp into alcohol by a slightly different fermentation pathway, but one that still produces alcohol. The "beer" is an acquired taste, but it can also be distilled into tequila (Brewers Association 2018).

APPLYING SCIENCE TO LIFE 7 Domesticating Microbes

When you think of domesticated organisms, what organisms come to mind? Cattle, dogs, or sheep, perhaps? Well, apparently brewers have also domesticated the yeasts they use for making beer. Beer was being made for a long, long time before we knew about microbes in our world. Unknowingly, brewers were selecting yeast strains that were fittest—able to thrive in the conditions of beer-making and that gave the beer the most preferred flavor. When yeast stop producing gas during the fermentation process, they sink. Brewers learned that if they transferred some of the sediment to a new batch, the new beer would be reliably like the previous fancied brews.

Because these yeast strains were being used all year for brewing beer, they didn't get out to "socialize" with other yeasts. Because of that reclusive, agreeable life in the brewing tanks, they didn't need their ability to interbreed (genetically mix) with other, wild strains of yeasts. Consequently, the genes for sex in this type of yeast are diminished or gone completely.

Although beer is brewed year-round, wines are made only at certain times of the year. It is interesting that wine-making yeasts are much more "social." The yeasts for winemaking are on the grape's skins, and so they have time in the wild to mix it up with other, savage yeast strains. Subtle changes in the yeasts can possibly make a difference in the flavor of the wine. Of course, most vintners know that many other factors influence the quality of their products too (Gallone 2016).

QRQ 20. Complete the following statements on two sources of ATP for muscles with vigorous physical activity: Skeletal muscles have a small amount of ATP available immediately. An enzyme can remove phosphate from ______ phosphate, stored in muscles, and add it to a(n) ______ molecule to make ATP.

QRQ 21. Complete the following statements on three sources of ATP for muscles with vigorous physical activity: With sufficient oxygen, muscles can break down sugar in aerobic respiration to yield more ATP. If oxygen levels drop enough, muscles can use anaerobic ______ of ______ to produce ______ acid. The process oxidizes ______ to NAD so the latter can return to glycolysis to pick up more electrons. With rest, ______ levels again rise in your muscles. Then the previous process can be reversed, and the 3-C sugars can be oxidized by the Krebs cycle.

QRQ 22. Yeasts can use ______ fermentation, which regenerates ______ for glycolysis and ATP production.

Image Credits

Fig. 7.1: Adapted from Copyright © by CNX OpenStax, (CC BY-SA 4.0) at https://commons.wikimedia.org/wiki/File:OSC_Microbio_08_02_Glycolysis.jpg.

Fig. 7.2: Source: https://commons.wikimedia.org/wiki/File:Animal_mitochondrion_diagram_en.svg.

Fig. 7.3a: Adapted from Copyright © by akane700, (CC BY-SA 3.0) at https://commons.wikimedia.org/wiki/File:Pyruvate_dehydrogenase_complex_reaction.PNG.

Fig. 7.3b: Adapted from Copyright © 2020 Depositphotos/VectorMine.

Fig. 7.4: Copyright © by Dw001 (CC BY-SA 4.0) at https://commons.wikimedia.org/wiki/File:Electron-TransportChainDw001.png.

Fig. 7.5: Source: https://commons.wikimedia.org/wiki/File:Creatine-Kinase.svg.

LAB #7

Laboratory Exercises

7.3. Laboratory on Food Inequality

7.3.1. Preliminary Work

In this lab, we examine dietary disparities in an urban situation. From the text's website, view the introductory video and "Food Inequality Utility Score;" download and open "*Food Inequality.nlogo.*"

1. Click the Info tab at the top of the page and read the section "What Is It?"
    a. ________________ As indicated in the section, which of the following are some of the important contributors to unhealthy dietary selections for low-income households? A. age, B. inequality in geographic distribution of stores selling healthy and reasonably priced food, C. low-income jobs, D. preferences for less healthy menus, E. snack foods, F. unappealing stores
    b. ________________ What factors does the model consider? A. distances to stores, B. income, C. the cost of food in each store, D. the degree to which a household prefers healthy foods, E. the degree to which a store changes its pricing scale, F. the habitual behavior of each household, G. upward mobility
    c. ______ Why is the cost of food relevant? A. everyone likes a bargain, B. expensive stores often sell a variety of healthy and unhealthy options, C. stores selling healthy foods are often more expensive, D. stores selling unhealthy foods often have sales on that food
    d. ________________ Select all factors why distance of a household to stores relevant: A. high-income households can afford food warehouse memberships, B. high-income households use more online shopping, C. low-income households are less likely to be able to afford transportation, D. low-income households are less likely to have access to good transportation
    e. ______ A household's habitual behavior is important because people tend to **1.** (A. drive the same routes each day, B. eat foods they recently ate, C. follow price-patterns of stores, D. shop at stores where they most recently shopped).

2. Read "How It Works" and "How to Use It."

a. Give the color (A. black, B. blue, C. brown, D. cyan, E. gray, F. green, G. orange, H. red, I. violet, J. white, K. yellow) of each model component:

_______ rich household

_______ poor household

_______ cheap and unhealthy store

_______ expensive and healthy store

_______ cheap and healthy store

_______ expensive and unhealthy store

b. ______ Where are poor households in the simulation's world? A. bottom, B. left, C. right, D. throughout, E. top

c. ______ Initially, where are the expensive and healthy stores? Use the same possible answers as in part b.

d. ______ to ______ Give the range of food preference values for high-income households.

e. ______ The parameter PREFERENCE-SHIFT represents the degree to which dietary education impacts food preference for **1.** (A. all households, B. high-income households, C. low-income households).

f. ______ PERCENT-HEALTHY-HUBS represents approximately the initial percent of stores that are **1.** (A. cheap and healthy in the poorer neighborhood, B. cheap and healthy, C. expensive and healthy).

g. ______ A tick represents a(n) (A. hour, B. day, C. week, D. month, E. year).

h. ______ In the model, a household shops at a store that yields the highest **1.** (A. distance score for that household, B. habitual-behavior score for that household, C. preference score for that household, D. price score for that household, E. utility score for that household).

i. ______ **1.** (A. Distance, B. Habitual behavior, C. Preference, D. Price, E. Utility) carries the greatest weight in calculation of a household-store utility score.

j. ______ Give the price score for a rich-household-cheap-store combination.

k. ______ Give the price score for a poor-household-expensive-store combination.

l. ______ Calculate the distance score, rounded to two decimal places, for a poor household and store that are ten units apart.

m. ______ Calculate the distance score, rounded to two decimal places, for a rich household and a store that are ten units apart.

n. ______ Give the household-store habit score, rounded to one decimal place, for a store in which the household shopped time-before-last.

o. ______ Calculate the preference score, rounded to one decimal place, for a healthy store when the household has a food preference value of 0.3.

p. ______ Calculate the preference score, rounded to one decimal place, for an unhealthy store when the household has a food preference value of 0.3.

q. ______ Give the color (A. black, B. blue, C. brown, D. cyan, E. gray, F. green, G. orange, H. red, I. violet, J. white, K. yellow) of the link when a household shops at a new store.

r. ______ Suppose a store with the least number of customers closes. What color does the patch become? Select from possible answers in part q.

s. ______ percent If ALLOW-STORE-CLOSE is ON, give the chance that when an unhealthy store closes a healthy store will open in its place after 26 weeks (ticks).

t. **1.** ______ **2.** ______ The simulation runs **1.** ticks, or about **2.** (whole number) years.

u. ______One simplifying assumption is that high-income households tend to prefer **1.** (A. cheaper, B. healthier, C. less healthy, D. more expensive) food than low-income households.

7.3.2. Laboratory Work

3. This question involves the situation where no diet education program exists and no cheap, healthy stores exist initially in the poor neighborhood. However, every 26 weeks, a store closes, and every 26 weeks, starting at 52 weeks, a store opens in the location of a previous grocery. Moreover, a new store could be of a different type than the original one.

a. ______ Make sure PREFERENCE-SHIFT and PERCENT-HEALTHY-HUBS are at their default values of 0 and ALLOW-STORE-CLOSE is ON. Increase the speed, press *setup*, and run the simulation. Observe carefully the animation. After the animation stops, export the interface (found under *File, Export, Export Interface...*), saving a picture of the interface. Also, export the output (found under *File, Export, Export Output...*) to save the output in a text file. Based on link color, which kind of household is more likely to change the store at which they shop from one time to the next, poor households (P) or rich households (R)?

b. ______ **1.** (A. Distance, B. Habitual behavior, C. Preference, D. Price) is most likely to cause this phenomenon.

c. ______ Complete the hypothesis: Poor households **1.** (L. are less likely to shop, M. are more likely to shop, S. have about the same likelihood of shopping) healthy compared to rich households.

d. ______ Complete the hypothesis: As compared with people from rich households, those from poor households **1.** (A. are more likely to travel longer, B. are more likely to travel shorter, C. travel about the same) distances to shop.

e. ______ Speed the simulation or uncheck *view updates*. After execution, check *view updates* and export the interface and output as described in part a, using different names. Repeat the process several times until each of the following situations has occurred: 1) stores on the left (poor area) remain all cheap and unhealthy (brown), while those on the right (rich area) remain all expensive and healthy (cyan); 2) at least one store on the left is cheap and healthy (green). This and subsequent parts relate to these situations. For situation 1, give the approximate proportions of rich households shopping at healthy stores: A. 0, B. 0.25, C. 0.50, D. 0.75, E. 1.0

f. ______ For situation 1, give the approximate proportions (A. 0, B. 0.25, C. 0.50, D. 0.75, E. 1.0) of poor households shopping at healthy stores.

g. ______ For situation 1, give the approximate average distance (A. 3, B. 6, C. 9, D. 12, E. 15) traveled to stores by the rich.

h. ______ For situation 1, give the approximate average distance traveled to stores by the poor. Use the same choices as in part g.

i. ______ For situation 2, does the average proportion of times poor households shop healthy go down (D) or up (U) as compared with situation 1?

j. Discuss how your findings support or refute the hypotheses in parts c and d.

4. This question involves assessing the impact on diet of varying numbers of healthy hubs alone in the poor neighborhood. Unlike the previous question, these experiments will have no store closures or openings, which will enable more accurate comparisons of the impacts of the number of hubs.

   a. ______ With PREFERENCE-SHIFT still being 0, make PERCENT-HEALTHY-HUBS 5 percent and turn off ALLOW-STORE-CLOSE. Keep pressing *setup* until the initial world has exactly one healthy, cheap store in the poor area. Speed the simulation or uncheck *view updates*. Export the output as described in question 3a. Repeat the experiment three times. Calculate the mean over all the experiments of the average proportion of times poor households shopped healthy and select your result, approximately: A. 0.00–0.01, B. 0.01–0.04, C. 0.04–0.07, D. 0.07–0.10, E. 0.10–0.17, F. 0.17–0.25, G. 0.25–0.40

   b. ______ Besides preference shift, what can impact the average proportion of times poor households shopped healthy in part a's experiment? Select all that apply: A. distribution of cheap healthy stores in poor area, B. distribution of expensive unhealthy stores in rich area, C. number of cheap healthy stores in poor area, D. number of expensive unhealthy stores in rich area

   c. ______ Repeat part a with PERCENT-HEALTHY-HUBS being 10 percent and exactly three healthy, cheap stores in the poor area.

   d. ______ Repeat part a, using the possible answers from that part, with PERCENT-HEALTHY-HUBS being 20 percent and exactly five healthy, cheap stores in the poor area.

   e. ______ Repeat part a, using the possible answers from that part, with PERCENT-HEALTHY-HUBS being 30 percent and exactly seven healthy, cheap stores in the poor area.

   f. ______ According to the model, when the number of cheap healthy grocery stores in the poor section increases from none to almost 30 percent, the number of poor people shopping healthy rises from almost 0 percent to about **1.** (A. 0.13 percent, B. 1.3 percent, C. 13 percent, D. 0.30 percent, E. 3.0 percent, F. 30 percent).

5. This question involves assessing the impact on diet of increasing dietary education alone for poor households. As in the last set of experiments, for more accurate comparisons, these experiments will have no store closures or openings.

   a. ______ With PERCENT-HEALTHY-HUBS being 0, make PREFERENCE-SHIFT 0.1 and turn off ALLOW-STORE-CLOSE. Speed the simulation or uncheck *view updates*. Export the output as described in question 3a. Repeat the experiment three times. Calculate the mean over all the experiments of the average proportion of times poor households shopped healthy and select your result, approximately: A. 0.000–0.005, B. 0.005–0.02, C. 0.02–0.03, D. 0.03–0.04, E. 0.04–0.1, F. 0.1–0.2

   b. ______ Repeat part a with PREFERENCE-SHIFT being 0.2.

   c. ______ Repeat part a with PREFERENCE-SHIFT being 0.3.

d. ______ Repeat part a with PREFERENCE-SHIFT being 0.4.

e. ______ According to the model, dietary education alone has **1.** (N. no, L. little, M. moderate, S. significant, I. indeterminate) impact on the healthy shopping habits of those in the low-income area.

6. This question involves assessing the impact on diet of increasing dietary education and the number of healthy hubs in the poor area. For more accurate comparisons, these experiments will have no store closures or openings.

   a. ______ Make PERCENT-HEALTHY-HUBS 5 percent and PREFERENCE-SHIFT 0.1 and turn off ALLOW-STORE-CLOSE. Keep pressing *setup* until the initial world has exactly one healthy, cheap store in the poor area. Speed the simulation or uncheck *view updates*. Export the output as described in question 3a. Repeat the experiment three times. Calculate the mean over all the experiments of the average proportion of times poor households shopped healthy and select your result, approximately: A. 0.00–0.01, B. 0.01–0.05, C. 0.05–0.12, D. 0.12–0.25, E. 0.25–0.35, F. 0.35–0.43, G. 0.43–0.60, H. 0.60–0.75, I. 0.75–0.85

   b. ______ Repeat part a with PREFERENCE-SHIFT being 0.2, PERCENT-HEALTHY-HUBS being 10 percent, and exactly three healthy cheap stores in the poor area.

   c. ______ Repeat part a with PREFERENCE-SHIFT being 0.3, PERCENT-HEALTHY-HUBS being 20 percent, and exactly five healthy cheap stores in the poor area.

   d. ______ Repeat part a with PREFERENCE-SHIFT being 0.4, PERCENT-HEALTHY-HUBS being 30 percent, and exactly seven healthy cheap stores in the poor area.

   e. ______ According to the model, healthy hubs along with dietary education has **1.** (A. practically no, B. a slightly greater, C. a moderately greater, D. a significantly greater, E. an indeterminate amount of) impact on the healthy shopping habits of those in the low-income area as compared with one of these interventions (healthy hubs or dietary education) alone.

7. Make another hypothesis and design an experiment to test the hypothesis. Run the experiment at least three times. Do the results support your hypothesis? Discuss the results.

MODULE

8

Hosting Unwelcome Guests—Noncellular Entities

8.1. Introduction

You may have heard about the Influenza Epidemic of 1918, which killed at least 50 million people worldwide. Viruses were virtually unknown early in the 20th century. Biologists were aware of some very small agents (not visible in the microscopes of the time), but they did not know much about their nature or their association with disease. Accordingly, in 1918, most scientists thought the influenza was caused by a bacterium. They were wrong. We now know much more about the virus strains that cause influenza, but we still have difficulty in preventing influenza epidemics. This section will introduce you to the world of viruses.

In this module, we focus on disease-causing agents that have no cellular structure, no metabolism of their own, and no means to replicate themselves without host-cell cooperation. Viruses are now well-known agents of disease, and we will concentrate on them in the section on spread of disease. We also briefly consider some other rather bizarre, acellular (without cell structure) agents—prions and viroids. Despite our emphasis here on noncellular agents, you can apply the same spread-of-disease principles to other, cellular agents, such as bacteria, fungi, and protozoa. Consequently, we will include some diseases that are caused by these agents.

APPLYING SCIENCE TO LIFE 1 Desperate Solution

For centuries, human cultures have employed curative treatments for infectious diseases. Sometimes they worked, and sometimes they did not. Those cultures might consider the successful treatments as "antimicrobial," if that term existed at the time. We in the West consider successful the antimicrobial or antibiotic treatments began in the 20th century. Salvarsan, an arsenic-containing compound, was the first effective treatment against the sexually transmitted disease syphilis, which is caused by a bacterium. However, the serendipitous discovery of penicillin in the 1920s catalyzed the discovery and isolation of a variety of antibiotic compounds. The successful extraction of penicillin took many years, and so it was not available until the 1940s for implementation in human patients. Utilizing this compound saved countless lives, and physicians thought they had finally found the "magic bullet" to kill pathogenic bacteria. Unfortunately, pathogenic bacterial strains became resistant to this drug within a decade. Following this development, new, more effective generations of penicillin and other types of antibiotics were developed. Unsurprisingly, pathogenic bacterial strains evolved resistance to these new antibiotics also. Now, no day passes where we do not hear that a new strain of some pathogen has become resistant to our current arsenal of antibacterial compounds (what we usually refer to as antibiotics). Additionally, viral diseases are *not* treatable with antibiotics, although some antiviral compounds are now in use.

Cystic fibrosis (CF) is an inherited disease wherein the body produces mucus that is unusually thick and sticky. Normally, mucus is more fluid and slippery. The thick mucus can clog the lungs and the pancreatic duct (tube carrying pancreatic secretions—digestive enzymes). CF patients have difficulty breathing. They may cough and wheeze, and the lungs are unable to clear out impurities and microbes. Consequently, they are very vulnerable to infections.

In 2017, a British teenager with CF received a lung transplant. This is drastic surgery, but she did survive the operation. To reduce the chance of transplant rejection, she was given immune suppressive drugs, which promoted an infection by a resistant bacterium (which was already in her body) related to tuberculosis. The bacterium attacked the surgical incision site, her liver, and her body surface. Almost out of hope, the family and team decided to try something experimental. Without such intervention, she had virtually no chance to survive. The treatment was a cocktail of viruses.

You might question the logic of using viruses to treat a bacterial infection. However, these viruses are not human, or even animal, viruses. These viruses were bacterial viruses, called *bacteriophages*, or *phages*. Phages only attack bacterial cells, and they do so very specifically. Consequently, you have to have a virus that would attack and kill the bacterium that was causing the problem. Her physicians contacted a scientist at the University of Pittsburgh and Howard Hughes Medical Institute (HHMI) in the United States. This scientist, Dr. Graham Hatfull, had the largest collection of known phages. Dr. Hatfull leads an HHMI program for undergraduate students where each student takes a phage, visualizes it using an electron microscope, sequences its nucleic acid, and screens bacteria to which it will attach and then kill (by lysis—cell bursting). So this alternative treatment results at least partially from the work of thousands of college freshmen and sophomores.

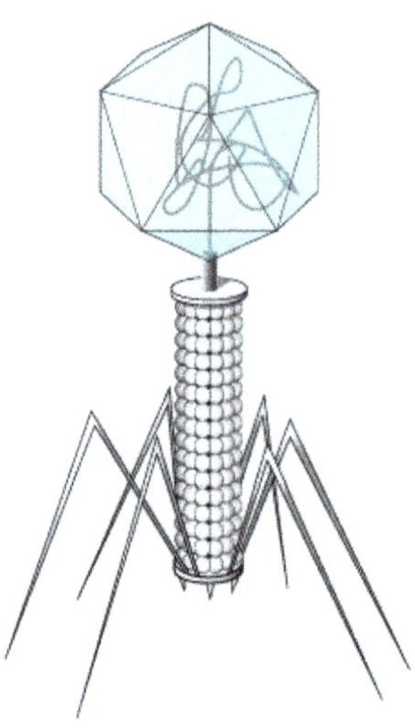

FIGURE 8.1 Bacteriophage structure

In Figure 8.1 is a cartoon of a bacteriophage's structure. Admire its complexity. Does it remind you of some sort of lunar lander? The blue-shaded head is formed by proteins that enclose the nucleic acid. The remainder of the phage functions for attachment and subsequent injection of the nucleic acid into the bacterial cell. Each phage type is very specific and will only bind to one bacterial strain. Even if it binds and enters the bacterial cell, it might not kill that cell. Effective phages will reproduce and cause the bacterium to burst, releasing more phage particles that infect more bacterial cells.

The research team in the United States found three phages that would attack the bacterium causing the teenager's infection. Two of them had to be modified slightly to make them more effective at killing the bacterium. Over a period of weeks, the infection began to fade. The patient gained weight and was eventually sent home. She still carries the infection, but it is being held at bay.

Despite this and a few other successes, phage therapy is not approved by the Food and Drug Administration (FDA) in the United States. However, there are now enough people requesting phage treatment that scientists may soon begin some clinical trials. Successful trials may eventually lead to FDA approval for more widespread use.

APPLYING SCIENCE TO LIFE 2 Viral News

The Democratic Republic of the Congo (DRC) has had its share of problems: contested presidential election, widespread violence from a multitude of armed groups, rampant corruption, weak governance and institutions. It is estimated that there are 4.5 million displaced people in the DRC. DRC is also a mineral-rich country that, if managed properly, could likely become a leading economic power in Africa. With all of its problems, it certainly didn't need another Ebola epidemic. However, in August 2018, the Ministry of Health in the Congo announced the country's tenth outbreak of the dreaded disease.

Ebola has always been a death sentence for any individual acquiring it. It is a brutal virus (see Figure 8.2) whose invasion prompted a hemorrhagic fever that, up to now, was almost certainly fatal. Combating outbreaks has been particularly challenging over the years, not only with all the lawless violence and absence of adequate public health response but also in the minds of the people. There were false rumors that Ebola didn't exist. Others equated the disease with a miserable death, where the dead were buried in pits and drenched with bleach by health workers in "spacesuits." Health care workers were attacked in places, mostly out of fear. With all the fear, families often hid those who were sick, preventing any possible treatment and promoting the spread of the disease.

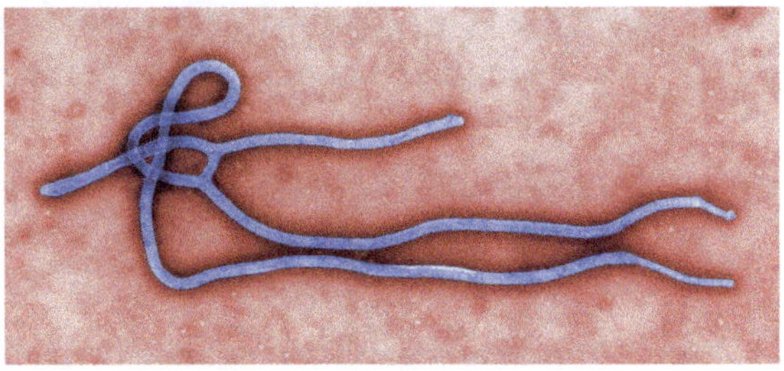

FIGURE 8.2 Transmission electron microscope photo showing Ebola virus (colorized)

Finally, a year later, the government and representatives from the World Health Organization and the National Institutes for Health announce the promising results from treatments using two vaccines, where it appears there is a 90 percent survival rate. Both treatments involved monoclonal antibodies (proteins that bind to selective targets to interfere with the viral invasion). Finally, the DRC has some good news to share with the world.

8.2. Essentials for Life

Do you see the fish in Figure 8.3? That fish is a stonefish, and it is one of the most poisonous fish in the world. The poison is produced by poison glands, located in the spines supporting the animal's fins. Stonefish are masters of camouflage. Note how well this specimen blends in with its surroundings. These fish are ambush predators, lying still along shallow sea bottoms. When a small fish or other unsuspecting prey swims into the vicinity of a stonefish, it quickly opens its oversized mouth and swallows the prey whole. The spines pose a particular problem for any predator that choses a stonefish for a meal. Furthermore, unsuspecting human waders, swimmers, and snorkelers can accidentally run afoul of those spines, which are stout enough to pierce a fin or boot. If the spine cracks, it releases a complex poison into the wound, which will be quite painful. Still worse, the toxin can interfere with heart function and breathing, which could lead to death. With the serious consequences that result from a stonefish encounter, distinguishing a living stonefish from the surrounding non-living rocks is quite important. What are the characteristics of a living organism that make it different from something non-living? Living organisms, no matter how complex, all share the following characteristics:

1. Cells are the basic structural and functional unit of all life forms on earth.
2. Living organisms are capable of reproducing, growing, and developing. Reproduction of the organism is based on cellular reproduction, which may include mutations and errors.
3. Living organisms metabolize, which includes breakdown and synthetic, chemical reactions. This metabolism is necessary for growth, maintenance, and reproduction. All organisms need energy, which may come from non-living sources—light (photosynthesis) or inorganic, chemical reactions (chemosynthesis). These organisms are called *autotrophs,* which includes photosynthetic organisms, like green plants, and chemosynthetic bacteria. Organisms that obtain their energy by breaking down organic molecules are called *heterotrophs.* These organisms, including humans, obtain these molecules from autotrophic organisms or other heterotrophic organisms.
4. Living organisms regulate their metabolism so as to maintain a constant internal state, called *homeostasis.*
5. Living organisms are sensitive to their environment, having the ability to receive environmental signals and respond to them. Your senses include smell, taste, sight, hearing, and touch. Some organisms are able to detect electrical or magnetic fields, light, or sound not detectable by human sense organs. With the reception of these environmental signals, living organisms may respond. Animals often act by moving toward or away from the signal. Movement of some sort also characterizes living things. So, if you smelled a bad odor, you might recoil from that odor. A hungry animal, spotting something to eat, would move toward that food. Although not moving very quickly, plants move toward light by growing in that direction.

FIGURE 8.3 Stonefish

Most human encounters with stonefish are unintentional, but prey, looking for something to eat, can easily stray into range of a quick-moving, voracious stonefish. If you examined a stonefish, you would be able to tell it was alive, based on the some of the above characteristics, hopefully, without endangering your life. On the other hand, you could examine some of the surrounding rock, after removing the organisms that have colonized it. The rock lacks cellular structure; is unable to reproduce, metabolize, or maintain homeostasis; and is unable to sense and respond to environmental signals. Therefore, the rock is non-living.

QRQ 1. Organisms that obtain their energy from non-living sources, such as photosynthesis or chemosynthesis, are called ______, while organisms that obtain their energy by breaking down organic molecules are called ______.

QRQ 2. A living organism's tendency to maintain a constant metabolic state called ______.

QRQ 3. Select characteristics of all living organisms: appearance, blood, cells, environmental sensitivity, eyes, grow, homeostasis, metabolism, reproduction, stability

8.3. Viruses—The Basics

Viruses lack cellular structure; are unable to reproduce, metabolize, or maintain homeostasis; and are unable to sense and respond to environmental signals. Because viruses do not meet these characteristics of living things, we think of them as infectious intracellular parasites. They are relatively small, measured in nanometers (1 nanometer < 25-millionth of an inch), and all known viruses require a *host cell* to allow them entry and support replication. Despite not being considered alive, they are still fascinating objects for study. Furthermore, viruses are of special interest to us because viruses cause disease in human beings, in domestic animals, and in crop plants.

There are many types of viruses, but they all are composed of a core of nucleic acid (DNA or RNA) and a protein coat, called the *capsid*, and some are surrounded by an *envelope* (membrane). There are viruses that will infect any organism on earth, including bacteria, protists (e.g., *Paramecium*), fungi, plants, and animals. The target organism or cell is usually quite specific, and replication is crucial for a virus to be successful. Virus replication is varied among the numerous types of viruses, but we will explore only a generalized virus replication cycle, as shown in Figure 8.4. The steps of viral infection follow:

1. **Binding** of virus to *specific* receptors on the exterior of the target cell using distinctive attachment glycoproteins on the capsid surface or in the exterior of its envelope, if present.
2. **Entry** into the cell, which may be by various means. Endocytosis is shown in the diagram.

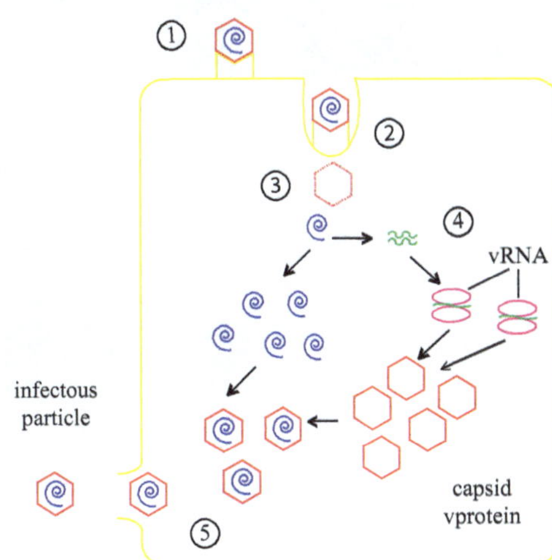

FIGURE 8.4 Generalized viral replication cycle

3. **Uncoating**, a dissolution of the capsid and release of the nucleic acids and other core molecules into the cytoplasm of the host cell.
4. **Replication** of the viral nucleic acid (generalized), accompanied by protein synthesis of enzymes and capsid proteins.
5. **Assembly** of new virus particles; viral DNA and RNA and enzymes are enclosed within new capsids.
6. **Release** of infectious virus particles.

Viruses have been classified traditionally by their shape (structure), the type of nucleic acid in their core, and sometimes by host preferences. Most viruses are so small that they must be viewed using electron microscopy. Recently, scientists in Brazil have found new viruses that are very large, large enough to view with a regular microscope. These particular viruses were collected from species of amoeba. They are not only physically large, but they also contain a huge genome—double-stranded DNA with about 1.5 million base pairs, sufficient to code for 1,276–1,425 proteins. They lack enzymes for their own metabolism, primarily missing the ability to make their own ribosomes, but the genes code for many of the proteins required for independent protein synthesis. Note the structure of the Tupanvirus in Figure 8.5. It has a capsid with an icosahedral symmetry and a large tail. You can also see fibers projecting from the viral surfaces.

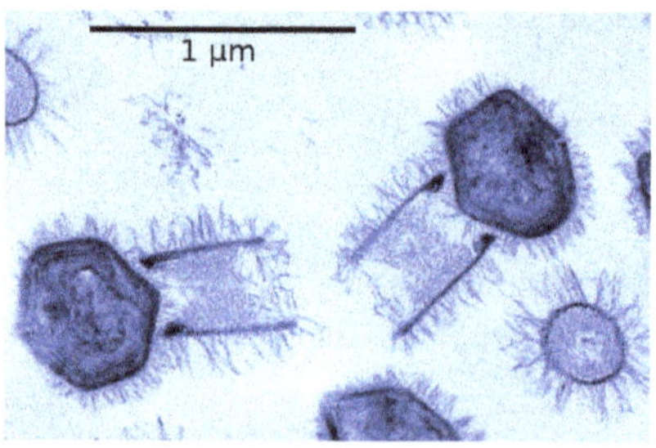

FIGURE 8.5 Tupanviruses

Common shapes of viruses are icosahedral, helical, complex, round, and bullet-shaped (see Figure 8.6). Classifications are also based on the nucleic acid—DNA, RNA; double-stranded, single-stranded—as well as the replication method. For instance, some viruses replicate in the cytoplasm and others replicate in the nucleus of the host cell—animal, plant, bacterium, protist.

One of the reasons scientists are so interested in viruses, besides simple curiosity, is that they are disease-causing agents. They can cause diseases in just about any cell-based organism you can name. Naturally, what attract the most attention are viruses associated with disease.

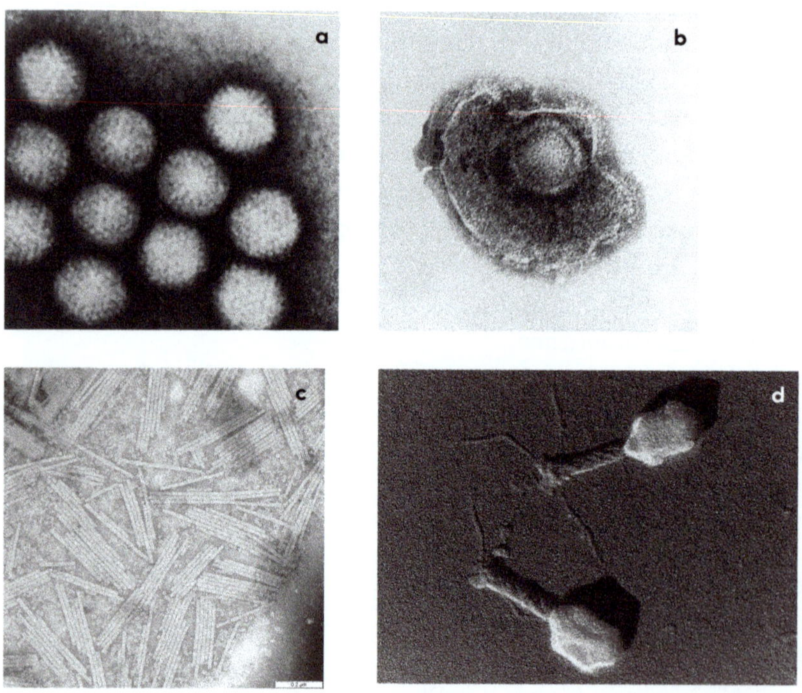

FIGURE 8.6 Common virus shapes a. Adenovirus, icosahedral. b. Varicella-zoster icosahedral, enveloped. c. Tobacco Mosaic viruses, helical. d. Bacteriophage, complex

QRQ 4. Select from the following the appropriate term for each description below: assembly, binding, capsid, entry, envelope, nucleic acid, protein synthesis, release, replication, uncoating

a. attachment of virus to specific receptors on exterior of target cell
b. copying of viral nucleic acid
c. core of a virus
d. dissolution of the capsid with release of core molecules into host's cytoplasm
e. enclosing of viral DNA and RNA and enzymes within new capsids
f. making of enzymes and capsid proteins
g. membrane surrounding some viruses
h. protein coat around virus core
i. virus goes into cell
j. virus goes out of cell to infect a new cell

QRQ 5. Give the general sequence that infection by a virus usually follows. Some letter(s) may not be used: A. assembly, B. binding, C. capsid, D. entry, E. release, F. replication/protein synthesis, G. uncoating

8.4. Epidemics and Spread of Disease

APPLYING SCIENCE TO LIFE 3 Tracking Down a Killer

On March 29, 2003, Dr. Carlo Urbani, an infectious disease specialist from WHO (World Health Organization), became a casualty of a disease he helped to contain. In early March, Dr. Urbani, 46, responded to an alert from a small, private hospital in Hanoi regarding a patient who presented with an unusual influenza-like virus. He recognized that this case represented some emerging disease, and he aggressively pursued a public health response that effectively checked the epidemic in Vietnam (Reilley et al. 2003). Later, that virus was identified as the causative agent for *SARS* (severe acute respiratory syndrome), and Vietnam's response has become a model for containing epidemics. Unfortunately, many of the victims who did not survive the disease were health care workers. One of those who did survive was Nguyen Thi Men, a pediatric nurse at the hospital. She emerged from a nine-day coma, still suffering but alive. While recovering, she stated that she hoped to return to nursing one day (Mydans 2003).

SARS is an example where quick and determined action by public health officials and organizations likely prevented a pandemic. Despite containing this disease, several hundred died, and there was tremendous economic loss in Asia and in Canada. History of human disease is replete with epidemics and pandemics. Epidemics of various types are still a real threat, and our modern lifestyles and pervasive technology both promote and deter them. Diseases can be spread easily from one country or continent to another by the rapid transportation and increasingly global commercial activities. On the other hand, our ability to communicate quickly and monitor on large and small scales can work to promote a rapid response to emerging diseases.

8.4.1. Fundamentals of the Spread of Infectious Diseases

In this section, we consider, using viruses, the nature and spread of *infectious diseases*. First, we will need to consider a few fundamentals about diseases and how they develop and are transmitted. Let's consider what we mean by *disease*. Actually, it is not that easy to define this term, but we'll start with the following:

> a pathological condition that exists in an organism, caused by an infectious agent, genetic abnormality, or environmental stress, that results in a characteristic set of signs and symptoms; the condition alters normal body function and disturbs normal homeostatic mechanisms

We can define an *infectious disease* as a disease caused by the entrance, establishment, and replication of a microorganism in a *susceptible host*. This microbe, termed a *pathogen*, potentially can be spread to other hosts. An *infection*, then, occurs with successful establishment (and usually reproduction) of a pathogen in the host.

Human beings have a large set of diseases caused by infectious microbial agents (e.g., bacteria, viruses, protozoa, fungi). Before we focus on the diseases and pathogens, let's also make certain we understand what we mean by the term *epidemic*. Most people consider epidemics to involve very large numbers of individuals who have a disease. Certainly, diseases like influenza or HIV can and do affect millions of people. However, to scientists who study epidemics (*epidemiologists*), there only needs to be a *significant increase* in the occurrence of a disease within a population at a certain time. For instance, in 2003 in several Midwestern states, there were unusual numbers of cases of monkeypox. Monkeypox is caused by a virus that is harbored by rodents, in this case, prairie dogs. Most of the 70+ cases were traced to some direct or indirect association with prairie dogs that had come in contact with small mammals imported from Africa (CDC 2003). If the epidemic affects various populations over a large area, we usually refer to the epidemic as a *pandemic*.

In studying outbreaks of infectious diseases, epidemiologists are especially interested in the *agent*, the *origin*, the *reservoir*, and how they are spread (*mode of transmission*). The *origin* is easily defined but not always that easy to determine. The origin is where an epidemic begins. With SARS, the origin was Guangdong Province, China, in November 2002, as that was when and where the first confirmed case occurred. Scientists pursued the source of this virus, which had made the jump to human hosts. They searched for the *reservoir*, an organism or location that provides a continual source of the infectious agent. Initially they discovered suspicious viruses in civets (catlike mammals) found in animal markets of southern China. However, with no sustaining evidence that civets were acting as a reservoir, scientists eventually focused on the most likely reservoirs—bats. The civets and other exotic animals in the marketplace probably only served as conduits for the spread of the virus. The viruses located in the bats showed considerable genetic diversity, and those viruses in the population that displayed a highly modified gene for the S-glycoprotein were able to become established in human hosts. (S-glycoprotein is a protein that has short chains of sugars attached to it and is found embedded in the outer covering [envelope] of the virus. The protein binds to a receptor on susceptible host cells, promoting invasion of the host cell by the virus.) (UCMP 2006; Li et al 2006).

The *agent* for this disease was eventually determined to belong to the coronavirus family, designated SARS-CoV (Mydans 2003). SARS-CoV (SARS-CoV-1) is a single-stranded RNA virus never seen previously in human beings. Disease symptoms appear after 3–17 days of incubation and include fever, headache, dyspnea (difficulty breathing), and dry cough, and X-rays display features of pneumonia. Most people recover after about two weeks. The virus is spread by respiratory droplets from infected individuals in close contact with susceptible individuals. These droplets can also be transferred from sewage or nonliving objects/surfaces (*fomites*) (Heymann 2013). Other diseases, like cholera (bacterial), can be spread from contaminated water or food. In cholera, water and food are considered *vehicles*. A vehicle is an inanimate source for indirect transfer of a disease pathogen. Generally, vehicles are liquids, gases, and solids that enter our bodies for our physiological maintenance. Examples include air, water, and food.

A number of human diseases have human reservoirs. It is easy to determine some human reservoirs, because they are acutely symptomatic. In other words, someone with an active cold, who is sneezing and sniffling, harbors millions of cold viruses that they can share with susceptible individuals. Then we have *carriers* of the disease agent. In some cases, although an individual is going to become ill, the person may be contagious before showing symptoms. There are *subclinical carriers*, who may display light or no symptoms and do not proceed to the clinical stage of the disease. *Convalescent carriers* remain infectious as they recover from a disease, and *chronic carriers* remain infectious even after they have recovered.

Many diseases have environmental reservoirs. Remember that a reservoir is a continuing source of infectious agents. Examples of such reservoirs include soil and water.

QRQ 6. A *significant increase* in the occurrence of a disease within a population at a certain time is called a(n) ______.

QRQ 7. A *significant increase* in the occurrence of a disease affecting various populations over a large area is called a(n) ______.

QRQ 8. Select from the following the appropriate term for each description below: agent, carrier, fomite, host, mode of transmission, origin, reservoir, vehicle

a. an inanimate source for indirect transfer of a disease pathogen
b. an organism or location that provides a continual source of an infectious agent
c. hamburger, if you got food poisoning from eating it
d. human who is susceptible to the influenza virus or who the virus has infected
e. influenza virus A/Hong Kong/1/68 (H_3N_2), cause of the 1968 influenza pandemic
f. nonliving objects/surfaces that can spread a virus
g. where an epidemic begins

QRQ 9. Select which of the following are fomites for transmitting a cold virus: aspirin, cough, fever, handkerchief, runny nose, sneeze, tissue

QRQ 10. Characterize each of the following kind of carrier using the descriptions below: chronic, convalescent, infected, infectious, recovered, subclinical, susceptible

a. displays light or no symptoms and does not proceed to the disease's clinical stage
b. remains infectious as he or she recovers from a disease
c. remains infectious even after recovery

PoI 1. In 2002–2003, the SARS epidemic (SARS-CoV-1 virus) was contained within a fairly short time. The COVID-19 (SARS CoV-2) pandemic in 2020 has seen a related virus that is very difficult to contain.

a. Compare the SARS-CoV-1 and SARS-CoV-2 viruses, including transmission and symptoms.
b. Explain the statistics that say the *case fatality rate* for COVID-19 is 6 percent (0.06) and *infection fatality rate* is 1 percent (0.01).
c. Give the meaning of each term and its significance to COVID-19: silent hypoxemia (or hypoxia), cytokine storm

PoI 2.

a. SARS is spread primarily through respiratory droplets. Determine how each of the following infectious diseases are transmitted: typhoid, hepatitis B, measles, polio, dengue, HIV, Lyme disease, giardiasis
b. Knowing the mode of transmission, determine the *portal of entry*, or method of getting into the body, for each of the diseases above.

PoI 3. Let's consider one disease more thoroughly—*cholera*. We read periodically about cholera epidemics. The causative agent is a bacterium, *Vibrio cholera*. The method of transmission is by contaminated water or food. The disease is characterized by severe diarrhea, and WHO estimates that there are two to three million cases per year. Most cases can be treated without antibiotics using rehydration. Think about this disease and how it is spread.

a. What kinds of factors influence the *incidence* (number of newly diagnosed cases) of cholera in a population?
b. Knowing factors that promote cholera spread, can you suggest interventions that might be useful in preventing an epidemic?

PoI 4. Find the following information on West Nile encephalitis.

a. What are its symptoms?
b. What is the agent?
c. What is the primary reservoir?
d. What is its mode of transmission?
e. What factors influence its transmission?
f. What interventions might lower the incidence of West Nile virus?

PoI 5.

a. Find examples of disease agents that are harbored in the soil.
b. What kind of disease is associated with each example you provide?

PoI 6.

a. Find examples of disease agents that are harbored in water.
b. What kind of disease is associated with each example you provide?

8.4.2. Viral Pathogenicity

Pathogenicity is generally defined as the capacity of an agent to cause disease. *Virulence*, although it is often used interchangeably with pathogenicity, is a term used to describe the severity of the disease.

Completely dependent on host cells for replication and metabolism, viruses are often described as intracellular parasites or mobile genetic elements. Their entire mission is to reproduce successfully, and to do that viruses must attach, transfer their nucleic acid into the target cell, take charge of needed metabolic machinery of the host cell, and then synthesize the necessary nucleic acids and proteins to make more of themselves. In the process, they will either kill the host cell or impair its normal function.

Virulence in viruses depends on the dose, strain, proper portal of entry, and a variety of host factors. Genes affecting the virulence of a viral strain often impact replication potential, spreading between or within the hosts, overcoming host defense, or producing toxic products. Viruses may kill cells or induce *apoptosis* (programmed cell death), disrupt necessary cellular functions (e.g., membrane transport, cell cycle), damage host cell chromosomes, or prompt cytokine (small proteins that affect cell-cell interactions or communications) release and other immune responses to infected cells.

PoI 7.

a. The influenza pandemic of 1918 killed, by some estimates, more than 50 million people. Why was this virus so virulent?
b. What other factors contributed to this virus causing a pandemic?
c. In 2009 there was another H1N1 influenza epidemic that killed thousands, but it never reached the numbers encountered in 1918. Can you determine factors that may explain this?

8.4.3. Hosting a Disease-Causing Microbe

We have been referring to *susceptible* individuals, implying that a portion of a population is capable of becoming *infected* and getting sick from a particular disease agent. Susceptibility to a disease varies greatly with individuals. Human beings possess some natural body structures or functions that help to deter the establishment of an infection. For instance, the skin and the mucous membranes that line the lumen (inside space) of your internal organs act as physical barriers to disease agents. You have ciliary-mucus mechanisms in your respiratory tract, acid in your stomach, plus several reflex actions (e.g., coughing, sneezing, vomiting) that are also protective. Moreover, you have various types of nonspecific components of your immune system that offer important lines of protection. For example, you have white blood cells (e.g., macrophages, dendritic cells) that can phagocytize (engulf) and kill bacteria and other infectious agents.

Your body also has an arm of the immune system that protects you particularly against many kinds of pathogens. When *antigens* (soluble alien substances or surface molecules of pathogens) enter your body, this system is capable of producing *antibodies* that bind *specifically* to the antigen. Furthermore, this system can generate special types of white blood cells that react with the antigen. Essentially, microorganisms and soluble antigens are not products of your body and will provoke your immune system to react to neutralize the challenge. It is not necessary to elaborate on the details of this system here, but each person has their own unique immune potential. These differences are translated into variable degrees of susceptibility to disease.

Scientists are very interested in susceptibility factors that are genetically determined. For instance, there is a small segment of the population that is HIV positive, but the disease does not progress to AIDS. Why? We are still not certain, although some researchers think that it is a result of a defective cell receptor that prevents the virus from binding to target cells as easily. In some cases, we understand lack of susceptibility, such as with the sickle cell trait and malaria. Although the defective protein causes the body to produce fragile red blood cells, these fragile cells do not allow the malaria parasite to complete its life cycle. This gives heterozygous individuals (with one normal copy of the gene and one abnormal recessive copy), who do not exhibit the sickle cell trait, an advantage in areas where malaria is endemic. As we find out more about our genomes, we are likely to find more mechanisms that make us either more or less susceptible to various infectious diseases.

Just as individuals vary in their susceptibility, they also vary in their ability to recover from illness. Sometimes this ability results from the degree of virulence of the infectious agent, and sometimes it is related to the state of the person's immunity or general health. Consequently, *recovery rate* differs with the person and/or with severity of the disease. In essence, your body's defenses fight with pathogens. If your defenses win, you recover. Even when your defenses overcome the challenge, the body may be permanently damaged. If your defenses don't win, you die. So you have three general outcomes that are possible at the end of an illness—complete recovery, some degree of physical/physiological impairment, or death. Assuming you survive, you are quite often left with *immunity* to the specific pathogen that made you sick. You may only be fully immune to a particular strain of pathogen, so another strain could make you ill. We see this variation in immunity often with the annual strains of influenza. With some diseases, you do *not* gain immunity. Syphilis is an example of a disease you can have over and over again.

QRQ 11. Select from the following the appropriate term for each description below: antibody, antigen, apoptosis, immunity, infected, pathogenicity, recovery rate, susceptible, virulence

a. capacity of an agent to cause disease
b. describes the severity of a disease
c. inability to become sick from a specific pathogen
d. part of the immune system that binds to a specific antigen
e. programmed cell death
f. soluble, alien substance or surface molecule of a pathogen
g. someone capable of becoming infected and getting sick from a particular disease agent

PoI 8. Dengue is a disease that has variable effects on a victim.

a. Describe the possible effects of dengue.
b. Is there a vaccine against dengue?
c. Can you have dengue more than once?

8.5. Other Noncellular Agents of Disease

8.5.1. Prions

Prions (proteinaceous infectious particles) are protein agents that are associated with various rare neurodegenerative diseases of animals, including human animals. These proteins are characterized as abnormally shaped versions of a normal neuronal protein, which can induce the normal protein into assuming the abnormal shape. The abnormal proteins then formed clumps, which disrupt normal functioning of neurons, particularly in the brain (see Figure 8.7). The most commonly occurring human disease is Creutzfeldt–Jacob disease (C-JD), and the best known diseases in nonhuman animals are bovine spongiform encephalopathy (mad cow disease), scrapie in sheep, and chronic wasting disease in deer. All of these diseases are characterized by neuronal loss and a spongiform (spongy) appearance of brain tissues, long incubation periods, and no obvious inflammatory response.

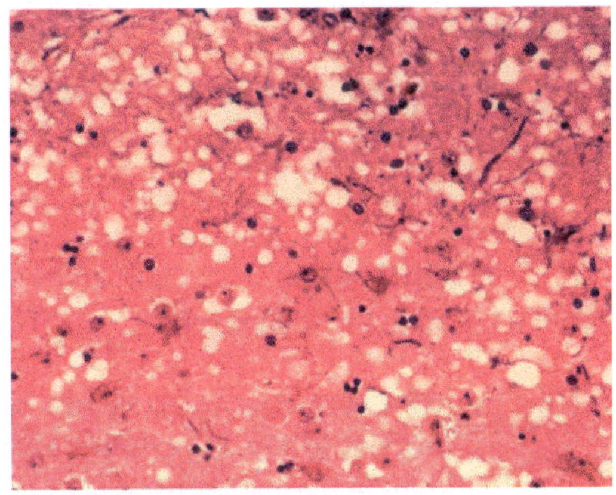

FIGURE 8.7 Brain section from a cow with bovine spongiform encephalopathy. Note spongy appearance (holes).

How are these disease agents spread? There are some forms of C–JD that seem to have a genetic influence (familial), but most cases are sporadic, normally caused by contamination from surgical equipment, particularly from corneal transplants. There is a variant C–JD that is acquired by eating contaminated meat. All forms of such human diseases are very rare, although with no really good treatment, and certainly no cure; prion diseases are always fatal. When mad cow disease is diagnosed in one cow, it will lead to the destruction of the entire herd.

After years of research, we are still learning about prions. A new disease, *multiple system atrophy*, has been connected with a new prion-like agent. In this case, they have identified the specific protein that is associated with this disease. The protein is in a number of tissue types and is abundant in the neurons of the brain. It is located at the tips of neurons, where chemicals (neurotransmitters) diffuse from one neuron to stimulate a nerve signal in the next neuron. Its specific role is being intensely investigated. The malfunction of this protein may be associated with a much more widely occurring disease, Parkinson's.

8.5.2. Viroids

Viroids are entities that have one living characteristic—they replicate, given a little help from a host plant cell. Composed of one small (250–400 nucleotides), single-stranded, circular piece of RNA, their base pairs do not code for any protein. They may invade the host leaf through aphids (insect *vector*), through mechanical injury, or from leaf contact with another plant that is contaminated. Viroids may attack some plants with economic significance, such as potatoes, avocados, citrus, and coconut palms. Some viroids replicate in the nucleus and others replicate in the chloroplasts. In either location, they primarily need the help of a host cell's RNA polymerases. A viroid infection may manifest itself mildly as inhibited growth, but it may also cause severe stunting, deformity, necrosis (plant parts darken and wilt), and chlorosis (lack of chlorophyll, visible as yellowing).

The question is, without any protein product, how does this pathogen induce disease? There is mounting evidence that the viroid, with the help of the host's enzyme system, operates by silencing sequences of the host's messenger RNA (mRNA). It does this using some of its own sequences to bind to sequences of the mRNA, targeting that sequence for degradation. This action effectively "silences" the host from gene coding that mRNA, as well as its protein product.

Little is known about viroids in animals. However, it is now thought that hepatitis D might be caused by a single-stranded RNA that does not seem to code for any proteins. This circular piece of RNA is termed a *virusoid*. They replicate and cause pathology only if a patient is coinfected with hepatitis B virus. Until we have more information, we will consider viroids plant pathogens.

QRQ 12. A proteinaceous infectious particle is a(n) ______.

QRQ 13. A(n) ______ is an entity that has one living characteristic—it replicates, given a little help from a host plant cell.

QRQ 14. A(n) ______ is an organism that transmits a disease agent from an infected host to a susceptible host. For example, a particular type of mosquito can bite someone with malaria and transfer the malarial parasite to a susceptible person.

QRQ 15. Associate each description or disease with either prion or viroid.

a. associated with various rare neurodegenerative diseases of animals
b. believed that it operates by silencing sequences of the host's messenger RNA
c. characterized by neuronal loss and a spongiform appearance of brain tissues, long incubation periods, and no obvious inflammatory response
d. composed of one small, single-stranded, circular piece of RNA
e. induces the normal protein into assuming the abnormal shape
f. mad cow disease
g. primarily needs the help of a host cell's RNA polymerases

Image Credits

Fig. 8.1: Copyright © by Database Center for Life Science (DBCLS) (CC BY 3.0) at https://commons.wikimedia.org/wiki/File:Bacteriophage.png.
Fig. 8.2: Source: https://commons.wikimedia.org/wiki/File:Ebola_virus_virion.jpg.
Fig. 8.3: Source: https://commons.wikimedia.org/wiki/File:Reef4432_(9718053032).jpg.
Fig. 8.5: Source: https://commons.wikimedia.org/wiki/File:Tupanvirus.jpeg.
Fig. 8.6a: Source: https://commons.wikimedia.org/wiki/File:Adenoviridae.jpg.
Fig. 8.6b: Source: https://commons.wikimedia.org/wiki/File:Varicella_(Chickenpox)_Virus_PHIL_1878_lores.jpg.
Fig. 8.6c: Source: https://commons.wikimedia.org/wiki/File:TMV_virus_under_magnification.jpg.
Fig. 8.6d: Copyright © by AFADadcADSasd (CC BY 4.0) at https://commons.wikimedia.org/wiki/File:Bacteriophage.jpg.
Fig. 8.7: Source: https://commons.wikimedia.org/wiki/File:Histology_bse.jpg.

LAB #8

Laboratory Exercises

8.6. Laboratory on the Spread of a Virus

8.6.1. Preliminary Work

In this laboratory, we consider "the transmission and perpetuation of a virus in a human population," where the dynamics of different diseases result in different outcomes. From the text's website, view the introductory video; download and open "*Virus Rev.nlogo.*"

1. Click the Info tab at the top of the page and read the sections "What Is It?" and "How It Works."
   a. Give the color (A. black, B. blue, C. brown, D. cyan, E. gray, F. green, G. orange, H. red, I. violet, J. white, K. yellow) of each model component:

      _______ healthy but susceptible to infection

      _______ sick and infectious

      _______ immune

   b. ______ Select the meaning of "carrying capacity": A. capacity of a population to carry a virus, B. maximum ability of a population to sustain a virus, C. maximum population size, D. maximum population size the environment can support indefinitely
   c. ______ For this model, give the carrying capacity.
   d. ______ For this model, give the state of a newborn: A. immune, B. infected, C. none, D. susceptible
   e. ______ Give a model-simplifying assumption stated in the subsection "Degree of Immunity": A. children have no immunity, B. degree of immunity varies, C. immunity is transient, D. immunity lasts a lifetime, E. no one can become immune
2. Read "How to Use It."
   a. ______ A tick represents a(n) **1.** (A. hour, B. day, C. week, D. month, E. year).
   b. ______ When switch STOP-AT-520 is on, the simulation will stop after about **1.** years and display output.

c. Table 8.1 displays information about three peaks that occur in a hypothetical graph of the number of those sick. Determine each of the following:

_______ MAX-SICK

_______ WHEN-MAX-SICK

_______ percent%MAX-SICK

TABLE 8.1. Hypothetical Peak Information in Graph of Those Sick

Week	Number Infected	Total Population
60	200	500
120	300	600
180	175	700

d. ______ Suppose initially 10 people are sick, and in the subsequent four weeks we have the following numbers of new cases: 2, 3, 1, 5. Give the cumulative total of those sick at the end of this time.

e. ______ percent If CHANCE-RECOVER is 25 percent, give the percent chance that an infection will end in death of the person.

f. ______ For transmission of the disease to occur, do an infected person and a susceptible person have to be in the same patch (S) or adjacent patches (A)?

g. ______ If the INFECTIOUSNESS slider is at 25 percent, an infected person must have approximately how many interactions with susceptible people to spread the virus?

h. ______ How many people are sick initially?

8.6.2. Laboratory Work

For each experiment below, run the simulation at least five times unless otherwise instructed. Labeling data appropriately, use a spreadsheet to record results and calculate averages. The following are default parameter values: STOP-AT-520 is ON, PEOPLE = 150, INFECTIOUSNESS = 65 percent, CHANCE-RECOVER = 75 percent, and DURATION = 20 weeks

3. In this question, we examine the spread of the disease for the default parameters, so use default slider bar values. Read the section "Things to Notice."

   a. To record data, start a spreadsheet called *Virus*, as in Table 8.2. Set up and run the simulation at least *five times* when the *virus does not disappear* from the population (i.e., when *SICK* does not become 0) and when the *first peak of illness is the highest* of the peaks (i.e. when MAX-SICK is height of the first peak). To obtain faster results, uncheck *view updates* or increase the simulation speed. Because tabs are separators for the data, we can copy the four output values at once and paste them into the spreadsheet, and the values will appear in different columns. Calculate and record the average of each column of data.

   _______ MAX-SICK

   _______ WHEN-MAX-SICK

_______ percent %MAX-SICK

_______ CUMULATIVE-SICK

TABLE 8.2. Start of Virus Lab Spreadsheet

Names				
Virus Lab				
	Question 3a			
	PEOPLE = 150			
	INFECTIOUSNESS = 65%			
	CHANCE-RECOVER = 75%			
	DURATION = 20 weeks			
	max-sick	**when-max-sick**	**%max-sick**	**cumulative-sick**

b. ______ Does the maximum number of sick usually occur before (B) or after (A) the total population reaches its carrying capacity?

c. ______ On average, approximately how many years (rounded to a whole number) does the simulation run before the number of sick reaches its first peak?

d. ______ T/F: Using the default parameters, an epidemic usually occurs.

e. ____________ Select all descriptions of how the number of healthy agents usually changes *before* the total population remains close to its carrying capacity: A. 0 for about the first half-year, B. alternates rising and falling, C. at some point falls below its initial number, D. mainly rises, though slowly at first, E. rapidly rises before leveling off, F. reaches a peak about when the population gets to its carrying capacity, G. rises slowly when it starts to increase

f. ____________ Even though the virus is present in the population, one explanation of why the number of healthy individuals rises initially is **1.** (A. births of susceptible individuals, B. deaths of sick, C, those that recover become susceptible again).

g. ____________ Select all descriptions of how the number of immune agents usually changes *before* the total population remains close to its carrying capacity. Use the same choices as part e.

h. ____________ Give all that could help explain why it takes so long for the number of immune to reach a significant level: A. initially not many people sick, B. number of sick people goes up and down, C. number of susceptible people goes up and down, D. takes about 20 weeks to recover

i. ______ Which population peaks first, sick (S) or immune (I)?

j. ______ One explanation of why the number of sick falls after an initial peak is that there are not as many **1.** (A. healthy, B. immune, C. people).

k. ______ Setup and run the simulation several times with the default slider values. Can you obtain a graph where the second wave of the number of sick is higher than the first wave, no (N) or yes (Y)?

l. **1.** ______ **2.** ______ **3.** ______ **4.** ______ Turn off STOP-AT-520. Repeat the simulation several times, allowing it to run beyond the time in which the total population stabilizes, say for 1,000 weeks. The graphs are **1.** (A. cyclic, B. decreasing, C. increasing), with the graph of the number of those who are **2.** (D. healthy, E. immune, F. sick) higher than the graph of the number of those who are **3.** (D. healthy, E. immune, F. sick), which is higher than the graph of the number of those who are **4.** (D. healthy, E. immune, F. sick).

m. Explain the results of part l.

4. In this question, we consider the impact of a higher initial population density. Use default parameters, except change the population density to 300 people.

   a. ______ Run the simulation. Do the shapes of the graphs generally change in comparison to those of question 3, no (N) or yes (Y)?

   b. ______ Do the relative positions of the graphs change from those of question 3, no (N) or yes (Y)?

   c. Repeat question 3a. In your spreadsheet, be sure to update the list of parameter values to indicate PEOPLE = 300. Give the means of the output.

      ________ MAX-SICK

      ________ WHEN-MAX-SICK

      ________ percent %MAX-SICK

      ________ CUMULATIVE-SICK

   d. **1.** ______ **2.** ______ **3.** ______ **4.** ______ With initially 300 instead of 150 individuals, the maximum number of sick at any one time is **1.** (A. about the same, B. larger, C. smaller) and occurs **2.** (D. about the same time, E. earlier, F. later), and the percentage of the population infectious at that time is **3.** (A. about the same, B. larger, C. smaller). Moreover, the cumulative number of sick is **4.** (A. about the same, B. larger, C. smaller).

   e. Explain these results in comparison to those of question 3, where the initial population density was half as much.

5. In this question, we consider the impact of how contagious the disease is. Use default parameters.

   a. ______ Complete the following hypothesis: As INFECTIOUSNESS increases, CUMULATIVE-SICK **1.** (A. decreases, B. increases, C. remains the same, D. varies) at week 520.

   b. **1.** ______ **2.** ______ **3.** ______ Set INFECTIOUSNESS to 10 percent, so that the disease does not spread easily, and turn off STOP-AT-520. Using the slider at the top of the screen, change the speed from normal to faster and allow the simulation to run beyond the time in which the total population stabilizes, say for 1,000 weeks. Repeat the simulation several times where the disease does not disappear. The graphs are **1.** (A. cyclic, B. decreasing, C. increasing), with the graph of the number of those who are **2.** (D. healthy, E. immune, F. sick) much higher than the other two graphs. The two lower graphs follow a similar track, but each peak in the graph of the number of **3.** (D. healthy, E. immune, F. sick) occurs before a peak in the other graph.

   c. ______ Repeat question 3a using default parameters, except INFECTIOUSNESS = 40 percent. In your spreadsheet, update the list of parameter values appropriately. In simulations where the disease does not die out, do significant numbers of people get sick before the population researches its carrying capacity, no (N) or yes (Y)?

d. Give the means of the output.

_______ MAX-SICK

_______ WHEN-MAX-SICK

_______ percent %MAX-SICK

_______ CUMULATIVE-SICK

e. **1.** ______ **2.** ______ **3.** ______ **4.** ______ When INFECTIOUSNESS is 40 percent, the graphs are **1.** (A. cyclic, B. decreasing, C. increasing). As time progresses, the graph of the number of those who are **2.** (D. healthy, E. immune, F. sick) is higher than the graph of the number of those who are **3.** (D. healthy, E. immune, F. sick), which is higher than the graph of the number of those who are **4.** (D. healthy, E. immune, F. sick).

f. **1.** ______ **2.** ______ **3.** ______ **4.** ______ Complete the following statement comparing these answers in part d (INFECTIOUSNESS = 40 percent) to those of question 3a (INFECTIOUSNESS is 65 percent) when an epidemic occurs and the highest peak is the first one: With INFECTIOUSNESS being 40 percent instead of 65 percent, so that the infection does not spread as readily, the maximum number of sick at any one time is **1.** (A. about the same [within 3], B. larger, C. smaller) and occurs **2.** (D. about the same time, E. earlier, F. later), and the percentage of the population infectious at that time is **3.** (A. about the same [within 3], B. larger, C. smaller). Moreover, the cumulative number of sick is **4.** (A. about the same [within 3], B. larger, C. smaller) at ten years.

g. What, if anything, surprises you about the answers to part e?

h. __________ Change INFECTIOUSNESS to 90 percent and repeat the experiment. However, consider results even if the maximum number of sick does not occur on the first peak. Only record the values for CUMULATIVE-SICK in your spreadsheet, and give the mean of these data here.

i. ______ Do the eventual relative positions of the graphs change from those in question 3l, when INFECTIOUSNESS was 65 percent, no (N) or yes (Y)?

j. __________ Have your spreadsheet plot CUMULATIVE-SICK versus INFECTIOUSNESS for INFECTIOUSNESS = 40 percent (question 5d), 65 percent (question 3a), and 90 percent (question 5h) and select all statements descriptive of the graph: A. almost straight, B. concave down strongly, C. concave up strongly, D. cyclic, E. decreasing, F. horizontal, G. increasing

k. ______ Do the results support your hypothesis in part a?

l. Explain your results.

m. ______ Uncheck *view updates* or increase the speed of the simulation. In situations when the disease does not die out, find a value for INFECTIOUSNESS to the nearest 5 percent (e.g., 5 percent, 10 percent, 15 percent, etc.) in which, eventually, the number of individuals in each state is approximately the same constant number.

6. In this question, we examine the impact of the duration that someone has the disease on the total number of people contracting the disease. Use default parameters.

a. ______ Complete the following hypothesis: As DURATION increases, CUMULATIVE-SICK **1.** (A. decreases, B. increases, C. remains the same, D. varies) at week 520.

b. ______ Change DURATION to one week and run the simulation several times. Does the infection usually die out, no (N) or yes (Y)?

c. Explain why.

d. ____________ With DURATION = 5 weeks, repeat the experiment of question 3a for situations where an epidemic occurs and the maximum number of sick at any one time occurs on the first peak. Record only CUMULATIVE-SICK. Give the mean of the five data values here.

e. ____________ Repeat part d using 10 weeks.

f. ____________ Repeat part d using 15 weeks.

g. ____________ Repeat part d using 40 weeks.

h. **1.** ______ **2.** ______ **3.** ______ **4.** ______ Complete the description of the graphs when DURATION is 40 weeks: The graphs are **1.** (A. cyclic, B. decreasing, C. increasing). As time progresses, the graph of the number of those who are **2.** (D. healthy, E. immune, F. sick) is higher than the graph of the number of those who are **3.** (D. healthy, E. immune, F. sick), which is much higher than the graph of the number of those who are **4.** (D. healthy, E. immune, F. sick).

i. ____________ Have your spreadsheet plot CUMULATIVE-SICK versus DURATION for DURATION = 5 weeks (question 6d), 10 weeks (question 6e), 15 weeks (question 6f), 20 weeks (question 3a), and 40 weeks (question 6g) and select all statements descriptive of the graph: A. almost straight, B. concave down strongly, C. concave up strongly, D. cyclic, E. decreasing, F. horizontal, G. increasing

j. Discuss the results in this question.

7. In this question, we consider scenarios appropriate for several diseases.

a. ______ In central Africa, Ebola is a deadly disease that is highly infectious but usually has a very short duration. Consider a similar disease with the following slider values: STOP-AT-520 = ON, PEOPLE = 150, INFECTIOUSNESS = 90 percent, CHANCE-RECOVER = 5 percent, DURATION = 2 weeks. Run the simulation several times. Does the disease ever reach the epidemic stage, no (N) or yes (Y)?

b. Discuss the results.

c. ______ The HIV virus causes AIDS. The virus endures a long time but has a very low infectiousness value. Without proper drug treatment, the recovery rate is extremely low. Consider a similar disease without drug treatment using the following slider values: STOP-AT-520 = OFF, PEOPLE = 150, INFECTIOUSNESS = 4 percent, CHANCE-RECOVER = 0 percent, DURATION = 99 weeks. Run the simulation several times. True (T) or false (F): For situations where the disease does not die out, the number of sick gradually grows and then oscillates, and the disease persists in the population.

UNIT III

Genetics/Genome

MODULE

9 Cell Reproduction and Genetics

9.1. Cell Division

Can you imagine a giant pumpkin (actually a type of squash) weighing 2,624.6 pounds? Well, this German fruit is currently the world record holder and weighs more than a Volkswagen Beetle! It is even harder to imagine that this behemoth started out as a single-celled *zygote* (fertilized egg). How did it grow to such a gigantic size? The answer to the giant's success is the process of cell division—and some tender care.

Cell division is the biological process by which a single parent cell divides into two or more usually identical daughter cells. In many prokaryotic organisms, such as bacteria, *binary fission* is a simple form of cell division in which the genetic material is replicated and positioned to each end of the cell prior to the division of the cell. Binary fission can be a powerful process, as evidenced by the speed at which some bacteria can quickly double their numbers. This is particularly concerning in the spread of bacterial infections and food poisoning. In eukaryotic organisms, such as plants, fungi, and animals, a cell duplicates its contents, including the genetic material, and usually splits into two identical cells. This process is described as the *cell cycle.* It is fundamentally important in the growth and repair in eukaryotes. Cancer can result when this process is not regulated and becomes out of control.

9.1.1. Developing an Understanding of Cell Division

In order to view cells, a microscope is essential, and the majority of biological historians credit the Dutch lens and spectacle maker *Zacharias Janssen* (circa 1585–1638) with the invention of the compound microscope. His scope literally opened up the world of cells to keen observers. The British scientist *Robert Hooke* (1635–1703) changed our view of life when he formally observed and gave cells their name in his classic work *Micrographia* in 1665. In fact, his work had a great influence on the Dutch surveyor, wine assayer, fabric merchant, and microscopist *Anthony van Leeuwenhoek* (1624–1723), known as the "Father of the Microscope." Although Leeuwenhoek did not invent the microscope, he popularized the use of the microscope by princes and paupers alike. His enthusiasm for his work is reflected in this quote about his observations of pond water: "the motion of most of these little animalcules in the water was so swift and varied—upward, downward, and round-about—that 'twas wonderful to see" (Dobnell, 1958). His observations and drawings were an inspiration for generations of microscopists.

During the 1800s, knowledge of the cell grew in leaps and bounds. In 1831, the British scientist *Robert Brown* described the nucleus, and other scientists unraveled the mysteries of the cell. During this century, the *cell theory* had its origins as well. After observing plants, in 1838, the German botanist *Matthias Schleiden* concluded that all plants are composed of cells, and, not to be outdone, the German zoologist *Theodor Schwann* stated in 1839 that all animals are made of cells. Thus, as the result of these two scientists, the cell theory was proposed. The first two tenets of the cell theory state that all living things are composed of cells and that the cell is the basic unit of structure and function of all living things. The third tenet of the cell theory was proposed in 1855 by the German physician and scientist *Rudolf Virchow.* It stated that all cells stem from preexisting cells. From the cell theory, there have emerged other important principles about how cells operate, including how they reproduce during cell division (National Geographic Society 2019).

APPLYING SCIENCE TO LIFE 1 A Special Place in the History of Biology: Rudolf Virchow

In the history of science and medicine, the name Rudolf Virchow stands as a giant. Virchow was a multitalented physician, pathologist, biologist, writer, and public health and social reformer. He is considered by many to be the most influential physician in history.

Rudolf Virchow was an eminent physician who rejected medical practices not based on science. In cellular pathology, Virchow pointed out that specific cells are responsible for disease, not the whole organism. This observation gave medical practitioners a more defined look at the cause and nature of disease and allowed physicians to provide more precise diagnosis, treatment, and prognosis of disease. In medicine, Virchow was the first person to describe disorders and diseases such as leukemia, spina bifida, thrombosis, and embolism. Furthermore, he was the first scientist to correctly recognize the origin of cancerous cells arising from normal cells.

Virchow added the final tenet to the cell theory, which was that cells stem from preexisting cells. Prior to his observation, many misconceptions regarding the origin of cells had flourished. Virchow was also a prolific writer, composing over 2,000 scientific papers and books that had a tremendous influence upon medical issues, public health practices, and society. He advocated social changes in medicine that are still used today.

Virchow coined the term *zoonosis*, or *zoonotic disease*. These diseases are those spread between animals and humans. Presently, scientists estimate that more than six in 10 infectious diseases are spread from animals to humans and three in four emerging infectious diseases in humans are spread by animals. Examples of zoonotic diseases include Ebola, rabies, Lyme disease, malaria, toxoplasmosis, ringworm, and hookworms.

Virchow's interest in zoonosis led to one of the strangest duels in history. In 1865, the Minister President of Prussia (Germany), Otto von Bismarck, declared that funding to the navy should be increased. Rudolf Virchow was the political leader of the Progressive Party and opposed Bismarck. At the time, the parasitic infection *trichinosis* was devastating the German populous, as poorly processed pork sausage was an important part of the diet. Trichinosis, a potentially fatal disease, is caused by the parasite *Trichinella spirilis*, a type of roundworm. Virchow noticed these tiny worms encysted in the muscles of infected people and concluded that the disease was spread by eating pork infected with these worms. He insisted that new reforms regarding the processing of food should be introduced. Bismarck refused to listen to Virchow's ideas, and a serious argument erupted. As a result of the argument, Bismarck challenged Virchow to a duel. A clever Virchow stated that he had the right to choose the weapon, and his choice would be sausage! In this case, one of the sausages would be prepared in sanitary conditions and the other in the typical unsanitary manner. Bismarck was given first choice in the meaty delights, and Virchow would eat the remaining sausage. Faced with this choice, Bismarck backed down, and Virchow was able to reform the sausage industry and end "the great sausage duel" (Schultz 2008).

Virchow was not always correct. He rejected the *germ theory of disease* (the idea that germs can lead to disease) and did not believe that the hand-cleaning process was necessary when delivering babies.

PoI 1. Rudolf Virchow was the first to use each of the following biological terms. For each, write a paragraph describing the term's meaning and its biological context:

a. neuroglia
b. parenchyma
c. osteoid

PoI 2. Rudolf Virchow was wrong about some biological concepts. For each listed here, find out what is correct:

a. He incorrectly stated that malaria was caused by rotting plant matter.
b. He incorrectly thought that typhus was caused by rotting animal matter.
c. He stated incorrectly that the newly discovered Neanderthal skeleton was actually a *Homo sapiens* and not ancient at all. He thought that the skeleton was disfigured by rickets and the head flattened by hard blows.

9.1.2. Binary Fission

Asexual reproduction results in the formation of new cells that are derived from a parent cell and do not involve the fusion of sex cells (*gametes*). There are several means of asexual reproduction, including budding, fragmentation, and making cuttings in plants (vegetative reproduction). However, *binary fission* is the simplest and most common strategy of asexual reproduction in bacteria and many one-celled protists, such as the amoeba.

Bacterial binary fission is relatively simple. The process of binary fission ultimately results in the formation of two identical cells with basically the same genetic information as the parent cell. First, the parent cell increases in size prior to binary fission. In the majority of bacteria, the genetic material is housed in a single circular DNA molecule, located in the *nucleoid* region and surrounded by cytoplasm (see Figure 9.1). The DNA molecule is tightly packed in this single chromosome around the middle of the cell. The chromosome begins to make a copy of itself and eventually forms two DNA molecules. These new molecules form new nucleoids and segregate to opposite ends of the cell. Finally, a *septum* (partition) forms, dividing the parent cell into two identical cells. These cells begin to function independently once the cell wall is completed. The actual division of the original cell into new cells is known as *cytokinesis*. Binary fission can occur very rapidly. Typically, the bacterium *Escherichia coli* can double its number every 20 minutes under optimum conditions. This rapid means of reproduction allowed the first prokaryotes to populate the barren earth nearly 3.5 billion years ago and permits pathogenic bacteria to overwhelm a victim rapidly. Periodic genetic mutations can change the *phenotype* (observable traits) of the bacteria and can bring about evolutionary change quickly. The ability of bacteria to resist antibiotics is a testimony to the power of these mutations and binary fission.

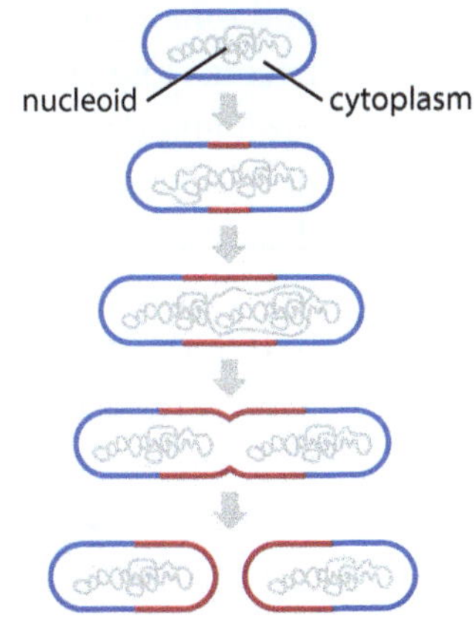

FIGURE 9.1 Binary fission in a bacterial cell

QRQ 1. Select all that apply to asexual reproduction: in many eukaryotic organisms, in many prokaryotic organisms, involves the fusion of sex cells, most common strategy is binary fission, most common strategy is binary fusion, results in the formation of new cells that are derived from a parent cell, results in two cells that have different genetic information than the parent cell, simple form of cell division

QRQ 2. Select all that are a part of cell theory: all cells stem from preexisting cells; cells divide; for all living things, the cell is the basic unit of function; germs can lead to disease; living things are composed of cells, as Robert Hooke stated; sexual reproduction involves fertilization of an egg; the cell is the basic unit of structure of all living things; viruses do not have cells

QRQ 3. Select from the following the appropriate term for each description below: cytokinesis, cytokinin, cytoplasm, gamete, nucleoid, nucleotide, nucleus, phenotype, schism, septum, zoonosis, zygote

a. prokaryotic cell's central region with no membrane that contains genetic material
b. division of an original cell into new cells
c. fertilized egg
d. observable traits
e. partition
f. sex cell

9.2. The Eukaryotic Chromosome

Whereas the prokaryotic chromosome and binary fission are relatively simple, cell division in eukaryotic organisms is more complex, involving many chromosomes. Our knowledge of chromosomes can be traced back to the development of *cytogenetics* (the relationship between chromosomes and inheritance) during the mid to late 19th century. Perhaps the first observation of what was to become known as chromosomes was made by *Karl Wilhelm von Nägeli* in 1842 in plant cells. *Walther Flemming* described "colored bodies," or chromosomes, while observing salamander embryological development. Later, *Theodor Boveri*, while working with the roundworm *Ascaris* and sea urchins, described events during cell division, and biologist *Walter Sutton* linked *chromosomal theory* (chromosomes are the vehicles of heredity) to the classic genetic works of Gregor Mendel. In the first half of the 20th century, the American geneticist *Thomas Hunt Morgan* and his colleagues linked genetic traits to the behavior of chromosomes. Today, chromosomal theory, as developed by these pioneers and others, is fundamental for modern genetics (O'Connor and Miko 2008).

Whereas a *prokaryotic chromosome* is usually a circular DNA molecule housed in the cytoplasm, eukaryotic cells have multiple *chromosomes*, each a compressed, single, uninterrupted molecule of DNA associated with proteins that function to compact the chromosomes. With so much genetic information in eukaryotic cells, the DNA far exceeds the size of the nucleus. Thus, the DNA is divided into several chromosomes, tightly compacted to fit into the nucleus. In cells other than sex cells, there are pairs of all chromosomes. Chromosomes play a fundamental role in ensuring that DNA is accurately copied and distributed to new cells. *Chromatin*, composed of approximately 60 percent proteins and 40 percent DNA, makes up the chromosomes. There are two types of chromatin in eukaryotic chromosomes. *Heterochromatin*, or inactive chromatin, is highly condensed and important in maintaining the structural and functional integrity of chromosomes, while *euchromatin*, or genetically active chromatin, is lightly condensed.

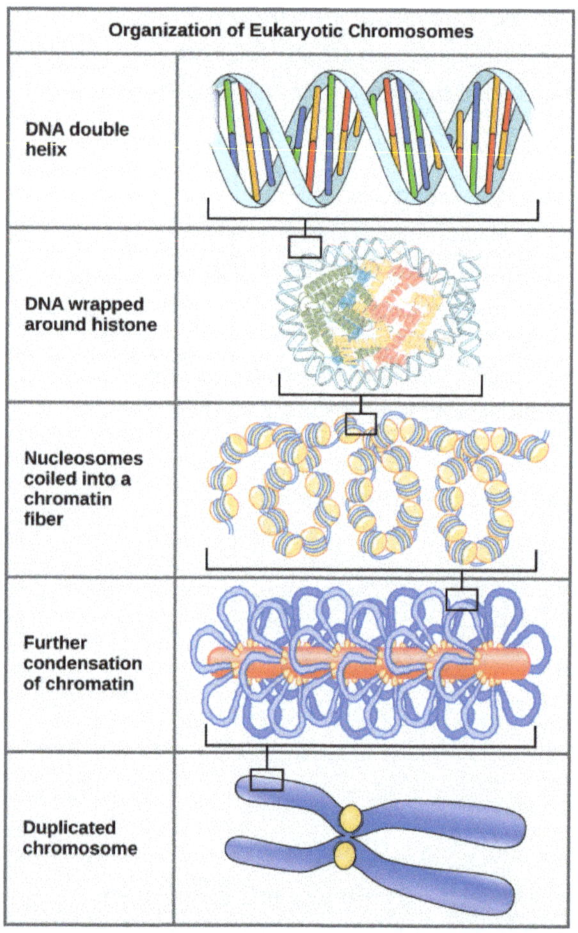

FIGURE 9.2 Organization of eukaryotic chromosomes

Humans have 46 chromosomes (23 pairs), each consisting of a single linear molecule of DNA. The length of DNA is far greater than the size of the nucleus. In order to fit within the nucleus, the DNA of each chromosome has to be tightly *compacted*, or packaged, in order to form a chromosome. For instance, the DNA molecule is wrapped around a core of eight *histone* proteins forming a *nucleosome*. These structures are "bead-like" in appearance and occur along the entire length of a chromosome. The nucleosomes may cluster, and chromatin may form looped domains to increase compaction (see Figures 9.2 and 9.3).

The size and shape of chromosomes vary in different species, but their fundamental anatomy remains consistent. Every chromosome possesses a constricted region of heterochromatin, or *centromere*, that divides the chromosome into arms of varying length (see Figure 9.3). The location of the centromere is used in classifying chromosomes and in describing the location of specific *genes* (units of heredity). The centromere also plays an integral role in the process of cell division. *Telomeres* are the regions of chromatin located at the terminal portion of a linear chromosome (see Figure 9.3). They are very important in the protection and stability of the chromosome. Telomeres are comprised of heterochromatin, featuring repeated sequences of DNA.

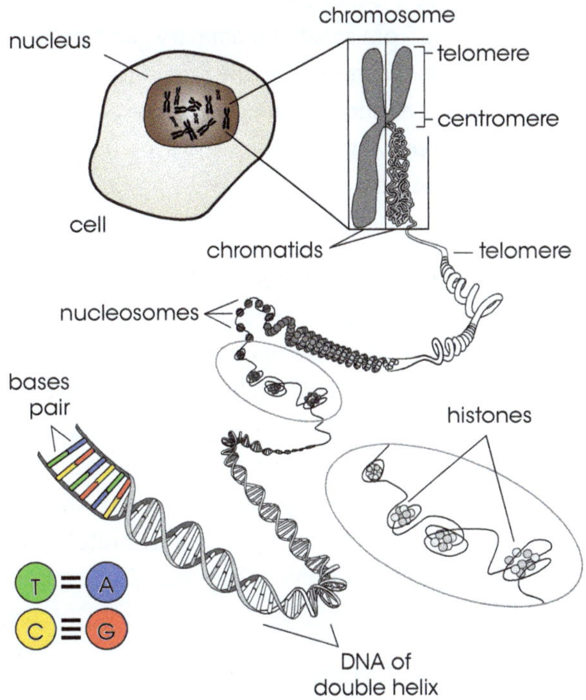

FIGURE 9.3 Basic chromosomal anatomy

APPLYING SCIENCE TO LIFE 2 Telomerase: The Fountain of Youth?

Sometimes it is hard for humans to accept the fact that we are mortal. Thus, throughout history there have been myths, stories of quests, and tall tales about the never-ending search for some magical elixir of life or fountain of youth that will extend our lives. In recent years, modern genetics is providing us with yet another search for the proverbial "fountain of youth."

Telomeres are structures that exist on the ends of chromosomes, serving to protect the DNA molecule beneath them from damage. They can be compared to the plastic caps at the end of a shoestring (aglet) that prevents fraying and damage to the shoestring. Originally, telomeres consist of approximately 100 to 1,000 repeats of the nucleotide sequence TTAGGG. However, after each division, the number of nucleotides is reduced, shortening the telomere. Cellular aging research has established that telomeres serve as a type of biological clock that, as cells divide, can bring on cellular *senescence* (inability to divide or grow). Scientists have discovered that many human cells can divide approximately 50 times before cells stop replicating and senescence occurs. The number of replications is called the *Hayflick limit*.

Enter *telomerase*, an enzyme that is capable of repairing telomeres. Certain stem cells and sex cells that must divide more than normal cells are influenced by telomerase. This enzyme serves as the mechanism that allows cells to undergo prolific divisions. Unfortunately, cancer cells also utilize telomerase to continue their divisions and form tumors. In order to understand how telomerase works in cell rejuvenation and cancer, it was necessary to map the telomerase gene. In 2013, international researchers at the University of Copenhagen announced the mapping of this all-important gene.

The mapping of the telomerase gene has led to a better understanding of cellular aging, cancer, and other disorders, such as Alzheimer's disease, type 2 diabetes, high blood pressure, and hardening of the arteries. Such studies may also lead to producing cells for transplantation, producing insulin, treating forms of muscular dystrophy and arthritis, and aiding in burn therapy and wound care. The understanding and treatment of many forms of cancer, including pancreatic, lung, and kidney cancer, could be aided by telomerase research. Already, researchers have devised methods to block telomerase in human prostate and breast cancer with promise. However, these therapies could impact wound healing, fertility, the immune response, and the formation of blood cells. Also, keep in mind that cloning can be impacted by telomeric aging. It is thought that the lethal complications of Dolly, the cloned sheep, were the result of telomeric aging from an adult mammary gland cell.

In the meantime, the scientific community will continue to make great strides in dealing with chronic illnesses and extending our life expectancy. Telomerase studies will certainly be an important contributor in improving our health and lives, but do not hold your breath about that magical "fountain of youth" just yet.

In cellular genetics, *ploidy* refers to the number of complete sets of chromosomes existing within a cell. A typical *sex cell* (*gamete*, or *sperm* and *ovum*) is considered *haploid* (N), since it contains just one complete set of chromosomes. The *haploid number* also reflects how many pairs of chromosomes are within a cell. In humans, the first 22 pairs are known as *autosomes*, and the two chromosomes making up the 23rd pair are known as the *sex chromosomes*. During reproduction, two haploid gametes unite to form a *diploid (2N)* cell, called the *zygote*. The zygote, as well as the *somatic*, or body, cells of most organisms, exists in the diploid state, containing two complete sets of chromosomes (one set from each parental gamete). The matching chromosomes in each pair are known as *homologous chromosomes*. The haploid number for human sex cells is 23, and the diploid number for cells other than sex cells is 46. The haploid and diploid numbers of the pineapple are 25 and 50, whereas they are 22 and 44 for the blue whale. Notice that the size and complexity of an organism is not determined by the number of chromosomes.

PoI 3. For each of the following organisms, give the diploid number (2N) and haploid number (N).

a. adder's tongue fern
b. corn
c. green pea
d. fruit fly
e. bald eagle
f. tiger
g. goldfish
h. horse
i. wooly mammoth
j. great white shark

Polyploidy is a heritable chromosomal state involving three or more copies of chromosomes in an organism. The condition is more common in fungi and plants than animals. Usually in animals, such as birds and mammals, polyploidy is detrimental to survival and, therefore, extremely rare. However, some invertebrates, fish, and amphibians exist in this state. Organisms can have many variations of polyploidy, including three, four, or even 64 sets of chromosomes. Botanists estimate that between 50 and 75 percent of flowering plants, 80 percent of grasses, and over 99 percent of ferns are polyploid. Plant polyploids include cotton, potatoes, and peanuts (4N); pansies (8N); and kiwis and chrysanthemums (6N). *Hybridization* (crossing different species), cell division complications (physical damage), exposure to chemicals, or other variables can lead to polyploidy. Polyploids may exhibit hybrid vigor (desirable or favorable traits), increasing the chance of desirable traits, and diversified gene function (how genes work), known as gene redundancy (multiple genes with the same function), leading to opportunities in evolutionary selection. However, polyploidy presents problems in cell division and difficulty accommodating the increased volume of genetic material in the nucleus (Woodhouse, Burkart-Waco, and Comai 2009).

APPLYING SCIENCE TO LIFE 3 The Amazing Human Chromosome Two

In the mid 1800s, Charles Darwin boldly predicted that humans and the great apes, including chimpanzees, bonobos, gorillas, and orangutans, share a common ancestor in geologic time. Through the years, comparative anatomy and fossils were the major evidence supporting Darwin's prediction. Bone for bone and muscle for muscle, scientists found no major difference between humans and the great apes. More recent genomic analysis has revealed that chimpanzees and humans vary by less than 2 percent of their DNA. Now to the bone-for-bone, muscle-for-muscle tenant, we can add molecule-for-molecule to support Darwin's conjecture.

Although genetic evidence overwhelmingly supported Darwin's idea, one problem existed. If humans and the great apes shared a common ancestor, why did humans have 46 chromosomes and the great apes have 48? Scientists hypothesized, perhaps humans have only 46 as the result of the fusion of two ape chromosomes. After much diligent testing and research, several scientific articles addressed the findings of the investigations. The scientists concluded that human chromosome two is the result of the fusion of two separate chromosomes that remain unfused in the great apes. Cytogeneticists have shown that regions of human chromosome two match separate chromosomes in great apes, specifically, chromosomes 12 and 13 of chimpanzees. Researchers have identified hypervariable regions of chromosome 2 that may have led to altered proteins unique to humans. Scientists have hypothesized that this fusion of chromosomes must have also involved a loss or rearrangement of part of the genetic material. Investigators also identified the exact location of a second inactivated centromere from another chromosome that is flanked by an abundance of a type of DNA-segmented duplication. Finally, although telomeres appear at the ends of chromosomes, on chromosome two, an extra telomere appears near the center of the chromosome. Research continues on this fascinating subject that will answer many age-old questions, including questions regarding our evolutionary split from the great apes (PBS 2007; Rejón 2017).

QRQ 4. Give a cell's diploid number if its haploid number is ______.

QRQ 5. How many pairs of chromosomes are in human somatic cells?

QRQ 6. (T/F) The number of chromosomes determines an organism's size and complexity.

QRQ 7. A prokaryotic chromosome is ______, while a eukaryotic chromosome is ______. Answer choices: usually a circular DNA molecule; a compacted, single, long, uninterrupted molecule of DNA; double-stranded DNA that is helix-shaped; a compacted, single, long, uninterrupted molecule of RNA; divided into arms, possibly of varying length; long, possibly interrupted molecule of DNA; a dendritic polymer

QRQ 8. Chromosomes are composed of the material ______, which is approximately 60 percent proteins and 40 percent DNA.

QRQ 9. The region at a terminal portion of a linear chromosome and containing repeated sequences of DNA is called the ______.

QRQ 10. A human cell can divide approximately ______ times before senescence

QRQ 11. Select from the following the appropriate term for each description below: centrale, centromere, chromosomal theory, diploid, euchromatin, haploid, Hayflick limit, heredity theory, heterochromatin, homologous, hybridization, mortality, nucleolus, nucleosome, ploidy, polyploidy, quiescence, senescence, somatic, telomerase

a. active chromatin
b. chromosomes are the vehicles of heredity
c. constricted region that divides a chromosome into arms of varying length
d. crossing different species
e. enzyme capable of repairing telomeres
f. heritable chromosomal state involving three or more copies of chromosomes
g. highly condensed; helps maintain chromosome's structural and functional integrity
h. number of chromosome pairs within a cell
i. inability of a cell to divide or grow
j. inactive chromatin
k. lightly condensed; houses genes
l. matching chromosome in each pair
m. number of complete sets of chromosomes existing within a cell
n. number of times a human cell can divide before death
o. segment of DNA coiled around a core of histones
p. type for a non-sex chromosome

9.3. The Cell Cycle

In addition to describing chromosomes, in the 1880s, *Walther Flemming* coined the term *mitosis* (*mitos* = "thread" in Greek) while observing the shape of chromosomes during cell division in salamander eggs. Botanist *Eduard Strasburger* also described this process in plants, and for years, the term *mitosis* was synonymous with cell division. However, today mitosis is recognized as a major part of the more complex cell cycle.

The cells of eukaryotic organisms pass through a programmed sequence of divisional events, called the *cell cycle*, which is essential in development, repair, maintenance, and growth. The cell cycle is regarded as the dynamic cellular process characterized by condensed chromosomes migrating to the central region of a cell and segregating into two identical daughter cells. Fundamentally, the cell cycle is controlled by chemical factors, including hormones, growth factors, and other signaling molecules. The frequency of cell division is highly variable in organisms. For example, in humans, skin cells and cells lining the digestive system are continuously dividing, and nerve and muscle cells do not divide at all in a mature human body. Regardless of the how often it occurs, the cell cycle has four distinct stages, G_1, S, G_2, and M, discussed below (see Figure 9.4).

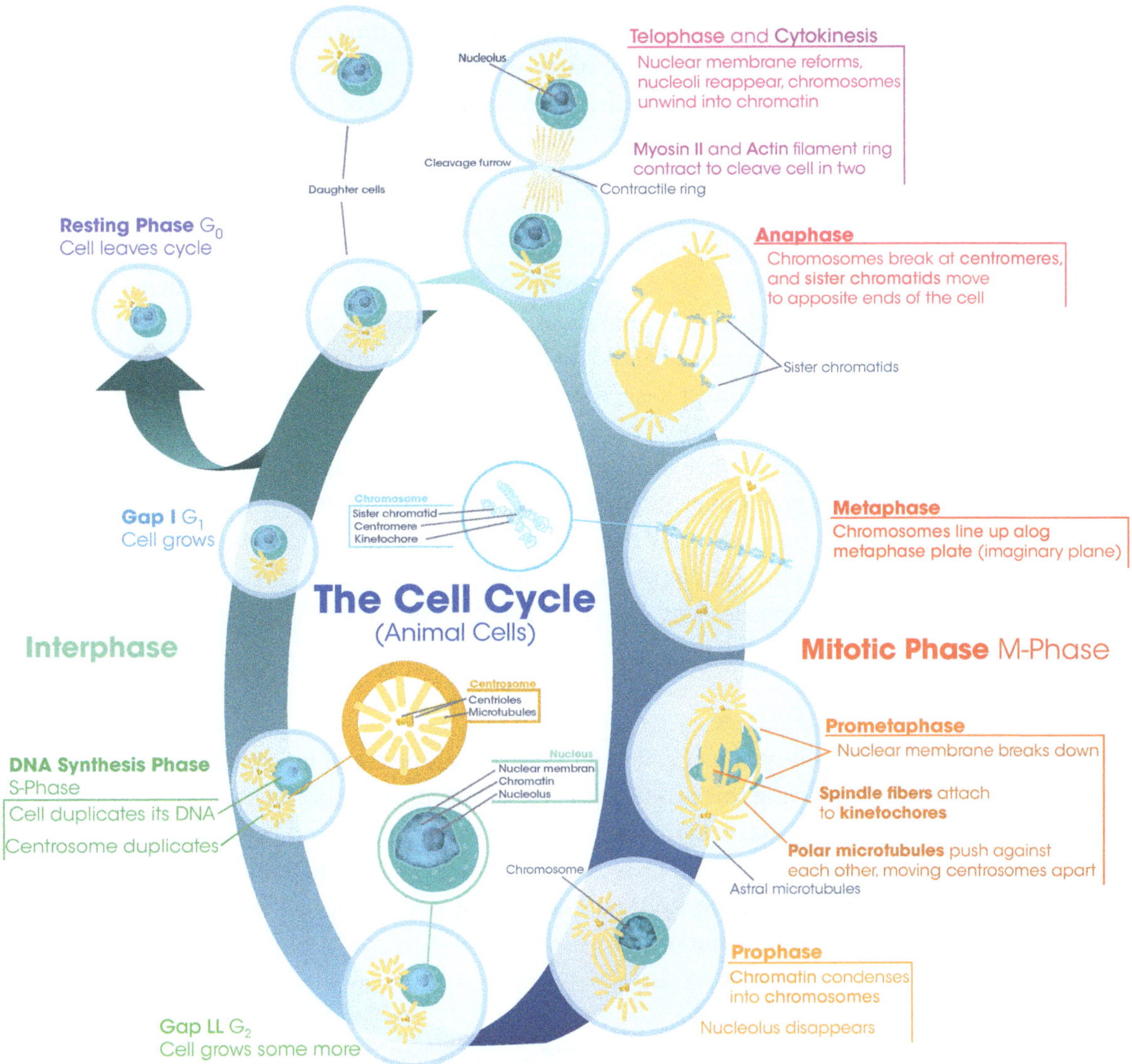

FIGURE 9.4 Cell cycle

9.3.1. Cell Cycle Phases 1–3: Interphase

Cell Cycle Phase 1: G_1 (Gap 1)

The *interphase* encompasses the first three phases of the cell cycle. In the early days of studying the cell cycle, this stage was known as the resting stage, because with limited observational power, no cellular activity was denoted. Presently, interphase is considered as extremely active, particularly at the molecular level. So many crucial preparations take place during interphase that a typical cell spends roughly 90 percent of its time in this phase. The first phase of the cell cycle and interphase, *G1* (*gap 1* phase) corresponds to the interval (gap) following the completion of mitosis to the beginning of DNA replication. During this stage, the new cell recovers from the previous division by increasing in size, forming new internal structures, and remaining metabolically active. Almost all animal cells have centrioles, which are two barrel-shaped structures made up of nine triplets of microtubules formed during G_1. They are self-replicating, and one of their functions is to organize the spindle fibers for mitosis (see Figure 9.4). During late G_1, a *checkpoint*, or restriction point, occurs that regulates the cell's entrance into the next phase of interphase. The *G1 checkpoint* ensures that there are enough nutrients, there is sufficient cell size, growth factors are present, and there is no DNA damage. One protein in particular, known as *p53*, can initiate *apoptosis* (programmed cell death) if a mutation is detected in G_1. If p53 is inhibited, perhaps a mutated cell, such as a cancer cell, is given permission to exit G_1 and proliferate. A large number of human cancers in which p53 is suppressed make this protein a prime candidate for cancer screenings. If a cell does not receive a signal to go into the next stage of interphase, the cell switches to a *quiescent*, or nondividing, state, known as G_0. At any given moment, most of the cells in your body are in G_0. In adults, nerve and muscle cells remain in this state. Other cells, such as liver cells, can stay in this state until cell division is needed, such as in tissue injury. Stem cells are particularly interesting because they have the capacity to differentiate into many cell types and are important in embryological development and tissue repair. These cells proliferate rapidly, with shortened G_1 and G_2 phases, but when they begin to differentiate, the gap phases lengthen. Control of these cell-cycle changes is an active area of investigation.

Cell Cycle Phase 2: S Phase

Provided that a cell successfully exits G_1, it enters the second phase of interphase, known as the *S phase*, or synthesis phase. During this phase, each chromosome makes a copy of itself (replicates) to produce two *sister chromatids* that remain attached at the centromere and begin to coil (see Figure 9.3). The function of replication of the DNA is to create enough genetic material for further divisions. Cell growth also continues during the S phase. The S phase occurs rapidly to lessen the chances of unpaired bases being damaged by mutagens.

Cell Cycle Phase 3: G_2 (Gap 2)

Following the S phase of interphase, the G_2, or gap 2 phase, begins. During G_2, the cell increases in size and makes copies of its organelles to prepare for division. In addition, the cell synthesizes essential molecules and continues its normal functions. To ensure that the cell is prepared to divide, the G_2 *checkpoint* occurs between the G_2 phase and the next phase, mitosis. Primarily, this checkpoint is responsible for checking for DNA replication problems and DNA integrity. If detected, the cell will attempt to correct a problem. However, if an error is not repairable, the process of apoptosis will be initiated, and the cell will be destroyed. This important event minimizes passing errors to future daughter cells. Although not visible, during G_2, the *condensation*, or coiling, of the chromosomes begins, and the centrioles duplicate.

9.3.2. Cell Cycle Phase 4: Mitosis (M Phase)

Mitosis is a common type of cell division in which a single cell, known as a *mother cell*, divides into two new, identical *daughter cells*. It is important for asexual reproduction and for cellular replication for growth, repair, and replacement in multicellular organisms. Mitosis is a dynamic, precise, and reliable process consisting of five distinct stages. However, keep in mind that these stages are part of a continuous process, not just stops in time (think of a video vs. a photograph). After successfully completing the G_2 checkpoint, the cell enters mitosis (*M*) (see Figure 9.5).

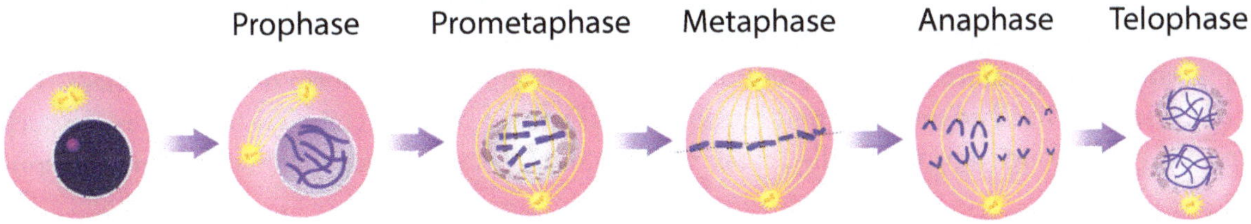

FIGURE 9.5 Mitosis

Mitosis Stage 1: Prophase

Prophase and *prometaphase* are the first stages of mitosis and are the most active and lengthy stages, accounting for as much as 70 percent of the time that a cell spends in the mitotic process. During prophase, the chromosomes progressively condense, an action that continues until metaphase. Driven by organizer proteins, the condensation makes chromosomal separation easier in later stages of mitosis. As the result of the condensation of the chromosomes, the sister chromatids become visible in prophase. The centriole pairs formed during G_2 now separate and begin their migration to opposite poles of the cell, and the *nucleolus* disappears. In addition, the other cellular components, such as the Golgi apparatus and endoplasmic reticulum, disperse, and the nuclear envelope begins to break down. During prophase, assembly of the *spindle apparatus,* composed of *microtubules* in association with the centrioles, begins.

Mitosis Stage 2: Prometaphase

During *prometaphase,* nuclear membrane breakdown into vesicles is completed. These fragments will be used to re-form the nuclear membranes of daughter nuclei. The spindle completes its expansion, forming a bridge from pole to pole, allowing the spindle access to sister chromatids. *Asters,* or microtubular extensions radiating from opposite ends of the centrioles, form and serve as an anchoring mechanism. The sister chromatids at their centrioles form *kinetochores,* or proteinaceous hooks, that attach to the spindle. Eventually, the sister chromatids facing opposite poles will be pulled toward the *equator,* or the center of the cell.

Mitosis Stage 3: Metaphase

In *metaphase,* the condensed sister chromatids are located at the *metaphase plate,* equidistant from the poles of the cell. The metaphase plate merely describes the location of the sister chromatids. At this stage, the sister chromatids are their most condensed and are easily observed. During metaphase, a *spindle checkpoint* (*M checkpoint*) takes place, ensuring that the centromeres of the sister chromatids are properly attached to the spindle microtubules in order for the next stage of mitosis to begin.

Mitosis Stage 4: Anaphase

The next stage, *anaphase,* is the shortest stage of mitosis, lasting perhaps only a few minutes. The sister chromatids separate, each becoming its own chromosome. The two new chromosomes move to opposite poles.

Mitosis Stage 5: Telophase

The final stage of mitosis is *telophase*. During this stage, the spindle breaks down into molecules that will be assembled to make cytoskeletal elements, and through a coalescence of membrane fragments, a nuclear envelope begins to form around each set of chromosomes. Also, in this stage the chromosomes begin to uncoil, becoming less condensed, and the nucleolus reappears. The Golgi apparatus, endoplasmic reticulum, and other cellular components re-form during telophase. The cell cycle is not finished until two new daughter cells form. In order to complete the process, the cell must divide through a process called *cytokinesis*. In animal cells, the formation of a *cleavage furrow*

QRQ 12. Select which of the following are reasons the cell cycle is important: development, repair, maintenance, growth

QRQ 13. List in order the four major phases of the cell cycle: G, G_0, G_1, G_2, C, C_1, C_2, T, M, S, S_0, S_1, S_2

QRQ 14. The cell cycle begins with ______, which encompasses the first three phases.

QRQ 15. A common type of cell division in which a mother cell divides into two new identical daughter cells is ______ (write out term, not letter).

QRQ 16. List in order the five stages of mitosis: anaphase, G_0, G_1, G_2, meiosis, metaphase, prometaphase, prophase, S, telophase

QRQ 17. Select all phases/stages/checkpoints that correspond to each situation listed below: anaphase, G_0, G_1 checkpoint, G_2 checkpoint, gap 1, gap 2, metaphase, prometaphase, prophase, S, spindle checkpoint, telophase

a. assembly of spindle apparatus begins
b. cell becomes longer
c. cell divides into two daughter cells
d. cell growth
e. cell increases in size and makes copies of its organelles to prepare for division
f. centriole pairs separate and begin migration to opposite poles of the cell
g. centrioles form in cells that have them
h. checks for DNA replication problems and DNA integrity before cell division
i. chromosomes progressively condense
j. coiling of chromosomes begins
k. completion of nuclear membrane breakdown into fragments
l. create enough genetic material for further divisions
m. each chromosome replicates to produce two sister chromatids
n. ensures centromeres are properly attached to spindle microtubules for the next stage of mitosis to begin
o. ensures sufficient nutrients and cell size, presence of growth factors, and undamaged DNA

p. new cell recovers from the previous division
q. nondividing state at the end of the cell cycle's first phase
r. nuclear envelope begins to form around each set of chromosomes
s. sister chromatids are their most compacted and easy to observe
t. sister chromatids attach to the spindle
u. sister chromatids line up along an imaginary plane, equidistant from cell's poles
v. sister chromatids separate, each becoming its own chromosome
w. spindle apparatus forms a bridge from pole to pole
x. spindle breaks down
y. two new chromosomes move to the poles

QRQ 18. Select from the following the appropriate term for each description below: apoptosis, central plate, checkpoint, chromatin, condensation, cytokinesis, kinetochore, metaphase plate, p53, sister chromatid, spindle apparatus

a. coiling up of the chromosomes
b. composed of microtubules in association with centrioles used to separate sister chromatids into daughter cells
c. during cell division, one of two longitudinal strands of a split chromosome
d. if suppressed, cancer can result
e. imaginary area, equidistant from cell poles, where sister chromatids move
f. process of a cell dividing into two daughter cells at the end of mitosis
g. programmed cell death
h. protein that can initiate cell death if a mutation is detected in G_1
i. proteinaceous hook that attaches a sister chromatid to the spindle apparatus
j. regulates the cell's entrance into the next phase

pinches off the cell into two new daughter cells, and in plant cells that are too rigid for a cleavage furrow to form, a cell plate separates the cell down the middle, forming the future cells.

9.3.3. Length of Cell Cycle

The amount of time that a cell spends in each stage of the cell cycle is dependent on the type of cell. Fruit fly embryonic cells can complete the cell cycle in as little as eight minutes, and budding yeast can complete the cycle in about 90 minutes. For rapidly proliferating human tissue, the cell cycle may last 24 hours, with mitosis itself lasting about an hour.

APPLYING SCIENCE TO LIFE 4 Asclepius and Dogs

Asclepius is the ancient Greek god of healing. His medical powers were supposedly so great that he could even bring the dead back to life. He is known for his daughters (Hygieia, the goddess of health and cleanliness, Aceso, the goddess of healing, and Panacea, the goddess of universal remedy), his medical staff, and his temples of healing that featured snakes and dogs.

Asclepeions were great temples of healing built in honor of Asclepius. More than 300 Asclepeions have been unearthed by archaeologists through the years. Many great physicians, including the "Father of Medicine," *Hippocrates* (460–370 BC), studied in these temples, which stressed cleanliness, proper diet, and rest (wouldn't that help us all). Welcoming patients into each temple was a statue or artwork featuring Asclepius holding a staff or rod with a snake wrapped around the core. Explanations for the snake's presence differ. However, many historians think that because snakes shed their skin, the snake represents rejuvenation (needed by most patients). Nonpoisonous snakes slithered across the floors of the Asclepions, chasing away evil spirits and easing the patients into a restful sleep (I bet!). This concept can still be seen in the modern symbol for medicine, known as the medical caduceus.

Asclepions also featured the presence of dogs. Myth has it that after his mother was burned to death, a dog looked after Asclepius, explaining his affinity for dogs. In the temple, some dogs served as guard dogs, and others were allowed to patrol the temples, licking the wounds of the injured. Have you ever noticed that your dog looks very concerned when you have a scratch and attempts to lick the wound? Dogs lick their wounds, and perhaps yours, to clean the region of the wound and to promote healing. The saliva of dogs is rich in *epidermal growth factors* (*EGFs*) that encourage cell division and healing. However, having a dog lick your wound is not such a good deal, considering other things dogs lick. Although not serving to transmit parasitic worms, as some believe, the saliva can pass on certain pathogenic bacteria. In modern medicine, artificial EGF, a polypeptide hormone, is used to accelerate healing of skin trauma, such as burns, and to promote healing in corneal transplants. How the ancient Greeks knew this makes for interesting speculation (Theoi Project 2020, Jahovic et al. 2004).

APPLYING SCIENCE TO LIFE 5 When the Cell Cycle Goes Wrong: Cancer

Simply put, *cancer* results from a total loss of control of the cell cycle—literally, cells gone wild. Throughout history, forms of cancer have appeared in plants, animals, and humans. Cancer has even been detected in ancient Egyptian mummies. The Greek Hippocrates first described cancer using the term *carcinos*, meaning crab, because the growing cancer had projections that reminded him of the legs of a crab. Later the Greek physician *Claudius Galen* (130–210 AD) described tumors as *oncos*, the Greek word for "swelling." This term eventually evolved into the term for the study of cancer, *oncology*. Cancer is a serious medical issue today, being the second leading cause of death in Americans (heart disease is first). In 2019, 1,762,450 new cases of cancer were diagnosed in the United States, along with 606,880 cancer-related deaths. As many as 15,338,988 Americans lived with cancer in 2019. Presently, it is estimated that one in three individuals will develop some form of cancer in their lifetime. Put this into perspective: consider visiting a crowded football stadium and counting, "one, two, three, one, two, three . . ." as the fans enter the facility. Only taking into account every third person to enter the stadium, think of the huge number of people!

More specifically, cancer is a term describing a group of over 100 related diseases that are characterized by uncontrolled growth of abnormal cells. The disease begins in a normal cell and continues developing in a group of other cells, becoming a *malignancy*. Cancerous cells have the ability to be *invasive* and spread to other parts of the body (*metastasize*) via the blood and lymph. Over time, cancer cells overcome some of the body's protective mechanisms, such as apoptosis, and can grow faster than their progenitors. A *sarcoma* is a cancer that begins in bone, cartilage, blood, fat, or muscle tissue, while a *carcinoma* is a cancer that develops in the skin or linings of the internal organs. Cancers that develop in the cells of the immune system are referred to as *multiple myelomas*, or *lymphomas*, and forms of *leukemia* originate in the blood. Cancers involving the brain are known as *central nervous system tumors*. Keep in mind that because mature nervous tissue cells do not divide, cancer in the nervous system is usually restricted to support cells, such as *glial cells* (*gliomas*). The five leading types of cancer that affect Americans are breast cancer, lung cancer, colorectal cancer, prostate cancer, and melanoma of the skin. Lung cancer is the leading cause of cancer deaths.

PoI 4. Describe each of the following types of cancer, including site, frequency in males, frequency in females, death rate, and other pertinent information or characteristics. The National Cancer Center website may be helpful.

a. breast
b. lung
c. prostate
d. malignant melanoma
e. pancreatic

APPLYING SCIENCE TO LIFE 6 Cancer, a Growing Concern

As any oncologist will confirm, many cancers cannot be attributed to a single cause. However, risk factors responsible for increasing the chances of cancer can be placed into several categories. Certain *biological factors* are related to an increased risk of developing cancer, including age, gender, and genetic makeup. *Lifestyle factors* can increase the risk of cancer and may include such things as use of tobacco and alcohol, exposure to UV light and smoke, and consumption of nitrates and nitrites. Some *organismic factors*, such as viruses (human papillomavirus, or HPV), bacteria, and fungi, can increase cancer risks. Several *environmental factors* are related to the increase risk of cancer, including exposure to radon and fine particulate matter. *Occupational factors*, such as the exposure to cancer-causing agents (*carcinogens*), have been directly related to the onset of cancer, including asbestos, formaldehyde, and benzene. Even some prescribed *medicines* can increase the chances of developing cancer, such as some hormones, certain antineoplastic (anticancer) agents, and immunosuppressant drugs.

PoI 5. Describe the risks of developing cancer from each occupation listed below.

a. farmer
b. construction worker
c. hairdresser
d. mining worker
e. plastics worker

A cell can become cancerous as the result of a *mutation* (misspelling of the genetic code or alteration of the genetic code) in a particular gene. (Mutations will be discussed in detail in later chapters.) Although cytogenetic studies of tumors indicate that some genes are more commonly mutated in some cancers than others, certain mutations possess similar mutational signatures. Perhaps a mutation will give a cancer cell a reproductive advantage by making the cell have a greater capability of outproducing normal cells or by damaging checkpoints or repair genes. If such a mutation occurs, future generations of the cells possess the same trait. Acceleration of the cell cycle and the inhibition of fundamental controls of the cell cycle increase the chance of cancer cells spreading. Thus, a malignant tumor or other forms of cancer will grow uncontrollably over time.

In recent years, a number of treatments have evolved that can be added to our arsenal to fight cancer. In addition, these treatments can be used in combination, such as radiation therapy and chemotherapy to fight certain cancers. Because there are many types of cancer, there are many kinds of cancer therapy. In some cancers, *surgical procedures* are utilized to remove cancerous tissue. For years, *radiation therapy* has been employed to target and destroy cancer cells. Many cancers can be stifled by the use of specific dosages and combinations of chemicals in *chemotherapy*. In recent years, *immunotherapy*, which aids the immune system in fighting cancer, has met success. *Hormone therapy*, presently used in breast and prostate cancers, utilizes hormones to stop or slow down cancer cell growth. Newly developed *targeted therapy* has been used to impact the instructions to cancer cells to divide, grow, and undergo metastasis. One of the newest weapons in the arsenal is *genetic precision medicine*, which employs changes in a patient's genetic makeup to counter cancer. Many more specific types of therapy are presently available or will be available in the future.

Cancer has a major impact on societies across the world, including the United States. Needless to say, every one of us has been touched by cancer in one way or another. It is imperative that our brightest minds be employed in cancer research across the globe to defeat this formidable enemy.

PoI 6. For each type of cancer listed below, describe at least two types of therapy, their target(s), advantages, and disadvantages.

a. breast
b. lung
c. prostate
d. malignant melanoma
e. pancreatic

9.4. Meiosis

Two of the most significant events in the history of life on Earth were the development of aerobic respiration and the evolution of sexual reproduction. Aerobic respiration provided organisms with copious amounts of energy to survive, and sexual reproduction provided the mechanism that led to diversity within a species, as well as the tremendous diversity of life on this planet (see Figure 9.6).

The process of *meiosis* is the fundamental mechanism underlying *sexual reproduction*. The offspring resulting from sexual reproduction, while maintaining many traits in common with their parents and siblings, possess unique traits particular to themselves. This genetic variation serves as the "spice of life" and the driving force of evolution. New variations may lead to a survival advantage. Although the mechanism of meiosis varies somewhat in eukaryotic organisms, in all cases, meiosis results in the formation of *gametes*, or sex cells, in contrast to cell division, which results in the formation of *somatic* (body, non-sex) cells. *Gametogenesis*, the process of forming gametes, utilizes meiosis to halve the chromosome number. Two sets of divisions occur, *meiosis I* and *meiosis II*. Each of these divisions is made up of prophase, prometaphase, metaphase, anaphase, and telophase stages. A sex cell produced through the meiotic process has a unique combination of parental traits, and when this cell unites with another sex cell, it results in a unique offspring.

FIGURE 9.6 Evolution of sexual reproduction led to diversity within a species, as seen in this litter of puppies.

In 1883, the Belgian cytologist Edouard van Beneden proposed that male gametes (sperm) and female gametes (ova) were haploid (N), as opposed to body cells that are diploid (2N). He further observed that the two haploid gametes fuse during *fertilization* to form a diploid *zygote*. The process of meiosis is sometimes referred to as *reduction division*, because it is the mechanism that reduces the diploid number of potential sex cells to the haploid number. Without this process of reduction division, the number of chromosomes would become unmanageable. Just think, in five generations of human beings, the number of chromosomes in the nucleus would be $46 \times 2^5 = 1{,}472$. Thus, meiosis serves to stabilize the number of chromosomes through the generations.

9.4.1. Spermatogenesis

Spermatogenesis is the meiotic process that ultimately results in the formation of haploid *spermatozoa* (*sperm*) (see Figure 9.7). When a male reaches puberty, a number of bodily changes occur, driven mostly by the hormone *testosterone*. The *testes* are responsible for production of sperm, and each testis is primarily composed of about 28 inches of highly convoluted *seminiferous tubules*. Once puberty is reached, *primordial germ cells* along the outer edge of the seminiferous tubules begin to divide by mitosis, forming diploid *spermatogonia* (*s. spermatogonium*). Each spermatogonium in turn undergoes mitosis, producing two diploid cells. One of the cells continues on as a spermatogonium, which perpetuates sperm production for nearly a lifetime. The other cell becomes the diploid *primary spermatocyte* that, through meiosis, will eventually form four haploid sperm cells.

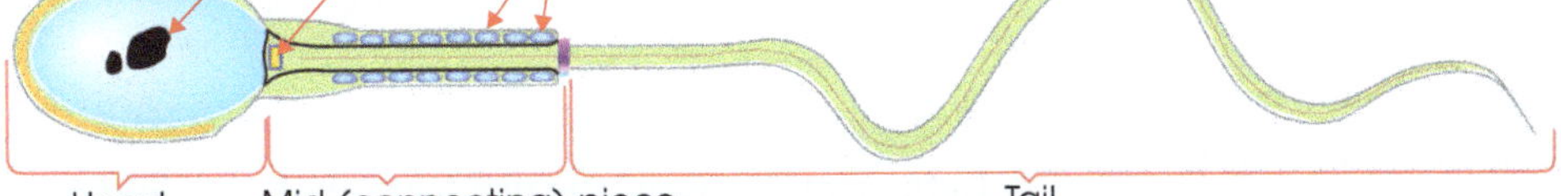

FIGURE 9.7 Human sperm anatomy

The primary spermatocytes are cells that are destined to enter the first meiotic division, meiosis I (Figure 9.8, Step 1). This first meiotic division is a complex process that results in the formation of a pair of *secondary spermatocytes*. Prior to *prophase I* of meiosis I, as in mitosis during the G_1 phase, all the chromosomes are duplicated and continue into the S and G_2 phases. During prophase I, homologous chromosomes find and align with each other. This pairing, or *synapsis*, occurs throughout the length of each pair of chromosomes, matching corresponding alleles. Each pair of the resulting complex is composed of two *chromatids*, linked at the centromere. A protein structure ensures that the homologous pairs are held together. Next, the homologous chromosomes can trade sequences of genetic material in a process known as *crossing over*. Multiple crossing-over events may take place in a pair of homologous chromosomes. The crossing over allows for the formation of new chromosomes with unique genetic combinations. This process is of the utmost importance in allowing for evolution and diversity. After crossing over, newly remixed homologous pairs, not individual chromosomes, line up during *metaphase I* of meiosis I. This process ultimately allows the formation of developing sperm cells with different combinations of *alleles* (alternative gene variants). Next, during *anaphase I* of meiosis I, homologous chromosomes, each made up of two chromatids, separate and move toward opposite poles. Finally, in *telophase I* of meiosis I, the cells separate. As a consequence of this division, the resulting two new cells retain two copies of each chromosome. At the completion of meiosis I, two *secondary spermatocytes* are formed (Figure 9.8, Step 2).

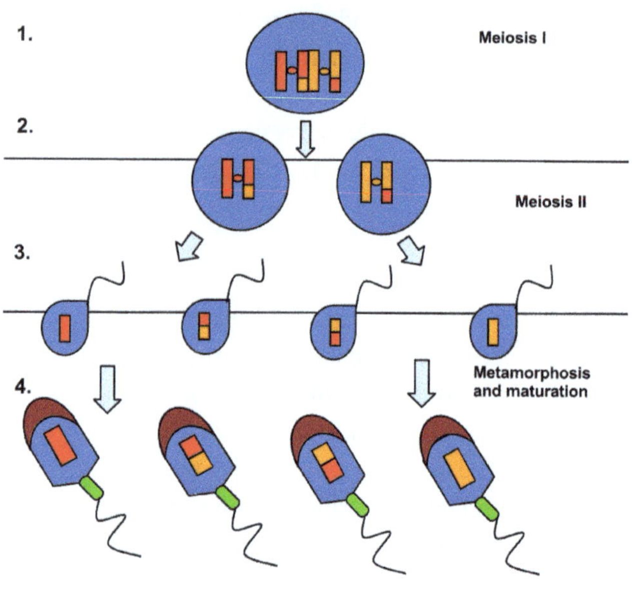

FIGURE 9.8 Generalized spermatogenesis

Shortly after the formation of the secondary spermatocytes, a period of interphase without the S stage (no DNA) begins meiosis II. *Prophase II* and *metaphase II* of meiosis II begin and are similar to the typical prophase and metaphase, with the exception that the newly remixed sister chromatids are lined up along the metaphase plate. In anaphase II, the sister chromatids are pulled toward the opposite poles, and in telophase II, the plasma membrane is reformed around four new haploid cells, known as *spermatids* (Figure 9.8, Step 3). Thus, each secondary spermatocyte gives rise to two spermatids. These newly derived cells undergo metamorphosis and form sperm cells (Figure 9.8, Step 4), each with a unique combination of genetic material. The process from spermatogonia to mature sperm usually takes about five weeks, and a typical human male produces between 100 to 200 million sperm daily throughout most of his life. Approximately 250–280 million sperm are released during ejaculation.

PoI 7. List several changes, physical and physiological, that occur in human males and females that are influenced by sex hormones.

9.4.2. Oogenesis

In the female ovary, the production of a haploid ovum results from the process of *oogenesis*. The basic blueprint of the process is similar to spermatogenesis but varies in several fundamental ways. In human beings, oogenesis begins during embryological development, as opposed to spermatogenesis, which begins during puberty. As early as the twelfth week of embryological development, primary oocytes are being produced in the female embryo. At birth, the number of primary oocytes is fixed, and they remain in prophase I of meiosis I. In contrast, primary spermatocytes are produced throughout the lifetime of a male. Since only one ovum is available for fertilization, excess genetic material is eliminated through *polar bodies*, whereas in sperm formation, every primary spermatocyte eventually forms four sperm. The female gamete, or *ovum*, is larger (10,000X) than the mobile sperm and contains organelles, enzymes, mitochondrial RNA (mRNA), and materials and structures needed to initiate and maintain metabolism and growth, as well as a haploid nucleus (see Figure 9.9).

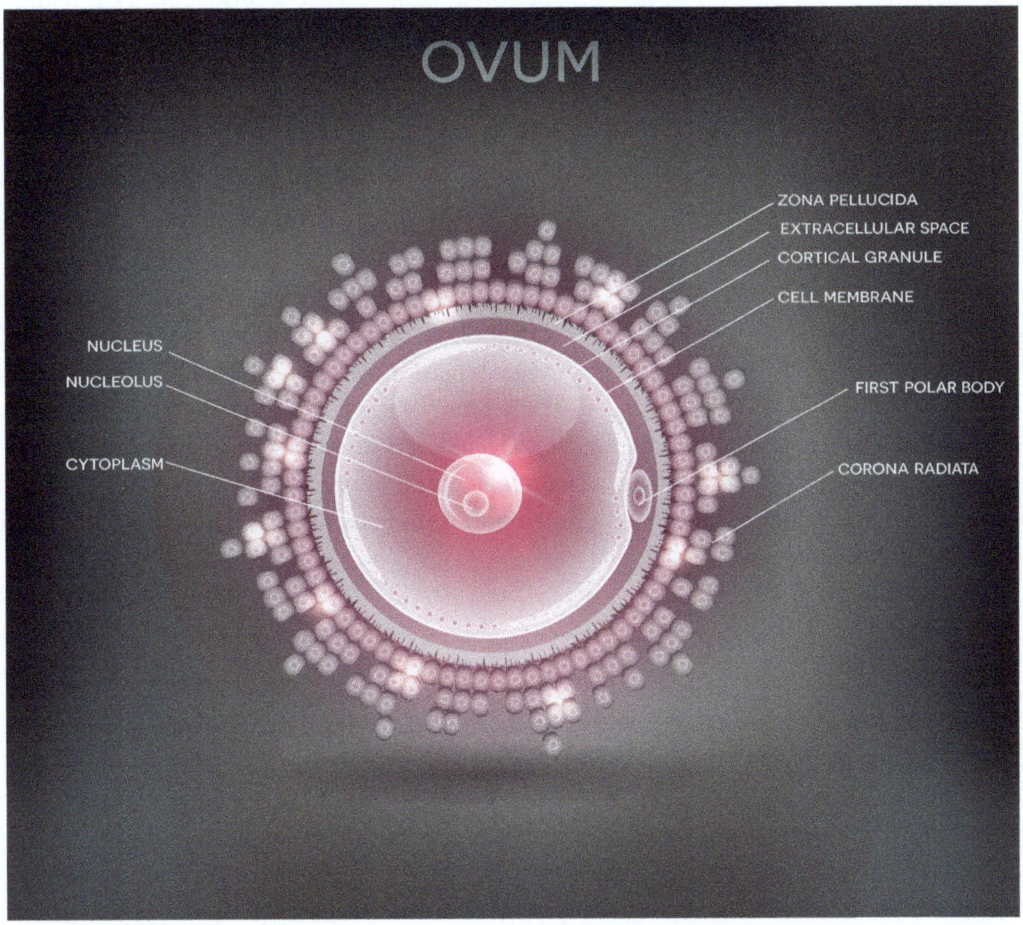

FIGURE 9.9 Human egg anatomy

In the ovaries, primordial germ cells give rise to diploid *oogonia*, which divide between the second and seventh month of embryonic development. The majority of oogonia die during this period, and the remaining oogonia become diploid *primary oocytes*, arrested in prophase I of meiosis I and still in the fetus. Meiosis will proceed during puberty and into the reproductive years. In some females,

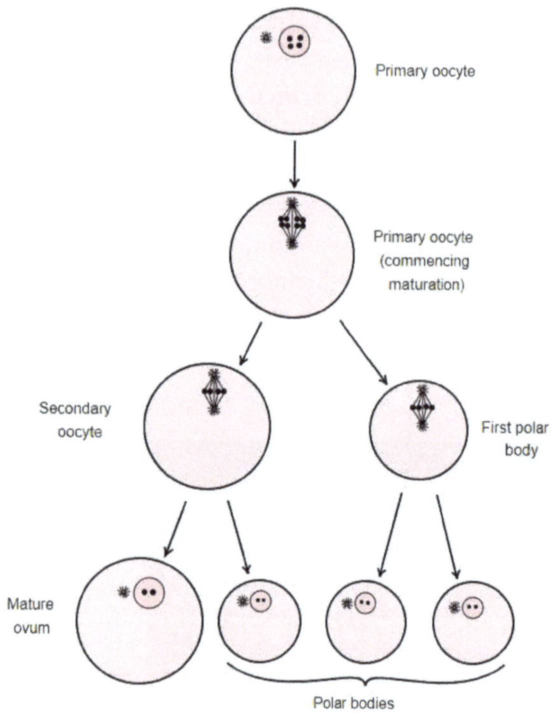

FIGURE 9.10 Generalized oogenesis

these primary oocytes can be meiotically inactive for 40 years. Only about 400 primary oocytes mature during the lifetime of a female.

At puberty, *follicle stimulating hormone (FSH)*, secreted by the anterior pituitary gland, stimulates a primary oocyte, residing in a *follicle* of the ovary, to complete meiosis I. In most instances, only one primary oocyte completes meiosis I monthly. As in the development of sperm, during meiosis I, the primary oocyte undergoes synapsis, tetrad formation, and crossing over. During telophase I, the primary oocyte has an unequal division, producing a *secondary oocyte* and a *first polar body* (see Figure 9.10). The secondary oocyte contains the majority of the primary oocyte's cytoplasm, cellular components, and one half of the genetic material. On the other hand, the polar body with very little cytoplasm or cellular constituents but half of the genetic material usually degenerates within several hours.

The secondary oocyte is released from the follicle during *ovulation* and travels to the *oviduct*, or *Fallopian tube*, where it awaits potential fertilization. A short time after forming, the secondary oocyte begins meiosis II but halts its development in metaphase II. If fertilization occurs, meiosis II resumes, yielding an *ovum* and a polar body (both haploid). The nuclei of the ovum and a sperm can fuse, forming a zygote, which has the potential of growing into a new organism. If fertilization does not occur, the secondary oocyte degenerates.

APPLYING SCIENCE TO LIFE 7 Puppies for Sale

Andrea is the proud owner of a beautiful buff-colored cocker spaniel named Sugar, and her husband Jeff is an equally proud owner of a playful Boston terrier named Chip. After much forethought, Andrea decided to mate Sugar to another buff cocker spaniel in order to have a litter of bouncy cocker puppies. After 63 days of gestation, the puppies were born, but Andrea was horrified. The litter consisted of two beautiful buff cocker spaniel puppies and three puppies slightly resembling Chip. What happened, and why was the litter mixed? It is obvious that old Chip was pretty sly and had mated with Sugar when no one was around. However, the mixed litter is a bit more complicated to explain. Unlike humans, which generally produce only one egg during ovulation, many animals, including dogs, are capable of multiple ovulation over a short period of time. Apparently, two of the eggs were fertilized by the cocker spaniel father and the other three eggs were fertilized by Chip. Anybody want a free puppy?

PoI 8. What are some of the evolutionary advantages of multiple ovulation?

QRQ 19. Are male and female gametes diploid or haploid?

QRQ 20. As spermatogenesis begins, ____ along the outer edge of the ____ begin to divide by ____, forming ____. Each undergoes ____, producing a ____ for perpetuation of sperm cell production and a ____. All of these cells are ____. Answer choices: diploid, haploid, meiosis, mitosis, primordial germ cell(s), seminiferous tubule(s), spermatocyte(s), spermatogonium/spermatogonia.

QRQ 21. Before meiosis I of spermatogenesis, ____. During prophase I of meiosis I, ____. In a crossing-over process, ____. During metaphase I, ____. During anaphase I, ____. In telophase I, ____. At the completion of meiosis I, ____. Answer choices: all chromosomes in a primary spermatocyte are duplicated, cells separate, homologous chromosomes find each other and align, homologous chromosomes separate and move toward opposite poles, homologous chromosomes trade sequences of genetic material, newly remixed homologous pairs line up, two diploid secondary spermatocytes are formed, two haploid secondary spermatocytes are formed.

QRQ 22. Answer with a number: In spermatogenesis, each primary spermatocyte gives rise to ____ spermatids, which mature into sperm cells.

QRQ 23. In oogenesis, primordial germ cells in a female embryo give rise to ____, which remain in ____. Beginning at ____, ____ stimulates a ____ to complete ____, producing a(n) ____ and a(n) ____ with excess genetic material. During ovulation, a(n) ____ travels to the ____ and begins ____. If fertilization occurs, this process completes, yielding a(n) ____ and a(n) ____, which can fuse with a sperm. Answer choices: birth, follicle stimulating hormone, meiosis I, meiosis II, oviduct, ovum/ova, polar body/bodies, primary oocyte(s), puberty, secondary oocyte(s), seminiferous tubule(s).

9.5. When Meiosis Goes Wrong

Meiotic errors in the gametes of humans are relatively common. In fact, meiotic anomalies in humans are more common than in any other animal. Presently, this phenomenon is under investigation. *Cytogeneticists* (scientists that study chromosome number, structure, appearance, and function) are developing detailed descriptions and causes of these disorders. The majority of maternal meiotic errors are due to an incorrect number of chromosomes in the egg, while the majority of paternal chromosomal problems are in the structure of the sperm.

9.5.1. Nondisjunction

The process of meiosis is complex, and occasionally errors occur. The most common meiotic error is *nondisjunction*, or the failure of the homologous chromosomes to separate (see Figure 9.11). Nondisjunction is relatively common in humans and can result in a fertilized egg having the incorrect

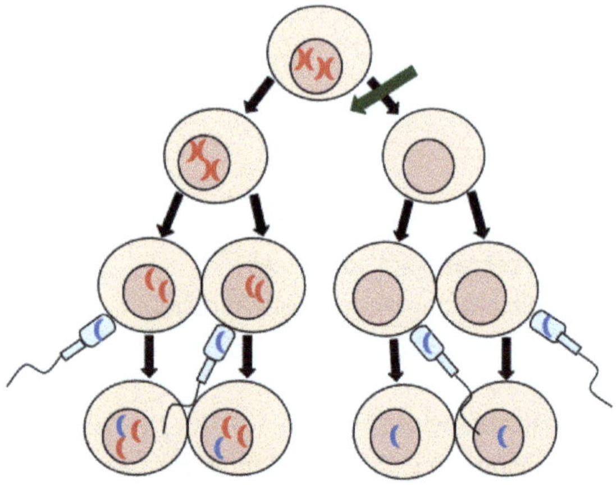

FIGURE 9.11 Nondisjunction occurring at meiosis I

PoI 9. If nondisjunction occurs at meiosis II, what results would occur (see Figure 9.11)?

number of chromosomes, or *aneuploidy*. *Karyotyping*, which displays the chromosomes, can easily be employed to identify errors in chromosomal number. Approximately 6 percent of newborns are aneuploid, and aneuploidy is present in about 60 percent of spontaneous abortions and 6 percent of stillbirths. The risk of nondisjunctional events increases greatly with the age of the mother. For women over 40 years old, the chance of a *trisomy* birth, where the infant has an extra set of chromosomes, is 35 percent, compared to 2 percent for a 25-year-old woman. Maternal nondisjunction is more common in errors involving non-sex chromosomes, while paternal nondisjunction is more likely in disorders involving the sex chromosome.

Nondisjunction can occur in either division of meiosis, with the most common error being the result of homologous chromosomes failing to separate during either anaphase I or II. If the error occurs during anaphase I, all four secondary oocytes will have an incorrect number of chromosomes, while if in anaphase II , two of the potential products will have the wrong number of chromosomes. The differences in whether nondisjunction occurs in meiosis I or II can have important clinical consequences. When an egg that has an extra set of chromosomes is fertilized by a normal sperm, a trisomy results. *Trisomy 21*, or *Down syndrome*, caused by an extra copy of chromosome 21, is the best-known trisomy. The traits of individuals with trisomy 21 vary greatly. Several of the physical traits common to individuals with trisomy include an enlarged tongue, slanting eyelids, poor muscle tone, short fingers, small hands, short neck, excessive flexibility, and short stature. Unfortunately, many people with trisomy 21 suffer from heart defects, intestinal problems, thyroid problems, hearing loss, and vision problems. They also may exhibit a number of developmental issues, including mild to moderate learning complications. Presently, cytogeneticists note three causes of trisomy 21, with approximately 95 percent of the cases associated with aneuploidy. One in 2,000 children of 20-year-old mothers has Down syndrome, while the chance increases to 1 in 30 for 45-year-old mothers.

Instead of an extra set of chromosomes, *monosomy* results when an individual has only one copy of a particular chromosome. A monosomy is considered lethal, with the exception of *Turner syndrome*, or XO. The disorder only affects females and results from a partially missing or missing sex chromosome. Eighty percent of Turner syndrome births are attributed to paternal nondisjunction.

APPLYING SCIENCE TO LIFE 8 Without Solution

On advice of her primary care physician and the school counselor, Olivia brings her 16-year-old son, Mark, to you, a human genetics specialist. She has long been concerned that he has some underlying undiagnosed medical condition that is affecting his appearance and his behavior. During your initial interview, she indicated that he often displays aggressive behavior, paranoia, inability to concentrate, and self-mutilation. He often is nervous and "out of sorts," and he doesn't sleep well. He has few friends, and he is periodically suspended from school for fighting. She has been unable to control his use of alcohol (which he began drinking at age 10) or cannabis (begun at age 12). His maternal grandfather suffered mental illness, which led to his suicide.

You note in a physical examination that he is very tall (6 feet 6 inches), has poorly developed muscle mass, and is obese or overweight (250 lb., BMI 30.4). You also note that he has unusually long arms and legs. His blood pressure and his heart rate were normal. His physical appearance, including excessive breast development and behavioral issues, suggested *Klinefelter syndrome*.

Cytogenetic studies confirmed this diagnosis, displaying a 47 XXY karyotype. All other physiological markers are within normal range (see Figure 9.12). Previous intelligence scores indicated a below-average intelligence. Klinefelter can be diagnosed using tests made during pregnancy, as described below, which prepares parents for the defect and allows them to consider their options.

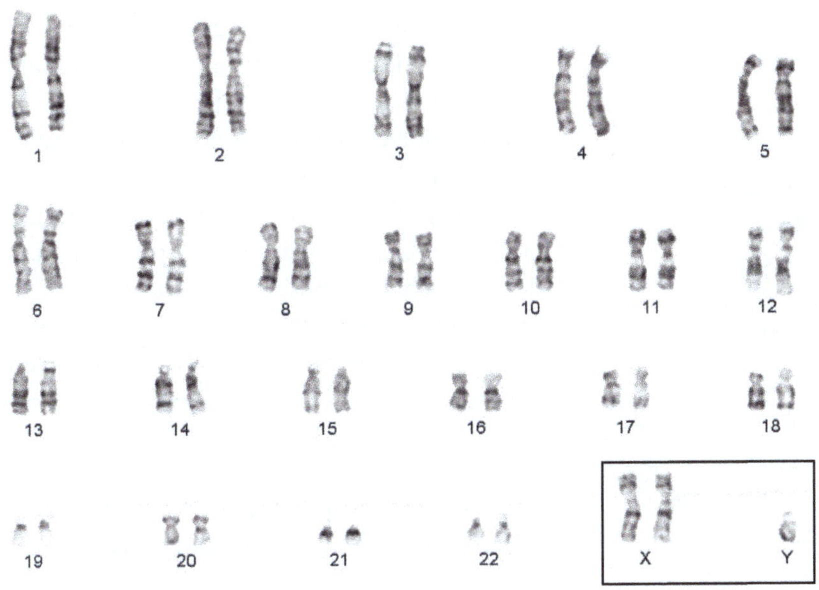

FIGURE 9.12 Klinefelter syndrome (XXY)

9.5.2. Structural Chromosomal Abnormalities

Other chromosomal errors, besides nondisjunction, include physical or structural abnormalities, such as deletions, inversions, duplications, and translocations, and many structural chromosomal abnormalities result in disease. These abnormalities result from the breakage or incorrect rejoining of chromosomal segments. The abnormalities can be denoted as *balanced* if the complete chromosome is represented or *unbalanced* when part of a chromosome is missing or added. Balanced abnormalities may not result in a genetic disorder. However, balanced states can result in disease when it interferes with the formation of a vital protein or proteins. Unbalanced arrangements can exist as deletions, duplications, and insertions. Other, more rare chromosomal abnormalities may result from errors in meiosis or cell division (IUPUI 2003).

A chromosomal *deletion*, which can affect any part of any chromosome, results from a breakage and loss of part of a chromosome. These abnormalities have the potential to cause severe and life-threatening genetic disorders.

PoI 10. For each of the following conditions, indicate the cause(s), traits, sex, incidence, complications, life expectancy, psychological issues, therapies, and possibility of producing viable offspring:

a. Patau syndrome, or trisomy 13
b. Edward syndrome, or trisomy 18
c. 47XXX syndrome, or trisomy X
d. 47XYY syndrome

PoI 11. For each of the following conditions, indicate the cause(s), traits, sex, incidence, complications, life expectancy, and possibility of producing viable offspring:

a. cry of the cat syndrome, or cri-du-chat syndrome
b. Williams syndrome, sometimes called the pixie syndrome
c. Prader-Willi syndrome
d. Angelman syndrome
e. Charcot-Marie-Tooth disease (CMT)

The cry of the cat syndrome, or *cri-du-chat*, is an interesting chromosomal deletion disorder that involves a deletion of part of the short arm of chromosome 5, denoted *5p-*. Infants with this disorder may exhibit a high-pitched noise that sounds similar to a cat's cry.

Another chromosomal deletion disorder, *Williams syndrome*, sometimes called the pixie syndrome, is the result of a deletion of part of chromosome 7, involving between 26 and 28 genes. Children with Williams syndrome remind many people of pixies, possessing large ears, a wide mouth, and a petite stature. They also have interesting personalities, exhibiting musical and singing talents, an engaging personality, and interest in entertaining. Recently, it was discovered that these individuals have three genes that make them seek the approval of other people. These three genes also exist in dogs, but not wolves. Some geneticists think that these genes made the domestication of the dog possible.

Many other chromosomal aberrations are responsible for a number of abnormalities. In the world of medical genetics, terms such as duplication, translocation, substitution, and inversion are linked to several genetic diseases. Understanding meiotic errors involving chromosome number, as well as structural damage of the chromosome, represents a challenge to modern cytogeneticists.

9.5.3. Detection During Pregnancy

If a genetic problem or meiotic error is suspected during a pregnancy, there are two primary methods that can be used to analyze and collect fetal fluids and cells, *amniocentesis* and *chorionic villus sampling*. During pregnancy, the developing fetus is surrounded by a clear yellow liquid, referred to as *amniotic fluid*. This fluid first appears within the amniotic sac that surrounds the embryo about 12 days after conception and persists until birth. The amniotic sac and fluid serve to protect, cushion, and insulate the developing fetus. The fluid also aids in muscle and lung development, support for the umbilical cord, and lubrication. Initially, the fluid is primarily composed of water, but after about 20 weeks of gestation, fetal urine replaces the water. Amniotic fluid also contains antibodies, nutrients, proteins, and hormones. The greatest volume of amniotic fluid (one quart) occurs around 36 weeks of development. When a mother's water breaks, the amniotic sac tears, and the fluid leaks out.

Administered between 15 and 20 weeks of pregnancy, amniocentesis is a commonly used prenatal diagnostic medical procedure used to determine chromosomal abnormalities, some fetal health issues, fetal blood type, fetal lung maturity, and sex. During amniocentesis, a needle is inserted into the abdomen and into the uterus to draw out less than one ounce of fluid from the amniotic sac. The route of the needle is guided through ultrasound. Once collected, the fluid is sent to the laboratory for analysis. The test cannot detect all of the genetic abnormalities but can measure *alpha fetoproteins* (proteins produced by fetal liver). Too much or two little may indicate a problem such as trisomy 21 or neural tube defects. Also, fetal cells can be collected for karyotyping, as will be discussed later. There is a small risk that amniocentesis can cause a miscarriage or an infection.

Usually performed between 11 and 14 weeks of pregnancy, *chorionic villus sampling* (*CVS*) is a prenatal testing procedure that is also used to diagnosis a number of congenital abnormalities. Chorionic villi are minute projections of tissue emanating from the fetal placenta that have the same genetic makeup as the developing fetus. It is not part of the umbilical cord that links the placenta to the fetus. In CVS, a small section of chorionic villus tissue is sent to the laboratory for testing. As in any medical procedure, there is some risks of miscarriage, limb defects, and infection.

Once a fetal sample is collected through amniocentesis or chorionic villus sampling, there are a number of laboratory procedures and staining methods that can be used to detect genetic abnormalities and disease. One of the best-known tests is *karyotyping* (chromosome examination), which pairs and examines chromosomes of a fetus or an individual. In this procedure, cells are grown under laboratory conditions and are arrested in the metaphase or prometaphase stage of the cell cycle, when the chromosomes are most condensed. The cells are then treated with a hypotonic solution that causes the cells and nuclei to burst, leaving the chromosomes. In turn, the chromosomes are then treated with a chemical fixative (preservative) and either photographed or placed on a glass slide. They are then arranged by size and centromere location in order to be counted, sorted, and analyzed (see Figure 9.13). Molecular staining of the chromosomes reveals chromosomal structural or numerical abnormalities with the chromosome.

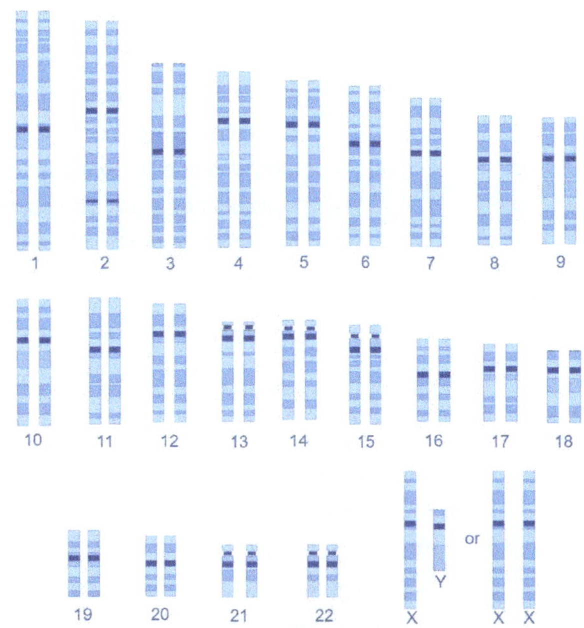

FIGURE 9.13 Karyotype of a normal human subject

APPLYING SCIENCE TO LIFE 9 Good News

Paul and Jennifer, a career-oriented couple, decided to have children a little bit later in life. Paul is 42 years old, and Jennifer is 40. They were aware that mothers of Jennifer's age have an increased risk of having a child with trisomy 21 (Down syndrome). Compared to the risk of a mother at 25 years old, being 1/1,340 births, the risk of a 40-year-old mother is 1/85 births. After consulting with their obstetrician, the couple decided to undergo *screening tests* for trisomy 21. The screening tests are not definitive, but they indicate if the baby is more likely to have trisomy 21. One screening test used during pregnancy, the *nuchal translucency test*, employs *ultrasound*. Ultrasound allows measurement of the tissue on the back of the baby's neck, known as the *nuchal fold*. If excess fluid is detected in this region, there is an increased chance of the baby having trisomy 21.

There are also several tests where the mother's blood is collected and analyzed. If unusual amounts of the hormone *human chorionic gonadotropin* (*HCG*) or the plasma protein *PAPP-A* are present, there is an increased chance for trisomy 21. Later in the pregnancy, unusual levels of *alpha-fetoprotein* (*AFP*) *estriol* and *inhibin A* can also be indicators of trisomy 21.

A more specific diagnostic test for trisomy 21 requires samples collected by amniocentesis or chorionic villus sampling for karyotyping. Presence of an extra 21st chromosome in the karyotype is indicative of a baby with trisomy 21. However, one must keep in mind there are some exceptions, as discussed earlier.

Fortunately, after screening and diagnostic tests, Paul and Jennifer's baby was diagnosed as negative for trisomy 21. However, if they are planning on more children, these tests must be repeated for each pregnancy.

QRQ 24. Select from the following the appropriate term for each description below: adjusted, amniocentesis, aneuploidy, balanced, CVS, deletion, karyotyping, monosomy, nondisjunction, trisomy, unbalanced

a. breakage and loss of part of a chromosome
b. chromosomal error where part of the chromosome is missing or added
c. chromosomal error where the complete chromosome is represented
d. failure of the homologous chromosomes to separate
e. incorrect number of chromosomes
f. individual has an extra set of chromosomes
g. individual only has one copy of a particular chromosome
h. prenatal diagnostic procedure analyzing amniotic fluid
i. test that pairs and examines chromosomes

Image Credits

LAB #9

Laboratory Exercises

9.6. Laboratory on Tumor

9.6.1. Preliminary Work

This laboratory relates to tumor growth and chemical treatment. From the text's website, view the introductory video; download and open "*Tumor.nlogo*." Write your answers in blanks to the left of the questions.

1. Click the Info tab at the top of the page and read sections "What Is It?" and "How It Works." For each of the following, fill the appropriate letter(s) into the blank.
   a. Give the color (A. black, B. blue, C. brown, D. cyan, E. gray, F. green, G. magenta, H. orange, I. red, J. white, K. yellow) of each model component.

      ________ stem cells

      ________ young transitory cells

      ________ "middle-aged" transitory cells

      ________ "old" transitory cells

   b. ______ (T/F): In mitosis, a cell divides, and the two daughter cells have the same number of chromosomes as the parent cell. (Refer to other sources as needed.)
   c. ____________ In the model for asymmetric mitosis of a stem cell, the daughter cells include what kinds of cells? A. stationary transitory cell, B. stem cell that replaces the parent, C. transitory cell that moves outward, D. two stem cells, E. two transitory cells
   d. ______ What kind of mitosis do young transitory cells exhibit, asymmetric (A), symmetric (S), both (B), or neither (N)?
   e. ____ For the model, select all that apply concerning the process that starts with symmetric division of the stem cell: A. cells die young in the secondary tumor, B. first stem cell moves at random, C. first stem cell remains static, D. daughter cell generates another tumor, E. happens more than once, F. happens only once, G. illustrates disease progression, H. illustrates metastasis, I. second stem cell moves at random, J. second stem cell moves to the right, K. two stem cells result

2. Read "How to Use It." Select the feature for each description. A. KILL STEM CELL, B. KILL TRANSITORY CELLS, C. KILL-MOVING-CELL, D. LEAVE-TRAIL, E. LIVING CELLS PLOT, F. SETUP

_______ causes display of each cell's path if ON

_______ creates two cancerous stem cells

_______ destroys young transitory cells (younger than 10 times steps)

_______ kills moving stem cell if KILL-MOVING-CELL is ON

_______ kills original stem cell if KILL-MOVING-CELL is OFF

9.6.2. Laboratory Work

3. This question involves observations of tumor growth in the model.

   a. ______ Read the first paragraph of "Things to Notice" under the Info tab. Follow the directions in the paragraph and let the simulation run until the cell count becomes constant. Give this cell count.

   b. ______ Read the second paragraph of "Things to Notice" and follow the directions. Does the system eventually reach equilibrium with cells being created at the same rate as they are dying, no (N) or yes (Y)?

   c. ____________ Read the last two paragraphs of "Things to Notice" and follow the directions. Next, stop the simulation and press *setup* again. Press *go* twice in quick succession so that the simulation displays two slightly separated stem cells and several transitory cells. Right-click on one of the red transitory cells toward the left, that is, one of the transitory cells divided from the stationary stem cell. At the bottom of the popup menu, select a turtle other than turtle 0 or 1, which are the stem cells, to *inspect*. A closeup of the area including this turtle appears. Note that each transitory cell is depicted with a black dot in the middle. Under the closeup, click *Watch*. How does the world on the left change? Select all that apply: A. a lighter circle appears around the moving stem cell, B. a lighter circle appears around the transitory cell, C. nothing, D. the background becomes darker

   d. ________________________ With the speed still *slower*, click *go* and, observing the world and close-up view of the area around the watched cell, select all that occur: A. Above the close-up view, the observed cell (turtle) is eventually indicated as dead, B. the closeup view eventually becomes solid gray, C. the observed cell changes color from red to pink to white to black, D. the observed cell is stationary, E. the observed cell moves outward, F. the observed cell never dies, G. when the observed cell dies, its surrounding circle disappears

4. This question involves a chemotherapy treatment that inhibits cell division by killing young transitory cells. However, stem cells are usually harder to kill. We will observe what happens as we kill the young transitory cells but not the stem cells.

   a. **1.** ______ **2.** ______ Complete the hypothesis: After a treatment, initially the number of cells **1.** (A. decreases, B. increases, C. stays approximately the same), but eventually, the number of cells will **2.** (D. continue to decrease, E. decrease sharply to 0, F. increase to a level less than the maximum, G. increase to the maximum, H. stay approximately the same).

   b. ______ Set up and run the simulation until the number of cells reaches equilibrium. Then perform this "chemotherapy" several times by pressing the *KILL TRANSITORY CELLS* button. Do your observations support your hypothesis, no (N) or yes (Y)?

c. ______ Can you eliminate the cancer with this technique, no (N) or yes (Y)?

d. Write a paragraph explanation of why or why not and what you observed.

5. This question involves a chemotherapy treatment that inhibits cell division by killing stem cells.

   a. ______ Complete the hypothesis: If we kill a stem cell, **1.** (A. its tumor will eventually disappear, B. its tumor will initially diminish but then start growing again, C. only the stem cell will die, D. the main tumor and metastasized tumor will disappear, E. the other tumor will disappear).

   b. __________ Continue the simulation at a slower speed until the number of cells reaches equilibrium. Then press the *KILL MOVING STEM CELL* button once. Select all that occurs: A. the cells in the original tumor continue to age but eventually die, B. the metastasized tumor disappears, C. the number of cells goes to zero, D. the number of cells reaches a new, slightly higher equilibrium, E. the number of cells reaches a new, slightly lower equilibrium, F. the original tumor disappears

   c. __________ Press the *KILL ORIGINAL STEM CELL* button once. Using the choices from part b, select all that occurs.

   d. ______ Do your observations support your hypothesis from part a, no (N) or yes (Y)?

   e. Write a paragraph explanation of why or why not and what you observed.

   f. Write a paragraph discussing the challenges of having chemotherapy strong enough to kill the cancerous stem cells.

6. This question steps through calculation of an appropriate concentration of a drug dosage.

   a. ______ *Concentration* of a drug in the plasma, the fluid that contains the blood cells, is the total amount of drug in the plasma divided by the volume of plasma: (amount of drug in plasma)/(volume of plasma). The amount of blood in an average adult is about 5 liters (L), while the amount of plasma is approximately 3 L. Suppose at a particular time, the amount of aspirin in an adult's body is 650 milligrams (mg). Express the plasma concentration of aspirin in mg/L, rounded to the nearest whole number.

   b. __________ Such concentrations are usually expressed in micrograms (µg) per milliliter (mL), or µg/mL. 1 milligram (mg) equals 1,000 micrograms (µg). Convert 650 mg to µg.

   c. __________ 1 liter equals 1,000 milliliters. Convert 3 L to mL.

   d. ______ We are multiplying both the numerator and denominator by 1,000 to convert mg/L to µg/mL. Will the plasma concentration of aspirin in mg/L from part a be the same number in µg/mL, no (N) or yes (Y)?

   e. ______ The *therapeutic range*, or *therapeutic window*, for a drug consists of concentrations between the least concentration of drug that is helpful and the largest concentration that is helpful without having dangerous or intolerable side effects. The aspirin's therapeutic range for an average adult is approximately 150 to 300 µg/mL. Is the concentration in part a within aspirin's therapeutic range, no (N) or yes (Y)?

MODULE

10 Fundamental Genetics

10.1. Introduction

For millennia, humans have been interested in what is today known as *genetics*—the study of heredity. Our ancestors have long crossbred plants and interbred animals to produce more offspring and those that were hardier, more productive, better-tasting, and so on. A walk through the fruit and vegetable section of a nearby supermarket is a testimony to the power of genetics. For example, consider the staple food product corn. How many kinds of corn can you name? For starters, you probably named yellow, white, sweet, field, and of course popcorn. Actually, there are perhaps as many as 1,000 different types of modern corn. Biodetectives have traced the origins of all these types of corn to central Mexico between 7,000 and 9,000 years ago. The ancestor to modern corn is thought to be a wild grass, known as *teosinte* (see Figure 10.1). By today's standards, teosinte is not particularly appetizing or palatable (Native Seeds n.d.). The ears are small, and the small kernels are covered by a thick, tough casing. In other words, chewing on an ear of teosinte would probably wear your teeth down quite easily. Through selective breeding, teosinte now serves as the precursor to modern corn varieties. Today, every plant and domesticated animal that we eat has a similar story. Genetics has certainly shaped our world.

FIGURE 10.1 Comparison of teosinte and modern corn

APPLYING SCIENCE TO LIFE 1 A Sad Story

Not many people have heard of the Russian botanist *Nikolai Vavilov* (1887–1943). He should occupy a special place in the history of biology, but unfortunately, as a result of jealousy and greed, his is a sad story.

As a young boy, Vavilov witnessed the Great Russian Famine in the winter of 1891, where half a million Russians perished from starvation and opportunistic diseases such as cholera. In the meantime, the Russian aristocracy dined on fine foods and exported grain for profit. This disaster served to spark Vavilov's pursuit of botany and agriculture and triggered his quest to eliminate hunger and starvation.

In 1911, despite antiquated farming strategies and equipment, Russia was the largest grain exporter on the planet. Working at a prestigious agriculture academy, Vavilov wished to revitalize Russian agriculture by applying the ideas of Mendel and Darwin and statistical analysis, combining them with the practical experience of successful farmers. During World War I, Vavilov began collecting seeds of commercially important plants in hopes of one day establishing a seed repository. This seed repository would be an important resource in Vavilov's quest to end hunger and starvation. With the success of the Russian revolution in 1917, he worked even harder to trace the genetic lineage of modern crops. In 1920, his hard work paid off, and he began his rise to dominance in Russian agriculture. He became obsessed with collecting seeds of commercially important plants and developing the story of their origins. As a result, his experiences made him one of the first experts in biodiversity. He collected more than 250,000 varieties of seeds in his worldwide Center of Origins endeavor, including sweet potatoes, cashews, tobacco, and corn from Central and South America; quinine (antimalarial) from Java; rice, sugar cane, black pepper, and mangos from India; carrots, grapes, and apples from Central Asia; and walnuts, peaches, and poppies from China. In 1926, he received the valued Lenin Prize for his contributions to science.

However, the following year, another Russian scientist began his rise to glory. Trofim Lysenko claimed that university science was not relevant in successful crop production and the works of Mendel and Darwin were rubbish. He was a firm believer in Lamarck's disproven ideas of use and disuse and acquired characteristics (discussed in our chapter on evolution). He was also extremely jealous of Vavilov's works. He pushed the idea of vernalization (a Lamarck theme), that soaking seeds in ice-cold water would make them more winter-hardy. Lysenko won the ear of the leader of Russia, Josef Stalin, an egotistical and unmercifully brutal dictator. Lysenko belittled and attacked Vavilov in order to discredit him. More importantly, Stalin appointed Lysenko director of the Institute of Genetics within the Russian Royal Academy of Sciences.

A debate with Lysenko sealed Vavilov's fate in 1939 when Lysenko used his pseudoscientific ideas in front of a prejudiced crowd to discredit Vavilov's valid science. In August 1940, Vavilov was picked up by a black sedan and taken to a prison managed by the secret police. While there, he resisted hours and days of brutal torture but finally broke, saying that he was a spy. He was abruptly sentenced to be shot. While he was sitting in his barren, cold jail cell in a weakened state of hunger, a group of guards dragged him away to escape the German forces invading Moscow in the early days of World War II. Vavilov was sent to another prison in the interior of Russia for safekeeping. Meanwhile, Vavilov's loyal laboratory colleagues continued his research throughout the siege on Leningrad. Although Stalin did not appreciate the research done in the seed laboratory, Hitler did. He ordered his forces to respect this prize. He even established a special tactical unit to take the seed bank and bring its riches back to Germany. Fighting rats and hunger, Vavilov's comrades starved, along with 800,000 citizens of Leningrad. However, out of respect for Vavilov, the researchers did not eat the precious nuts, seeds, and grains stored within. In late January 1943,

Nikolai Vavilov starved to death in a dreary prison. Oh, yes, what happened to Lysenko? Vavilov's nemesis continued his grip on Russian genetics until the 1960s. Finally, he was removed from power by Premier Nikita Khrushchev for his pseudoscientific strategies and his role in the loss of millions of lives. This one man set Russian genetics back decades. With the rise of epigenetics (which will be discussed later), Lysenko's discredited ideas are unfortunately being resurrected in some circles.

Today, the successor to Vavilov's seed bank, known as the Svalbard Global Seed Vault, located between Norway and the North Pole, houses over one million sample seeds and is a standing tribute to one of the unsung heroes of science. So the next time you visit a supermarket, think of Nikolai Vavilov (Baranski 2014).

PoI 1.

a. What is the purpose of the Svalbard Global Seed Vault?
b. What is its capacity?
c. How are the seeds stored?
d. What is the global need for such a facility?
e. Are there other seed vaults located across the planet?

Watch the news, read a newspaper, or thumb through a science-and-news-related magazine, and you will find that hardly a day passes without new breakthroughs in genetics. Genetics is a thriving discipline of the biological sciences that incorporates scientific observations, sophisticated techniques, and powerful statistics. It has the capacity to help us understand evolution, improve crops, develop medicines, and diagnose and cure disease. Unfortunately, in the wrong hands, genetics can become a potent bioweapon. It is our duty as responsible global citizens to see that science and genetics serve to benefit humankind.

Not only has genetics shaped everything around you, it also shapes *you*, from the color of your eyes to your ability to resist disease. The subdisciplines of *human* and *medical genetics* have grown exponentially in recent years. The Human Genome Project and other endeavors have given us a new understanding of our place in the living world, an understanding of inheritance, a new view of disease, and a hope to cure devastating genetic disorders.

APPLYING SCIENCE TO LIFE 2 From Spirits to an Understanding

Imagine living in a world before the age of medicine and genetics. Your newborn daughter seemed perfectly fine at birth, but after several months, she began displaying unusual symptoms. At the age of five, your child was profoundly physically and mentally challenged, unable to speak or feed herself or walk, and possessed a strange musky odor. She displayed muscle rigidity, skin and bladder problems, and stunted growth. What could be the cause of this horrible condition?

Through the years, many physicians examined the child and could offer only hollow explanations about the sickness, accompanied by a series of mystical chants and incantations. Predominant explanations were that the young girl was possessed by evil spirits or that the gods were angry. Such conclusions may be drawn when people feel powerless. Through science, we can now face such problems with logic and confidence.

In 1938, Professor Ivar Asbjørn Følling, after conducting intense research, concluded that the disorder that plagued such children for centuries was the result of their genetics. He and his colleagues concluded that these children possessed too much of an amino acid called phenylalanine in their urine and blood. As a result, these children had a metabolic defect in their way of processing phenylalanine, a disease now called *phenylketonuria (PKU)* (Mayo Clinic 2020, "Phenylketonuria").

Have you ever noticed warnings on a diet cola or candy for phenylketonurics, to be aware that this product contains phenylalanine? Check it out! Further genetic studies concluded that the inheritance of PKU was simple. If both parents of an affected child were in normal health, each parent was a carrier for the defective gene. This defective gene could be passed to the offspring. Children possessing two defective genes would in turn have PKU. Later in this chapter, we will address specific terminology for this inheritance pattern.

In the 1950s, science took the next great step in countering this genetic anomaly by reducing the amount of phenylalanine in the diet. If PKU was detected a few weeks after birth, a diet low in phenylalanine circumvented the deleterious symptoms of PKU. However, this special diet was not effective if started one or two months after birth. Usually, the restrictive diet was lifted in the late teen years after mental and physical development was completed. As a result, early detection of PKU became a medical priority. In 1963, a simple blood test was developed to test for phenylalanine. Today, this test is given at birth, and a horrible genetic disorder has been put against the ropes.

10.2. Gregor Mendel

For centuries, farmers understood the basics of crossbreeding plants and animals in order to attain desirable results. Also, people noticed family resemblances within particular families. How could this be? Despite these fundamental observations, many invalid mechanisms influencing inheritance were put forward by scientists. About the same time Charles Darwin was working on his monumental *On the Origin of Species*, *Gregor Mendel* (see Figure 10.2), a humble Austrian monk, was establishing the fundamentals of inheritance in a monastery garden in what is now the Czech Republic.

PoI 2. Unfortunately, it was discovered that mothers spared the manifestations of PKU could still give birth to children exhibiting the irrecoverable effects of PKU.

a. Why did this happen?
b. What has been done to circumvent this problem?

Gregor Mendel had an intense interest in mathematics and science and planned on becoming a teacher or professor. However, he did not pass the test to join his desired occupation, so he entered the rural St. Thomas monastery, where he eventually became an abbot, to serve as a substitute teacher. While in the monastery from 1856 to 1863, Mendel began growing an experimental garden featuring the garden pea. He grew thousands of pea plants, focusing on seven physical characteristics, plant height, pod form, pod shape, flower color, flower position, seed shape, and seed color (see Figure 10.3). Throughout his experiments, Mendel recorded detailed notes on his work. Mendel was looking for a predictive theory of breeding that could predict exactly the results of crossing plants of contrasting characteristics (Biography.com Editors 2020).

FIGURE 10.2 Gregor Mendel (1822–1884)

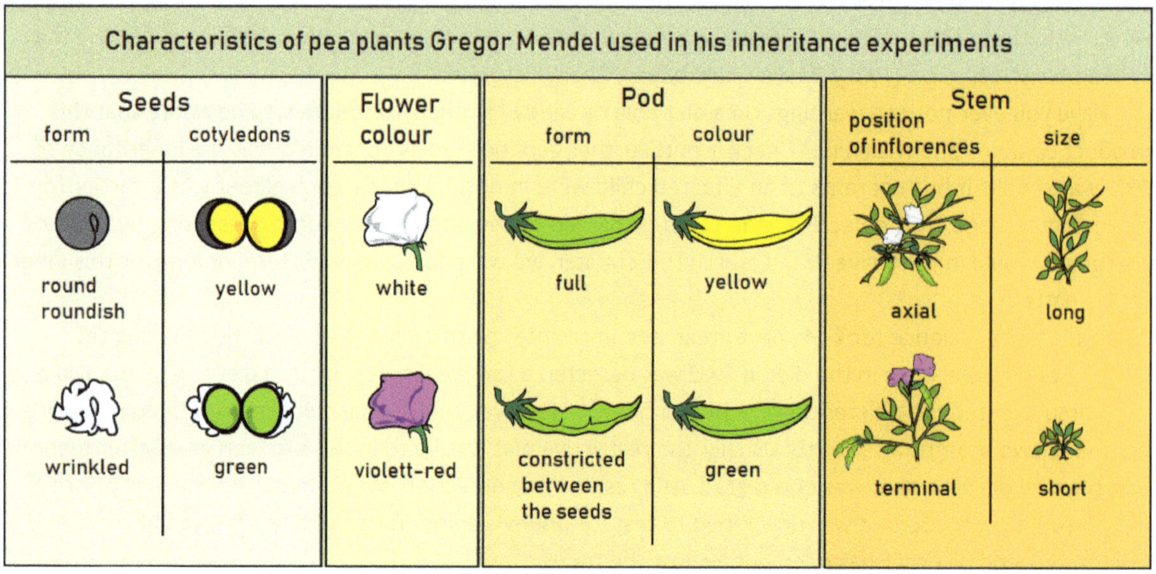

FIGURE 10.3 The seven basic traits of the peas that Mendel studied

For example, when Mendel crossed a yellow pea plant with a green pea plant, all of the resulting peas were yellow in color. Thus, the yellow trait masked the expression of the green trait. He called yellow pea a *dominant trait* and the trait masked by the dominant trait (green pea) the *recessive trait.* He discovered that if he crossed the progeny of this cross, it would yield a three yellow-pea to one green-pea ratio. Thus, as Mendel postulated, these traits were influenced by invisible factors that caused predictable outcomes. These factors are now known as *genes* (units of heredity). In 1866, Mendel published his findings in the work *Experiments on Plant Hybridization.* Interestingly, the humble monk had published his monumental work in a relatively obscure journal, so few initially took any notice of it. Poor Mendel died in 1884, not knowing that his ideas spawned a critically important field of biology— *genetics.* In the early 1900s, three scientists independently rediscovered and verified Mendel's work, opening the door to a new field of science. Today, Gregor Mendel is known as "the Father of Genetics."

Through his research with pea plants, Gregor Mendel revealed the basic laws of inheritance. He deduced that his distinct invisible factors (genes) occurred in pairs, one from each parent. He also introduced the concept of dominant and recessive factors. Mendel proposed several important laws of inheritance: 1.) *Law of Segregation*: Every organism has two *alleles* (possible forms a gene may take), one from each parent. During the production of gametes, or sex cells, the alleles separate so that each gamete has only one allele. (This process is meiosis, covered in the previous chapter.) 2.) *Law of Independent Assortment*: Every trait is inherited independently of every other trait. For example, a plant with a red flower does not necessarily have yellow peas. These traits are inherited independently. 3.) *Law of Dominance*: In an organism with different alleles, the dominant allele will be expressed. Of course, modern genetics presents many exceptions to these original concepts. In fact, if Mendel were to read a modern genetics text, the exceptions to this law would shock him.

To develop an understanding for solving genetics problems, it is necessary to define some terms. As stated earlier, a dominant trait, represented with an uppercase letter, overshadows the expression of the recessive trait, designated as a lower-case letter. For example, in peas, the yellow pea trait (Y) is dominant to the green pea trait (y). The term *genotype* refers to the genetic makeup of an organism. A plant with yellow peas can be *homozygous* (same alleles) *dominant* (YY) or *heterozygous* (different alleles) (Yy). A plant with green peas is referred to as being *homozygous recessive* (yy). The homozygous condition may be termed pure and the heterozygous state to be a carrier. The term *phenotype* refers to the physical description of the traits expressed—the plant produces a phenotype of yellow peas or green peas.

QRQ 1. The last name of "the Father of Genetics" is ______.

QRQ 2. The genetic makeup of an organism, such as Rr, is its ______, while physical description of the trait(s) expressed, such as red flower, is its ______.

QRQ 3. Characterize each of the following allele pairs as heterozygous, homozygous dominant, or homozygous recessive:

a. RR
b. Rr
c. rr

QRQ 4. According to one of the first laws of inheritance, in an organism with different alleles, the ______ allele will be expressed.

QRQ 5. (T/F) According to one of the first laws of inheritance, every trait is inherited independently of every other trait.

QRQ 6. Every organism has (number) ______ (term) ______, or possible form(s) of a gene.

10.3. Punnett Square

Two British biologists, *Reginald Punnett* and *William Bateson*, played important roles in establishing the field of genetics. Punnett is perhaps best known for developing the *Punnett square*, a simple diagrammatic device that can be used to predict the potential genotypes and phenotypes resulting from parents of known genetic makeup. For example, suppose a father has dominant (A) and recessive (a) alleles for a gene, and so does the mother. A child gets one allele from each parent. With two possibilities from each parent, there are 2 x 2 = 4 possible zygotic combinations, as the following Punnett square illustrates:

	A	a
A	AA	Aa
a	aA	aa

Thus, genotypically, one out of the four possibilities (ratio 1:4) is homozygous dominant, AA; two are heterozygous, Aa or aA (ratio 2:4); and one is heterozygous recessive, aa (ratio 1:4). Phenotypically, three out of the four possibilities (AA, Aa, aA) result in the dominant trait (ratio 3:4), while one (aa) exhibits the recessive trait (ratio 1:4).

It is conventional to construct a Punnett square by placing the male alleles along the top of the square and the female alleles along the side of the square. Each potential zygotic combination is placed in the intersection (square) of the alleles. To determine the number of squares in a Punnett square, use the formula 4*n*, where *n* represents the number of traits. One trait, such as pea color, would use 4^1, or 4 squares, while pea color and flower color would involve 4^2, or 16 squares, and so on.

10.3.1. Practice Problems

1. In the gene for pea-plant flower color, a red-flower allele (R) is dominant to the allele for a white flower (r). If a homozygous dominant male flower (RR) is crossed with a homozygous recessive white flower (rr), predict the genotypic and phenotypic ratio of the potential offspring.

	R	R
r	rR	rR
r	rR	rR

Genotypic ratio: 4:4 red flowers

Phenotypic ratio: 4 red flowers:0 white flowers

2. In humans, having normal phenylalanine metabolism (P) is dominant to the allele for having phenylketonuria (p). If a homozygous dominant man (PP) and a heterozygous woman (Pp) decide to conceive a child, predict the genotypic and phenotypic ratios of the potential offspring.

	P	P
P	PP	PP
p	pP	pP

Genotypic ratio: 2:4 homozygous dominant normal, 2:4 heterozygous normal

Phenotypic ratio: 4:4 normal: 0:4 phenylketonuria

3. If you were a genetic counselor knowing that cystic fibrosis is a recessive genetic disorder, how would you explain to normal parents using a Punnett square what happened to their child born with cystic fibrosis?

	C	c
C	CC	Cc
c	cC	cc

Genotypic ratio: 1:4 homozygous normal, 2:4 heterozygous normal, 1:4 homozygous recessive cystic fibrosis

Phenotypic ratio: 3:4 normal, 1:4 cystic fibrosis

Both parents had to be heterozygous (Cc) for cystic fibrosis.

GtM 1. In the gene for watermelon shell design, the allele for having a solid-colored shell (S) is dominant to the allele for having a striped shell (s). If two heterozygous watermelons are crossed, predict the genotypic and phenotypic ratio of the resulting offspring.

GtM 2. If you have a cleft chin, like Tom Brady, John Travolta, Vanessa Hudgens, and Sandra Bullock, you possess a dominant trait. In humans, having a cleft chin (C) is dominant to having a smooth chin (c). If a man that is heterozygous and a woman that is homozygous recessive decide to have a child, predict the possible genotypic and phenotypic ratio of their potential offspring.

FIGURE 10.4 A polydactyl cat from Ernest Hemingway's home in Florida

GtM 3. Many of the famous Hemingway cats are polydactyl (have extra digits), sometimes called mouse catchers (see Figure 10.4). The allele for polydactyly (P) is dominant to the allele for the normal number of digits (p). If a male homozygous dominant cat mates with a female heterozygous cat, predict the genotypic and phenotypic ratio of the potential offspring. By the way, polydactyly is a dominant trait in humans, too.

QRQ 7. Fill in numbers: The number of squares in a Punnett square that addresses all seven traits Mendel investigated in his pea plants is ______ raised to the power ______.

QRQ 8. The gene for hairline has a dominant allele (W) for widow's peak and a recessive allele (w) for straight hairline. Suppose a man with Ww and a woman with ww plan to have children. Develop a Punnett square before continuing.

a. Does the father have a widow's peak?
b. Does the mother have a widow's peak?

Predict the following ratios for potential offspring:

a. Genotypic ratio for ww
b. Genotypic ratio for wW
c. Phenotypic ratio for straight hairline

10.4. Beyond Mendel

Although Mendel's basic ideas are brilliant, scientists soon discovered that his simplistic patterns of inheritance do not hold true in all cases. Not all inheritance abides by simple dominant-recessive rules, and a single gene does not determine each heritable trait, as once thought. These scientists were and are amazed at the dynamics of gene expression.

10.4.1. Incomplete and Codominance

Have you ever noticed that some plants have red flowers, white flowers, and pink flowers? These flowers are displaying *incomplete dominance,* in which the heterozygous condition results in an intermediary phenotype. In this case, a plant with red flowers (RR) crossed with a plant with white flowers (rr) will result in offspring with pink flowers. Other examples of incomplete dominance include the color of the fruit in eggplants, tail length in some varieties of dogs, and familial hypercholesterolemia (excess blood cholesterol) in humans (Fossella 2001).

GtM 4. If two Rr pink plants are crossed, what will be the potential genotypes and phenotypes of their offspring?

GtM 5. Give the blood type for each of the following genotypes.

a. AA
b. BB
c. OO
d. AO

Another genetic mechanism with dominant-recessive exceptions is *codominance,* where both alleles are expressed in the heterozygous condition. In codominance, the phenotype is not an intermediate between the two traits expressed, but both phenotypes are exhibited. A classic example of codominance in humans is the *ABO blood group,* where both A and B alleles can be expressed in the heterozygous condition, resulting in the blood type AB. However, types A and B are both dominant to type O. Thus, genotype BO results in blood type B phenotype. Multiple alleles of a particular gene can be expressed on the surface of red blood cells, which is the basis of the blood types A, B, AB, and O.

Do you know your blood type? The ABO blood group is not the only parameter evaluated when you get your blood typed or crossmatched. However, the ABO group is best known, and if a person receives the wrong blood type through a transfusion, it can be life-threatening. The exact mechanisms, involving antigens and antibodies, are a bit too complex for this reading but are a great topic for further investigation. Tables 10.1 and 10.2 are useful summaries for understanding blood types

TABLE 10.1. Prevalence of ABO blood types (Digg 2001)

Phenotype (Blood type)	% in US	Genotype
A	42%	AA, AO
B	10%	BB, BO
AB	4%	AB
O	44%	O

TABLE 10.2. Blood-Type Donation Possibilities, with + and – Indicating Rh^+ and RH^-, Respectively (American Red Cross 2020)

Blood Type	Can Donate To	Can Receive From
A+	A+, AB+	A+, A–, O+, O–
A–	A+, A–, AB+, AB-	A–, O–
B+	B+, AB+	B+, B–, O+, O–
B–	B+, B–, AB+, AB–	B–, O–
AB+	AB+	Anyone
AB–	AB+, AB–	AB–, A–, B–, O–
O+	O+, A+, B+, AB+	O+, O–
O–	Anyone	O–

10.4.2. Range of Expression

Penetrance is an all-or-nothing mode—the gene is expressed or not. However, if an organism has the genotype to express a particular trait, sometimes the trait is not expressed. This phenomenon is called *incomplete penetrance,* which may occur as the result of environmental or genetic variables. For example, having extra digits, or polydactyly, is a dominant trait that is not expressed in some people with the gene. The lower degree of penetrance may obscure the dominant inheritance pattern of this trait.

PoI 3. Another important blood cell surface protein is the *Rhesus factor* (Rh+ and RH–).

a. What is the RH factor?
b. What is the natural function of the RH factor?
c. Why is this factor important to pregnant women?

Besides discrete expressions of allelic combinations, many traits exhibit *variable expressivity,* or differences in the observed effects of a particular genotype. Often, such expression exists as a continuum at the phenotypic level rather than being expressed or not expressed. This is evident in genetic disorders, such as *neurofibromatosis.* One in 3,000 individuals have this autosomal dominant genetic disorder. Some with neurofibromatosis only express a lightly pigmented "café-au-lait" spot that resembles spilled milky coffee, while others develop thousands of disfiguring neurofibromas on their bodies. A number of genetic and environmental variables can influence expressivity of this gene.

Siamese cats exhibit variable coat patterns as the result of *environmentally induced variation* (temperature), causing the warmer regions of their body to be light-colored and the cooler regions of their body to be dark-colored. Variable age of onset can also influence expressivity as well, such as in certain forms of diabetes. *Modifier genes* can also influence expressivity of another gene, as evidenced by the size and location of piebald (the solid color with colored patches) spots on dogs, such as beagles (Griffiths et al. 2000).

Besides a continuum at the phenotypic levels, we can have a situation of one gene having multiple phenotypic effects, called *pleiotropy*, from Greek, meaning "many turnings." Such a gene can impact a number of seemingly unrelated systems. *Marfan syndrome*, an autosomal dominant genetic disorder that primarily affects the body's connective tissues, is a classic example. Some individuals hardly display any symptoms, while others show a host of them. One symptom, aortic enlargement (expansion of the main blood vessel that carries blood away from the heart) can be life-threatening. Other symptoms include tall and thin body type, long legs and arms, long fingers, flexible joints, flat feet, a curved spine, vision problems, and overgrowth of the ribs that produces a misshapen sternum. Historians suggest that perhaps Abraham Lincoln had Marfan syndrome. Other genetic disorders that are considered pleiotropic include cystic fibrosis, sickle cell anemia, phenylketonuria, and schizophrenia (Stearns 2010).

10.4.3. Not One, But Many

In the new world of genetics, it seems nothing is simple. Fortuitously, the traits that Mendel investigated in pea plants were either dominant or recessive. However, today we know that a particular phenotypic type of an organism can be influenced by several genes, a condition called *polygenic inheritance*. Geneticists today realize that over 180 genes are responsible for the variation in height of humans. Human height is a *continuously varying trait* that, if displayed for a population, forms a bell-shaped curve. However, external variables, such as diet, can also influence the expression of height. Skin color is an example of a polygenic trait that involves three genes. Of course, exposure to the sun and certain metabolic disorders can skew these patterns. Perhaps the best-known polygenic trait in humans is eye color (color of the iris), which ranges along a continuum from the lightest of blue to the darkest of brown. Brown eye color is the most common eye color in humans. Generally, blue and green eyes are restricted to people of European ancestry. Eye color is directly related to the amount of a brown pigment known as *melanin* deposited in the frontal regions of the iris. At least two major genes and several minor genes are responsible in the variation of eye color from blue to hazel to brown (JoVE n.d.).

Often, one gene masks or influences the expression of another gene, called *epistasis*. Many times, epistatic events can produce a novel phenotype, such as in the color of coats of Labrador retrievers (Labs). In Labs, the allele for producing a black coat (B) is dominant to the allele for producing a brown coat (b). Labs with brown coats (bb) are commonly referred to as chocolate Labs. In these dogs, another gene (extension gene) influences whether (E) or not (e) a particular type of pigment will be deposited in the shaft of the hair (BE = black; bbE = brown). If a Lab homozygous recessive for the extension gene (ee), it will display a yellow coat color, resulting in yellow Labs (Miko 2008).

The environment can play a part in the combined expression(s) of genes too. *Multifactorial traits* reflect the action of multiple genes (polygenic) that interact with one or more environmental factors to alter the phenotype of an organism. One example is *coronary artery disease (CAD)*, the leading cause of death in industrialized nations. Coronary artery disease involves a narrowing and blockage of the coronary arteries, caused by a buildup of cholesterol-containing deposits (plaque) in these arteries that supply the heart with blood, nutrients, and oxygen. CAD is certainly influenced by several genes, but it is also influenced by lifestyle and environmental factors. For example,

PoI 4. If type 2 diabetes runs in your family, what can you do to reduce the impact of this multifactorial disease?

an individual that is genetically prone to coronary artery disease can be severely affected by diet. In other words, watch those greasy pork chops and cheeseburgers! Other examples of multifactorial disorders include obesity, some types of cancer and arthritis, Alzheimer's disease, and hypertension (Children's Wisconsin n.d.).

QRQ 9. Classify each situation listed below: codominance, continuously varying trait, environmentally induced variation, epistasis, incomplete dominance, incomplete penetrance, multifactorial trait, penetrance, pleiotropy, polygenic inheritance, variable expressivity

a. continuum of expression at phenotypic level
b. dominant trait sometimes not expressed
c. multiple genes and environmental factors influence phenotype
d. one gene having multiple phenotypic effects
e. one gene masks or influences expression of another
f. phenotype is intermediate between two traits
g. plot of phenotype for a population forms bell-shaped curve
h. several genes influence phenotype
i. temperature influences expression of a gene
j. two phenotypes, both exhibited

10.4.4. Mosaicism

Have you ever noticed that some celebrities, such as Jane Seymour, Keifer Sutherland, Kate Bosworth, Demi Moore, and Dan Akroyd, have two distinctly different colored eyes? Some historians even describe Alexander the Great as having one brown eye and one blue eye! *Heterochromia* results when an individual's irises are different colors. Although heterochromia can be the result of a number of genetic and physical causes, one cause is a genetic occurrence known as *mosaicism*. A mosaic individual manifests two different genotypes that developed from a single fertilized egg, not to be confused with a chimera, which develops from two different fertilized eggs. Varied fur patterns in mice, cats, and other mammals are also related to mosaicism.

APPLYING SCIENCE TO LIFE 3 Of Calico and Tortoiseshell Cats

Did you know or ever wonder why most Calico cats are female? This phenomenon, plus the highly variable coat pattern of white, orange, and black fur patterns, makes them interesting genetic examples. Calicos are distinguished from tortoiseshell cats with their black and orange fur patterns (see Figure 10.5).

FIGURE 10.5 A calico cat

In cats, the gene responsible for white fur is located on an autosomal chromosome, and the gene responsible for black or orange fur is on the female X chromosome. The smaller male Y chromosome does not possess the gene for pigmentation, and the vast majority of male cats are black; orange; black and white; or orange and white, because they only possess one color gene on the X chromosome. In calico cats, white fur is marked by various-sized black and orange patches of color, while tortoiseshell cats have little or no white fur. So what is going on?

In the 1960s, British geneticist *Mary Lyon*, while studying color patterns in mice, helped answer the question of color patterns in calico and tortoiseshell cats by describing *X-inactivation*, or the *Lyon hypothesis*. Mammalian females, including humans, have two X chromosomes, but one remains inactive. In fact, the inactive X chromosome can be identified by a compact structure within the nucleus called a *Barr body*, which was identified by Murray Barr in 1948. In cats, if the X chromosome that carries the gene for black fur gets inactivated, that cell will create orange fur, and if the X chromosome that carries the gene for orange fur is inactivated, that cell will create black fur. Because the X chromosomes are inactivated at random throughout the body, the pattern on each calico and tortoiseshell cat will be unique.

One in 3,000 calico or tortoiseshell cats is male. These special cats are the result of a trisomic event, in which they have received two X chromosomes and a single Y chromosome (XXY) and are usually sterile. Humans have a similar condition, known as Klinefelter syndrome (LeMieux 2016).

PoI 5. Discuss genetic reasons for color patterns of each of the following:

a. tabby cats
b. Siamese cats

QRQ 10. A condition where two different genotypes develop from a single fertilized egg is called ______.

10.4.5. Sex Linkage

Humans possess two distinct kinds of sex chromosome, the *X chromosome* and the *Y chromosome*. It is thought that the X chromosome contains about 5 percent of the total DNA of a cell, and that translates to well over 1,000 genes. In contrast, the physically much smaller Y chromosome contains approximately 70 genes, but it is the Y chromosome that is active in determining the sex of an individual. In particular, the *SRY gene* (sex-determining region of Y) on the Y chromosome is responsible for the development of male genitalia and male sex characteristics. In the vast majority of humans, XX individuals are female, and XY individuals are male. However, in rare instances, as the result of a number of anomalies, some XX individuals may develop as male and some XY individuals may develop as female (NIH 2020, "X Chromosome"; NIH 2020b, "Y Chromosome").

Although females have two copies of the X chromosome, one X chromosome is switched off during embryological development to ensure that the female does not produce twice as many proteins. This developmental insurance is known as *dosage compensation*. Which X chromosome is inactivated is the result of a random process (see above, "Of Calico and Tortoiseshell Cats").

A gene that is located on the X or Y chromosome is considered *sex-linked* and is designated *X-linked* or *Y-linked*, respectively. Sex-linked genes may have nothing to do with sexual differences and may influence diverse characteristics, such as vision, glucose metabolism, the muscular system, and the skin. Because females and males inherit a different number of X chromosomes, different patterns of inheritance result. For example, with only one X chromosome, we say that males are *hemizygous* for any X-linked allele, and the dominant or recessive allele will express itself. In females, if an allele is recessive, she will only express that allele in the homozygous recessive condition. Red-green color-blindness is a common X-linked genetic disorder that affects more males than females. The same is true concerning Duchenne muscular dystrophy, affecting approximately 1 in 3,500 males born in the United States. The disease features continued muscular weakness over time, and affected individuals rarely live past 25 years old.

Perhaps the best-known example of an X-linked genetic disorder is *hemophilia*. This disorder affects the production of a single protein that is essential in the blood-clotting cascade. If left untreated, cuts can be deadly, as the victim can bleed to death. Not only cuts can cause problems, but bumps, bruises, and even regular movements can be painful and perhaps deadly. Because of *consanguinity* (inbreeding), the royal families of Europe (England, Spain, Germany, Russia) were riddled with hemophilia in the late 1800s and beyond. Remarkably, one person, *Queen Victoria of England*, was the initial carrier of hemophilia, which spread throughout the royal family tree (see Figure 10.6). One well-known victim was Alexi Tsarevich of Russia. Born into Russian royalty in 1901, it is said that the walls of the royal palace shook with his desperate cries until he was exiled and executed during the Russian revolution in 1918 (Mayo Clinic 2020, "Hemophilia").

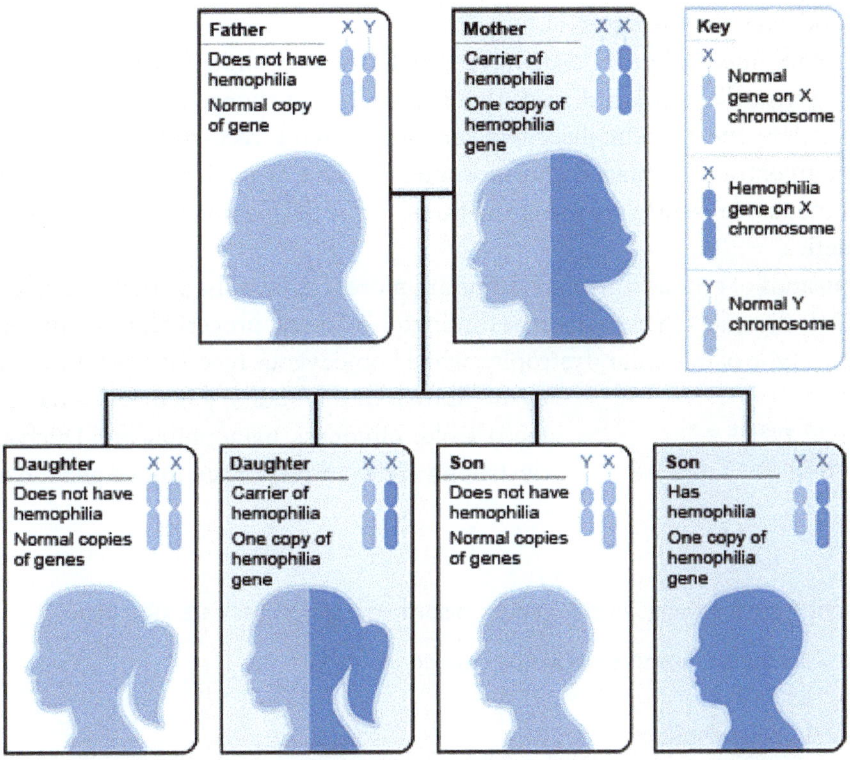

FIGURE 10.6 Inheritance pattern of hemophilia from female carrier

QRQ 11. (T/F) Sex-linked genes are always expressed sexually.

QRQ 12. Are XY individuals almost always female or male?

QRQ 13. Select from the following the appropriate term for each description below: consanguinity, dosage compensation, hemizygous, hemophilia, heterochromia, SRY gene, Y-linked

a. having one copy of a gene
b. one X chromosome switched off during embryological development
c. sex-determining region of Y chromosome
d. X-linked genetic disorder causing problems with blood-clotting

10.5. Medical Genetics

One of the most fascinating and important fields of the medical sciences is *medical genetics*. Careers in this field include researchers, technicians, counselors, and physicians. In recent years, a new emphasis in medicine has been developing an understanding of the genetic basis of disease. *The National Center for Biotechnology Information* and the *National Institute of Health* maintain a genetic database that is very helpful to anyone interested in genetics. One of its most valuable tools is the *Online Mendelian Inheritance in Man* (*OMIM*) site (NCBI 2020). Updated daily, OMIM presents the most comprehensive and recent information regarding all known genetic disorders in humans (over 8,600) and over 16,000 genes.

In recent years, medical geneticists have concluded that almost every disease has a genetic component. A *genetic disorder* is a disease that is caused by a change in the DNA sequence (*mutation*) of a particular gene. Genetic disorders can be dominant, recessive, polygenic, multifactorial, resulting from an extra or damaged chromosome, induced by environmental factors, sex-linked, or even spontaneous. Presently, we are at the threshold of developing a much better understanding of human genetics.

Of the thousands of genetic disorders, some are more common than others. Several homozygous dominant genetic disorders in humans are Huntington's chorea, progeria (premature aging), Marfan syndrome, and a type of muscular dystrophy, while homozygous recessive genetic disorders include cystic fibrosis and hereditary hemochromatosis (iron overload). Color blindness, hemophilia, and Duchenne muscular dystrophy are examples of sex-linked genetic disorders.

QRQ 14. A(n) ______ of a gene is a change in its DNA sequence.

PoI 6. Describe the following unusual genetic disorders and how they are inherited:

a. Jumping Frenchmen of Maine disorder
b. Ichthyosis
c. Hypertrichosis
d. Michelin tire baby
e. Achoo syndrome

PoI 7. Identify the chromosome or chromosomes, mode of inheritance, and basic characteristics associated with the following inherited disorders:

a. Tay-Sachs disease
b. Thalassemia
c. Ehlers-Danlos syndrome
d. Albinism
e. Sickle cell anemia
f. Achondroplasia

APPLYING SCIENCE TO LIFE 4 CF, from Chaos Comes Clarity

One evening during dinner, newlyweds Bryan and Jena began discussing the future, including starting a family. The couple became ecstatic about the possibility of becoming parents. In the excitement, Bryan became very anxious when he recalled the chaos of living in a family with the genetic disorder *cystic fibrosis (CF)*. Having CF, Bryan remembered the countless doctor visits, scary hospital stays, painful digestive problems, hours of rigorous treatments, and the fight merely to breathe. Most of all, he remembered how his older brother, Shaun, gallantly fought and succumbed to the fatal disease with dignity and courage. Bryan asked, "How can I bring a child into this world to face such pain and suffering?" Also, he recalled that 97–98 percent of males with CF are infertile. Silently, Bryan faced the chaos of CF once again.

Today, in the United States, over 30,000 people have cystic fibrosis. Approximately 1,000 new cases of CF are diagnosed yearly, usually before the age of two. Traditionally, CF is considered a chronic, progressive genetic disorder whose incidence is estimated at 1 in 2,500 live births in Caucasians of Northern European decent. Nearly 1 in 30 Caucasians are carriers of CF mutations, whereas the carrier frequencies for other ethnic groups are much lower. The occurrence of CF is equal in both males and females. However, females with CF have more serious symptoms than males.

Located on the long arm of chromosome 7 is the *cystic fibrosis transmembrane conductance regulator (CFTR)* gene. This gene serves to make the CFTR protein, which controls the movement of salt and water in and out of bodily cells. In the normal state, the CFTR protein acts as a type of membranal channel that allows cells to release chloride and other ions into the extracellular environment. In individuals with CF, the CFTR protein is defective, and the cells do not release their chloride ions. In turn, the abnormal cells draw water out of the surrounding mucus. The result is an improper salt balance within the cells and the formation of a thick, sticky mucus that can clog or impede the function of various organs and structures, such as the lungs, reproductive organs, and pancreas. The anomaly also causes the patient to produce copious amounts of sweat. Although there are as many as 1,700 mutations that can cause CF, 70 percent of CF patients have the Δ508 mutation. The anomaly consists of a mutation at position 508 of a 1,480-nucleotide sequence that results in the deletion of three nucleotides from the CFTR gene. The loss of these nucleotides causes the CFTR protein to lack a single amino acid known as phenylalanine.

A day in the life of Bryan is a very chaotic event, as he deals with CF and the defective CFTR protein. Because of the treatment routine and the high level of discipline required, this endeavor can be overwhelming and at times requires dedication, vigilance, and discipline. First thing in the morning, his blood glucose level is checked and recorded and the following oral medications are administered: GERD medicine, multivitamins, vitamin D3, calcium, medicine that allows chloride ions to pass through the channel pore and increases the activity of the CFTR protein, two antimicrobials, an appetite stimulant, a mucus-thinning agent, and allergy medicine. Pancreatic enzymes to aid in digestion are administered according to the nutritional content of the consumed meal (up to eight capsules per meal). Also, during breakfast, Bryan counts the amount of carbohydrates in the food consumed, adjusts his fast-acting insulin shot, and administers a shot of long-acting insulin. After breakfast, he performs airway clearance therapy. This involves inhaling a bronchodilator, nebulizing medicines for thinning pulmonary secretions and antimicrobials, and wearing an airway clearance vest, which shakes the lungs to break up the mucus.

FIGURE 10.7 Bryan and Jena, now the proud parents of the happy and healthy Rhys

Before lunch, Bryan performs routine blood glucose testing. His lunch, like all meals, should be high in calories, fats, carbohydrates, and proteins. During the meal, Bryan counts the amount of carbohydrates and adjusts his insulin shot. The amount of carbohydrates and fat content is important in order to administer the proper dosage of pancreatic enzymes. Before dinner, blood glucose level is checked and recorded, and carbohydrates are counted before many of the same drugs as are taken at breakfast are administered. Healthy snacks are encouraged for CF patients. Any snacks throughout the day require the use of pancreatic enzymes and insulin. For example, a Snickers bar requires two pancreatic enzymes, along with two units of fast-acting insulin. Before bedtime, inhalers and nebulizers are used while airway-clearance-vest therapy is performed.

Eventually, from the chaos of cystic fibrosis came clarity. A new treatment that addressed the genetics of the disorder, not the symptoms; better medicines; and better therapies improved the quality of Bryan's life. The couple visited a fertility institute and learned how science has pierced the veil of infertility, despite CF. Thanks to assisted reproductive therapies, Bryan and Jena are now the proud parents of a healthy baby boy named Rhys, and Bryan's brave brother's final words, "Stay positive and love your life," will influence another generation (see Figure 10.7).

10.6. Epigenetics

In recent years, a new field of genetics has become increasingly important—*epigenetics*, which investigates the changes in organisms caused by modification of gene expression rather than alteration of the genetic code itself. These changes to the genome are called *epigenetic changes*. All of the chemicals that have been added to the genome (intentionally or accidentally) to regulate the expression of genes comprise the *epigenome*. These chemicals are not part of the DNA itself but remain attached to the DNA molecule. These epigenetic changes can be passed on to future generations. In addition to certain chemicals, scientists are discovering that a number of other sources, such as diet, pollution, age, and lifestyle, have the potential to alter the epigenome by turning particular genes off or on and influencing the production of important proteins. For example, *DNA methylation* is a type of epigenetic activity in which small molecules of compounds, called methyl groups, attach to the DNA, turning off genes (WIE 2019).

APPLYING SCIENCE TO LIFE 5 Starvation and Epigenetics

The "Dutch Hunger Winter" lasted from November 1944 to the spring of 1945 during World War II. The cause of this famine, responsible for over 20,000 deaths, was a German blockade of Holland, ending the import of supplies. People resorted to eating grass, pets, and even tulip bulbs. As the result of excellent health care record-keeping, incredible studies have been made on the survivors of this disaster.

Epidemiologists and geneticists have discovered that pregnant women were particularly vulnerable to the impact of the famine. If a mother suffered malnutrition *only* during the first trimester of the pregnancy (at end of famine period), the baby was born within normal size parameters. However, extended studies have shown that years later these children suffered a number of medical issues, including diabetes, obesity, high triglyceride levels, and mental health issues (schizophrenia). If a mother was well-fed prior to conception but suffered malnutrition during the entire pregnancy, their babies were born small. Years later, when these babies became adults, they were small in stature and thin but suffered none of the issues of the other population. Even more surprising, scientists determined that when children of the mothers that were malnourished (in the first trimester or throughout their pregnancies) had their own offspring, their children (grandchildren of the malnourished women) developed diabetes and other medical issues later in life. Scientists began to wonder what was responsible for such results.

Investigators have determined that certain genes of the babies that were malnourished during the first trimester received epigenetic methylation tags. This event and these changes in the genes were determined to be responsible for the health issues of the population. As future studies are conducted, the power of environmental influences on heredity will be revealed (Schultz 2010; Zimmer 2018).

It is awesome to think how epigenetic changes have influenced evolution. Certain changes are probably responsible for the success of some species and the demise of others. These changes are also responsible for making us unique. In recent years, epigenetics has been linked to several diseases, including some cancers, Prader–Willi syndrome, Angelman syndrome, autoimmune diseases, leukemia, neurodegenerative disorders (schizophrenia), and perhaps addiction.

In discussions about genetics, we often hear about the continuing debate involving nature versus nurture. As we learn more about this subject, new light shows that genes and environmental factors are part of a fully interactional system, and neither nature nor nurture can totally stand alone.

QRQ 15. All of the chemicals that have been added to the genome (intentionally or accidentally) to regulate the expression of genes constitute the ______.

QRQ 16. (T/F) Chemicals to regulate the expression of genes become part of the DNA itself.

QRQ 17. ______ is the area of genetics that investigates changes in organisms caused by modification of gene expression rather than genetic code.

Image Credits

Fig. 10.1: Source: https://commons.wikimedia.org/wiki/File:Teosinte_and_Modern_Corn_Comparison_(3745571067).jpg.
Fig. 10.2: Source: https://commons.wikimedia.org/wiki/File:Gregor_Mendel_portrait.jpg.
Fig. 10.3: Source: https://commons.wikimedia.org/wiki/File:Gregor_Mendel_-_characteristics_of_pea_plants_-_english.png.
Fig. 10.4: Copyright © by Averette (CC BY 3.0) at https://commons.wikimedia.org/wiki/File:Polydactyl-cat.jpg.
Fig. 10.5: Copyright © by Caspian blue (CC BY-SA 3.0) at https://commons.wikimedia.org/wiki/File:Calico_cat_in_La_Coruna_of_Spain-01.jpg.
Fig. 10.6: Source: https://commons.wikimedia.org/wiki/File:Hemophilia_01.gif.

LAB #10

Laboratory Exercises

10.7. Laboratory on ProbLab Genetics

10.7.1. Preliminary Work

In this laboratory, we will explore Mendelian genetics with dominant and recessive alleles of genes using probability. For each of the following questions, fill the appropriate letter(s) or value(s) into the blanks. From the text's website, view the introductory video and "ProbLab Genetics Diagram;" download and open "*ProbLab Genetics Rev.nlogo.*"

1. Click the Info tab at the top of the page and read sections "What Is It?," "Pedagogical Note," and "How It Works."

   a. Match each component with its possible color(s) in the model: A. black, B. blue, C. brown, D. cyan, E. gray, F. green, G. orange, H. red, I. violet, J. white, K. yellow

   _______ fish body color

   _______ fish fin color

   _______ dominant allele for body color in model

   _______ recessive allele for fin color in model

   b. **1.** ______ **2.** ______ Complete the following statement using A. dominant genes, B. genotype, C. phenotype, D. recessive genes. Fish appearance (coloration of body and fin) indicates the fish's **1.**, while a four-block of genes determining the body color and fin color represents the fish's **2.**

   c. ___________ Select all that the top row of a four-block has: A. alleles that determine body and fin color, B. alleles that determine body color, C. alleles that determine fin color, D. genetic code for body and fin color, E. genetic code for body color, F. genetic code for fin color, G. two letters representing alleles that determine body color, H. two letters representing alleles that determine fin color

   d. ______ With two alleles—G (dominant) and g (recessive)—for the gene determining body color, give the number of possible two-allele sequences, such as GG, for fish body color.

e. Answer the following questions concerning two-allele sequences for the gene determining body color with alleles G (green, dominant) and g (gray, recessive):

_______ number of sequences resulting in a green body

_______ percent of possible sequences resulting in a green body

_______ percent of possible sequences resulting in a gray body

_______ probability of green-bodied fish

_______ probability of gray-bodied fish

f. Similarly, answer the following questions considering the two alleles, B (blue, dominant) and b (black, recessive), for the gene that determines fin color:

_______ probability of fish having blue fin

_______ probability of fish having black fin

g. **1.** _______ **2.** _______ From each parent, an offspring gets **1.** (1, 2, 3, 4) allele(s), selected at random, from **2.** (E. each column, F. each row, G. the four-block).

h. ______ With two possible colors for the body and two possible colors for the fin, give the number of possible phenotypes.

i. ______ If there were three possible colors for the body and five for the fin, give the number of possible phenotypes.

j. ______ With two genes determining body color and two genes determining fin color, give the number of possible genotypes.

k. ______ Suppose an animal has one gene with two alleles determining body color, another gene with two alleles determining fin color, and another gene with two alleles determining eye color; give the number of possible genotypes.

2. Read "How to Use It." Select the model component for each description: A. 4-BLOCK DISTRIBUTION, B. AVE BLOCK, C. FISH TOTAL, D. GO, E. GO ONCE, F. GRAY-BLUE, G. GREEN TOTAL, H. LIFE-SPAN, I. MATE-WITH, J. NUM-FISH, K. PERCENT FISH BY PROPERTIES, L. REVEAL GENES, M. STOP-1000

_______ Designates mating rules

_______ Displays the mean number of green alleles

_______ Establishes number of years (time-steps) a fish lives

_______ Graphs numbers of each phenotype

_______ Halts the simulation on tick 1,000

_______ If pressed and a fish is selected, the fish's four-block is displayed

_______ If pressed and a mating pair is selected, the parents' and offspring's four-blocks are displayed

_______ Initial number of fish

_______ Run simulation for one time-step

_______ Shows histogram of the number of green alleles in the four-blocks

_______ Shows how many fish have green bodies

_______ Shows how many fish have gray bodies and blue fins

10.7.2. Laboratory Work

Use a spreadsheet to record results and calculate averages.

3. This question considers the initial phenotypes and genotypes of fish.

a. **1.** _______ **2.** _______ **3.** _______ **4.** _______ **5.** _______ Fish in this model have one gene with a dominant allele (G) and recessive allele (g) determining fish body color and another gene with a dominant allele (B) and recessive allele (b) controlling fish fin color. Complete the following hypothesis: Suppose we add a large number of fish to the world without pressing *go*. The initial percentage of green (or blue) fish should be about **1.** (A. 0, B. 25, C. 50, D. 75, E. 100) percent. Moreover, the most prevalent phenotype should be **2.** (F. gray-black, G. gray-blue, H. green-black, I. green-blue), while the phenotype with the least number of animals should be **3.** (F. gray-black, G. gray-blue, H. green-black, I. green-blue). The numbers of fish having the other two phenotypes should be **4.** (J. about the average of the greatest and least phenotypes, K. approximately equal, L. not equal). Moreover, the value of AVE BLOCK should be approximately **5.** (M. 1, N. 2, O. 2.5, P. 3, Q. 4).

b. In the model, have NUM-FISH be less than 25, press *setup*, and observe the histogram. The height of the first rectangle, whose base goes from 0 to 1, gives the total number of four-blocks having zero dominant alleles, or the number of fishes having no green (G) and no blue (B) alleles (i.e., g and b alleles only, or four-block ggbb). With base from 4 to 5, the height of the last rectangle indicates the number of fishes having four (all) dominant alleles (four-block GGBB). The monitor AVE BLOCK and the red line indicate the average number of dominant alleles in a four-block. Then, repeat the process changing NUM-FISH to be about 50, 75, 100, 200, 500, and 1,000. In a spreadsheet with your name(s) at the top, for NUM-FISH being 1,000, record the nine values in the upper right monitors in a column, labeling the area "Question 3b" and using the order below. Have the spreadsheet calculate the percentage of each type of fish.

_______ percent green

_______ percent gray

_______ percent blue

_______ percent black

_______ percent green-blue

_______ percent green-black

_______ percent gray-blue

_______ percent gray-black

_______ Give AVE BLOCK

_______ Do the results support your part a hypothesis, no (N) or yes (Y)?

c. P(__________ body) × P(__________ fin) The probability of two *independent events* (events that do not depend on one another) is the product of their individual probabilities. That is, suppose *A* is an event that has a probability *P(A)* of occurring, and *B* is an event with probability *P(B)*. If *A* and *B* are independent, then the probability of both *A* and *B* occurring, *P(A* and *B)*, is *P(A)* × *P(B)*. For example, green body and blue fin are independent events; body coloration has nothing to do with fin coloration. Thus, *P*(green body–blue fin) = *P*(green body and blue fin) is the product of which two probabilities?

d. In your spreadsheet, have another column with header "Theoretical Probability." Fill in the theoretical probabilities by totals for green body, gray body, blue fin, and black fin, as listed in answers to questions 1e and 1f. Using the formula for the probability of independent events *A* and *B*, *P(A* and *B)* = *P(A)* × *P(B)*, have the spreadsheet calculate the *probabilities* of green body–blue fin, green body–black fin, gray body–blue fin, and gray body–black fin. Give the theoretical probabilities using all decimal places needed. Your observed percentages, expressed as decimal numbers, from part b should approximately agree with these theoretical probabilities.

_______ green

_______ gray

_______ blue

_______ black

_______ green-blue

_______ green-black

_______ gray-blue

_______ gray-black

4. This question involves inheritance.

a. ______ Change NUM-FISH to 10 by moving/clicking the slider or change the value as follows: Right-click on NUM-FISH, select *Edit*, change *Value* to 10, and click OK. Change the speed to *slower* and press GO. Why does the fish population disappear so quickly? A. not enough blue-fin fish, B. not enough fish find each other to mate, C. not enough green-body fish, D. not enough green body–blue fin fish, E. too many dominant alleles, F. too many recessive alleles

b. **1.** ______ **2.** ______ **3.** ______ **4.** ______ Change NUM-FISH to be 30 and select *reveal genes*. Advance one step at a time, looking for a situation where two fish of the same body color have a child that is of a different body color by clicking on yellow squares. The body color of the two parents was **1.** (A. black, B. blue, C. gray, D. green), while the body color of the child was **2.** (A. black, B. blue, C. gray, D. green). Each parent had alleles **3.** (E. G and G, F. G and g, G. g and g) for body color, while the child had alleles **4.** (E. G and G, F. G and g, G. g and g) for body color.

For the rest of this question, when asked to select a phenotype, employ the following lettering: A. gray-black, B. gray-blue, C. green-black, D. green-blue.

c. **1.** ______ **2.** ______ Complete the hypothesis: When MATE-WITH is "Any Fish," NUM-FISH = 300, and LIFE-SPAN = 5, after 1,000 ticks, phenotype **1.** (A, B, C, D) is most likely to dominate, and on the average, AVE BLOCK will be close to **2.** (E. 0, F. 1, G. 2, H. 3, I. 4).

d. Turn on switch STOP-1000. Also, set NUM-FISH to be 300 with LIFE-SPAN being 5 and MATE-WITH being "Any Fish." Uncheck *view updates* or speed up the simulation. Have "Question 4d" and appropriate headers in your spreadsheet. For 10 times, press *setup* and *go*, recording in your spreadsheet AVE BLOCK, the frequency of each phenotype, and the dominant phenotype. Have your spreadsheet calculate the means and standard deviations of AVE BLOCK and the phenotypes.

_______ AVE BLOCK mean

_______ mean for green-blue phenotype frequency

_______ mean for gray-black phenotype frequency

_______ dominant phenotype (A, B, C, D)

_______ Do your experiments support your hypothesis from part c, no (N) or yes (Y)?

1. _______ **2.** _______ **3.** _______ Your data support the following: When MATE-WITH is "Any Fish," NUM-FISH = 300, and LIFE-SPAN = 5, after 1,000 ticks, phenotype **1.** (A, B, C, D) is most likely to dominate, while other phenotypes **2.** (E. may, F. do not) occur, and on average, *AVE BLOCK* is close to **3.** (0, 1, 2, 3, 4).

e. **1.** _______ **2.** _______ Complete the hypothesis: When MATE-WITH is "Same Body," NUM-FISH = 300, and LIFE-SPAN = 5, after 1,000 ticks, phenotype **1.** (A, B, C, D) is much more likely to dominate, and on average, AVE BLOCK is close to **2.** (0, 1, 2, 3, 4).

f. Keeping all other switches and sliders the same, change MATE-WITH to be "Same Body" and repeat part d.

_______ AVE BLOCK mean

_______ mean for green-blue phenotype frequency

_______ mean for gray-black phenotype frequency

_______ dominant phenotype

_______ Do your experiments support your hypothesis from part e, no (N) or yes (Y)?

1. _______ **2.** _______ **3.** _______ **4.** _______ Your data support the following: When MATE-WITH is "Same Body," NUM-FISH = 300, and LIFE-SPAN = 5, after 1,000 ticks, phenotype **1.** (A, B, C, D) is most likely to dominate, while phenotype **2.** (A, B, C, D) is the next most likely to occur, and the other phenotypes **3.** (E. may, F. do not or are extremely unlikely to) occur, and on average, AVE BLOCK is close to **4.** (0, 1, 2, 3, 4).

Why do these answers make sense?

g. **1.** _______ **2.** _______ Considering parts d and f, complete the following hypothesis concerning the situation where a fish can only mate with another fish of the same body and fin types: When MATE-WITH is "Same Both," NUM-FISH = 300, and LIFE-SPAN = 5, after 1,000 ticks, phenotype **1.** (A, B, C, D) will dominate, and on average, AVE BLOCK will be close to **2.** (0, 1, 2, 3, 4).

h. Keeping all other switches and sliders the same, change MATE-WITH to be "Same Both" and repeat part d, except run the experiment only five times.

_______ AVE BLOCK mean

_______ mean for green-blue phenotype frequency

_______ mean for gray-black phenotype frequency

_______ Do your experiments support your hypothesis from part g, no (N) or yes (Y)?

1. _______ **2.** _______ **3.** _______ Your data support the following: When MATE-WITH is "Same Both," NUM-FISH = 300, and LIFE-SPAN = 5, after 1,000 ticks, phenotype **1.** (A, B, C, D) **2.** (E. is the only one to survive, F. will dominate, but the other phenotypes occur), and on average, AVE BLOCK will be **3.** (0, 1, 2, 3, 4).

MODULE

11 Life's Instructions—It's All about DNA

Ignorance more frequently begets confidence than knowledge: it is those who know little, and not those who know much, who so positively assert that this or that problem will never be solved by science.

—Charles Darwin

11.1. Introduction

In recent years, DNA has become a buzzword and a part of modern culture. Statements like, "It's in my DNA," are commonly heard on the streets and in the media. The distinctly iconic double helix of DNA appears everywhere from cartoons to cosmetics labels to medicine boxes, and even on your pet's food can. DNA is also featured in science fiction movies, such as *Jurassic Park*, *Gattaca*, *Splice*, *X-Men*, and *Rampage*. In this century, the double helix has served as an important tool in catching bad guys in the television series *NCIS*. The history, chemical nature, and positive and negative potential of DNA technology are explored in documentaries and discussed by world and national leaders. Yet, with all of this hype and consideration, the majority of the public does not really understand this marvelous molecule and its importance in evolution, diversity, human health, and life itself. They know little about the potential of DNA and how people can use DNA to benefit or negatively impact the planet and humankind. The purpose of this chapter is to introduce you to the story of DNA, discuss its chemistry, help you understand how DNA works, and aid you in making wise decisions regarding DNA-related issues.

Why all the buzz about *DNA* (*deoxyribonucleic acid*)? It is an incredible molecule that possesses all of the chemical instructions necessary to build an organism and maintain its existence. DNA and its chemical cousin, *RNA* (*ribonucleic acid*), serve as the most fundamental units of heredity in living things and viruses. Within the chemistry of DNA can be found a universal set of chemical instructions known as the *genetic code*. Ultimately, within the code are the instructions for first making all of life on Earth, including bacteria, slime molds, roses, sea stars, and humans (see Figure 11.1). However, the code also allows for individuality within the species. It serves as the common thread from the origin of life throughout that great tree of life and from generation to generation. DNA is certainly the molecule of our past, present, and future.

FIGURE 11.1 Life on Earth, a testimony to the power of DNA

APPLYING SCIENCE TO LIFE 1 Discovery of DNA, Its Structure and Function

For such a hot topic, DNA discovery had very humble origins. The Swiss physician and researcher *Friedrich Miescher* graduated from medical school in 1868, but he was too shy for clinical practice, so he began a research career studying the chemistry of tissues. His focus for study was pus, an excellent source of white blood cells, which are cells that help defend the body against infections. After developing good techniques for isolating these cells, he began to study the contents of their nuclei. From his work, he described a new substance, rich in carbon, hydrogen, oxygen, nitrogen, and phosphorus, residing in the nucleus. He called this substance *nuclein*. Involving students in his research, Friedrich investigated the mysterious nuclein from various sources for the remainder of his career. When he died in 1895, the structure or function of nuclein was still unknown (DNA Learning Center 2011). Nuclein was deoxyribonucleic acid (DNA), a macromolecule found in all cells, as well as in some viruses. During the first half of the 20th century, many scientists tried to determine the structure of DNA. Scientists successfully discovered much about the chemistry of the molecule, but the correct structure was not elucidated until 1952 by Watson and Crick.

There remained the question of what role DNA played in the cell. Most scientists thought DNA too simple to contain the information necessary to guide the development of complex living systems. Protein, however, contained an array of amino acids that a majority of scientists thought could provide ample instructions to be the genetic material.

For many years, scientists tried to resolve the question. In 1952, Alfred Hershey and Martha Chase released results from some experiments they conducted with bacteria and bacteriophages (viruses) that were very convincing. Remember, phages are made up of a core of nucleic acid surrounded by a coat of protein. We now know that phages attach to the cell wall of bacteria and inject their DNA inside, where it is replicated. Only the DNA enters the bacterial cell. Phage proteins are synthesized from instructions from the replicated phage DNA. However, no one knew these details prior to 1952. In one test, Hershey and Chase grew bacterial cultures and phages whose protein coats were labeled with radioactive sulfur. In another test, they cultured bacteria with phages whose DNA was labeled with radioactive phosphorous. In each case, after growth, they separated the phages from the bacteria. Results showed that only the bacterial cells with the DNA from the phage showed radioactivity. This confirmed that DNA must contain the information for the production of new phages and, therefore, the hereditary material. Figure 11.2 illustrates the basics of these experiments.

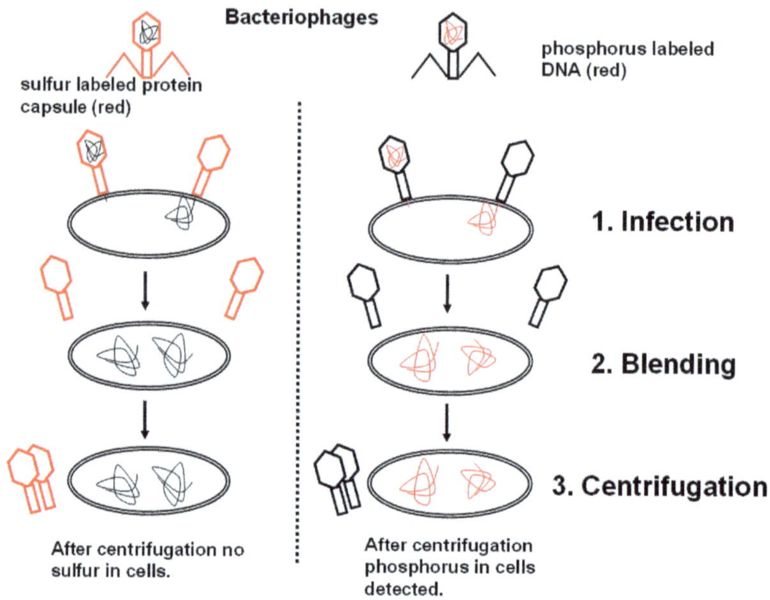

FIGURE 11.2 Hershey-Chase experiment

PoI 1. The following investigators contributed to our understanding of genetics. For each, describe when and what the basic contribution was:

a. Archibald Garrod
b. Frederick Griffith
c. Oswald Avery, Colin MacLeod, and Maclyn McCarty
d. George Beadle, Edward Tatum, and Joshua Lederberg
e. Phoebus Levene

11.2. Crick and Watson

Following World War II, building on studies that revealed the basics of DNA chemistry and function, scientists investigated the three-dimensional molecular architecture of DNA. Today, we credit *Francis Crick* and *James Watson* for uncovering the molecular architecture of DNA, but the true story is much more complicated, with many twists and turns rivaling that of a spy novel (see Figure 11.3). Little did they know that once determined, this discovery would usher in a new age of genetics.

Although Crick and Watson were eager to discover the molecular configuration of DNA, their accomplishment would not have been made possible without the essential contributions of several of their contemporaries. Keep in mind that the relationships among these individuals was not ideal, and egos, quirks, and some deviousness sometimes undermined good science. *Linus Pauling*, esteemed biochemist and double Nobel Prize laureate, was one of the early investigators of the architecture of DNA. However, he was not completely convinced of the role of DNA in inheritance and directed most of his efforts to investigating the role of proteins in heredity. Although his triple-stranded DNA model proved incorrect, many of his suggestions regarding the chemistry of DNA were invaluable. Escaping the terror in Nazi Germany, another biochemist, *Erwin Chargaff*, joined the faculty at Columbia University in New York. He observed

FIGURE 11.3a Francis Crick

FIGURE 11.3b James Watson

that the DNA of all species appeared to have the same four nucleotides, although each had varying sequences. His best-known observation, known as *Chargaff's rule,* stated that the ratios of bases adenine (A) to thymine (T) and cytosine (C) to guanine (G) were 1:1. The observations of these two scientists were invaluable to Crick and Watson's model development.

In the 1950s, the technique of *X-ray diffraction crystallography* would yield more crucial information about the three-dimensional architecture of DNA. At Kings College, London, Professor *Maurice Wilkins's* lab overcame a number of technical problems to produce the first X-ray images of DNA. These crude images suggested that DNA had a helical configuration. Enter one of the heroines of science, *Rosalind Franklin,* a colleague of Wilkins and an expert X-ray crystallographer (see Figure 11.4). Her images were much clearer than Wilkins's images because she used a form of DNA whose base pairs are almost perpendicular to the helix's axis. In 1953, Rosalind Franklin left DNA to work on the tobacco mosaic virus at another laboratory, but she sadly died of ovarian cancer, at the age of 37, in 1958. The cancer was likely caused by the lengthy exposure to X-rays in her research. Many biological historians speculate that if Franklin had lived and continued her DNA studies, she would have been awarded a Nobel Prize.

James Watson and Francis Crick were latecomers to the quest to determine the structure of DNA. In October 1951, the pair of scientists met at the University of Cambridge in England and immediately "hit it off." Prior to World War II, Crick was interested in physics, and during the war he helped develop radar and magnetic mines. Following the war, he returned to school to study biochemistry. Crick was very talkative, with a booming voice and a very entertaining

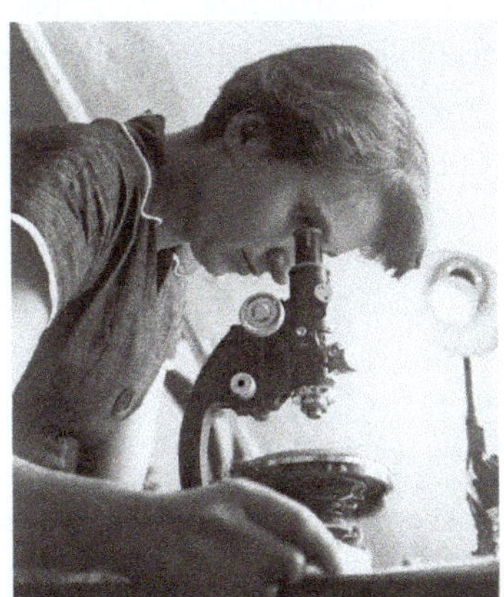

FIGURE 11.4a Rosalind Franklin

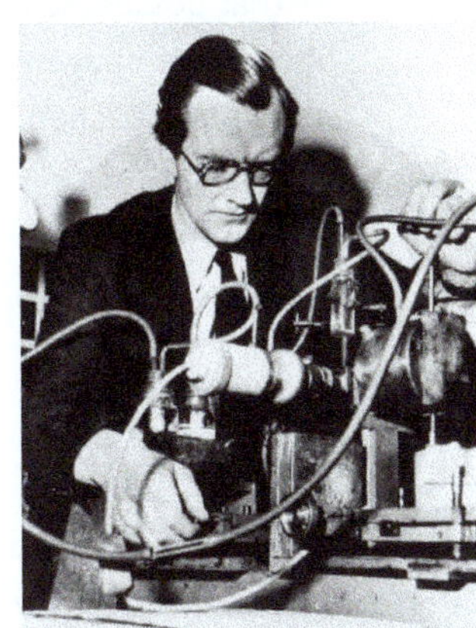

FIGURE 11.4b Maurice Wilkins

personality. With a strong interest in ornithology, James Watson, assertive and ambitious, received his PhD in genetics from Indiana University at the age of 22. Crick and Watson excelled in taking information established by other scientists to construct three-dimensional models. In 1951, they constructed models of DNA using cardboard cutouts representing the known components of DNA, as if they were piecing together a jigsaw puzzle. The scientists knew that the backbone of DNA was made from alternating deoxyribose sugar and phosphate molecules, to which adenine, cytosine, thymine, and guanine molecules were attached. They also were aware that the molecule was probably helically shaped, but initial models were flawed. The biggest breakthrough in their understanding of the molecular architecture of DNA came from the X-ray crystallography work of Rosalind Franklin. Her image of DNA, revealed without her knowledge to Watson and Crick, was just what they needed to complete their model. Watson realized that DNA was a *double helix*, with the nucleotides (A-T, G-C) held together by internal hydrogen bonds in an antiparallel configuration, and the resulting model proved accurate. Even better, the nucleotide configuration matched Chargaff's rule exactly!

In the prestigious scientific journal *Nature* on April 25, 1953, an article entitled "A Structure for Deoxyribose Nucleic Acid" appeared and changed biology forever. Two other articles also appeared in the journal, one authored by Rosalind Franklin and her collaborators and the other authored by Maurice Wilkins and his collaborators. Ironically, the work of Crick and Watson was primarily based upon their work. In 1962, the Nobel Prize in Medicine was awarded to Crick, Watson, and Wilkins "for their discoveries concerning the molecular structure of nucleic acids and its significance for information transfer in living material." According to the rules, a Nobel Prize must be awarded to no more than three *living* scientists. Rosalind Franklin had died of cancer in 1958 and was not eligible. However, Maurice Wilkins did mention her in passing in his acceptance speech. The discovery of the DNA double helix forever transformed genetics (Pray 2008).

APPLYING SCIENCE TO LIFE 2 A Physicist Who Inspired Geneticists

It is hard to believe that an Austrian quantum physicist and Nobel Prize winner, *Erwin Schrödinger*, had a tremendous influence on a generation of geneticists, including Crick, Watson, and Wilkins. In 1943, Schrödinger gave a lecture on the molecular nature of the gene. He inferred that it had to have a chemistry that possessed a nonrepetitive structure, be compact enough to fit in a cell, and pass along some kind of code. He imagined it stretching along the length of a chromosome fiber. Sound familiar? This lecture was later published in a book entitled *What Is Life: The Physical Aspect of the Living Cell* that influenced many young geneticists (Cobb 2013).

11.3. The Nuts and Bolts of DNA and RNA Structures

Although DNA and RNA seem remarkably simple, made of only sugar, phosphate, and nitrogen bases, they are incredibly large molecules composed of the elements carbon, hydrogen, oxygen, nitrogen, and phosphorus. Where did all of these elements come from? Get ready—the stars! As the astronomer *Carl Sagan* once said, "The nitrogen in our DNA, the calcium in our teeth, the iron in our blood, the carbon in our apple pies were made in the interiors of collapsing stars. We are made of star stuff"

(Sagan 1980). That's right. Life on Earth is based on carbon, formed in the death throes of ancient stars as three helium atoms fused to form a carbon atom. For example, the carbon atom in a molecule of your own DNA made its way through space and time to our humble planet. Recycled through the eons, it may have been a part of an ancient sea scorpion, a dinosaur, or another person in history.

We described briefly the structure of nucleic acids in section 5 of module 3, but hopefully you remember that DNA and RNA are *polymers* (macromolecules comprised of repeated subunits that are structurally similar) and that each nucleic acid chain is made up of a succession of subunits called *nucleotides*. A nucleotide is a *monomer* (one subunit of a chain of subunits in a polymer), and each nucleotide monomer is made up of a five-carbon *sugar* (pentose), a *phosphate* (PO_4), and a *nitrogen base* (nitrogen-containing molecule, with the chemical properties of a base). In DNA, the sugar is *deoxyribose*, and in RNA, the sugar is *ribose*. As Figure 3.19 shows, each phosphate binds covalently to two sugars. This chain of sugars and phosphates forms the backbone of the DNA, and each nitrogen base links to one sugar. In each type of nucleic acid, DNA or RNA, there are four different nitrogen bases. In both DNA and RNA are the nitrogen bases *adenine* (*A*), *guanine* (*G*), and *cytosine* (*C*). In DNA, the fourth nitrogen base is *thymine* (*T*), and in RNA, the fourth nitrogen base is *uracil* (*U*) (see Figure 11.5).

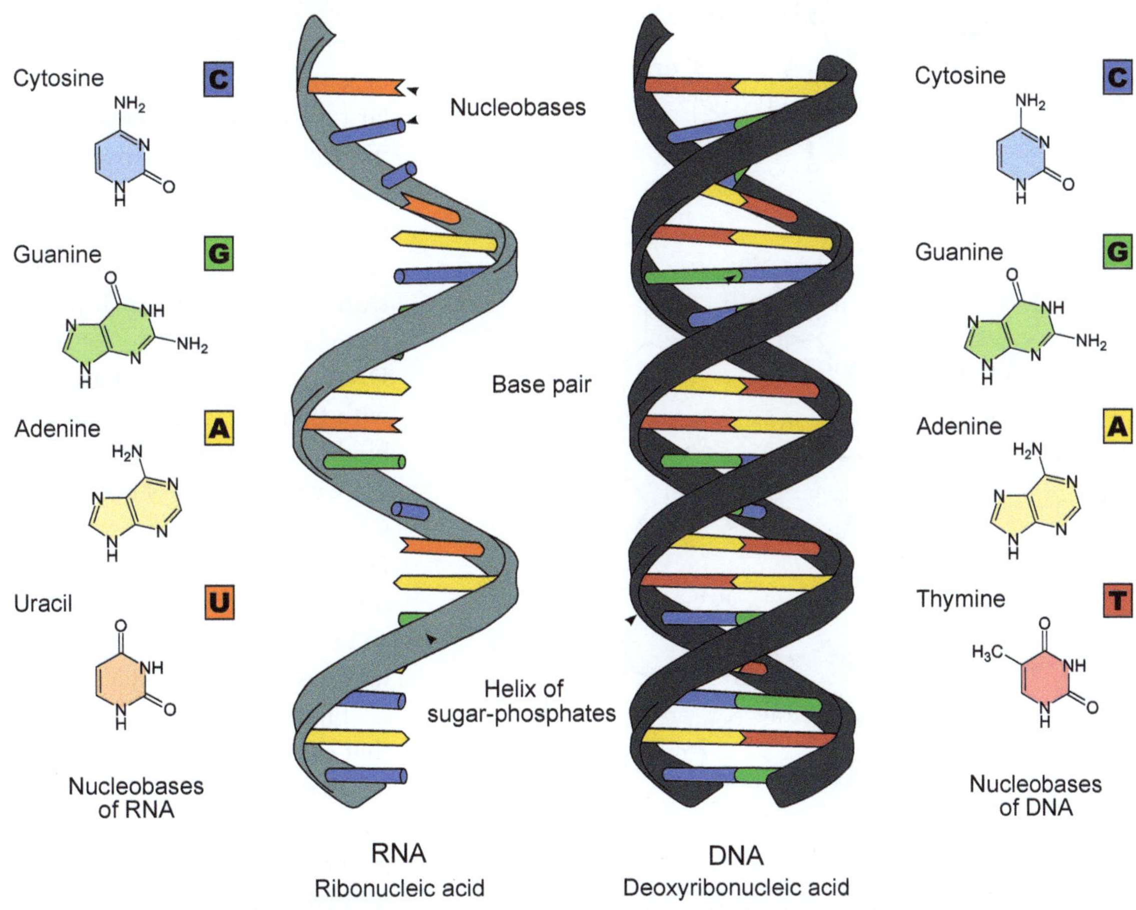

FIGURE 11.5 Structures of RNA and DNA, each with accompanying nitrogen bases

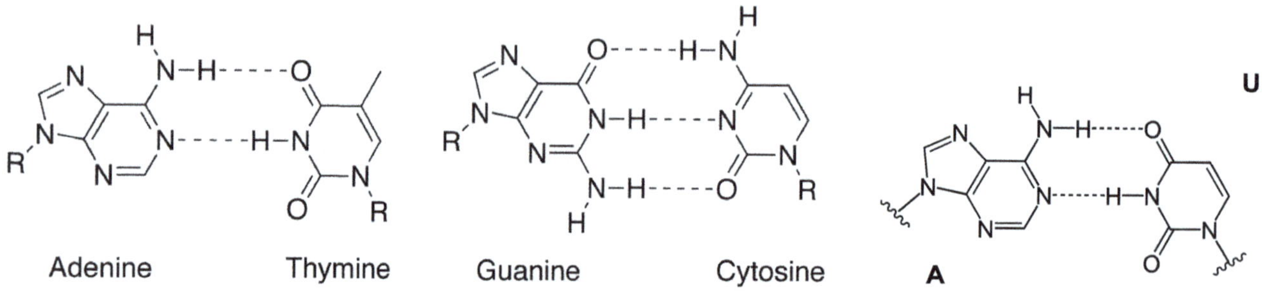

FIGURE 11.6 Base pairings: a. A-T in DNA. b. G-C in DNA or RNA. c. A-U in RNA

Figure 11.6 displays the chemical representations of the five nitrogen bases. It is the order of these nitrogen bases that gives rise to the genetic code of DNA and the intermediary codes of RNA. Except in certain viruses, DNA is double-stranded, and RNA is single-stranded. Picture DNA as a coiled or twisted ladder, where the sugar-phosphates are linked to form the side rails and the rungs are formed by the base pairs bonding between the two side-rails. The double stranding in DNA is maintained by the *hydrogen bonds* formed between the nitrogen bases of opposing strands. Each nitrogen base will *pair* only with a particular type of base, a *complementary nitrogen base*, in another nucleic acid strand. This *base-pairing* is essential, not only for the structure of DNA but also for DNA replication and transcription and translation involving RNA. In DNA, adenine always base-pairs with thymine, and cytosine only binds with guanine. Thus, A and T are complementary bases, as are C and G. With the transcribing of DNA into RNA, or with translation, adenine base pairs with uracil so that A and U are complementary bases in these situations.

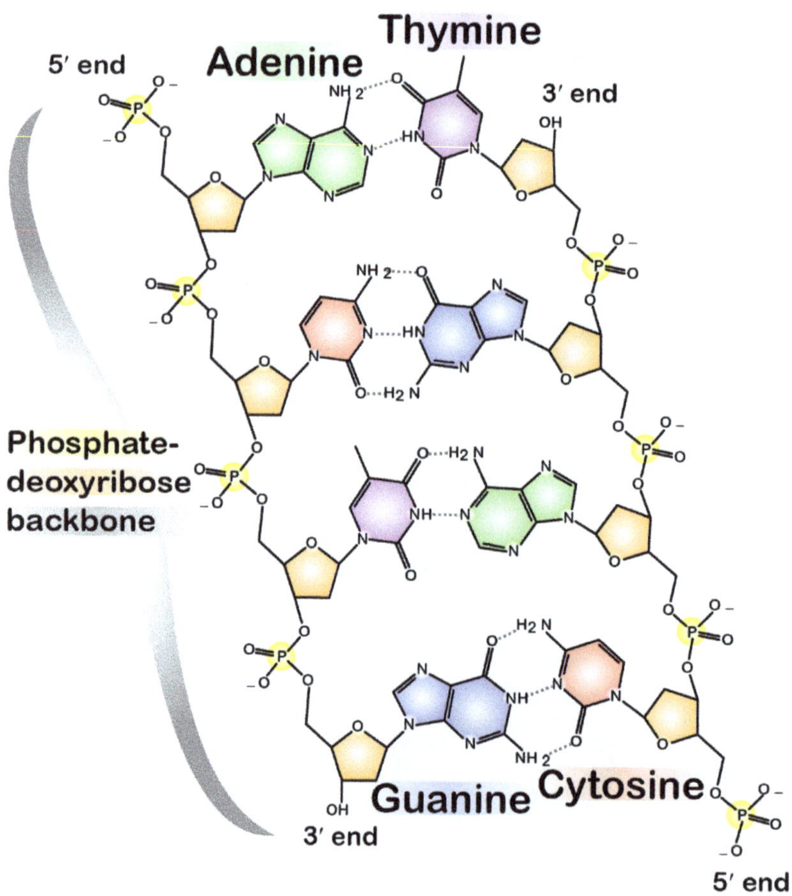

FIGURE 11.7 Short segment of double-stranded DNA with hydrogen bonds between base pairs

Figure 11.7 shows a cartoon that summarizes the fundamental structure of DNA. Note the base pairing, and also that the two strands run in opposite directions, termed *antiparallel*. Furthermore, each strand has what are called the

5' end and the *3' end*. The 5' end denotes the end of the nucleotide where the phosphate attaches to the sugar, while the 3' end indicates where the –OH interacts with the phosphate of the next nucleotide.

QRQ 1. The scientists given primary credit for the discovery of the chemistry of DNA were ______ and ______ (last names, alphabetical order).

QRQ 2. Each nucleotide in a nucleic acid is made up of a 5-carbon ______, a(n) ______, and a nitrogen ______.

QRQ 3. A(n) ______ bond is formed between the nitrogen bases of opposing strands in DNA.

QRQ 4. The sugar is ______ in DNA and ______ in RNA.

QRQ 5. Which is double-stranded, DNA or RNA?

QRQ 6. Select all that are bases in DNA and/or RNA: adenine, adesine, cytocil, cytosine, guanine, guasine, thycil, thymine, uracil, urasine

QRQ 7. Give the complementary bases for T, C, and U.

QRQ 8. Where the phosphate attaches to the sugar end of a nucleotide is the ______, while where the phosphate interacts with -OH is the ______ end.

11.4. DNA Replication

Whatever the biological occasion (growth, replacement, repair), whenever a cell undergoes cell division, DNA must make a copy of itself with extreme fidelity. *Replication* is the biological process by which a double-stranded DNA molecule makes an exact copy of itself during cell division. Genetic information resides in the sequence of the nucleotides found in the DNA of chromosomes. It is essential for the daughter cells to have an exact copy of that genetic information from the parent cell.

The process of DNA replication is a highly evolved, efficient, accurate, and complex molecular mechanism that ensures the continuity of the genetic material from one generation of cells to the next. In eukaryotic organisms, like ourselves, a variety of enzymes are the primary drivers of DNA replication, which occurs in the cell's nucleus. In a human being with 100 trillion cells, every somatic (body) cell possesses 46 chromosomes, each with its own unique DNA molecule. Approximately 3.2 billion nucleotide pairs reside in the chromosomes, with the smallest chromosome having about 50 million nucleotides and the largest approximately 260 million nucleotides. If you could type 200 characters/min for eight hours a day, every day (and you lived long enough!), it would take you over 180 years to type all the As, Ts, Gs, and Cs of the nucleotides in DNA. These vast numbers are represented by between 20,000 and 25,000 genes in the human genome. How long do you think that copying this takes to complete? DNA replication can occur within one of your cells in just a few hours with very few errors.

Each parental strand of the double helix can serve as a *template* for replication (see Figure 11.8). In eukaryotic cells, the process initiates prior to the S phase of interphase, but the actual replication of DNA takes place during the S phase. Because eukaryotic chromosomes are so large, there are multiple *points of origin* along a chromosome where the processes of replication begin. The number of points of origin can vary depending on the needs of a cell. Each point of origin is only used once in a cell cycle. DNA replication begins with an enzyme called *DNA helicase* opening up the double helix, exposing the single-strand sequences that will be copied. Literally, the DNA helicase serves to "unzip" the molecule by breaking the hydrogen bonds holding the complimentary bases together. This action can be seen with an electron microscope. The two separate strands, which begin to form along a *replication fork*, or point of separation, in turn serve as templates for building DNA. New strands are *always* synthesized in the 5' to 3' direction, beginning at the 3' end of their template strands. Using the strand template whose 3' end is exposed, one new strand, called the *leading strand*, is formed toward the replication fork, adding nucleotides one at a time continuously. Thus, being antiparallel to the original template, the leading strand is synthesized from its 5' end to its 3' end. The other strand, known as the *lagging strand*, grows away from the replication fork in a series of pieces, base-pairing nucleotide by nucleotide. With the template's 5' end exposed, the new antiparallel strand is formed in a backtracking fashion, a segment at a time, in the lagging strand's 5'-to-3' direction. Synthesis of the leading and lagging strand occurs concurrently in both strands at a similar rate.

In the leading strand, the starting point (*initiation*) for the synthesis of DNA begins as the enzyme *primase* builds a short piece of RNA, called the *primer* (5–10 RNA nucleotides long). In turn, the primer binds to the end of the leading strand of the replication fork, where replication starts. Another enzyme, known as *DNA polymerase*, travels along the original strand and selects, places, and bonds complementary DNA nucleotides to the original, *only* adding a nucleotide to the 3' end

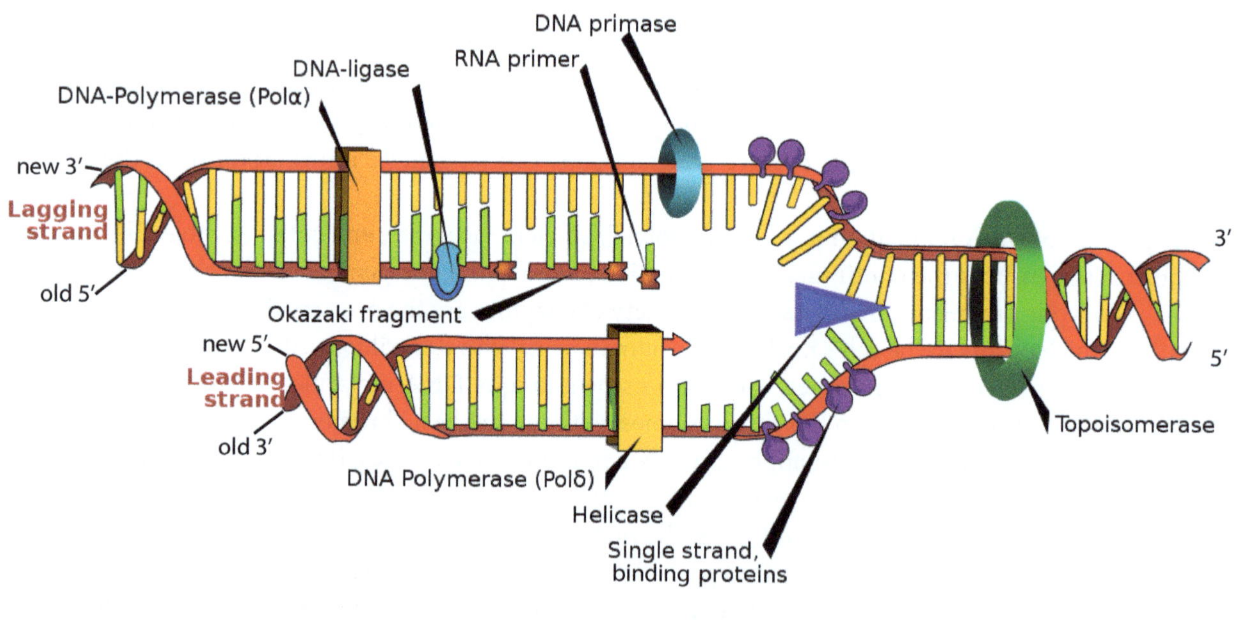

FIGURE 11.8 DNA replication

of the previous nucleotide, thus building the new strand. In human cells, about 50 nucleotides are added per second in a particular fork as the molecule is elongated.

The orientation of the lagging strand is different from that of the leading strand. Because of the antiparallel nature of the DNA molecule, the lagging strand-building occurs in segments, proceeding away from the replication fork within each segment. As the fork opens further, new template is revealed, new primer is laid down, and replication proceeds. Consequently, the DNA in the lagging strand is produced in segments, called *Okazaki fragments*. These fragments are usually less than 200 nucleotides long in eukaryotic cells, such as that of humans. The enzyme *ligase* joins the segments in order to complete the building of the lagging strand.

Once the complementary bases are matched up, an enzyme known as *exonuclease* strips away the primers that were replaced by complementary DNA nucleotides. Replication forms two new molecules of DNA, with each new strand complementary in sequence to the parent strand. Following replication, DNA coils up into the familiar double helix.

As stated earlier, DNA replication is not perfect, and errors periodically occur. Errors are rare, because DNA polymerase *proofreads* its work as it builds nucleotides. If the polymerase detects an error, it can remove the incorrect nucleotide and replace it with the proper nucleotide. Not all errors are detected by polymerase, which can result in a *mismatch error*. Other enzymes can often correct the problem. However, sometimes the error remains in place, resulting in a range of often serious problems. Additionally, cells are constantly exposed to agents called *mutagens* that can also alter or damage the DNA, such as chemicals, radiation, and UV light.

PoI 2. Describe the following human disorders as related to faulty repair enzymes:

a. Xeroderma pigmentosum
b. Bloom's syndrome
c. Ataxia-telangiectasia
d. Cockayne syndrome

QRQ 9. The number of chromosomes in every human somatic cell is ______.

QRQ 10. New strands are always synthesized in the ______' to ______' (numbers) direction.

QRQ 11. The ______ (Answer choices: lagging, leading) strand forms toward the replication fork, adding nucleotides ______ (Answer choices: one at a time, in a series of pieces).

QRQ 12. Select from the following the appropriate term for each description below: DNA polymerase, exonuclease, helicase, initiation, ligase, mismatch, mutagen, Okazaki fragment, primase, primer, replication fork, replication

a. agent that can alter or damage DNA
b. enzyme that builds a primer in DNA synthesis
c. enzyme that helps open up the double helix
d. enzyme that joins segments to complete the lagging strand
e. enzyme that selects, places, and bonds complimentary DNA nucleotides
f. enzyme that strips away primers
g. segment produced in the forming lagging strand
h. point of separation in the "unzipping" DNA molecule
i. short piece of RNA that gives the starting point for DNA synthesis

11.5. DNA (Gene) to RNA (Message) to Protein (Function)

11.5.1. Introduction

What do the following have in common: lion claws, feathers, spider webs, hair, hemoglobin, antibodies, enzymes, and egg whites? The answer is that they are all comprised of *proteins,* complex macromolecules that perform a variety of dynamic duties for cells. Proteins are important in structure, transport, growth, cellular communication, motion, maintenance, defense, immunity, storage, and physiological processes. Thousands of proteins are at work within your body and comprise about 12 percent of your mass. Proteins are made up of hundreds or thousands of smaller molecular units known as *amino acids.* In living things, there are 20 different types of amino acids that can be combined to build a protein under the control of genes (see Figure 3.14).

After scientists learned of DNA's structure and general function, biochemists began to ask an important question. How does a gene specify how to build a protein? Crick thought that perhaps the sequences of the bases residing in DNA served as a code that could specify which amino acids should be linked to build a specific protein. However, research soon indicated that amino acids are not directly built from DNA. After all, DNA resides in the nucleus, and proteins are assembled on ribosomes residing in the cytoplasm. Two French biologists, *François Jacob* and *Jacques Monod,* hypothesized that RNA molecules serve as the link between the DNA residing on genes in the nucleus and the ribosomes in the cytoplasm. Once their hypothesis was tested and accepted, Francis Crick articulated what is now the "central dogma" for the flow of biological information: information goes from DNA (gene) to RNA (message) to protein (function). This concept has proven applicable in all life forms from bacteria to the blue whale; it is a means in which genes were literally brought to life (Science Learning Hub 2011; Scitable 2014). Let's examine the differences between DNA and RNA in Table 11.1.

TABLE 11.1. Differences between DNA and RNA

	DNA	RNA
Name	Deoxyribonucleic acid	Ribonucleic acid
Structure	Two strands comprising a double helix composed of nucleotides	Single strand composed of nucleotides
Size	Long molecule constituting a single linear chromosome	Relatively short molecule of variable length
Location	In nucleus	In nucleolus, ribosome, and cytoplasm
Bases with Links	Adenine (A) – Thymine (T) Guanine (G) – Cytosine (C)	Adenine (A) – Uracil (U) Guanine (G) – Cytosine (C)
Stability	Generally stable; can be damaged by UV light and certain chemicals	More reactive under certain conditions; not as susceptible to UV light but more susceptible to enzymes
Evolution	Thought to be derived from RNA	Thought to give rise to DNA

Principal Types	**Messenger RNA (mRNA):** Carries transcribed genetic information from the DNA in the nucleus to the ribosome in the cytoplasm **Transfer RNA (tRNA):** Carries newly assembled amino acids to the ribosome **Ribosomal RNA (rRNA):** Located in the ribosome and noncoding

11.5.2. Transcription

The process of *protein synthesis* involves the building of a protein from amino acids. In cells, genetic information is encoded in certain sequences of nucleotides of DNA, or a *gene*. These genes can vary in length from a few hundred to more than two million bases. The bases are responsible for directing synthesis of proteins from amino acids. For example, in the genetic disorder cystic fibrosis, the CFTR protein is made up of 1,480 amino acids. A mutation in this sequence can result in cystic fibrosis. The gene serves as a physical as well as functional unit of heredity. Some genes carry the biochemical instructions written in the nucleotides to build specific proteins; others do not.

Protein synthesis consists of two fundamental events, *transcription* and *translation*. Ultimately, transcription and translation are responsible for *gene expression*, which involves converting genetic information stored in the DNA of a gene into a protein. In eukaryotic cells, transcription occurs in the nucleus. In transcription, a base sequence of DNA serves as a template to assemble a complementary segment of *messenger RNA (mRNA)*. Only one strand of the DNA, known as the *template strand*, is transcribed. Not all of the genes in a chromosome are transcribed at once, because proteins known as *transcription factors* ensure that only certain genes are activated to undergo transcription. The enzyme *RNA polymerase* has several duties in this process. RNA polymerase is responsible for unwinding the specific segment of DNA to be transcribed and temporarily breaking the hydrogen bonds between the opposing bases in the DNA. It also attaches a sequence of bases, known as a *promoter*, where the process is to begin. Promoters are very short and consist of mostly Ts and As, called the *TATA box*. The process of transcription begins after the site of the promoter, with new complementary ribonucleotides, available in the nucleus, being linked to their DNA template with hydrogen bonds. As stated in Table 11.1, uracil replaces thymine in RNA. This process continues until reaching a *transcription termination sequence*, when the promoter and the RNA polymerase detach from the template and the two strands of DNA rejoin. Depending on the needs of the biological system, this process can be very active or occur periodically.

In order for the mRNA to become mature, a few more steps are needed. Within the gene's DNA nucleotide sequence, there are coding portions, called *exons* (expressed sequences), and noncoding portions, known as *introns* (intervening sequences). Eventually, exons serve to code for amino acids. It is important that the introns are removed before mature mRNA forms. Otherwise, a faulty protein may result. These introns can consist of tens to thousands of base pairs and are sprinkled throughout the gene. At one time, introns were considered mere "genetic junk," but we know now that introns have regulatory and translation functions important to cellular success. RNA-protein complexes, known as *spliceosomes*, chemically remove introns from the complex and splice the ends of the exons together. After the removal of the introns, the RNA molecule matures into mRNA.

Once mature, the mRNA exits the nucleus through the nuclear pores. Arriving in the cytoplasm, the mRNA attaches to a ribosome, which makes proteins. The bases of mRNA work in triplets, or *codons*, and a codon represents a three-base code for a specific amino acid (see Figure 11.9). With each of the three RNA codon positions being one of four bases, there are $4 \times 4 \times 4 = 64$ possible three-base sequences, or codons, in the genetic code. However, each triplet of mRNA specifies one of only 20 amino acids. Led by the American biochemist *Marshall Nirenberg* and others, scientists finally revealed the genetic code in the late 1960s. Just like writing a sentence, proper punctuation is necessary in writing the genetic code. Of the 64 codons, 61 of the codons specify a specific amino acid. The codon AUG codes for the amino acid methionine and signifies *start* for building a protein. The AUG sequence is synonymous to beginning a sentence with a capital letter. The codons UAA, UAG, and UGC are synonymous to ending a sentence with a period and serve to *stop* or terminate the building of a protein. The genetic code is *degenerate*, with more than one codon specifying an amino acid. For example, CGU, CGC, CGA, and CGG all code for the amino acid arginine. Francis Crick noticed that, with a few exceptions, the first two nucleotides are the same in each codon and initiate coding specificity (Greenwood 2018). Examining the chart below, what exceptions can you find?

First Base	Second Base: U		C		A		G	
U	UUU	Phe	UCU	Ser	UAU	Tyr	UGU	Cys
	UUC		UCC		UAC		UGC	
	UUA	Leu	UCA		UAA	STOP	UGA	STOP
	UUG		UCG		UAG		UGG	Trp
C	CUU	Leu	CCU	Pro	CAU	His	CGU	Arg
	CUC		CCC		CAC		CGC	
	CUA		CCA		CAA	Gln	CGA	
	CUG		CCG		CAG		CGG	
A	AUU	Ile	ACU	Thr	AAU	Asn	AGU	Ser
	AUC		ACC		AAC		AGC	
	AUA		ACA		AAA	Lys	AGA	Arg
	AUG	Met/Start	ACG		AAG		AGG	
G	GUU	Val	GCU	Ala	GAU	Asp	GGU	Gly
	GUC		GCC		GAC		GGC	
	GUA		GCA		GAA	Glu	GGA	
	GUG		GCG		GAG		GGG	

FIGURE 11.9 The genetic code

11.5.3. Translation

The process of translation, which involves building a protein from the mRNA sequence on ribosomes using tRNAs, is one of the most complex processes within a cell. The site of translation is the ribosome, which is responsible for decoding the message in the codons and building a specific protein from amino acids. The steps of translation (corresponding to circled white numbers in Figure 11.10) are summarized below:

> **GtM 1.** Give the amino acid for which each codon codes:
>
> a. UUU
> b. ACG
> c. CAG
> d. AUC
> e. GUC
>
> **GtM 2.** List the codons that code for each amino acid:
>
> a. Valine
> b. Tryptophan
> c. Alanine
> d. Threonine
> e. Aspartic acid

Initiation (process not shown in diagram)

1. Transcribed mRNA moves into the cytoplasm, where its start codon (AUG) binds to a ribosome's Site P.
2. In the cytoplasm, there are a variety of *transfer RNA* (*tRNA*) molecules, each of which has two functional regions: an *anticodon*, formed by three nucleotides, complementary to a specific codon of mRNA, and a region carrying the amino acid specified by this anticodon.
3. The *start* tRNA, with anticodon UAC and carrying the specified amino acid methionine, base-pairs with the *start* codon (AUG) of mRNA. This binding occurs at Site P of the ribosome.

Elongation

Repeat the following until the ribosome encounters a *stop codon* (UAA, UAG, UGA):

4. tRNA, with an anticodon and carrying the next amino acid, base-pairs with the next codon at Site A.
5. A peptide bond forms between the newest amino acid in the growing amino acid chain at Site P and this next amino acid at Site A (dotted line).
6. tRNA from Site P (minus its amino acid) is released. The ribosome moves (to the right in the diagram), transporting the tRNA, now with an additional amino acid, from Site A to Site P and revealing a new codon at Site A.
7. Released tRNA can attach to a new amino acid molecule for reuse.

Termination

8. The ribosomal complex is dissembled, releasing the new polypeptide.

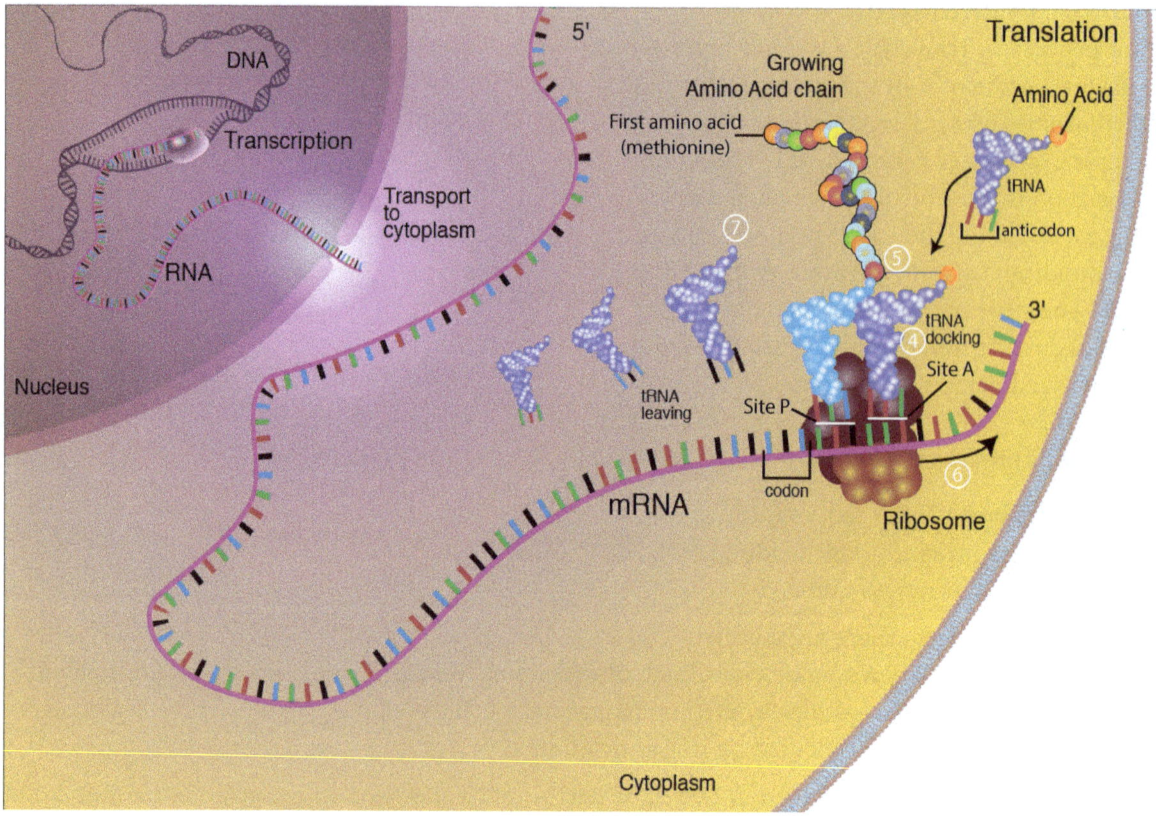

FIGURE 11.10 Process of translation

PoI 3. Give the sources for the following components required for translation:

a. amino acids
b. tRNA
c. rRNA
d. proteins for enzymes and for ribosome structure

QRQ 13. The "central dogma" for the flow of biological information is that information goes from ______ (gene) to ______ (message) to ______ (function).

QRQ 14. Give the number of

a. bases in a codon
b. amino acids

QRQ 15. The first subprocess of protein synthesis is ______ and the second is ______.

QRQ 16. The coding portion of a gene's DNA nucleotide sequence is called a(n) ______, while the noncoding portion is a(n) ______.

QRQ 17. List in order the initial steps of protein synthesis:

a. mRNA assembled as complement to a strand segment of DNA
b. promoter and the RNA polymerase detach from template
c. RNA polymerase attaches promoter to DNA
d. RNA polymerase unwinds a specific segment of DNA
e. spliceosomes chemically remove introns
f. spliceosomes splice ends of exons together
g. two strands of DNA rejoin

QRQ 18. List in order the final steps of protein synthesis:

a. mRNA attaches to a ribosome in cytoplasm
b. mRNA exits nucleus
c. protein built one amino acid at a time
d. stop codon encountered; polypeptide released
e. tRNA attaches to a specific amino acid
f. tRNA brings amino acid–matching anticodon to ribosome
g. tRNA formed as anticodon to a codon of mRNA
h. tRNA travels to cytoplasm

11.6. Control of Gene Expression

At one point in your development, you were only one cell—a fertilized egg, or zygote. Now look at yourself! How did you get here from only one cell? Every one of your cells was derived from that first cell, and except for your gametes, all of your trillions of somatic cells contain DNA that is essentially the same. Wait a minute—Don't you have about two hundred different types of cells? If every somatic cell came from cell divisions from that first cell, where did you get skin, muscle, blood, and nerve cells? They don't look much alike, do they? You have cells that produce hormones, like human growth hormone, which specialized cells in your anterior pituitary gland produce. Your muscle cells contain the same DNA as your anterior pituitary, but muscles don't secrete growth hormone. Likewise, adrenal glands don't produce myosin, one of the proteins needed for muscle contraction. So cells of different types have different structures and functions.

Following fertilization, the zygote that became you began dividing, but as the number of cells increased, there were many unseen changes occurring. The process of *differentiation* (specialization of cells) had begun. Differentiation results in the production of different cell types, all of which were derived from the zygote. What you can't see immediately is that differentiating cells develop different patterns of gene expression. Some genes are actively expressed in all the cells, such as genes that code for enzymes used for many metabolic activities. Sometimes such genes are described as "housekeeping genes." However, as cells become more specialized, they become increasingly different, as certain genes are expressed while others are not. Each type of cell now expresses altered phenotypes, which are visible at some level.

Bacteria are responsive to external factors, like nutrients. If a novel food source becomes available, the bacteria usually respond by altering gene expression to ramp up the metabolic machinery to utilize that food source. Cells in multicellular organisms are also sensitive to signals, often originating from other cell types, that may influence gene expression in target cells. For example, growth hormone from the pituitary has been shown to bind to surface receptors of target cells and promote the production of *collagen* (structural protein found in skin, muscle, connective tissues, etc.) in tendon and muscle cells, the targets (Doessing et al. 2010).

Cellular regulation of gene expression may occur at various levels. Although some genes may only produce RNAs, like tRNA, assume that for this discussion, gene expression is equivalent to protein synthesis.

1. **Transcriptional control.** The most common level of gene regulation occurs at the transcriptional level. In eukaryotic cells, there is an assortment of protein *transcription factors* (*TFs*) that, when activated, modify expression of certain genes. Most genes, except for housekeeping genes, are off, so most regulatory factors are positive (activators). For instance, in Figure 11.11 is a diagram of the region of DNA, some distance from the gene, containing *enhancers*, to which *activator proteins* can bind. If they do bind, the enhancers will interact with other transcription factors to form a complex that binds to the promoter region, which can bind with and place the RNA polymerase on the promoter to begin that gene's transcription.

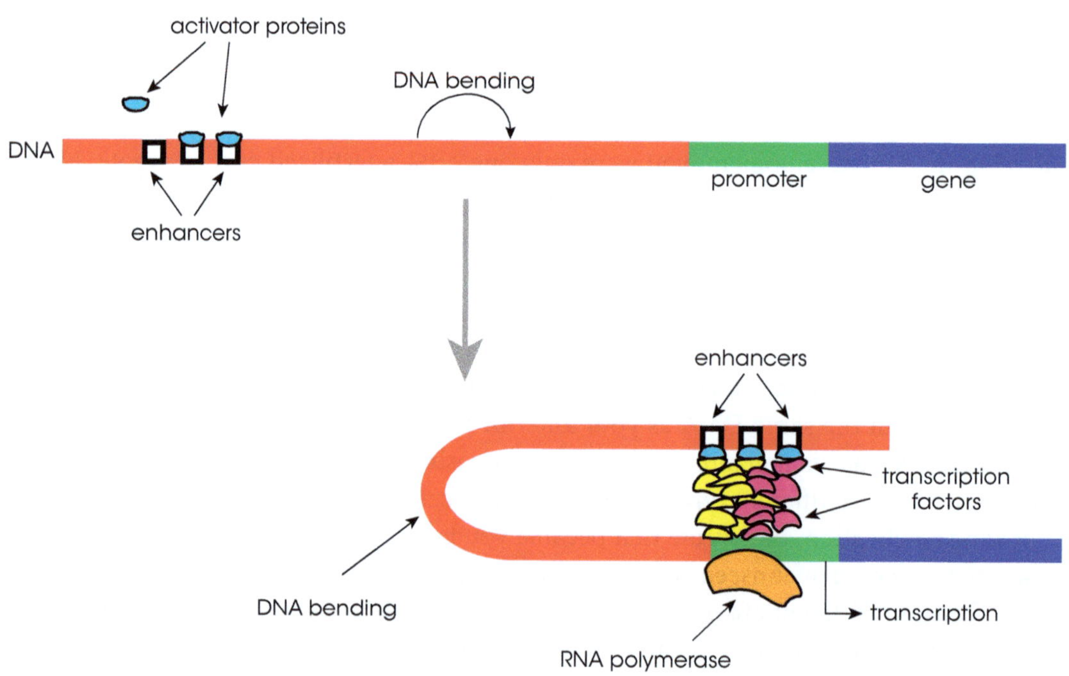

FIGURE 11.11 Region of DNA containing enhancers, remote from the gene, which bind to activator proteins that can bind to transcription factors

APPLYING SCIENCE TO LIFE 3 For the Sake of a Nail (Transcription Factor): ESRRB

For the want of a nail the shoe was lost, For the want of a shoe the horse was lost,
For the want of a horse the rider was lost, For the want of a rider the battle was lost,
For the want of a battle the kingdom was lost, And all for the want of a horseshoe-nail.

(Ancient proverb)

By analogy, *ESRRB* is a transcription factor (TF) that binds to the promoter and enhancer regions of various genes, sensitive to the hormone estrogen. When TF is defective, the targeted genes are not transcribed properly. The genes are responsible for proper development of the inner ear, so the defect can lead to a form of deafness (Lee et al. 2011).

2. **Processing pre-mRNA.** *Alternative splicing* (removal of introns) of the pre-RNA transcript can also help to determine which genes are expressed. In the female fruit fly, a transformer gene (*tra*) is spliced to yield a functional protein that is important for development of the female phenotype. In males, the *tra* gene has an alternative splicing that produces a nonfunctional protein (see Figure 11.12).

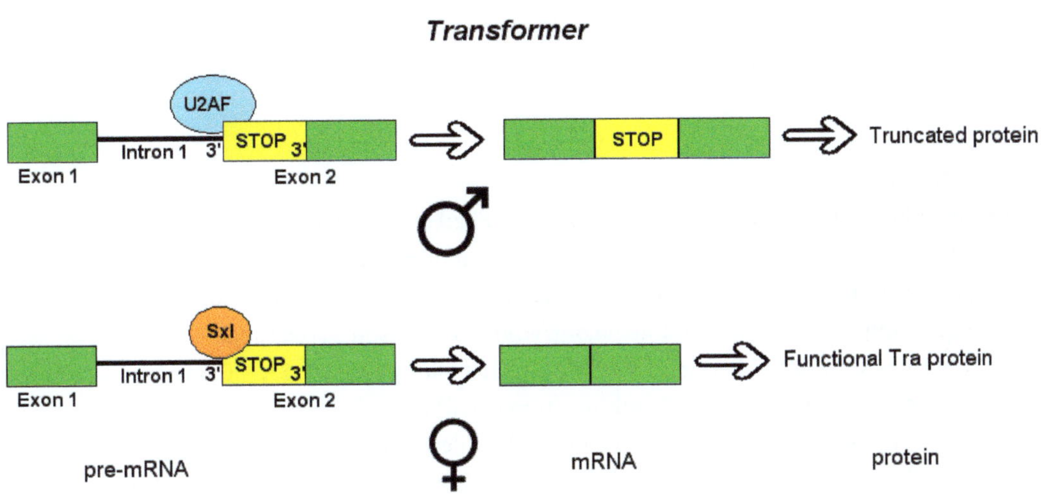

FIGURE 11.12 Alternative splicing in fruit fly

3. **Chromosomal condensation.** Remember how the chromosomes coil up tightly (condense) during mitosis? This tight coiling inhibits access to the gene segments by enzymes, thereby obstructing RNA production and subsequent protein synthesis (gene expression).
4. **Epigenetic modifications.** These are heritable changes in the gene that do not involve a change in the nucleotide code. For instance, adding a chemical group (i.e., $-CH_3$ [methyl]) to a nucleotide base in the promoter region generally inhibits transcription of that gene.

Besides the above examples, there are other stages of protein synthesis where gene expression can be controlled. These include the following:

1. **Varying the stability of mRNA.** Often related to nucleotide sequence, chemical modifications, and structural factors.
2. **Translation regulation.** Faster or slower affects levels of the gene product in the cell.
3. **Life span of proteins.** Rate of breakdown of synthesized proteins varies; the mechanisms are too complex to present here.

With so many working parts in a molecule of DNA, surely things can go awry from time to time. These misspellings of the DNA within a gene are known as *mutations*. Even if the gene itself is normal, there are many other factors that influence the expression of the gene. In module 16 we will discuss the causes, biology, and types of mutations in the biological world. We will also witness the power of mutations in evolution, speciation, and genetic disease.

QRQ 19. Select from the following the appropriate term for each description below: activator, differentiation, enhancer, epigenetic, mutation, splicing, transcription factor

a. heritable modification not involving a nucleotide code change
b. protein that can bind to an area of the DNA, making transcription more likely
c. protein that, when activated, modifies expression of a certain gene
d. specialization of cells

Image Credits

Fig. 11.2: Source: https://commons.wikimedia.org/wiki/File:Hersheychaseexperiment.gif.

Fig. 11.3a: Copyright © by Wellcome Images (CC BY 4.0) at https://commons.wikimedia.org/wiki/File:-Francis_Crick_lecturing_at_Cambridge_University,_n.d_Wellcome_L0042633.jpg.

Fig. 11.3b: Source: https://commons.wikimedia.org/wiki/File:Watson,_james.jpg.

Fig. 11.4a: Copyright © by MRC Laboratory of Molecular Biology (CC BY-SA 4.0) at https://commons.wikimedia.org/wiki/File:Rosalind_Franklin_(retouched).jpg.

Fig. 11.4b: Source: https://commons.wikimedia.org/wiki/File:Maurice_H_F_Wilkins.jpg.

Fig. 11.5: Copyright © by Sponk (CC BY-SA 3.0) at https://commons.wikimedia.org/wiki/File:Difference_DNA_RNA-EN.svg.

Fig. 11.6a: Source: https://commons.wikimedia.org/wiki/File:AT_base_pair_jypx3.png.

Fig. 11.6b: Source: https://commons.wikimedia.org/wiki/File:GC_base_pair_jypx3.png.

Fig. 11.6c: Source: https://commons.wikimedia.org/wiki/File:Base_pair_AU.svg.

Fig. 11.7: Source: https://commons.wikimedia.org/wiki/File:DNA_chemical_structure.svg.

Fig. 11.8: Source: https://commons.wikimedia.org/wiki/File:DNA_replication_en.svg.

Fig. 11.10: Source: https://www.genome.gov/genetics-glossary/Translation.

Fig. 11.11: Adapted from Copyright © by Philippe Hupé, (CC BY-SA 3.0) at https://commons.wikimedia.org/wiki/File:Role_of_transcription_factor_in_gene_expression_regulation.svg.

Fig. 11.12: Copyright © by Agathman (CC BY 3.0) at https://commons.wikimedia.org/wiki/File:Transformer_splicing.gif.

LAB #11

Laboratory Exercises

11.7. Laboratory on Lac Operon

11.7.1. Preliminary Work

In this laboratory, we investigate gene expression using a model of *E. coli*'s lac operon. From the text's website, view the introductory video; download and open "*GenEvo 1 Genetic Switch Rev.nlogo.*"

1. Click the Info tab at the top of the page and read the section "What Is It?" Consulting other sources as needed, select from the following the best term for each description below: A. DNA, B. enzyme, C. gene, D. genetic switch, E. lac operon, F. LacI, G. lactose and glucose, H. LacY, I. LacZ, J. mRNA, K. operator, L. operon, M. permease , N. polymerase, O. positive feedback loop, P. promoter, Q. protein (simple), R. regulatory protein, S. repressor, T. RNA, U. RNAP, V. terminator, W. transcription
   - **a.** ______ biomolecule that is a chain of amino acids and performs many critical functions
   - **b.** ______ contiguous section of a chromosome that encodes information to build a protein or an RNA molecule
   - **c.** ______ controls gene expression (i.e., acts to switch genes on and off) by influencing regions of DNA that are transcribed
   - **d.** ______ double-stranded nucleic acid that contains the encoded information for the manufacture of all the proteins a cell needs
   - **e.** ______ enzyme that digests lactose molecules inside the cell to produce energy
   - **f.** ______ enzyme that catalyzes the formation of new DNA or RNA from an existing DNA or RNA, respectively
   - **g.** ______ enzyme that promotes the formation (synthesis) of mRNA from DNA
   - **h.** ______ intermediary, single-stranded nucleic acid carrying instructions from DNA that specifies the amino acid order of proteins
   - **i.** ______ molecule that conveys genetic information from DNA to a ribosome
   - **j.** ______ molecule that stimulates or prevents the transcription of proteins
   - **k.** ______ process in which a product of an activity causes more of that activity to occur

l. ______ protein that aids the transport of a specific molecule through the membrane

m. ______ protein that can bind to an operator and inhibit the expression of one or more genes by preventing transcription of the genes into mRNA

n. ______ protein that facilitates transport of lactose through the membrane into the cell

o. ______ protein that serves as a catalyst for a biochemical reaction in a living organism

p. ______ region of DNA to which a repressor protein can bind; portion of lac between the promoter and the operon genes

q. ______ repressor protein that binds to the operator region of the DNA, not allowing RNAP to pass along the gene, thus stopping protein synthesis

r. ______ responsible for regulating the synthesis of proteins required to conduct the uptake and digestion of lactose in *E. coli.*

s. ______ segment of DNA with adjacent genes under the control of one promoter; transcription of the segment occurs together, making a single mRNA; and expression of the genes occurs at the same time or not at all

t. ______ subsequence of DNA indicating where transcription of a gene is to begin

u. ______ subsequence of DNA that causes RNA polymerase to stop transcription

v. ______ sugar(s)

w. ______ synthesis of RNA from a single strand of DNA

2. Read "How It Works" up to "Important Proteins."

a. ________________ In the model, what is preferred as an energy source?

b. ____________ Select all that occur when no lactose is in the cell's surrounding environment: A. genetic switch is at an off steady-state, B. LacI prevents (mostly) the bacteria from producing the proteins, C. LacI readily binds to the operator site of the DNA, D. no glucose is in the cell, E. relatively little LacI is produced, F. relatively little LacY and LacZ are produced

c. ________________ Give the sequence of events, in order, that occurs when lactose is introduced to the outside environment, causing a positive feedback loop: A. LacIs cannot bind to the DNA operator site and thus cannot stop protein synthesis, B. lactose molecules enter into the bacterial cell through LacYs, C. more lactose can enter the cell, D. more LacYs are produced, E. newly formed LacYs get inserted into the cell membrane, F. some lactose molecules in the cell bind to LacIs

d. ______ Under what circumstance is the genetic switch turned on? A. glucose and lactose in the environment, B. glucose but no lactose in the environment, C. lactose but no glucose in the environment, D. no glucose and no lactose in the environment

e. ______ (T/F) When glucose is present, the rates of LacY and LacZ production are reduced significantly.

3. Read the remainder of "How It Works" and "How to Use It."

   a. Match each component with its color visualization in the model: A. black, B. blue, C. brown, D. green, E. gray, F. light red patch, G. light red rectangle, H. light red triangle, I. orange, J. purple, K. purple with gray dot, L. white

      _______ cell when the number of LacY is greater than 10

      _______ cell when the number of LacY is less than or equal to 10

      _______ LacI

      _______ LacI-lactose complex

      _______ lactose

      _______ LacY inside cell

      _______ LacY on cell wall

      _______ LacZ

      _______ operator

      _______ operon

      _______ promoter

      _______ RNAP

      _______ terminator

   b. ______ Give the number of LacY (or LacZ) produced with each transcription.

   c. ______ When the energy of the cell is **1.** (number) times its initial value, the cell splits into two daughter cells, each having half the amount of energy as the original cell.

   d. ______ When GLUCOSE? is ON, energy of a cell increases at a constant rate of **1.** units/tick.

11.7.2. Laboratory Work

4. This question involves observation of the cell with glucose, but no lactose, present. Set GLUCOSE? to ON and LACTOSE? to OFF. Press *setup*, change the speed to *slower*, and click *go*. Stop the simulation as needed to make observations.

   a. ______ What energy level triggers cell division?

   b. Match each situation to its visualization: A. additional light red rectangles and triangles appear in the cell, B. gray pentagon going through light red patch into the cell, C. light red patch appears on the cell wall, D. purple shape with jagged edge fits into orange area of strand, E. purple shape with jagged edge inside the cell moves around, F. starting at green area, brown blob travels down strand to gray area

      _______ a repressor preventing transcription

      _______ transcription completed

      _______ transcription

      _______ transport of lactose

c. ______ Ideally, the genetic switch should be off when lactose is not present, so that transcription cannot occur. Is that always the case, no (N) or yes (Y)?

d. ______ (T/F) Immediately after transcription, the energy level drops a little before rising again.

e. ______ (T/F) The LacZ number increases by 10 after a transcription.

f. ______ (T/F) The LacZ number is cut in half after cell division.

g. ______ (T/F) Cell division time is shorter after a transcription.

5. This question involves observation of the cell with lactose but no glucose present, so that we have a positive feedback loop.

   a. Before changing the settings, adjust the speed to *normal*, then press *setup* and *go*, allowing the simulation to run for about 500 ticks, and then stop the simulation. We will use the graphs to this point as a basis for comparison. Set GLUCOSE? to OFF and LACTOSE? to ON. Press *setup*, change the speed to *slower*, and click *go*. Stop the simulation as needed to make observations. Match each situation below to its visualization listed here: A. additional light red rectangles and triangles appear in the cell, B. gray pentagon going through light red patch into the cell, C. gray pentagons inside the cell move around, D. gray pentagons outside the cell move around, E. light red patch appears on the cell wall

   ______ lactose immediately after turning LACTOSE? to ON

   ______ lactose getting into the cell

   b. ______ Observe the behavior of the genetic switch by considering the change in the LacZ production. Is LacZ production more (M) or less (L) than when no lactose was in the system?

   c. ______ Is the cell division time more (M) or less (L) than when no lactose was in the system?

   d. ______ Uncheck *view updates* or increase the speed of the simulation. Run the simulation for several thousand more ticks. Does the cell's energy ever fall below the lowest values when no lactose is present, no (N) or yes (Y)?

   e. ______ Stop the simulation and check *view updates*. Does any LacI not bound in a LacI-lactose complex exist, no (N) or yes (Y)?

   f. ______ In this situation, can transcription occur, no (N) or yes (Y)?

   g. 1. ______ 2. ______ 3. ______ 4. ______ 5. ______ Starting with the following first step of one LacY molecule inside the cell, indicate the order of the steps of the lac-operon positive feedback loop in the presence of lactose: The presence of lactose in the cell results in more lactose being in the cell, facilitated by the presence of LacYs in the cell, resulting in more LacYs being in the cell.

      a. Bound LacIs are not available to repress transcription.

      b. Lactose molecules pass through LacY(s) into the bacterial cell.

      c. Some lactose molecules bind with LacIs to form LacI-lactose complexes.

      d. Some LacYs are inserted into the cell membrane.

      e. Transcription results in more LacYs that can help transport lactose into the cell and more LacZs that can digest lactose to produce energy.

MODULE

12

Biotechnology—Transforming the Future

An exact determination of the laws of heredity will probably work more change in man's outlook on the world, and in his power over nature, than any other advance in natural knowledge that can be foreseen.

—William Bateson (1902)

12.1. Introduction

Biotechnology is a discipline grounded in the biology of organisms, cells, and subcellular components, where scientists utilize biological processes to fashion new technologies and tools (see Figure 12.1). The products from these efforts can be applied to research, medicine, agriculture, and industry, where the outcomes are intended to advance human lives and the earth. Biotechnology is not really a new field, because people have applied it for thousands of years. We have long used artificial selection in agriculture to improve crops and domestic animals. Much of what comes from farms results from these endeavors. Plants have long been cultivated for dyes and medicine. Also, many ancient cultures learned to make alcohol and cheese using bacteria and yeast.

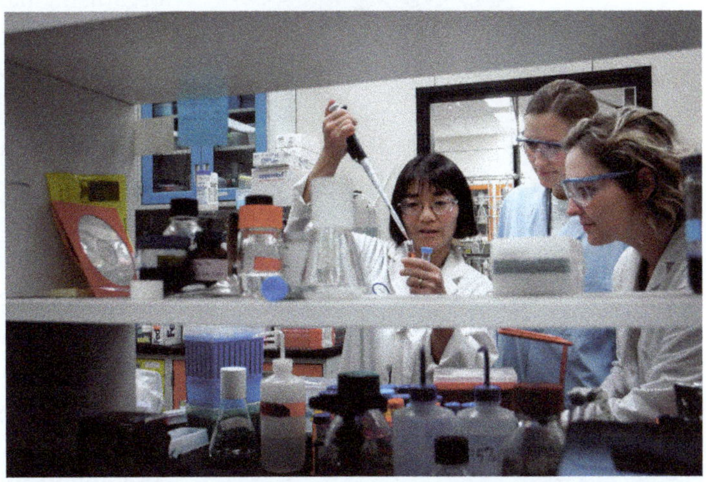

FIGURE 12.1 Scientists working in a biotechnology laboratory

History is replete with our use of biotechnology in medicine. An excellent example of such applications is Edward Jenner's development of a smallpox vaccine. A country doctor, *Edward Jenner* developed his vaccine,

PoI 1. List and describe the following biotechnology companies, providing a brief description of their mission and the primary product or products of each:

a. Seattle Genetics
b. Alexion
c. Celltrion
d. Gilead Sciences

PoI 2. We have sequenced the genomes of many pathogenic bacteria, parasites, domestic plants, and domestic animals. Defend this sequencing.

FIGURE 12.2 The grizzly bear's genome is known.

which has saved millions of lives, using material from cowpox lesions. With the recent advancements in molecular biology, biotechnology has mushroomed. Now part of our daily lives, it asserts tremendous impact upon industry, medicine, agriculture, forensics, and the environment. Biotechnology companies throughout the world provide many jobs to laymen as well as scientists.

Any new technology generates fears with an air of excitement. When Jenner developed the vaccine, many members of the public feared people might develop cow parts. In the same sense, today vaccines, cloning, and stem cell technologies generate many irrational fears. On the other hand, many of these technologies give great hope for diagnosing and treating diseases, confronting pandemics, understanding evolution, and enacting bioremediation of the environment. As global citizens, we must attempt to develop a rational understanding of these tools, with caution, to ensure their proper application.

To understand some of the applications of biotechnology, we must differentiate among confusing terms. *Genomics* refers to a field that studies the total genetic complement of an organism. It employs a number of technologies to sequence and analyze the genetic makeup of an organism. As a result of the genetic revolution, organisms from papayas to the great grizzly bears have been genetically sequenced (see Figure 12.2). In addition, several extinct organisms have been sequenced, including the mammoth and Neanderthal man.

Ultimately, gene expression is revealed as proteins. The *proteome* is the protein makeup of a cell, tissue, organ, or organism, making *proteomics* the discipline employing sophisticated analytic instruments for the comprehensive study of proteins. Proteomics can yield valuable information regarding protein complement of a system, how proteins are involved in biochemical pathways, and how various proteins interact. Important applications include vaccine production, understanding disease, and pharmaceutical research.

Bioinformatics uses biology, statistics, biochemistry, and computer science to study, store, retrieve, and analyze vast amounts of biological data, including genetic information. In genetics, bioinformatics improves our understanding of sequence analysis, genomics, proteomics, phylogeny, evolution, inheritance, and disease. It is a rapidly developing discipline and will provide many incredible research and job opportunities in the future.

QRQ 1. Select from the following the appropriate term for each description below: bioinformatics, biostatistics, biomics, biotechnology, genetics, genomics, proteome, proteomics

a. discipline utilizing biological processes to fashion new technologies and tools
b. discipline utilizing biology, statistics, biochemistry, and computer science to study, store, retrieve, and analyze vast amounts of biological data
c. protein makeup of a cell, tissue, organ, or organism
d. study of the total genetic complement of an organism

12.2. Cloning

Cloning refers to developing genetically identical individuals from an original biological source. Unfortunately, in recent years, cloning has become a buzzword in nonscientific circles and has led to many misconceptions. However, cloning is a powerful agricultural and medical tool.

Although science fiction movies would have you believe that cloning is a new process created in some laboratory, cloning has been around for billions of years. Cloning occurs naturally in some species of bacteria, protists, fungi, plants, and animals through asexual reproduction. In plants, *vegetative propagation* is used in cloning plants through cuttings and grafting. In strawberries, the original plant produces runners that give rise to new plants or clones. Sprawling stands of quaking aspen trees are the most widely distributed tree species in North America. These trees can reproduce sexually, but asexual propagation occurs though shoots and sucker formation. While enjoying a seedless navel orange, explain where new trees come from? Cloning can also occur in some animal species through budding, as in *Hydra*. Several species of animals, such as bees, ants, and aphids, can undergo a type of cloning known as *parthenogenesis*. In this process, usually under harsh conditions, an unfertilized ovum can develop into a clone. Identical twins in humans and other animals are also examples of clones. In this instance, a fertilized egg splits, forming two embryos with nearly identical DNA. However, the offspring are different from either parent, and each may display different phenotypes.

There are three distinct types of artificial cloning: gene cloning, reproductive cloning, and therapeutic cloning. In *gene cloning* (*DNA cloning*), copies of specific genes are produced from the DNA. Gene cloning is extremely valuable in studying the structure and function of specific genes. This procedure extracts and inserts the gene from an organism being studied into a carrier, or *vector*, such as a virus or bacterium. Once the gene is successfully inserted into the vector, it is encouraged to reproduce within the vector, making multiple copies of that particular gene. Currently, valuable uses of gene cloning include the study and analysis of normal gene function and the development of artificial genes. Biopharmaceuticals benefit greatly from gene cloning. Various proteins, such as human insulin and human growth hormone, are produced from bacteria in this fashion. Gene cloning is one of the underlying processes governing gene therapy, which will be discussed later (NIH 2017).

To the general public, the most intriguing aspect of cloning involves *reproductive cloning* of animals. Animal cloning first appeared in the news in 1996 when *Ian Wilmut* and his colleagues at the Roslin Institute in Scotland successfully cloned the first animal, a sheep known as Dolly. Since then, over 20 different species of animals have been cloned. Perhaps the most famous clones include Mira the goat and her sisters, cloned in 1998; Ombretta, an endangered European mouflon sheep cloned in 2000; Carbon Copy (CC), the first cloned cat in 2001; Trakr, a hero dog from 9/11 cloned into five pups in 2009; and Samantha, Barbara Streisand's dog cloned into two pups in 2018. If you are thinking about cloning your pet, be ready to pay over $50,000.

In reproductive cloning, the nucleus is removed by micropipette from an unfertilized egg of a target animal, leaving behind the plasma membrane and cytoplasm. As a result, the egg is *anucleated* and lacks primary genetic material. Next, a mature diploid body cell is taken from a tissue culture or frozen tissue of the donor animal (one to be cloned). The nucleus of this cell is removed by micropipette and inserted into the anucleated egg. This egg is then stimulated, in many cases by an electrical current, to prompt division. If cell division begins, it develops into a preimplantation embryo, which is then implanted into the uterus of the surrogate mother to complete embryological development. The animal produced from this process will have the genetic makeup of the donor animal (see Figure 12.3). The procedure is much easier said than done, often involving multiple attempts before success. If multiple clones are made, they will differ physically, physiologically, or behaviorally, because the genetic material is not the only determinant of these characteristics. Other variables will also influence the outcome, such as uterine environment and nutrients and, after birth, parental influence, society, and personal experiences.

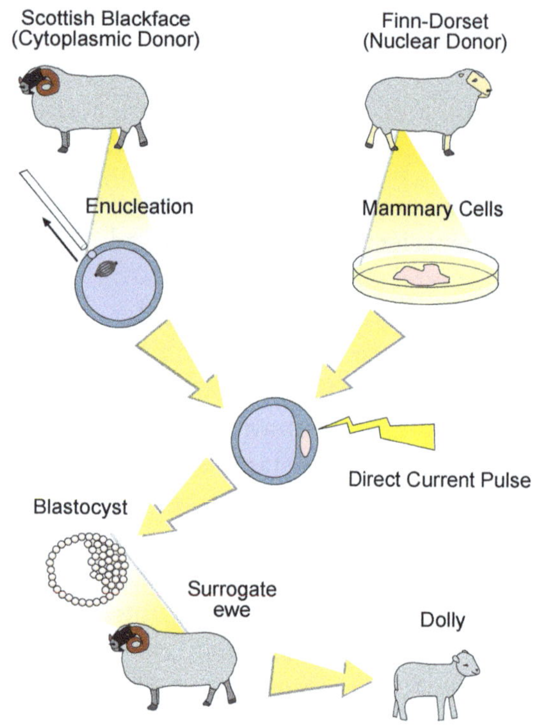

FIGURE 12.3 The making of Dolly the Sheep

Therapeutic cloning is an important constituent of *regenerative medicine* that involves replacing or regenerating cells and tissues in order to restore normal function. Therapeutic cloning uses stem cells for replacing or repairing damaged tissues or organs in a target animal. *Stem cells* are undifferentiated cells that have the potential to give rise to specialized cells through differentiation. Therapeutic cloning begins in the same way as reproductive cloning, except stem cells are removed from oocytes seeded with the patient's cells and used to grow specific types of cells for the patient (see Figure 12.4). Important in growing skin, corneal tissue, bone, cartilage, and muscle tissue, therapeutic cloning shows much promise in the repair of nerves, Parkinson's disease, Duchenne muscular dystrophy, diabetes mellitus, and organ

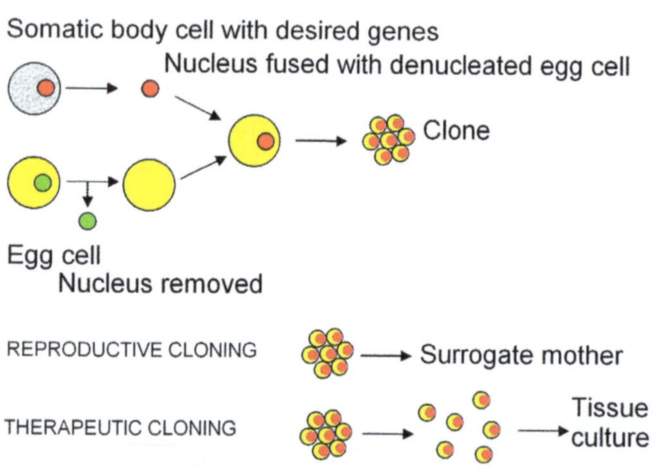

FIGURE 12.4 Reproductive and therapeutic cloning techniques

transplantation. Moreover, since therapeutic cloning uses the patient's cells, there is no risk of immune rejection (Murnaghan 2020).

QRQ 2. In ______ cloning, an egg's ______ is replaced. ______ cloning uses stem cells with a similar process.

QRQ 3. (T/F) Stem cells are specialized cells.

12.3. Recombinant DNA (rDNA)

Recombinant DNA technology has become an incredibly valuable biotechnology tool. A gene removed from a donor eukaryotic organism that codes for a desired protein is inserted, using enzymes, into a vector, commonly a bacterial *plasmid* (small, circular bacterial DNA that is separate from the main bacterial chromosome). This recombined DNA in the cloning vector is the *recombinant DNA* (*rDNA*). Cultured bacteria take up the plasmids, which replicate with their hosts. The gene, being cloned thousands of times, can produce large quantities of the protein product (see Figure 12.5).

In 1978, one of the most stunning applications in early recombinant technologies was development of a recombinant microorganism (bacterium *E. coli*) that could produce human insulin. *Insulin* is a hormone responsible for the regulation of glucose metabolism. Individuals with insulin deficiencies or metabolic problems in which the body does not respond properly to insulin, with either cause resulting in high blood sugar, have *diabetes*. Previously, insulin procured from the pig or cow pancreas was processed and administered. Although either form of insulin works, neither is as compatible and efficient as using human insulin, and both are more expensive. Almost simultaneously, researchers cloned the gene for human growth hormone. In 1982, Genentech and Eli Lilly marketed recombinant insulin, Humulin, the first

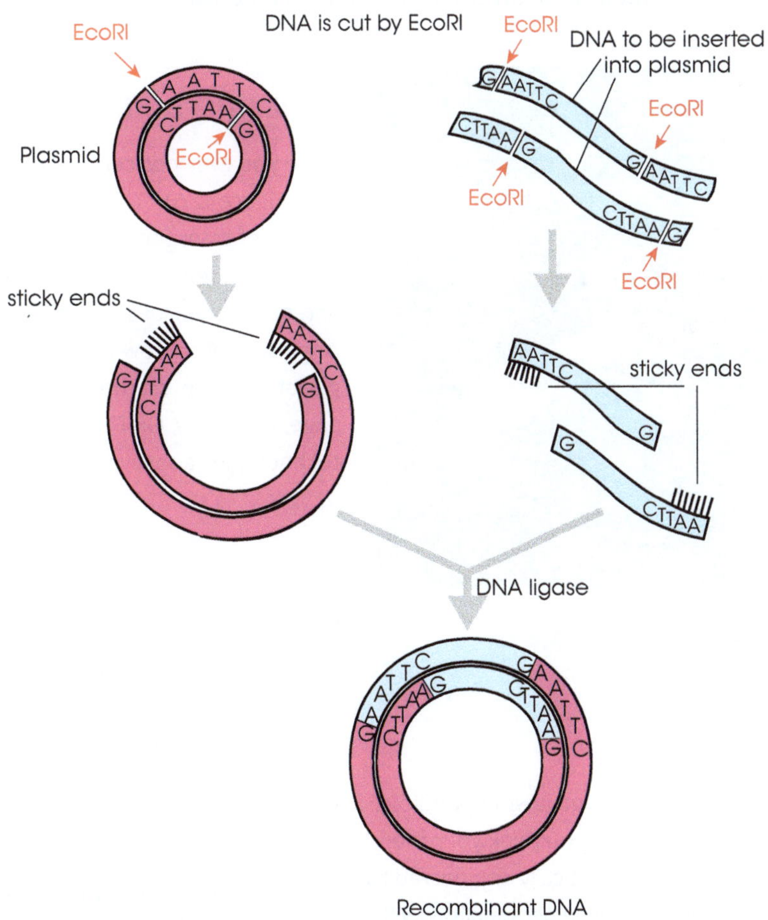

FIGURE 12.5 Basic process of creating recombinant DNA

recombinant pharmaceutical approved for use in the United States. Industrial techniques allow for *E.coli* to be grown in vast cultures and large volumes of insulin to be collected, purified, and inspected for patient use. With the incidence of diabetes on the rise due to dietary and lifestyle changes, such production will become crucial (Baeshen et al. 2014).

New recombinant techniques have developed in recent years that are commonly employed in agriculture, industry, environmental projects, and medicine. Recombinant DNA has been and is used to produce transgenic organisms, knock-out and knock-in organisms, and genetically modified organisms, which we will discuss next.

Transgenic organisms are intriguing and constantly in the news. These organisms are the result of placing the genes from another species in a host organism to obtain a desirable result. For example, we can get silk from goat milk engineered to have spider silk and cotton resistant to a moth larva (bollworm) from the plant modified to express a bacterial protein. These examples reiterate the commonality of the genetic code. Literally, an oak tree's cells can read your DNA. The world of agriculture and medicine utilize many more practical transgenic organisms, including rice that makes beta-carotene, a precursor in the production of vitamin A; vaccine-producing bananas that make vaccines easier to administer; sheep that have a missing human clotting factor in their milk; and human antibodies in mice (Genetics Generation 2015).

FIGURE 12.6 GloFish

APPLYING SCIENCE TO LIFE 1 The Amazing GloFish

Have you ever noticed the absolutely stunning *GloFish* in a local pet store (see Figure 12.6)? They dazzle us with their beautiful colors. GloFish are the results of transgenic recombinant technologies that involve inserting bioluminescent genes from marine organisms, like jellyfish, into zebra fish, betas, and other tropical species. Originally green because of their transgenic origins, GloFish now are available in several constructs, such as cosmic blue, galactic purple, starfire red, electric green, moonrise pink, and sunburst orange. Currently, GloFish are the only animal designed by recombinant DNA technologies approved for sale in the United States, except California.

Another application related to recombinant techniques is developing knock-out and knock-in organisms. In a *knock-out* organism, a specific gene has been inactivated or removed and an artificial piece of DNA inserted so that the gene of interest is rendered nonfunctional. As the result of these techniques, researchers can determine what that gene's function might be and thereby develop a better understanding of mutations and the relationship between an organism's genotype and phenotype. Although a number of plant and animal species have been the subject of knock-out techniques, the mouse is the most common. Knock-out mice have been successfully employed in the study of Parkinson's

PoI 3. Although it would be illegal and unethical, do you think your friend could be genetically engineered to have a "glowing personality"? Why or why not?

disease, obesity, diabetes, heart disease, arthritis, substance abuse, and many other conditions. Experimental mice in which the cystic fibrosis transmembrane conductance regulator gene (CFTR) has been knocked out have been good indicators of how this gene is involved in determining the phenotype.

In *knock-in organisms*, such as mice, a normal allele is targeted and replaced by another homologous allele at the same site. Thus, in studying genetic disorders, the normal gene in a mouse can be replaced by a defective gene in order to see manifestations of this gene in animal models. The technique has been important in developing mouse models of human diseases, such as skin disorders, Huntington's chorea, and cystic fibrosis.

For years, living organisms have been modified, such as sweeter corn and faster horses, through artificial selection techniques. In recent years, new techniques and advances in genetics have opened the door for a more far-reaching and precise control of the modification of organisms. Another buzzword related to the application of this new genetics is *genetically modified organism* (GMO). The DNA of these transgenic, knock-out, or knock-in organisms has been modified as the result of genetic engineering or transgenic techniques. Genetically modified organisms possess new genetic combinations that do not occur in nature. Using the common genetic code, genes from one organism can be inserted into another organism to obtain desired results. Future applications may involve vitamin enrichment in grains, new vaccines, and more efficient biofuels. Understandably, this new power provokes some concerns. Are we altering evolution and the environment? What are the physiological and metabolic consequence of altering an organism's genome? Will there be horizontal gene transfer to other organisms? How will such alterations affect human beings?

In agriculture, we are benefiting from GMOs with increased crop yields, improved nutrient composition, enhanced resistance to disease and pests, faster maturing rates, and resistance to frost and drought. Also, in animals, salmon have been modified to grow faster and larger and cattle have been altered to have increased resistance to mad cow disease. *Genetically modified foods* are a hot topic in public, in the courts, and in government. Some positive consequences of genetically modified foods are more nutritious and tasty food, decreased use of pesticides, disease- and drought-resistant food sources, faster growing plants and animals, and foods for medicinal uses. However, concerns focus on evolutionary incompatibility with our diet, possible harmful genetic consequences, and unexpected consequences such as allergies. As of now, these concerns have proven unfounded. The US Food and Drug Administration (FDA) vigorously studies genetically modified foods before they can be sold.

APPLYING SCIENCE TO LIFE 2 Challenges to Feeding the World

Beginning in the 1950s, a change in agricultural practices transformed farming across the globe—the *Green Revolution*. This change was prompted by a plant scientist from the United States who worked in Mexico for years developing new disease-resistant, highly productive strains of wheat. *Dr. Norman Borlaug* and his team from the United States and Mexico worked tirelessly generating genetically modified strains of wheat that increased yields and resistance to a fungal disease called wheat rust. During his work and travels, Dr. Borlaug was overwhelmed at the poverty and starvation that plagued many developing nations, moving him to spread new agricultural practices, beginning in Mexico and India. He hoped that these new agricultural practices would increase the food available to the poor, making them better able to overcome poverty and malnutrition. Furthermore, with more farmland productivity, he hoped to reduce the rate of deforestation to generate areas for crop cultivation.

There is no doubt that the early phases of the Green Revolution resulted in increased crop yields. In fact, Dr. Borlaug won the Nobel Peace Prize in 1970 for his dedicated work. However, other fallouts have generated fierce critics of such agricultural practices.

PoI 4.

a. What were the positive consequences from this shift in agricultural practices? Who were the main beneficiaries?
b. What were the negative consequences from this shift?
c. How might biotechnology help not only to improve crop yield but to reduce the negative results?
d. What are the pros and cons of GMO crops?

APPLYING SCIENCE TO LIFE 3 The Papaya: A Victory for GMOs

Almost everyone enjoying a refreshing papaya does not realize that this luscious fruit was almost wiped out. Beginning in the 1940s, an aphid-transmitted virus, known as the papaya ringspot virus (PRSV), began attacking papaya plants in Hawaii. Leaves yellowed and blistered; trunks had spotty, oily streaks; and young infected trees did not produce fruit. In the 1990s, the virus seriously impacted the papaya industry, but in 1998, papaya became the first commercial genetically modified plant. Because of the introduction of this Rainbow papaya, today the papaya business is booming and a testament to the use of GMOs (Azad, Amin, and Sidik 2014).

QRQ 4. A ______ mouse has had a particular gene inactivated or removed.

QRQ 5. The ______ DNA technique is used to produce large quantities of a protein. The typical vector in the technique is a bacterial ______.

QRQ 6. The acronym for an organism in which the DNA has been modified so that it possesses a new genetic combination not occurring in nature is ______.

QRQ 7. A(n) ______ organism, such as a GloFish, is the result of placing the genes from another species into a host organism to obtain a desirable result.

12.4. Polymerase Chain Reaction

In the 1980s, a revolutionary new genetic technique, *polymerase chain reaction* (PCR), was developed by the American biochemist *Kary Mullis*. Today, PCR serves as one of the fundamental processes used in gene sequencing and other genetic endeavors. In the past, significant amounts of DNA were needed in order to conduct molecular and genetic analysis. PCR provides scientists with a fast and relatively inexpensive way to perform this once challenging task.

PCR has become a common molecular laboratory tool that is used to *amplify*, or make multiple (millions of) copies of, a small segment of DNA for analysis. The source of DNA can be from a faulty gene, DNA left at a crime scene, or any source the researcher wants to investigate. Being automated, the process can be completed in just a few hours in an efficient and precise manner. In order to perform PCR, six key components are necessary: 1. *thermocycler*, which alters the temperature for denaturing (breaking bonds in) DNA and synthesizing new DNA; 2. *DNA* that is to be copied; 3. *primers*, which initiate synthesis; 4. supply of *nucleotide bases*; 5. *Taq polymerase* from a high-temperature-loving bacterium to form new strands of DNA by adding bases; 6. materials or instruments for *DNA isolation*. The process of PCR then can be accomplished in three major steps (see Figure 12.7):

1. **Denaturation** separates DNA into two single strands by exposing DNA to high temperatures to break the hydrogen bonds between the nucleotide pairs.
2. **Annealing** (binding of primers to complementary sequences on each strand) occurs as the temperature cools. As in replication, the primers serve as the starting point of DNA synthesis. Because complementary strands run in opposite directions, one primer is the forward primer, and the other is the reverse primer. Annealing results in two single strands of DNA that are ready to be copied.
3. **Extension** (adding nucleotides complementary to the template) uses *Taq polymerase* under higher temperatures. At the completion, two new strands of DNA are produced. The process is repeated using the new and original DNA as templates for making more DNA. This process results in millions of copies in a short period of time.

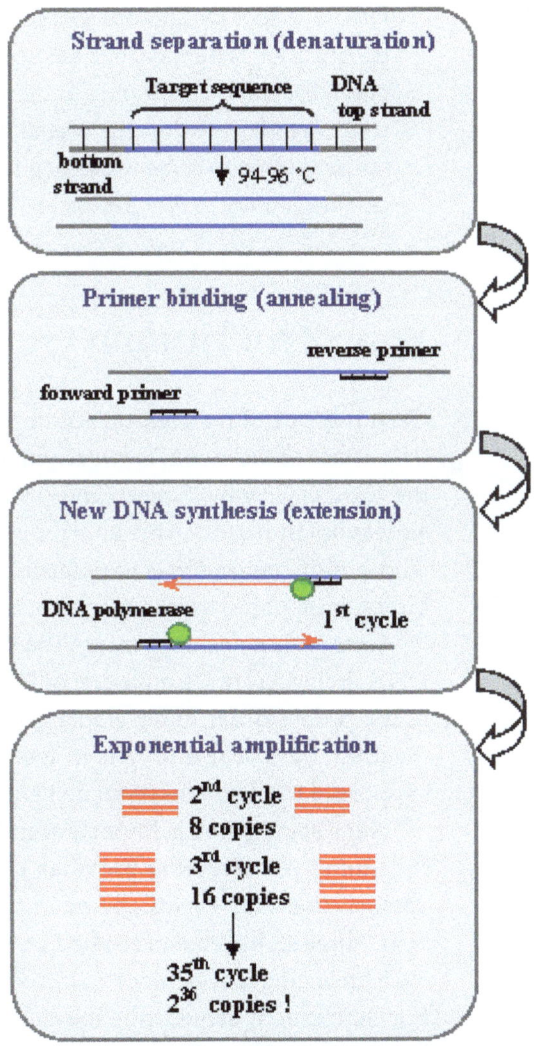

FIGURE 12.7 Summary of PCR methodology

PCR is used in a broad range of applications. Agricultural scientists employ PCR to develop and test genetically modified foods. In forensics, the process of making a genetic "fingerprint," or profile, of an individual from a sample of

DNA involves PCR. The chances of two people having the same "fingerprint" is minuscule, so the legal system employs such evidence to convict the guilty and exonerate the innocent. Also, legally, PCR is a mainstay in paternity DNA testing, important in determining the rights and responsibilities of a father. With small amounts of DNA, sequencing can be completed in about two hours. One of the greatest applications of PCR comes in the biomedical industry to understand the genetic basis of disease, catalog and diagnose disease, identify genetic disorders, and search for treatments and cures for a disease. For example, PCR can be used to amplify the genetic material of SARS-CoV-2 if it is present in mucus acquired for testing, confirming the virus is present. Also, PCR was a workhorse powering the Human Genome Project (NIH 2015, "Polymerase Chain Reaction").

Select the appropriate terms to complete the statement: ACR, amplifying, alleling, annealing, denaturation, extension, nucleotides, PCR, polymer, polymerase, primers

QRQ 8. Select the appropriate terms to complete the statement: ACR, amplifying, alleling, annealing, denaturation, extension, nucleotides, PCR, polymer, polymerase, primers. **a.** ______, or **b.** ______ chain reaction is used in **c.** ______, or making millions copies of, a small segment of DNA. The steps, in order, are **d.** ______ (separating DNA into two strands), **e.** ______ (binding of **f.** ______ to each strand to serve as starting points of DNA synthesis), and **g.** ______ (adding **h.** ______ complementary to each template). Then, the process is repeated using the new and original DNA.

12.5. The Human Genome Project

Certainly one of the greatest accomplishments in the history of humankind was placing humans on the surface of the moon. However, in terms of challenge, effort, cooperation, and computational power, the *Human Genome Project (HGP)* stands on an equal footing. One of the most ambitious scientific endeavors in history, this enormous project was a collaborative international scientific endeavor whose ultimate goal was to determine the nucleotide sequence of the entire human genome (three billion base pairs) and to map functional genes. Begun in 1990, the HGP was completed in 2003.

Gene sequencing determines the order of the nucleotides that constitute a gene in a molecule of DNA. Being so long, a molecule of DNA has to be sequenced in short segments and then reassembled in the proper order to develop a genome. Sequencing is similar to putting a complex jigsaw puzzle together, but breakthroughs in technology, automation, and more efficient methods, many with support of the HGP, have made gene sequencing faster, more reliable, and less expensive.

Gene sequencing is important in helping find specific genes more easily and quickly. Presently, the *Cancer Genome Atlas* provides valuable data on more than 30 types of cancer and the genomic changes that may eventually be important in finding cures. Other disorders, such as diabetes and heart disease, have been studied in a similar fashion.

Sequencing is also important in determining the genome and how it works in numerous organisms. For example, by sequencing the causative agent for COVID-19, an RNA virus of the Coronaviridae family of viruses, scientists have a better understanding of its origins and how to develop a successful vaccine. As in cellular genomes, sequencing technologies will give scientists the ability to study regulatory genes and introns as well. Comparative genome studies are also shedding light on viral evolution (NIH 2020, "DNA Sequencing").

Through the years, better technologies have evolved, making exploring the human genome faster, more accurate, and less expensive. One by one, genes were identified, such as the first gene associated with Parkinson's disease in 1996 and the gene map of chromosome 7, completed in 1997. By 2003, the HGP was essentially completed, finishing five years ahead of schedule (NIH 2020, "NHGRI History").

Of course, the project spawned many ethical questions, such as, "Can genes be patented?" The United States Supreme Court ruled in 2013 that "naturally occurring genes" could not be patented.

PoI 5. List five well-known genetic diseases or traits associated with five chromosomes of your choice.

Through the years, the list of accomplishments of the HGP have been astounding. Interestingly, with modern technologies, the human genome can now be sequenced in as little as three days, with a cost of less than one thousand dollars. As the direct result of the HGP, the complete genome of *Homo sapiens* has been sequenced, and important genes responsible for disease and genetic disorders have been identified. The HGP has opened the door for sequencing the genome of other organisms that can be useful in molecular medicine, waste control, energy sources, other biotechnology subjects, and evolutionary studies.

APPLYING SCIENCE TO LIFE 4 Every Gene Has a Story

The human genome is filled with nearly 25,000 genes, shaped through eons of time and evolution, each with its own unique and intriguing story. Let's take a quick look at two intriguing genes and their stories.

Located on the short arm of chromosome 7, consisting of nearly 12,500 base pairs, resides the gene with perhaps the coolest name—the *sonic hedgehog gene* (*SHH*), named after the cartoon character. Based on work conducted in the 1950s investigating the formation of limbs in chickens and other animals and the discovery of the sonic hedgehog gene in fruit flies, geneticists came to understand SHH. Common throughout the animal kingdom, one of the gene's major responsibilities in vertebrates, such as salamanders, chickens, and humans, is the regulation and timing of cell growth and specialization. The SHH gene is fundamental in the positioning, shaping, and makeup of the body, including the proper development of the spinal cord, eyes, and the two brain hemispheres. More than 100 mutations of the SHH gene are responsible for *nonsyndromic holoprosencephaly*, a disorder characterized by abnormal development of the face and brain affecting about 1 in 10,000 newborns. The manifestation of this disorder is highly variable, from the brain not separating into two hemispheres to the presence of one central eye and perhaps a tubular nasal structure located above the eye to a cleft palate (NIH 2020, "SSH Gene").

Another gene, found on the short arm of chromosome 7 and consisting of over 600,000 base pairs, is the *FOXP2* gene. This gene is important in the formation of the *forkhead fox P2* protein, essential for normal brain development and function, including vocalization, such as speech in humans. Reports in 2007 noted that the FOXP2 was even found in our Neanderthal relatives (NIH 2020, "FOXP2 Gene").

Our genome consists of many intriguing genes, many shared with other life forms. These genes are part of our genetic toolkit, and each has a unique story.

12.6. Gene Therapy

In the early 1960s, the American molecular biologist *Joshua Lederberg* stated, "it is only a matter of time ... before polynucleotide sequences can be grafted by chemical procedures onto a virus DNA," literally envisioning the world of *gene therapy*. By the late 1980s, despite great potential, gene therapy development was stalled by ethical and other issues. Finally, progress on gene therapies is being made.

For genetic disorders such as sickle cell anemia and cystic fibrosis, gene therapy can replace a faulty gene with a normal one. Introducing a normal gene requires a vector, usually a virus, customized to deliver the new gene into affected cells. Such viruses are modified so they are not pathogenic. The vector can be directly injected into the patient, given intravenously (IV), or via the patient's cells exposed to the vector in the laboratory and returned to the patient. If the treatment is successful, the corrected gene will oversee the production of a functional protein (see Figure 12.8).

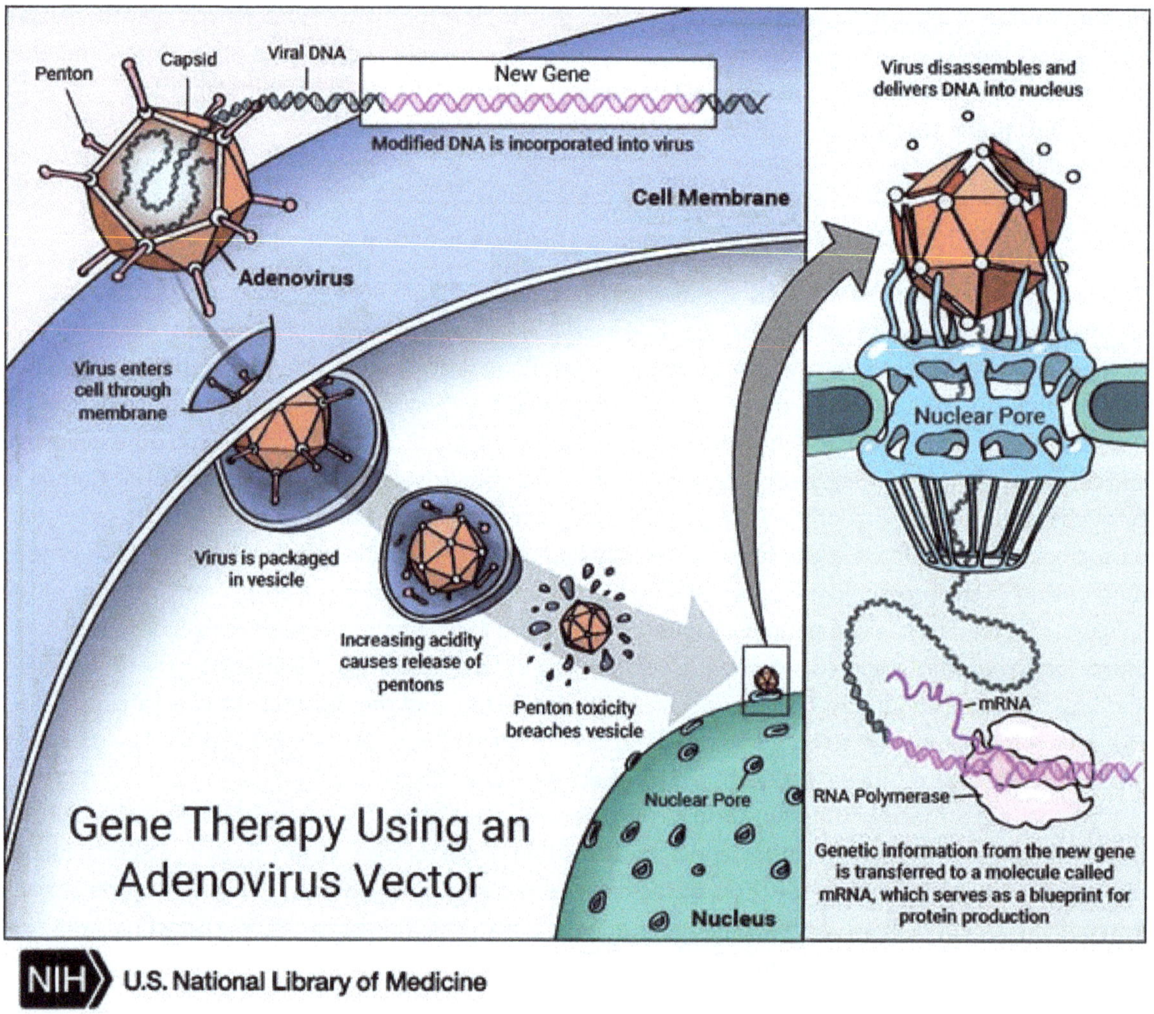

FIGURE 12.8 Using a virus in gene therapy

Today, gene therapy is focused on targeting somatic cells, not sex cells, so that the therapy does not influence future generations (NIH 2020, "Gene Therapy").

Although offering great promise, gene therapy can cause serious health risks and, in some cases, unpredictable results. The National Institutes of Health (NIH), US Food and Drug Administration, and researchers make every effort to ensure the safety of this technology. In order to be used in a clinical trial, the protocol for the trial is rigorously scrutinized by the NIH Office of Biotechnology Activities, the NIH Recombinant DNA Advisory Committee, an Institutional Review Board, and an Institutional Biosafety Committee. No federal funding is permitted for gene therapy involving sex cells. These safeguards have been designed to protect patients and the public. Gene therapy also raises ethical issues in society that must be addressed. Across the nation, hundreds of gene therapy projects are legitimately investigating the treatment for conditions that burden human health.

PoI 6. Describe several ethical issues that may stem from gene therapy.

APPLYING SCIENCE TO LIFE 5 Tragic Setback for Gene Therapy and Human Health

After much promise and success in other trials, another milestone in gene therapy was perhaps just around the corner. It began in 1993, when two pediatricians, Mark Batshaw and James Wilson, began to experiment with the power of gene therapy to cure children that had an X-linked single-gene mutation responsible for a liver disorder called *OTC* or ornithine transcarbamylase. The OTC gene is responsible for the production of an enzyme that is critical in the breakdown of certain proteins. Malfunction of this gene can result in fatigue, migraines, forgetfulness, delirium, loss of appetite, and seizures, with many patients not surviving beyond childhood.

Batshaw and Wilson's approach was radical compared to other gene therapy attempts, because they proposed modifying a cold virus to carry the corrected OTC gene. The modified virus particles would deliver a corrected sequence directly into liver cells, causing them to function properly. Their early experiments with mice were successful. However, experiments with monkeys were not as successful, with many monkeys developing acute immune responses. Batshaw and Wilson deleted many of the virus genes that evoked an immune response and reduced the dosage before human trials. In 1997, the doctors sent their request to the Recombinant DNA Advisory Board for the therapy to be used in human subjects. After debate, the request was approved with the recommendation that the trial should be given to normal volunteers or patients with mild cases of the disorder.

In September of 1999, an 18-year-old young man with a mild case of OTC enrolled in the experimental program. Even with a lighter version of OTC requiring a restricted diet, he had had several life-threatening episodes of OTC. On September 13, despite his having abnormal amounts of ammonia in his blood, the physicians decided to continue the procedure. The night after the infusion, Jesse's temperature rose to 104° F, and the next morning he was jaundiced, indicating liver failure. On September 17, Jesse was pronounced dead from a severe immune reaction to the therapy. As a result of the investigation into Jesse's death, a moratorium was placed on clinical gene therapy trials. It would take another decade, using stricter protocols with new delivery vehicles, before gene therapy would make a comeback (Mukherjee 2016).

12.7. CRISPR

Soon, *Clustered Regularly Interspaced Short Palindromic Repeats* (*CRISPR*) will serve as the foundation for revolutionary new genome editing technologies that will allow scientific researchers to make ultraprecise modifications to DNA. The road to CRISPR technologies began in 2011, when two researchers, *Jennifer Doudna* and *Emmanuelle Charpentier*, began investigating the ability of bacteria to defend themselves from viral attack. Viruses are like molecular pirates that use their DNA or RNA to commandeer the genetic mechanisms of a host cell, like a bacterium. Doudna and Charpentier's work was based on the findings of French scientists in the early 2000s working in the cheese and yogurt industry, which is highly dependent on healthy bacteria. These scientists described a defense mechanism against viral attack found in some bacteria. The bacteria uses a piece of RNA encoded in their genome that recognizes a particular genetic sequence in the viral DNA and tag it for attack. The tag is recognized by a bacterial enzyme called *Cas9*, and the enzyme is tasked to destroy the viral genome.

Doudna and Charpentier reasoned they could develop an efficient *gene editing* system, called CRISPR. Such a system could be used to recognize a faulty gene, like those in Tay–Sach's disease, and scissor out the faulty sequence in that gene, then weld in place an intact copy of the unaffected sequence, editing and correcting the defect. In order to function, CRISPR/Cas9 technologies would have to be applied to human embryonic stem cells. Because stem cells are responsible for forming a plethora of cells, a genetic disorder, theoretically, could be corrected at the molecular level. However, these corrections would also influence potential sex cells and could alter evolution, ushering in an age of *designer babies*. Parents could literally choose from a variety of phenotypes and create their ideal baby. Eliminating genetic disorders is a noble goal, but deciding on a child's potential height or color of eyes presents an ethical dilemma. In 2015, backed by Doudna and others, a moratorium was placed on gene editing and other gene-altering techniques in the human clinical setting.

CRISPR/Cas9 technologies have unimaginable applications in genetic engineering, agriculture, industry, and biomedical research. In medicine, it can be used to develop transformative gene-based medicines for patients with serious diseases, such as cancer, malaria, and HIV. CRISPR/Cas9 could also disrupt genes in pathogens, such as human papillomavirus (HPV), Zika virus, and many harmful bacteria. Interestingly, CRISPR/Cas9 technologies could be employed to save species at the brink of extinction or to resurrect extinct species. Moreover, with these technologies, we could produce more healthy animals and plants and eradicate a variety of pests. Of course, like any new technology (smallpox vaccine and use of stem cells), many individuals are leery of CRISPR. Hopefully, scientists and a scientifically literate society will see that such technologies are used to benefit the planet and humankind (Vidyasagar 2018).

APPLYING SCIENCE TO LIFE 6 Malaria on the Run?

The term *malaria*, from medieval Italian, has ancient Roman origins and literally means "bad air." The Romans believed the disease was caused by foul and putrid fumes that emanated from swamps. A variety of ancient manuscripts described what appeared to be malaria, and the ancient Indians described malaria as the "king of diseases." In the 1990s, scientists discovered the malarial antigen in ancient Egyptian remains, and malaria was probably responsible for the death of Alexander the Great at the age of 32. Malaria made its way to the United States and infected millions of people, including George Washington and Abraham Lincoln. The *Centers for Disease Control and Prevention* (*CDC*) was founded primarily because of malaria. Today in the United States, malaria has been somewhat controlled by understanding its life cycle, using effective drugs, and employing mosquito control and other prophylactic measures. However, the wrath of malaria is still prevalent, primarily in Africa.

Malaria is a mosquito-borne tropical disease that is still a serious and potentially fatal disease with symptoms such as high fever, chills, abdominal and muscle pain, and enlarged spleen. The causative agent is a one-celled parasite belonging to the genus *Plasmodium* with a highly evolved, complex life cycle, and the *vector* (transmitting agent) for malaria is a female mosquito of the genus *Anopheles* (see Figure 12.9).

FIGURE 12.9 ***Anopheles*** **mosquito engorged with human blood**

Prior to the coming of age of science and reasoning in the 19th century, malarial outbreaks were attributed to bad air, contaminated water, demons, and spirits. In the 1880s, the French physician *Louis Laveran* was responsible for revealing the life cycle of *Plasmodium*. Another scientist, *Camillo Golgi*, differentiated between the four species of *Plasmodium* and described the asexual stage of malaria. Two veterans from the development of the understanding of the virus that causes yellow fever and its link to mosquito transmission now entered the stage. *Patrick Manson*, a Scottish physician, and his student *Ronald Ross*, a British physician, veterans of mosquito-borne illness research, determined that the *Anopheles* mosquito is the vector of malaria and outlined the complete life cycle of *Plasmodium*.

It was known for centuries that the bark of the *cinchona tree* was an effective agent against malaria. In the 1820s, two French pharmacists and chemists, *Pierre Pelletier* and *Jean Caventou*, isolated and developed the antimalarial drug *quinine* from this remarkable plant. Quinine immediately became the favored therapy for the dread disease and is still an important weapon in our arsenal against malaria. Unfortunately, *Plasmodium* is developing a resistance to quinine, but scientists have and continue to develop other medicines.

During the COVID-19 pandemic, many people thought that the antimalarial drug *chloroquine* would be effective against the coronavirus, but, unfortunately, it was not. Although developed by German chemists after World War I, chloroquine did not come into popular use until World War II because of a quinine shortage. After World War II, the *World Health Organization* (*WHO*) and the US Army jointly developed *mefloquine*, which has proved effective against *Plasmodium*. In the 1970s, Chinese scientists rediscovered the herb *qing-hao*, used to treat fevers nearly 2,000 years ago, and from it developed the medicine *artemisinin*, a main line of defense against drug-resistant malaria in many parts of Asia.

Not only is the malarial parasite becoming resistant to drugs, the *Anopheles* mosquito is or is becoming resistant to pesticides. These trends could lead to a substantial increase in the incidence of malaria and malarial deaths.

According to the World Malaria Report 2019, conducted by the WHO, in 2018 there were 228,000 cases of malaria worldwide and an estimated 405,000 deaths, costing over 2.7 billion dollars in malaria control and elimination. Most of the cases occurred in Africa, disproportionately affecting children under five years old and accounting for 67 percent of malarial deaths worldwide. In the United States, approximately 2,000 cases of malaria are diagnosed per year. The vast majority of these cases are associated with travelers returning from malaria-infested regions and immigrants. As malarial and pesticide resistance rises and ease of world travel grows, it is reasonable to assume that these numbers will increase in the future (WHO 2019; NAP 2004).

How will we successfully counter the spread of malaria in the future? Should we continue to use and tweak traditional methods of using antimalarial drugs and pesticides, or is there a better way? Scientists are investigating using the power of biotechnology to edit the genes of the *Anopheles* mosquitoes in an attempt to counter the spread of malaria. Such editing will be accomplished by developing a *gene drive*, or suite of genes to be inserted into the mosquito, that will be passed on to future generations. Presently two strategies using CRISPR are being advanced for developing a gene drive. In the first method, *Anopheles* mosquitoes are driven to extinction by rendering them sterile, and in laboratory experiments, positive results have been observed. However, this strategy may be hard to employ in a broad area, and it raises many ethical issues. The alternative modification method involves rendering the mosquitoes incapable of spreading the disease. This gene drive would incorporate one of two genetic strategies, either to cause a mosquito to kill the *Plasmodium* within or expel the *Plasmodium* without injection into humans. Hopefully, one of these strategies will help curb this deadly disease (Bier and Sober 2020).

The CDC highly recommended that travelers to countries that are at high risk for malaria visit the site to learn more about the disease, its biology, helpful maps, prevention, risk assessment, and what to do if you contract malaria.

12.8. The Future

Undoubtedly, improving existing and developing new biotechnologies will continue to occur at an unprecedented pace. Fields such as gene editing, molecular medicine, genetic screening, preventative medicine, pharmacogenomics, energy technologies, and bioremediation will burgeon in the future.

Gene editing, allowing geneticists to alter the DNA of an organism, has been used since the 1980s. Using *molecular medicine* research, medical professionals will be able to address the fundamental causes and remedies of a plethora of diseases, instead of just treating the symptoms. Futuristic molecular testing will make a diagnosis more efficient and rapid. In addition, *genetic screening* will become a valuable tool in understanding a patient's health and wellness. Physicians, anticipating potential multifactorial maladies, such as high blood pressure, will aid the patient in developing a lifetime wellness plan. *Precision medicine* and *pharmacogenomics*, using genetic information to predict how patients will respond to specific medications, will reshape medicine. Also, many companies offer individuals relatively inexpensive gene-testing kits or services that can yield valuable information about ancestry, family history, and health profiles. (Before purchasing, research several brands thoroughly, investigating scientific validity, consumer reviews, periodic upgrades of data, and privacy policies.)

Besides our own genome, understanding the genome of certain methane-producing microorganisms could revolutionize the conversion of biomass into more efficient and cheaper fuels. *Bioremediation*, or the use of custom-designed microorganisms to treat pollution and contaminated soil and water, will become an important tool in maintaining a healthy environment and in aiding in recovery of the environment and living things from disasters, such as oil spills (see Figure 12.10). New energy technologies and bioremediation can improve *sustainability* by developing systems and strategies to meet our needs without jeopardizing the earth's health.

Biotechnology has the potential to address a number of medical, industrial, and environmental needs in the future—that is, if we use it wisely! On the other hand, if used incorrectly, it could become the "stuff" of nightmares. Let's do it right!

FIGURE 12.10 Heavily oiled pelicans waiting to be cleaned after an oil spill

PoI 7. Discuss the pros and cons of genetic engineering.

QRQ 9. The gene editing system, called ______, is used to recognize a faulty gene and cut out the faulty sequence in that gene.

QRQ 10. The ______ ______ Project, completed in 2003, determined the nucleotide sequence of the entire human genome and mapped functional genes.

QRQ 11. The use of custom designed microorganisms to treat pollution and contaminated soil and water is called ______.

Image Credits

Fig. 12.1: Copyright © by Idaho National Laboratory (CC BY 2.0) at https://commons.wikimedia.org/wiki/File:Biotechnology_Research_(3617772958).jpg.

Fig. 12.3: Source: https://commons.wikimedia.org/wiki/File:Dolly_clone.svg.

Fig. 12.4: Copyright © by Wikibob (CC BY-SA 3.0) at https://commons.wikimedia.org/wiki/File:Cloning_diagram_english.png.

Fig. 12.5: Adapted from Source: https://commons.wikimedia.org/wiki/File:Recombinant_DNA.JPG.

Fig. 12.6: Copyright © by Ccm272 (CC BY-SA 4.0) at https://commons.wikimedia.org/wiki/File:Glofish_shark_purple.jpg.

Fig. 12.7: Source: https://www.ncbi.nlm.nih.gov/probe/docs/techpcr/.

Fig. 12.8: Source: https://ghr.nlm.nih.gov/primer/therapy/procedures.

Fig. 12.9: Copyright © by Dunpharlain (CC BY-SA 4.0) at https://commons.wikimedia.org/wiki/File:Anopheles_mosquito_engorged_with_human_blood.jpg.

Fig. 12.10: Copyright © by International Bird Rescue Research Center (CC BY 2.0) at https://commons.wikimedia.org/wiki/File:Gulf-Oiled-Pelicans-June-3-2010.jpg.

LAB #12

Laboratory Exercises

12.9. Laboratory on DNA Protein Synthesis

12.9.1. Preliminary Work

This laboratory explores DNA protein synthesis and the effects of mutations. From the text's website, view the introductory video; download and open *"DNA Protein Synthesis Rev.nlogo."*

1. Consulting other sources as needed, select from the following the best term for each description below: A. A-C and T-G, B. A-G and T-C, C. A-T and C-G, D. codon, E. DNA, F. gene, G. mRNA, H. nucleotide, I. promoter, J. protein (simple), K. reading frame, L. ribosome, M. RNA, N. transcription, O. translation, P. tRNA

   a. ______ bases that can form a base pair in DNA

   b. ______ biomolecule that is a chain of amino acids and that performs many critical functions

   c. ______ binds to a codon of messenger RNA and carries resulting amino acid to a ribosome

   d. ______ contiguous section of a chromosome that encodes information to build a protein or an RNA molecule

   e. ______ double-stranded nucleic acid that contains the encoded information for the manufacture of all the proteins a cell needs

   f. ______ intermediary, single-stranded nucleic acid, carrying instructions from DNA that specifies the amino acid order of proteins

   g. ______ molecular complex that synthesizes a sequence of amino acids for the protein from an mRNA sequence

   h. ______ molecule made up of a sugar, a phosphate, and a nitrogen base forming the basic subunit of DNA or RNA

   i. ______ molecule that conveys genetic information from DNA to a ribosome

   j. ______ process of synthesizing a protein from a DNA sequence's instructions into a sequence of amino acids

   k. ______ sequence of consecutive, nonoverlapping triplets in DNA or RNA, for example, with nucleotide sequence GGATTACCA: GGA TTA CCA or GAT TAC or ATT ACC

l. ______ sequence of three nucleotides in RNA that specifies an amino acid or stops the synthesis of a protein

m. ______ subsequence of DNA indicating where transcription of a gene is to begin

n. ______ synthesis of RNA from a single strand of DNA

2. Click the Info tab at the top of the page and read sections "What Is It?" and "How It Works."

   a. ______ In the model, what does a colored polygon represent? A. DNA, B. gene, C. mRNA, D. nitrogen base, E. phosphate group, F. tRNA

   b. ______ In the model, what does a yellow circle represent? Use the choices from part a.

   c. ______ Give the letter abbreviation for the nitrogen base that is in mRNA and tRNA but not DNA.

   d. ______ Give the start codon in all capital letters that signals the start of transcription for mRNA.

   e. ______ From the following list, give the nucleotide sequence that is *not* a stop codon: ATT, ATC, ACT, ACC

   f. ______ (T/F) If the start codon begins at location 3 and a stop signal begins at location 18, then the gene is between locations 3 and 17, inclusive, in this model.

   g. ______ (T/F) If the start codon begins at location 4 and a stop signal begins at location 18, then the gene is between locations 4 and 17 (inclusive) in this model.

   h. ______ Give the model step to release a transcribed mRNA molecule in the original DNA.

   i. ______ Give the model step to release an amino acid chain from that attached to the tRNA chain in the original cell.

3. Read "How to Use It." Select from the following the appropriate term for each definition below: A. 1-TRANSCRIBE, B. 10-SHOW PRODUCTION SUMMARY, C. 2-RELEASE, D. 3-TRANSLATE, E. 4-RELEASE, F. 5-REPLICATE THE ORIGINAL DNA, G. INITIAL-DNA-STRING, H. INSTRUCTION #, I. MUTATION-TYPE, J. NEXT-INSTRUCTION, K. USER-CREATED-CODE

   a. ______ indicates kind of starting DNA string, such as "1 long gene" or "from user-created-code"

   b. ______ moves the text in the instruction window forward one "page"

   c. ______ releases an mRNA from the nucleus into the ribosome of the original cell

   d. ______ reports all proteins produced

   e. ______ sets mutation to deletion, insertion, or substitution

12.9.2. Laboratory Work

4. This question goes through a modeled DNA protein synthesis.

   a. ______ Set INITIAL-DNA-STRING to "1 long gene." Press SETUP. Read instruction #1 of the instruction window and press NEXT INSTRUCTION, reading the next instruction. Give the stop codon using all capital letters and no blanks.

   b. ____________ Give the last four bases in the gene using all capital letters and no blanks.

   c. ____________ Advancing with NEXT INSTRUCTION, read instruction #3 and follow the instructions. Give the sequence of the last four bases in the mRNA molecule using all capital letters and no blanks.

d. ____________ Read and perform instruction #4. In this step, mRNA is released from the nucleus into the **1.** of the cell.

e. ______ Different codons sometimes result in the same amino acid. For example, amino acid L-Threonine, abbreviated Thr, is formed from any of the following RNA codon sequences: ACU, ACC, ACA, ACG. Give the number of times this amino acid will occur in the protein. Be sure to consider only codons, searching from every third base by looking at positions 1, 4, 7, and so on.

f. ______ Give the codon for the first occurrence of Thr.

g. ______ The codon for the second occurrence of Thr is ACU. Give the corresponding triplet in the DNA.

h. ______ Read and perform instruction #5. Including the start codon, there are 39 bases in the gene and in the mRNA. Thus, how many subsequences of three nucleotides are there in the gene?

i. ______ Read and perform instruction #6. How many amino acids are in the protein?

j. ______ Read instruction #7, which is applicable if the sequence contains two or more genes. Our current sequence contains only one gene, so we continue to the next step. Make sure MUTATION-TYPE is set to "substitution." Read and perform instruction #8. Read instruction #9, noting the information in the Command Center at the bottom of the screen. Find the mutation, which is a substituted base in green. Are there any other changes in the replicated DNA, which is in a new cell, no (N) or yes (Y)?

k. ______ Press 5-REPLICATE THE ORIGINAL DNA button several times, each time noting the substituted base and the length of the gene. Even with a mutation, sometimes you may get the same protein as the original one, because the substitution does not occur within the gene or because different codons sometimes result in the same amino acid. Does the length of the gene in the replicated DNA ever change, no (N) or yes (Y)?

l. ______ Read and perform instruction #10. How many proteins are made from the replicated gene with mutation? Perform steps 5 through 9 several times, noting that we usually obtain a protein sequence different than the original one.

m. ______ Change MUTATION-TYPE to "deletion." Press 5-REPLICATE THE ORIGINAL DNA button several times, each time noting the deleted base and the length of the gene. Does the length of the gene in the replicated DNA ever change, no (N) or yes (Y)? Perform steps 5 through 9 several times, noting the variation in different protein sequences.

n. ______ Change MUTATION-TYPE to "insertion." Press 5-REPLICATE THE ORIGINAL DNA button several times, each time noting the inserted base and the length of the gene. Does the length of the gene in the replicated DNA ever change, no (N) or yes (Y)? Perform steps 5 through 9 several times, noting the variation in different protein sequences.

o. ______ Select the possible situation that *cannot* occur in replication of the original DNA when we have one mutation, or mark A if all situations can occur: A. all these situations can occur, B. different resulting protein, C. gene starting earlier in the sequence, D. longer gene, E. multiple genes, F. no gene, G. same gene, H. same resulting protein, I. shorter gene

5. Perform other DNA protein synthesis examples, such as changing the INITIAL-DNA-STRING or adding your own USER-CREATED-CODE.

UNIT IV

Evolution/Natural Selection

MODULE

13 Evolution of Evolutionary Thought

13.1. Introduction

February 12, 1809, was an important day in the history of the 19th century and centuries to follow. Two men that had unprecedented influence on the worlds of politics and biology were born on the same day an ocean apart. The 16th president of the United States, Abraham Lincoln, was born into a poor family in rural Kentucky, and Charles Robert Darwin, the icon of evolutionary thought, was born into a wealthy family in Shrewsbury, England (see Figure 13.1). The accomplishments of Abraham Lincoln are well known and legendary. However, to many, the life and times of Charles Darwin are shrouded in misunderstanding.

FIGURE 13.1 Charles Darwin's wedding portrait circa 1839

"There is grandeur in this view of life, with its several powers, having been originally breathed into a few forms or into one; and that, whilst this planet has gone cycling on according to the fixed law of gravity, from so simple a beginning endless forms most beautiful and most wonderful have been, and are being, evolved."

On the Origin of Species 1859

To many, the term *evolution* simply means a change over time or the change in a species over geologic time. However, in early biological thought, it was derived from the word *evolvere* or *evolution*, meaning "to unfold," such as a pupa within a chrysalis unfolding into a butterfly. Not until the 1800s did the term approach today's meaning. Even Darwin only used the term once in his definitive work *On the Origin of Species*. Contemporary definitions of evolution address a change in the gene frequency of a population over time.

13.2. Evolution Prior to Darwin

Humans have always been intrigued by the origin of the universe and life itself. However, in the days before the emergence of science, these questions could only be answered with mysticism and misconception. Early civilizations, such as those in Egypt, China, Mesopotamia, and the Americas, developed fascinating and wonderful stories addressing how the universe began and life emerged. Science came of age in ancient Greece, and early Greek scientists set the stage for evolutionary inquiry. However, it is the Greek philosopher and teacher Aristotle (384–322 BC) that is credited with developing many early evolutionary concepts. Many of these ideas would be refined and rewritten by numerous scientists, including Darwin.

Aristotle's passion was marine biology, and his numerous experiences gave him great insight into the world of life. In evolution, however, he believed that living things, especially lower forms, such as invertebrates, could appear through *spontaneous generation* from nonliving matter. Thus, fleas could arise from specs of dirt, horsehair worms could form from fallen horse hairs, mice could originate from meat or cheese wrapped in a rag and placed in a dark corner, and maggots could be derived from rotten meat. Of course, today, as the result of the work of Pasteur and others, spontaneous generation is just a discourse in biological history.

APPLYING SCIENCE TO LIFE 1 That Is Hard to Believe!

Have you ever seen a freshwater eel in a lake or pond near your home? Perhaps you took for granted that this elongated fish was born in this environment. Our ancestors believed that these creatures underwent spontaneous generation and arose from the slime on the bottom of a pond or lake. After all, they are very slimy.

FIGURE 13.2 Freshwater eel

Today, science reveals that there are two noteworthy species of freshwater eels. The European eel (see Figure 13.2), is an inhabitant of western Europe, and the American eel is found mostly on the coasts of North America and even eastern South America. Distinctions between the two species are based on the number of vertebrae and molecular markers. Eels also have unbelievable life cycles. After living in their pond or lake for as many as 25 years and reaching sexual maturity, if possible, they migrate to the ocean and eventually to the Sargasso Sea in the Atlantic Ocean. Here, they mate and then die from exhaustion. The immature eels undergo five stages of development and make their journey back to their respective continents, including many out-of-the-way places (USFW 2019; Bittel 2015). No wonder our ancestors thought they came about through spontaneous generation!

Another idea championed by Aristotle and other Greeks was *fixity*, or *immutability*, *of species*. This concept asserts that species were placed on the earth in their original form and have remained unchanged over time. This idea has been countered by modern evolutionary evidence, such as fossils and molecular data, but in some philosophies remains a fundamental belief.

PoI 1. Discuss other examples of what some thought were the results of spontaneous generation.

Aristotle developed a concept presented by Plato known as the *Great Chain of Being*, which represented a hierarchy from inanimate matter to God based on certain characteristics, one characteristic being the type of soul an organism possessed. Plants had vegetable souls, humans had rational souls, and gods and God had an immortal soul. Certainly, this characteristic is immeasurable, but you may think twice about eating that green salad! Today, Darwin's Tree of Life, which will be discussed later, is a more scientific attempt to represent the unity of life on Earth. Following Aristotle, a multitude of scientists contributed to evolutionary thought. Basic strides in comparative anatomy, paleontology, geology, astronomy, inheritance, and philosophy set the stage for the arrival of Charles Darwin.

APPLYING SCIENCE TO LIFE 2 Lamarck: Soldier, Scientist, and Visionary

To many, the name Jean-Baptiste Lamarck (1744–1829) is only associated with the "law of use and disuse" or the "theory of acquired characteristics," discredited ideas of heredity. However, Lamarck should be associated with many accomplishments in zoology and being a forerunner in the quest for understanding evolution. He also contributed to the sciences of meteorology, paleontology, and astronomy. After a distinguished career in the French army, Lamarck studied botany and medicine. His celebrated botany book, *Flore Française*, brought him great acclaim in scientific circles and led to academic promotions. Lamarck provided biology with one of the early definitions of species. He eventually became a professor of invertebrates (a term that he coined), where he did important work on crustaceans, arachnids, mollusks, and annelids. His work on evolution began in the early 1800s. He wisely commented, "Time and favorable conditions are the two principal [sic] means which nature has employed in giving existence to all her productions." In his book *Philosophie Zoologique*, he wrongly stated that evolution was a process of increasing complexity and perfection and was not driven by chance. His law of use and disuse argued that, depending on whether an organism uses or disuses a specific structure, it would either become strengthened or disappear in subsequent generations. His giraffe example still remains a classic of Lamarckism. If a giraffe stretched its neck to acquire leaves, its neck would become longer. In turn, its offspring would inherit a longer neck, and continued stretching over time would result in a giraffe population with long necks. His ideas were attacked by many, and much of his great work remains forgotten. Lamarck died in 1829 discredited, blind, and impoverished (McCarthy 2020, "Jean-Baptiste Lamarck"; PBS 2019; UCMP 2019, "Jean-Baptiste Lamarck").

QRQ 1. The term ______ means change over time, including change in the gene frequency of a population over time.

QRQ 2. Match the from following erroneous evolutionary terms to the descriptions below: Great Chain of Being, Great Chain of Characters, immutability of species, instantaneous generation, rigidity of life, spontaneous generation

a. hierarchy from inanimate matter to God based on things such as the type of an organism's soul
b. incorrect idea that species were placed on the earth in their original form and have remained unchanged over time
c. supposed production of living things from nonliving matter

13.3. Contributions of Charles Darwin

13.3.1. Before the Islands

Charles Darwin's father, Robert, once said of his son, *"You care for nothing but shooting, dogs, and rat-catching, and you will be a disgrace to yourself and all your family"* (*Darwin 1958*). Little did he know that his son would be considered one of the most significant contributors to science and biology in history and would be buried with honors at Westminster Abbey.

As a young man, Darwin enjoyed natural history and the outdoors, engaging in hunting, horseback riding, and collecting seashells and fossils. He was considered by his teachers and father an "ordinary boy below the common standard" (Darwin 1958) Although not a strong student, at the age of 16, Darwin was sent to Edinburgh to carry on the family tradition of becoming a physician. Finding the lectures intolerably dull and the surgeries ghastly, he dropped out of medical school within two years. However, during this time, he developed an intense interest in beetle collecting and zoology.

Upon his return home, Darwin and his father decided that he would pursue a career in divinity at Christ's College, Cambridge. He completed his prerequisite bachelor of arts degree in 1831, and during his stay, Darwin's interest in biology and geology grew immensely. His scientific talents and "enlarged curiosity" were noted by Professor John Stephens Henslow, a prominent botanist and reverend. Henslow recommended young Darwin to Captain Robert Fitzroy of the scientific sailing vessel the HMS *Beagle* to become the ship's naturalist and the captain's gentleman companion. Darwin's father was totally against his son taking the two-year position, because he feared the dangers of the trip and had to finance his son's journey. However, a resourceful Darwin finally received permission to be a member of the expedition. Upon his interview with Fitzroy, Charles almost did not become a member of the crew because of his political ideas and the shape of his nose. Fitzroy believed that Darwin had the round nose of a commoner.

On December 1831, the HMS *Beagle* left England with the 22-year-old naturalist, Charles Darwin, for a two-year scientific journey that was to circumnavigate the globe (see Figure 13.3). Although the voyage actually took five years, 1831–1836, Darwin commented, *"The voyage of the Beagle has been by far the most important event in my life and has determined my whole career"* (Darwin 1958)

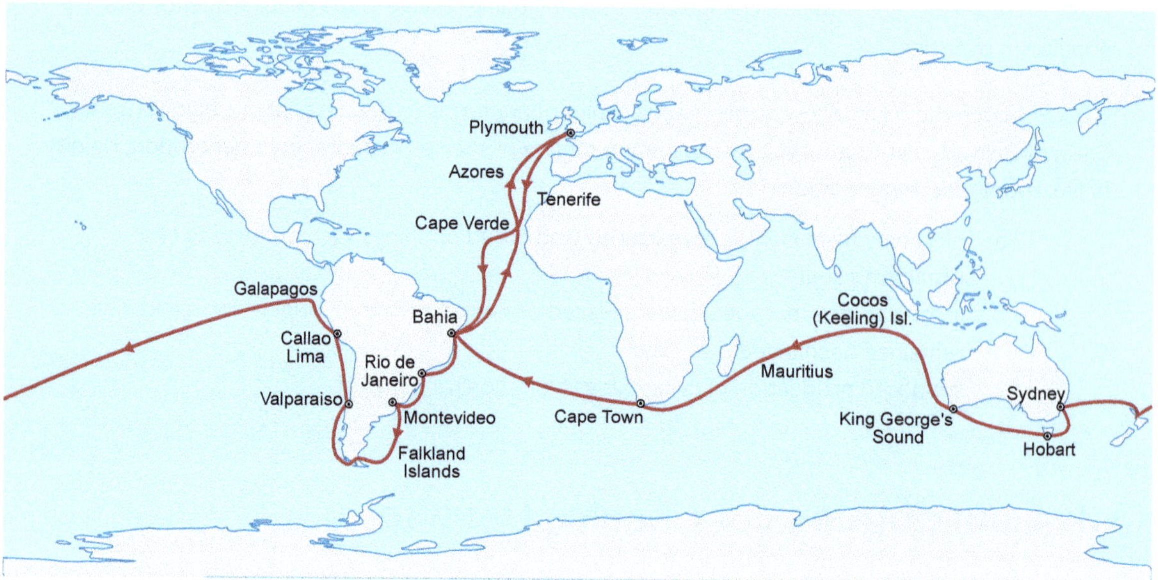

FIGURE 13.3 Voyage of the *Beagle*, 1831–1836

Immediately after entering the open ocean, young Darwin became seasick, a malady that haunted him for the rest of his journey. As the result of his discomfort, the naturalist took advantage of every stop to explore and collect specimens on land. In his book *The Voyage of the Beagle*, published in 1839, and other published works, Darwin brought his great adventure to life and tantalized the curiosity of readers. Among his most vivid accounts were the diversity and wonder of the tropical rainforest of Brazil and the giant fossil mammals found along the seashore of Uruguay. He also wrote about the indigenous people in Tierra del Fuego, riding with gauchos in the pampas, the flightless rheas of Patagonia, surviving an earthquake in Chile, and collecting marine fossils high in the mountains of the Andes. His work and observations of coral atolls in the Pacific catapulted him into being one of the leading scientists of his time. Darwin became extremely ill while visiting the Andes. It is thought that he contracted *Chagas disease*, caused by an endoparasite from the bite of the kissing bug. He suffered from the disease for the rest of his life, with bouts of weakness, vomiting, headaches, and depression (Brent 1981; Darwin 2005; Wyhe 2019; Russ 1998; Stark 2009; Thomson 2009; Van White 2008).

13.3.2. Galapagos Islands

Throughout his travels, Darwin was a creationist and believed in the *immutability of species* (species do not change). He referred to "centers of creation" when describing species in new lands, such as the *Galapagos Islands*. However, he slowly began to question the ancient Greek concepts of evolution, including fixity of species. Perhaps in the Galapagos Islands, he began to turn more toward scientific explanations, as evidenced by his statement from years later: "Hence, both in space and time, we seem to be brought somewhat near to that great fact—that mystery of mysteries—the first appearance of new beings on this earth"(Darwin 2005).

The *Galapagos Islands*, or Enchanted Islands, are a series of volcanic islands in the Pacific Ocean about 600 miles west of Ecuador. Virtually untouched for centuries, although now besieged with tourists, the islands continue to serve as a living laboratory for evolution. Darwin visited the Galapagos in September 1835 and stayed for five weeks.

While in the Galapagos, he observed and collected many unique and interesting animals. Darwin was particularly impressed by the island's giant Galapagos tortoises and even took the opportunity to ride one. He was stunned to find that each island had its own unique species.

APPLYING SCIENCE TO LIFE 3 Lonesome George

One of the best-known Galapagos animals was the giant male Pinta Island tortoise Lonesome George. In the 1800s, sailors, whalers, sealers, and pirates exploited these slow-moving animals for their meat, and as a result, he was the last of his species and served as an icon for conservation. At the time of his death, it is estimated that Lonesome George was over 100 years old. George remained celibate for his 40 years in captivity, not interested in mating with similar species. After his death, geneticists discovered Pinta-like tortoises on another island. In analyzing Lonesome George's DNA, scientists discovered that, like all tortoises, he possessed genes for longevity, a healthy immune response, and fighting cancer. Although all that physically remains of George is his stuffed body at the Fausto Llerena Breeding Center on Santa Cruz Island, his DNA lives on, and some scientists hope possibly to clone him one day.

Darwin was also intrigued that marine iguanas returned to the sea for feeding, counter to their land-borne cousins. Other animals that impressed Darwin were the penguins, flightless cormorants, magnificent frigate birds, blue-footed boobies, sea lions, and Sally Lightfoot crabs. Later, Darwin concluded that many organisms were unique to a particular island and similar species occupied exclusive niches on each island.

APPLYING SCIENCE TO LIFE 4 Sally's Light on "Her" . . . Claws

Why is the Sally Lightfoot crab so brightly colored (see Figure 13.4), when so many other crabs have muted coloration that seems to help them avoid being eaten? As it turns out, against the dark volcanic rock, these crabs get away with the bright coloration that makes them so appealing to potential mates because they are quick, nimble, and agile movers who are good at escaping predators. Hence, the name "Sally Lightfoot" seems fitting.

FIGURE 13.4 Sally Lightfoot crab

Despite all the spectacularly distinctive organisms, three species of mockingbird and several species (13–15) of obscure finches would eventually have the greatest impact on Darwin's thoughts on natural selection. He found it hard to believe that the mockingbirds were so varied on the

FIGURE 13.5 Beak variation of several Galapagos finches

different islands. At the time, he still believed in centers of creation, not that natural forces caused their diversity. Although he overlooked the finches at first, later he realized that the variation of morphology (study of form), beak structure, and behavior was key to understanding how species form. He observed that some finch beaks were used to crush seeds, other beaks were used to eat insects, and one finch used its beak to hold a twig to dig insects from a stem (see Figure 13.5). In 1837, he concluded, "If there is the slightest foundation for these remarks, the zoology of archipelagoes will be well worth examining; for such facts would undermine the stability of species" (Darwin 2005). Today *Darwin's finches* are a classic example of *adaptive radiation,* or how organisms may diversify into new species as a result of changing environments, isolation, or the availability of new resources (AMNH, 2020, Wyhe 2019, National Geographic 2019b, Van White 2008).

QRQ 3. (Answer choices in parentheses) Charles Darwin traveled as a naturalist on the HMS ______ (*Ark Royal, Beagle, Britannia, Queen Elizabeth, Vulcan*) from ______ (England, France, Nova Scotia, the United States) around the world. Before the trip, he believed in ______ (Great Chain of Being, Great Chain of Characters, immutability of species, instantaneous generation, rigidity of life, spontaneous generation). However, upon observing life, particularly in ______ (Brazil, Chile, the Galapagos Islands, the Philippine Islands), he gradually changed his ideas.

QRQ 4. How organisms diversify into new species as a result of changing environments or the availability of new resources is called ______ ______. A classic example of this mechanism involves ______, which Darwin studied, observing their different beak structures, behavior, and so on.

PoI 2. Write about or discuss threats to the Galapagos Islands' fragile ecosystem.

13.3.3. After the Islands

Charles Darwin returned from the adventure of a lifetime in October 1836. His reputation as a scientist preceded him, and his observations were sought by many. Wasting no time, Darwin began cataloging specimens and conducting research. In addition, he began writing prolifically on his experiences and a variety of subjects. He also married his cousin, Emma Wedgewood, in 1839. The couple had ten children and would stay married until his death in 1882.

The Darwin family lived in Kent, England, in a beautiful country home that featured gardens, a greenhouse, and numerous areas for scientific research, recreation, and family pets. Darwin conducted a variety of intense research projects, including work on earthworms, barnacles, bees, pigeons, and insectivorous plants. He was a prolific letter writer who exchanged letters with more than 2,000 people. Today, 15,000 letters written by or to him are still preserved. He also wrote many articles and books, including *The Voyage of the Beagle*, *The Descent of Man*, and of course, *On the Origin of Species by Means of Natural Selection.*

Darwin kept private notebooks for many years that were going to be part of his "big book" on species. He kept them secret because he wanted to develop a solid case for what he knew were groundbreaking ideas. Darwin knew that his concepts were threatening to many. Despite being urged to publish by his colleagues, Darwin was reluctant. Then, quite unexpectedly, in June 1858, he received a letter from the naturalist *Alfred Russel Wallace* (AMNH 2019; Wyhe 2019; National Geographic 2019; Van White 2008).

APPLYING SCIENCE TO LIFE 5 Who Was Alfred Russel Wallace?

Alfred Russel Wallace (1823–1913) was a Welsh naturalist, explorer, collector, surveyor, and geographer. After a brief career as a teacher, Wallace teamed up with fellow naturalist Henry William Bates to explore and map the Amazon region. Bates is best known for his study of *mimicry*, or the ability of an animal to imitate another in order to survive. He illustrated that harmless, edible species, such as certain butterflies, occasionally produced similar color patterns as distasteful or toxic species. After Wallace and Bates went their separate ways, Wallace amassed a tremendous collection of South American specimens. Unfortunately, his collection was lost when his ship caught fire and sank on the way back to England. Wallace escaped with his life and was set adrift in a life raft for ten days before he was rescued.

During 1854–1862, Wallace began exploring the Malay Archipelago, a vast group of islands that extends from southeastern Asia to Australia. He collected 125,600 specimens, including over 1,000 new species. He noticed a remarkable pattern in the distribution of animals in the archipelago. He proposed a gerrymandered imaginary line dividing the area into two biologically distinct regions, later called *Wallace's Line*. The fauna west of the line was represented by Indomalayan species, such as elephants, apes, and tigers, and those located east of the line represented Australian species, such as marsupials and cockatoos. Wallacea is a transition zone between the two and is very biodiverse, including animals such as the Komodo dragon. He later contributed to biogeography by describing six (now eight) distinct *biogeographic realms* (ecozones) defined by climate, altitude, and terrain. For example, the continental United States is comprised of the Nearctic realm and a fragment of the Neotropical realm.

While living in a simple hut in Indonesia and suffering from a malarial-induced fever, Wallace conceived the basic concept of natural selection in a eureka moment. He wrote a brief letter to Darwin outlying his ideas, such that the fittest individuals of a species survived and, as a result, passed their advantageous traits on to their offspring. Wallace's letter awakened Darwin and motivated him to seriously consider publishing his work. The legacy of A. R. Wallace is often overshadowed by Darwin's achievements. However, today, any discussion of Darwin should also acknowledge Wallace, and he should be applauded for his many accomplishments (McCarthy 2020, "Alfred Russel Wallace"; National Geographic 2019, "Alfred Wallace"; Beccaloni and Smith 2015).

In June 1858, the content of Wallace's letter shocked Darwin. It was as if the letter contained the topics and substance of Darwin's private notebooks and the "big book." Wallace had come to the same conclusions about natural selection that Darwin had been privately writing about for 20 years. Perplexed, Darwin did the honorable thing and collaborated with Wallace on a paper that was presented to the Linnaean Society in July 1858. The paper was met with no acclaim. After the paper, Darwin began writing *On the Origin of Species by Means of Natural Selection,* which was published in December 1859. *On the Origin of Species* was considered by many a direct threat to predominant religious beliefs, spawning much animosity. Others considered it a great accomplishment and the most important and influential book ever written. Great debates erupted about a book that only mentioned monkeys once, and not in an evolutionary context. Eventually, most intellectual circles accepted the book, which shifted the paradigm of evolution toward a more modern approach, based on evidence.

13.3.4. Darwin's Thoughts

Thomas Henry Huxley, staunch defender of Darwinism after reading *On the Origin of Species*, proclaimed, "I believe that Mr. Darwin's theory may explain many things, and throw a great light upon numerous questions" (Huxley 2019). The ideas generated in the book are relatively simple, which we can break down into several concepts.

FIGURE 13.6 Gray wolf, ancestor of all dogs

Long before Darwin, humans purposely bred for desirable traits in plants and animals. Thus, through the centuries, we have bred for sweeter corn, faster horses, and more adorable puppies. This variation under domestication is known as *artificial selection*, or *selective breeding*. Agriculture has particularly reaped the benefits of this powerful tool. The potential of this concept was reinforced in Darwin as the result of his pigeon-breeding observations and experiments. Think about it—in approximately 15,000 years, the domestication of the gray wolf has led to well over 200 pure breeds of dog and many more hybrids (see Figure 13.6). Today, dogs have the widest range of variation of any mammal. Keep in mind that variation is due to heredity, and that these variations can add up over time.

APPLYING SCIENCE TO LIFE 6 Taming of the Fox

Although beautiful, foxes generally do not make good pets. However, as the result of the work of a Russian geneticist, Dmitry K. Belyayev, in the 1950s, an elite group of silver foxes were rendered very dog-like through artificial selection (see Figure 13.7). He divided the foxes into three classes, based on *flight distance*. In his experiment, Class III foxes were scared of humans, Class II foxes tolerated humans, and Class I foxes demonstrated a degree of tameness. The geneticist selectively bred for Class I kits. After just six generations, a *domesticated elite* was developed that sought human attention and licked their handlers. After nearly 40 generations, many of the foxes not only displayed a dispositional change but lost their "foxy" pelage, developed floppy ears and a turned-up tail, and vocalized like dogs (Dugatkin and Trut 2019; Jones 2016; Goldman 2010, 2018). Now that is fascinating!

FIGURE 13.7 Silver fox, subject of Belyayev's genetics experimentation on domestication

At the hands of humans, artificial selection has resulted in incredible changes within species over the last few thousand years. In contrast, nature has a 3.5-billion-year head start, and the diversity and wonder of the living world is the result of *natural selection*. Both Darwin and Wallace envisioned the process, but Darwin was much more meticulous and detailed in working out the details. The basic premise of natural selection is simple. Within a population, variations of many degrees exist. Some of the variations may result in an individual's improved ability to survive and reproduce. In succeeding generations, the population will gradually become more suited for survival. Eventually, the accumulation of these advantageous traits can result in new species. Darwin considered this his "great fact one" of his most important contributions to science, and it provided him with valuable insight into the sculpting of species. Today, the resistance of insects to pesticides, bacteria to antibiotics, and the variation in Darwin's finches are examples of natural selection in action.

In nature, there is a great propensity to *overproduce* in order to assure survival. This tendency is apparent in the number of acorns produced by an oak tree, the number of spores produced by a mushroom, and the number of eggs laid by a salmon. For instance, a female bullfrog may produce up to 5,000 eggs, but after consumption of some eggs, environmental fluctuations, and predators having their way with other tadpoles, few frogs make it to adulthood. This phenomenon was described by many biologists prior to Darwin, including Aristotle. While traveling onboard the HMS *Beagle*, Darwin noted this occurrence and included it in his ideas on evolution. The overproduction of offspring ensures the survival of enough offspring to carry on the species.

Organisms struggle to survive every day. Consider when class is over, and you only have a short time to get something to eat. Suddenly, you find yourself rushing for that elusive fast-food goodie. You have now entered Darwin's world of *struggle for existence*. He really liked this term, which served as the title of chapter 3 in *On the Origin of Species*. Darwin also referred to this concept as "the war of nature" being red in tooth and claw. He pointed out that nature could be very brutal as organisms compete to survive. For example, a parasitic ichneumon wasp possesses a long ovipositor (a stinger is a modified ovipositor) that is used to place wasp eggs into an unsuspecting caterpillar. The larval wasps in turn eat the caterpillar from the inside out. Darwin also pointed out that plants and animals also have to compete with their own kind for shelter, food, mates, and so on.

As a result of the struggle for existence, the metaphor *survival of the fittest* has become associated with Darwin's basic ideas. However, the exact term was not Darwin's but was introduced by the philosopher Herbert Spencer in 1864. Darwin used the term in the fifth edition of *On the Origin of Species* in 1869. Unfortunately, many people misunderstand survival of the fittest and think that it means strongest, fastest, or smartest, instead of the members of a particular species that are best suited for the environment. For example, bower birds are native to New Guinea and Australia. In order to attract a mate, the male bower birds build an elaborate nest, adorned in seashells, glass, plastic, stones, leaves, flowers, berries, and colorful objects. (See "Bird Seduction Techniques" [BBC 2015] for an awesome brief video on bower bird behavior.) If a male does not meet the requirements

PoI 3. Discuss each of the following examples of natural selection/survival of the fittest:

a. blind fish in underground rivers and lakes
b. insects that are resistant to a pesticide
c. sharks that are dark on top and light on bottom

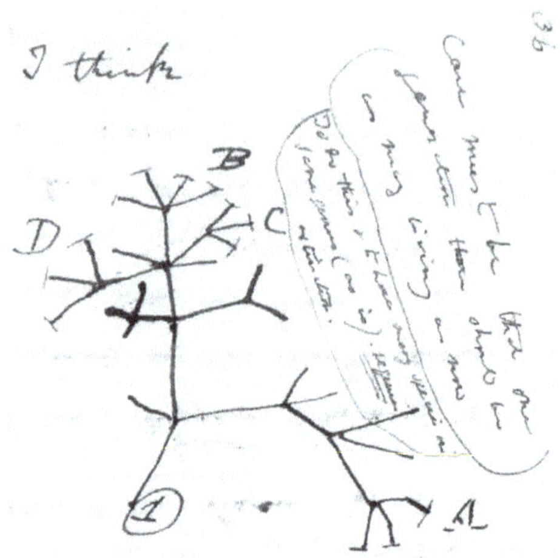

FIGURE 13.8a Darwin's Tree of Life

of the female, he in turn is selected against in the world of survival of the fittest. Luck may also be factored into the term "fittest," since being at the right place at the right time has a lot to do with survival.

Another of Darwin's important evolutionary tenets is *descent with modification*. It can be represented by the *tree of life*, in which all living things have descended from a *common ancestor* (Figure 13.8). Today many scientists agree that life on Earth is descended from the *last common universal ancestor* (*LUCA*) that theoretically existed as early as four billion years ago. Scientists use tools of comparative anatomy, paleontology, molecular biology, and the various fields of genetics to build the ever-branching tree of life and reinforce descent with modification.

Charles Robert Darwin died on April 19, 1882, from cardiac complications, in the arms of his loving wife. He was to be buried in a small churchyard along with his children, but his friends lobbied to have him buried in a place of honor at Westminster Abbey. Darwin is buried only a few feet away from Sir Isaac Newton.

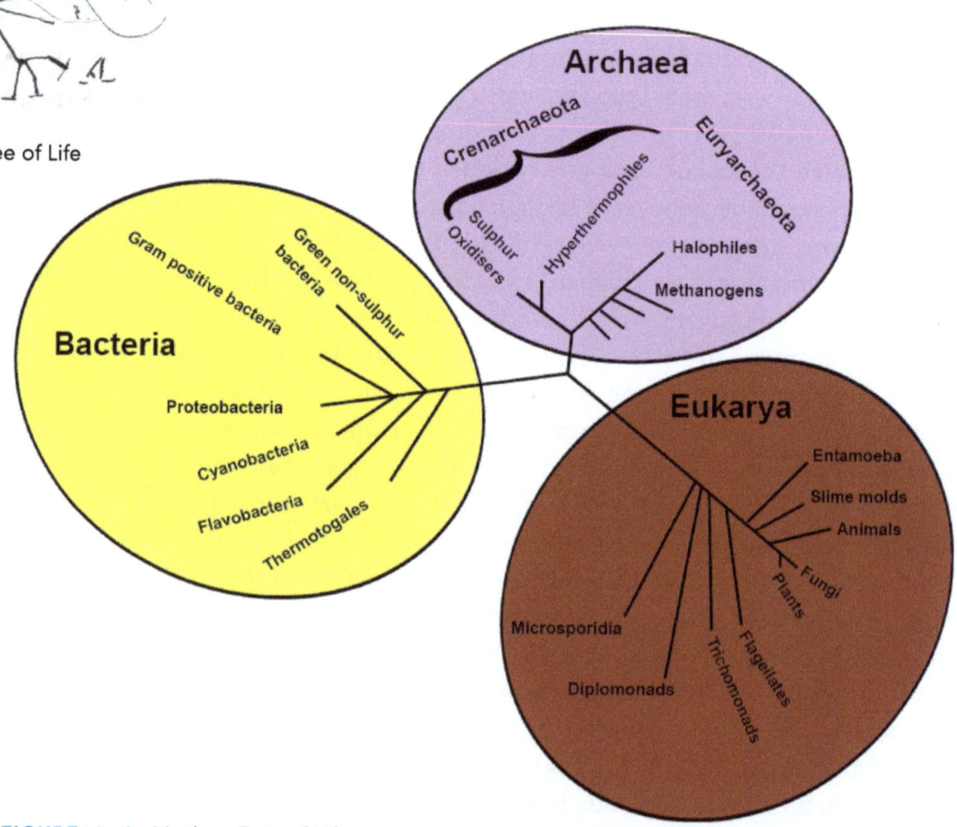

FIGURE 13.8b Modern Tree of Life

APPLYING SCIENCE TO LIFE 7 Darwin and the Origin of Whales

Darwin speculated about the origin of whales (*cetaceans*), stating that he thought that whales evolved from bears that became ever so aquatic. Darwin said, "I can see no difficulty in a race of bears being rendered, by natural selection, more and more aquatic in their structure and habits, with larger and larger mouths, till a creature was produced as monstrous as a whale" (Darwin 1859). However, he had little fossil evidence to support his idea and did not include the concept in formal writings. Today, Darwin has been vindicated by a plethora of fossils that illustrate that cetaceans did evolve from land animals, but not bears. Modern paleontology, comparative anatomy, and molecular biology reveal that extinct *artiodactyls* (even-toed ungulates) gave rise to Order Cetacea (see Figure 13.9). Thus, extinct hippo-like mammals are the ancestor to whales, porpoises, and dolphins (Switek 2010; Thewissen 2009; Thewissen 2014; UCMP 2019, "Evolution of Whales"). What a whale of an evolutionary tale! Check out the Smithsonian's evolution animation on whales (Smithsonian Institution 2019).

FIGURE 13.9 Extinct Artiodactyl, an ancestral whale

Today other aquatic mammals demonstrate similar paths to evolution. Modern *pinnipeds* (seals, sea lions, etc.) share a common ancestry with otter-like creatures, and members of the order *Sirenia* (sea cows and manatees) share a common ancestry with elephantine animals.

Modern evolutionary biology has seen countless advancements since Charles Darwin. He would be amazed by new breakthroughs in geology, genetics, embryology, paleontology, and anthropology. Considering that his work was done in the mid-1800s, he was amazingly insightful with many of his basic assertions.

QRQ 5. The ability of an animal to imitate another in order to survive is called ______.

QRQ 6. Select all who conceived the basic concept of natural selection in the early stages of the study of evolution: Aristotle, Charles Darwin, Robert Darwin, John Stephens Henslow, Jean-Baptiste Lamarck, Alfred Russel Wallace

QRQ 7. (T/F) In nature, there is a great propensity to overproduce in order to assure survival.

QRQ 8. Match the following concepts to the descriptions below: artificial selection, descent with modification, family tree, favorable selection, last common universal ancestor, least common denominator, natural ancestor, natural selection, natural variation, survival of the fittest, survival of the finest

a. some naturally occurring variations may result in an individual's ability to survive and reproduce, so that in succeeding generations, the population will gradually become more suited for survival, eventually resulting in a new species
b. most recent population from which life on Earth descended
c. purposely breeding plants or animals for desirable traits
d. represented by the tree of life, in which all living things have descended from a common ancestor
e. tendency of members of a species that are best suited for the environment to survive

13.4. General Principles of Evolution

In 1973, the evolutionary biologist *Theodosius Dobzhansky* profoundly stated, "Nothing in biology makes sense except in the light of evolution" (Dobzhansky 1973). Every field of the biological sciences, from anatomy to zoology, just like every living thing and every cell, is shaped by evolution. Evolutionary biology has given us great insight into the origin, history, distribution, and unity of life, as well as development, adaptation, behavior, ecology, and the future of life on this planet and perhaps other worlds. Today, evolutionary medicine is growing in leaps and bounds as scientist explore the origin of disease, the immune system, and new ways to treat various diseases. Evolution is truly the unifying principle of the biological sciences. Although the study of evolutionary biology is quite complex, there are some general characteristics of evolution that will simplify your understanding:

- **Life on Earth and the characteristics of living things are the products of evolutionary forces.** Everything in the biological world, from minute viruses to the great blue whale, is the product of evolutionary forces. Even the cells in that majestic redwood or tiny shrew are all shaped by evolution.
- **All degrees of variation exist within populations in nature.** Many times, these differences are obvious, such as the diversity of cats in a neighborhood. Other times, the variation is subtle and may be best observed at the molecular level.
- **Evolution does not necessarily occur in a gradual manner.** Darwin implied that evolution occurs in a slow and gradual manner as species accumulate small changes over vast periods of time. The small changes can in turn result in *speciation* (the formation of different species). This can be seen in the evolution of the horse and the whale. However, in 1972, Stephen Jay Gould and Niles Eldredge proposed the theory of *punctuated equilibrium* that explains how organisms undergo *stasis* (little change) for long periods of time followed by a *punctuated leap* (rapid change over a relatively short period of geologic time). This change is usually associated with environmental changes or small population size. This idea complements and does not negate Darwin's ideas. Examples can be seen in the evolution of certain butterflies and mollusks.

- **Populations evolve, not individuals.** This is illustrated by several devastating genetic disorders in humans. In both sickle cell anemia and cystic fibrosis, those individuals that are *heterozygous* (carriers) for these disorders have an evolutionary survival advantage against malaria and enteric bacterial diseases (salmonella, cholera, etc.), respectively. Individuals without copies of the sickle cell or cystic fibrosis gene may succumb to such a disease. The presence of carriers in a population ensures the population survives.
- **Evolutionary changes are opportunistic and not goal-oriented.** Since humans are an extremely goal-oriented species, it is difficult to imagine a world driven by randomness. Thus, when examining the evolutionary process, we tend to place ourselves as the epitome of evolution. In actuality, as part of the whole, humans are shaped by the same laws and tenets (gravity, genetics, natural selection) that rule the universe and affect our biology.
- **In examining the traits of living things, not everything is what it first appears to be.** What is the function of spines in cacti? The initial answer would be for defense, and certainly, spines are an important defensive structure. In actuality, they are modified leaves that serve to reduce evaporation and conserve water (see Figure 6.6). Needles in pines and spruces are also evolved to conserve water. How about the plates on the dorsal side of a stegosaurus, which was a type of dinosaur? The plates were actually an essential element in thermoregulation. Thus, in studying organs, the biologist or paleontologist must be a bit of a detective, or paleodetective.
- ***Extinction*, or the termination of a biological species, is part of the evolutionary process.** While studying the extinction of organisms, both prehistoric and in his world, Darwin illustrated that the same selective pressures that steer some species to extinction shape others to adapt, evolve, and survive. Three types of extinction are recognized in evolutionary biology and are considered in detail in the next module. Well over 90 percent of all organisms that have ever lived on Earth have become the victim of extinction.

In the next two modules, we will consider scientific evidence for evolution and two interesting products of evolution, adaptations and diversity.

QRQ 9. Select from the following the appropriate concept for each description below: extermination, extinction, fixation, heterogous, heterozygous, homozygous, punctuated equilibrium, punctuated equipoise, punctuated leap, speciation, speciesation, stasis

a. carrier
b. formation of different species
c. little change
d. rapid change over a relatively short period of geologic time
e. termination of a biological species
f. theory that organisms can undergo little change for long periods of geologic time followed by rapid change

QRQ 10. Indicate true or false for each:

a. Evidence of evolution can be observed in changes an individual experiences.
b. Evolution always takes a long time and occurs gradually.
c. Evolution can be random or purposeful.
d. Evolutionary changes only occur opportunistically.
e. Many levels of variation occur in a population.
f. Over 90 percent of all organisms that have ever lived on Earth have become extinct.
g. Some organisms never have evolved.
h. Some traits appear to be useful only for one thing but are also useful for something else.

PoI 4. Discuss examples of punctuated equilibrium.

Image Credits

Fig. 13.1: Source: https://commons.wikimedia.org/wiki/File:Charles_Darwin_by_G._Richmond.jpg.

Fig. 13.2: Copyright © by National Institute of Korean Language, (CC BY-SA 2.0 KR) at https://commons.wikimedia.org/wiki/File:Freshwater_eel.jpg.

Fig. 13.3: Copyright © by Sémhur (CC BY-SA 4.0) at https://commons.wikimedia.org/wiki/File:Voyage_of_the_Beagle-en.svg.

Fig. 13.5: Source: https://commons.wikimedia.org/wiki/File:Darwin%27s_finches_by_Gould.jpg.

Fig. 13.8a: Source: https://commons.wikimedia.org/wiki/File:Darwins_first_tree.jpg.

Fig. 13.8b: Copyright © by Samsara (CC BY-SA 3.0) at https://commons.wikimedia.org/wiki/File:Tree_of_life.svg.

Fig. 13.9: Copyright © by Nobu Tamura (CC BY-SA 4.0) at https://commons.wikimedia.org/wiki/File:Ambulocetus_new_NT_small.jpg.

LAB #13

Laboratory Exercises

13.5. Laboratory on Antibiotic Resistance

13.5.1. Preliminary Work

Antibiotics can promote animal growth and treat or control animal diseases but can lead to *antibiotic-resistant bacteria* (*ARB*) reservoirs and contamination of animal food products for humans. Unnecessary antibiotic prescriptions for people can also contribute to the evolution of antibiotic-resistant strains of bacteria. In this lab, we examine the development of antibiotic resistance in humans from agricultural and prescription antibiotic use. From the text's website, view the introductory video; download and open "*Antibiotic Resistance.nlogo*."

1. Click the Info tab at the top of the page and read "What Is It?"
    a. ________________ Select the main uses of antibiotics in animals: A. calm, B. control disease, C. discourage dehydration, D. promote fertility, E. promote growth, F. treat disease
    b. ________________ Give main cause(s) of antibiotic resistance in humans that the model considers: A. antibiotic prescriptions for humans, B. contamination of animal food products for humans, C. dispersal through groundwater, D. lack of immunity in animals, E. lack of quarantine, F. nonprescription drugs for humans, G. spreading of infection between animals
    c. ________________ Select all that apply to people in the susceptible category: A. can become amplified because of medical antibiotic use, B. can become colonized, C. can become exposed, D. can become exposed due to agricultural antibiotic use, E. can become exposed due to background exposure, F. can become susceptible, G. high load of antibiotic-resistant bacteria, H. low load of antibiotic-resistant bacteria, I. no load of antibiotic-resistant bacteria, J. not contagious, K. somewhat contagious, L. very contagious
    d. ________________ Using possible answers from part c, select all that apply to people in the exposed category.
    e. ________________ Using possible answers from part c, select all that apply to people in the colonized category.
    f. ________________ Using possible answers from part c, select all that apply to people in the amplified category.

g. In Figure 13.10, draw arrows from one category to another indicating where a person in one category can become a person in another category.

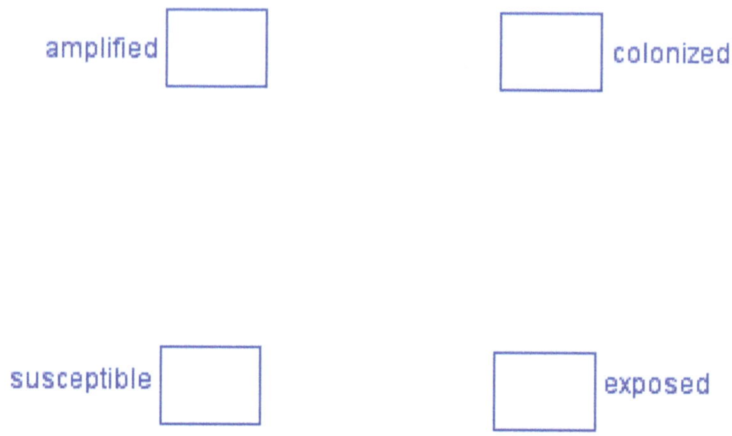

FIGURE 13.10 Categories of humans relative to antibiotic-resistant bacteria (ARB)

2. Read "How It Works."

   a. ______ Initially, how many people are colonized in the model?

   b. Match each component with its model color: A. black, B. blue, C. brown, D. cyan, E. green, F. magenta, G. orange, H. pink, I. red, J. white, K. yellow

      ______ susceptible

      ______ exposed

      ______ colonized

      ______ amplified

   c. ______ Can the number of people change in the model, no (N) or yes (Y)?

   d. **1.** ______ **2.** ______ **3.** ______ **4.** ______ Complete the meaning of CONTAMINATED-EXPOSURE-RATE: the fraction of **1.** (A. amplified, B. colonized, C. exposed, D. susceptible) population changing to the **2.** (A. amplified, B. colonized, C. exposed, D. susceptible) state each **3.** (E. day, F. week, G. month, H. year) due to **4.** (I. agricultural, J. normal, K. prescription) antibiotic use.

   e. **1.** ______ **2.** ______ **3.** ______ **4.** ______ PRESCRIPTION-RATE is the fraction of **1.** (A. colonized, B. colonized and amplified, C. exposed, D. exposed and colonized) population changing to the **2.** (E. amplified, F. colonized, G. exposed, H. susceptible) state each **3.** (I. day, J. week, K. month, L. year) due to **4.** (M. agricultural, N. normal, O. prescription) antibiotic use exposure.

3. Read "How to Use It."

   a. ______ TIME-TIL-PRESCRIPTIONS is the time at which prescriptions **1.** (A. are effective, B. are prescribed by the doctor for AAU, C. are prescribed by the doctor for ARB, D. begin, E. end).

b. ______ If TIME-TIL-CONTAMINATED is 730, give the value of CONTAMINATED-EXPOSURE-RATE at 365 ticks.

c. ______ Using default values, how many *days* does the simulation run?

d. **1.** ______ **2.** ______ **3.** ______ We can use the mean total infected over the last simulated three years to approximate the **1.** (A. cumulative prevalence, B. degree of infection, C. equilibrium prevalence) for **2.** (D. all, E. epidemic, F. nonepidemic) situations where the populations **3.** (G. are diminishing, H. are increasing, I. have become fairly steady, J. vary).

e. ____________ Select all that apply to the infection's cumulative prevalence: A. measure of total infection impact over the simulation length, B. proportional to area under the curve for total infected, C. sum of the total infected at each tick, D. total infected at simulation end

13.5.2. Laboratory Work

Throughout the experiments, leave the NUMBER-PEOPLE slider at 1,000 and the YEARS slider at 10. Express numeric answers as integers (whole numbers), with no decimal points. Perform each experiment at least five times, preferably more.

4. This question uses the default settings for *medical antibiotic use (MAU)* and *agricultural antibiotic use (AAU)* exposure. Thus, PRESCRIPTION-RATE is 0.003 and CONTAMINATED-EXPOSURE-RATE is 0.001, while TIME-TIL-PRESCRIPTIONS and TIME-TIL-CONTAMINATED are 0. The results will serve as a basis for comparison in other questions.

a. **1.** ______ **2.** ______ **3.** ______ **4.** ______ Press *setup* and *go*. After observing the animation for a while, uncheck *view updates* or speed the simulation. The graph of susceptibles **1.** (A. decreases, B. increases) **2.** (C. quickly, D. slowly) at first, then **3.** (E. falls slowly, F. falls steeply, G. rises slowly, H. rises steeply) before **4.** (I. decreasing, J. increasing, K. leveling off).

Start a spreadsheet, noting the question number (Question 4) and the slider values for PRESCRIPTION-RATE, CONTAMINATED-EXPOSURE-RATE, TIME-TIL-PRESCRIPTIONS, and TIME-TIL-CONTAMINATED. Have the following columns: Mean total infected over last three years, rounded to the nearest integer; time at which 5 percent of the population, i.e., 50 people out of 1,000, is infected (exposed, colonized, or amplified); cumulative prevalence

b. ______ Give the largest infected population: (A. amplified, B. colonized, C. exposed, D. susceptible)

c. ______ Using the "Infected Populations" graph, estimate the equilibrium for the total infected, and answer the following: Does the "Mean total infected over last three years" (in the output window) approximate this equilibrium, no (N) or yes (Y)?

d. ______ Run the simulation at least four more times (*five times in all*), recording the results in your spreadsheet. Have your spreadsheet calculate the means for each column. Give the average of the mean total infected over last three years.

e. ______ Give the mean time (# ticks, or days), expressed as a whole number, at which 5 percent of the population is infected.

f. ____________ Give the mean cumulative prevalence.

5. This question involves having no MAU or AAU exposure. Thus, change the PRESCRIPTION-RATE and CONTAMINATED-EXPOSURE-RATE to 0.

a. ______ Complete the hypothesis: With no MAU or AAU exposure, the prevalence of antibiotic resistant bacteria in humans will be **1.** (A. close to zero, B. large, C. moderate, D. variable).

b. ______ Run the simulation at least five times, recording "Mean total infected over last three years." Specify the average of these recordings: A. 0–1, B. 1–2, C. 2–5, D. 5–10, E. 10–15

c. ______ Do the results support your hypothesis, no (N) or yes (Y)?

6. This question involves varying levels of medical antibiotic use (MAU) exposure. Set the CONTAMINATED-EXPOSURE-RATE slider to 0.001, while TIME-TIL-PRESCRIPTIONS and TIME-TIL-CONTAMINATED are 0.

a. **1.** ______ **2.** ______ **3.** ______ Complete the following hypothesis: As the rate of antibiotic prescriptions increases, the severity of epidemics **1.** (A. decreases, B. decreases then increases, C. increases, D. increases then decreases, E. remains about the same, F. varies widely), an epidemic usually occurs **2.** (G. at about the same time, H. at widely varying times, I. earlier, J. later), and over time, **3.** (K. about the same number of, L. many fewer, M. many more) people become infected.

b. ______ For PRESCRIPTION-RATE = 0.0015, which is half the amount in question 4, run the simulation at least five times in which an epidemic occurs, recording the results in three columns of your spreadsheet. Note the question number (Question 6b) and the values for PRESCRIPTION-RATE and CONTAMINATED-EXPOSURE-RATE. Calculate the means of each column. Give the whole number average of column 1, the mean total infected over last three years.

c. ______ Repeat part b for PRESCRIPTION-RATE = 0.006, which is double the amount in question 4. On your spreadsheet, note the question number and values for PRESCRIPTION-RATE and CONTAMINATED-EXPOSURE-RATE.

d. ______ Using results from questions 6b, 4d, and 6c, graph the mean total infected over the last three years versus PRESCRIPTION-RATE (in order, 0.0015, 0.003, 0.006 on the horizontal axis). As MAU exposure increases, the prevalence of ARB **1.** (A. decreases, B. decreases then increases, C. increases, D. increases then decreases, E. stays about the same, F. varies widely).

e. ______ For PRESCRIPTION-RATE = 0.0015 (part b), give the mean time at which 5 percent of the population is infected (whole number).

f. ______ For PRESCRIPTION-RATE = 0.006 (part c), give the mean time at which 5 percent of the population is infected.

g. ______ Using results from 6e, 4e, and 6f, graph the mean time at which 5 percent of the population is infected versus PRESCRIPTION-RATE (in order, 0.0015, 0.003, 0.006 on the horizontal axis). As MAU exposure increases, the time at which epidemics strike occurs **1.** (A. at about the same time, B. at widely varying times, C. earlier, D. later).

h. ____________ For PRESCRIPTION-RATE = 0.0015 (part b), give the mean cumulative prevalence (whole number).

i. ____________ For PRESCRIPTION-RATE = 0.006 (part c), give the mean cumulative prevalence (whole number).

j. ______ Using results from questions 6h, 4f, and 6i, graph the mean cumulative prevalence versus PRESCRIPTION-RATE (in order 0.0015, 0.003, 0.006 on the horizontal axis). As MAU exposure increases, over time **1.** (A. about the same number of, B. many fewer, C. many more) people become infected.

k. ______ Do the results from parts d, g, and j support your hypothesis from part a, no (N) or yes (Y)?

l. Write a paragraph about what your simulations indicate about the impact of prescription antibiotic use on epidemics of antibiotic-resistant bacterial infections in humans.

7. This question involves varying levels of agricultural antibiotic use (AAU) exposure. Set PRESCRIPTION-RATE to 0.003, while TIME-TIL-PRESCRIPTIONS and TIME-TIL-CONTAMINATED are 0.

   a. **1.** ______ **2.** ______ Complete the following hypothesis: As the rate at which antibiotics are prescribed for animals increases, the severity of epidemics in the humans **1.** (A. decreases, B. decreases then increases, C. increases, D. increases then decreases, E. stays about the same, F. varies widely), and the epidemic usually occurs **2.** (G. at about the same time, H. at widely varying times, I. earlier, J. later).

   b. ______ For CONTAMINATED-EXPOSURE-RATE = 0.0005 (5.0E-4 on slider bar, which represents 5.0×10^{-4}), which is half the amount in question 4, run the simulation at least five times in which an epidemic occurs, recording the results in three columns in your spreadsheet. Note the question number and the values for PRESCRIPTION-RATE and CONTAMINATED-EXPOSURE-RATE. Calculate the means of each column. Give the average (as a whole number) of column 1, the mean total infected over last three years.

   c. ______ Repeat part b for CONTAMINATED-EXPOSURE-RATE = 0.002, which is double the amount in question 4. On your spreadsheet, note the question number and the values for PRESCRIPTION-RATE and CONTAMINATED-EXPOSURE-RATE.

   d. ______ Using results from questions 7b, 4d, and 7c, graph the mean total infected over last three years versus CONTAMINATED-EXPOSURE-RATE (in order 0.0005, 0.001, 0.002 on the horizontal axis). As AAU increases, on average, the prevalence of ARB **1.** (A. decreases, B. decreases then increases, C. increases, D. increases then decreases, E. stays about the same, F. varies widely).

   e. ______ For CONTAMINATED-EXPOSURE-RATE = 0.0005 (part b), give the time at which 5 percent of the population is infected (whole number).

   f. ______ For CONTAMINATED-EXPOSURE-RATE = 0.002 (part b), give the time at which 5 percent of the population is infected (whole number).

   g. ______ Using results from questions 7e, 4e, and 7f, graph the time at which 5 percent of the population is infected versus CONTAMINATED-EXPOSURE-RATE (in order 0.0005, 0.001, 0.002 on the horizontal axis). As AAU increases, the average time at which epidemics strike occurs **1.** (A. at about the same time, B. at widely varying times, C. earlier, D. later).

   h. ______ Do the results from parts d and g support your hypothesis from part a, no (N) or yes (Y)?

   i. **1.** ______ **2.** ______ Complete the following statement clarifying your observations: Agricultural antibiotic use **1.** (A. does, B. does not) change the equilibrium prevalence dramatically but does cause antibiotic resistance to invade **2.** (C. a little later, D. a little sooner, E. at about the same time, F. much later, G. much sooner).

j. ______ Considering your data from questions 6d and 7d and according to this model, which has the largest impact on the prevalence of antibiotic-resistant bacteria in people, medical antibiotic use (M) or agricultural antibiotic use (A)?

k. Write a paragraph about what your simulations indicate about the impact of agricultural antibiotic use on epidemics of antibiotic-resistant bacterial infections in humans.

8. This question involves delaying AAU exposure until after MAU exposure. Set PRESCRIPTION-RATE to 0.003 and CONTAMINATED-EXPOSURE-RATE to 0.001, while TIME-TIL-PRESCRIPTIONS is 0.

a. **1.** ______ **2.** ______ Complete the following hypothesis: As agricultural use of an antibiotic is delayed, on average, the time at which an epidemic begins **1.** (A. remains approximately the same, B. varies, C. will be earlier, D. will be later), and over a 10-year period, the total number of infected people will be **2.** (E. about the same, F. less, G. more, H. unpredictable).

b. ______ Set TIME-TIL-CONTAMINATED to 365 days (one year) and run the simulation at least five times in which an epidemic occurs, recording the results in your spreadsheet. Note the question number and the values for PRESCRIPTION-RATE, CONTAMINATED-EXPOSURE-RATE, and TIME-TIL-CONTAMINATED. Omit collecting data for mean total infected over last three years. Calculate the means of each column. Give the time at which 5 percent of the population is infected.

c. ______ Repeat part b for TIME-TIL-CONTAMINATED to 730 days (two years).

d. ______ Using results from questions 4e, 8b, and 8c, graph the mean time at which 5 percent of the population is infected versus TIME-TIL-CONTAMINATED (in order 0, 365, 730 days on the horizontal axis). As TIME-TIL-CONTAMINATED increases, the mean time at which 5 percent of the population is infected **1.** (A. decreases, B. decreases then increases, C. increases, D. increases then decreases, E. stays about the same, F. varies widely).

e. __________ For TIME-TIL-CONTAMINATED = 365 days (part b), give the average cumulative prevalence (whole number).

f. __________ For TIME-TIL-CONTAMINATED = 730 days (part c), give the average cumulative prevalence (whole number).

g. ______ Using results from questions 4f, 8e, and 8f, graph the average cumulative prevalence versus TIME-TIL-CONTAMINATED (in order 0, 365, 730 days on the horizontal axis). As TIME-TIL-CONTAMINATED increases, what happens to the cumulative prevalence? (A. decreases, B. decreases then increases, C. increases, D. increases then decreases, E. stays about the same, F. varies widely).

h. ______ Do the results from parts d and g support your hypothesis from part a, no (N) or yes (Y)?

i. Considering emergence of antibiotic-resistant bacteria, write a paragraph making a recommendation about when agricultural use for an antibiotic should begin relative to its medical use for humans. What advantage(s) do you anticipate incurring from your recommendation? Support your recommendation with your data.

MODULE

14

Evidence for Evolution

14.1. Introduction

The evidence for evolution is universal in the living world, from the minute bacteria to the majestic redwood trees to the very cells in your little toe, all are a testimony to the power of the evolutionary process. This evidence is compelling, undeniable, and genuine. Every discipline of the biological sciences, like every cell, is a testimony to the power of evolution. Evolutionary concepts are based firmly on more than 150 years of careful scientific examination, testing, and observation. Through the years, the overwhelming evidence for evolution has grown extensively and now reinforces our understanding of life on Earth.

APPLYING SCIENCE TO LIFE 1 Neanderthals?

Homo neaderthalensis, our closest extinct relative, lived from 400,000 to 40,000 years ago in Europe and southwestern to central Asia. The first Neanderthal fossil was uncovered in Germany in 1856. Neanderthals were characterized by having thick bones; low brow ridges; an oval-shaped skull with a receding jaw; large face; braincase set back in a long skull; space behind the wisdom teeth; a broad, projecting nose; and an occipital bun or rounded knot at the back of the skull. New DNA evidence suggests that *Homo sapiens* interbred with Neanderthals, and new DNA tests confirm that many individuals of European descent have Neanderthal DNA. Several traits of modern humans, including immunity to Eurasian pathogens, red hair, fair skin and freckles, rosy cheeks, wide fingers and thumbs, and increased risk of type II diabetes, can be traced to Neanderthal ancestry (Hendry 2018).

14.2. Paleontology

Unfortunately, many people think that *paleontology* is just the study of dinosaurs and fossils. In reality, paleontology incorporates numerous disciplines, including geology, biology, ecology, anthropology, archaeology, climatology, and computer science, in order to develop an understanding of life in the past. It is divided into several subdisciplines, including taphonomy (how fossils form) and vertebrate paleontology. Some paleontologists, called paleoecologists, reconstruct communities and ecosystems that existed at specific time periods. For example, paleoecologists have determined that the plants and animals living in the Antarctic 53 million years ago were subtropical. That world included palm trees!

Look closer at rocks—many of them may contain fossils. A *fossil* is the remains or signs of organisms that lived in the past. Technically, in order to be a fossil, a specimen must be at least 10,000 years old. To paleontologists, fossils are valuable because they tell a story of *deep time* (time through the ancient eras of Earth history), ancient life, climate, geography, and ecology. They serve as valuable clues in deciphering the past, understanding the present, and predicting the future.

Finding a fossil requires familiarity with the geologic history and organisms that lived there through geologic time and a bit of luck. For example, Figure 14.1 presents a diagram of geologic history's primary *eras*, subdivided into *periods*, along with their times in millions of years ago (*MYA*) and some of their organisms that became fossils. A better chance exists for fossil formation if an organism possesses hard parts, such as shells, teeth, and bones. The fossil record of ancient mollusks is excellent because they possessed shells. Although sharks have skeletons of cartilage, their "teeth" (actually placoid scales) fossilize easily. Another factor that increases the chance of fossilization is quick burial. An organism covered by sediment, ash, or some other source is protected from scavengers and oxygen that cause damage or loss of the specimen. Fossil fish and fossil fern fronds, such as those found in Wyoming and Pennsylvania, respectively, were formed in this way. Studying geological formations and fossils gives paleontologists clues about the ancient geography, ecology, living organisms, and evolutionary trends for a region. In addition, some fossils are used as museum specimens, in jewelry, for conversation pieces, and even in the search for oil.

Fossil formation can occur in a myriad of other ways too. Here, we only describe major examples. Ancient mammoths have been preserved in ice and found in Alaska and Siberia. In many cases, these animals from millions of years ago are so well-preserved that scientists can study the contents of these creatures' stomachs. Thanks to the book and movie *Jurassic Park*, more people are aware that fossilized tree sap (*amber*) encases animals, feathers, pollen, and so on, preventing them from deteriorating (see Figure 14.2a). Organisms that once lived in caves or very dry environments have been *mummified* into fossils after they were desiccated (dried out). *Sedimentary fossils* are embedded in rocks formed from mud, sand, and silt (see Figure 14.2b, c). Limestone is an example of rock containing a wealth of sedimentary fossils, including microskeletons of one-celled organisms, as well as corals, mollusks, animals related to sea stars, and trilobites (marine arthropods). The gravel in your driveway may contain many sedimentary fossils—look around! *Coal*, a type of sedimentary rock, is actually plant material that has fossilized as the result of time, heat, and pressure. In Hollywood, tar pits have revealed the La Brea fossils, including animals like the saber-toothed cat and the camel. The tar pits formed from crude oil that seeped to the earth's surface through fissures. Tar, which

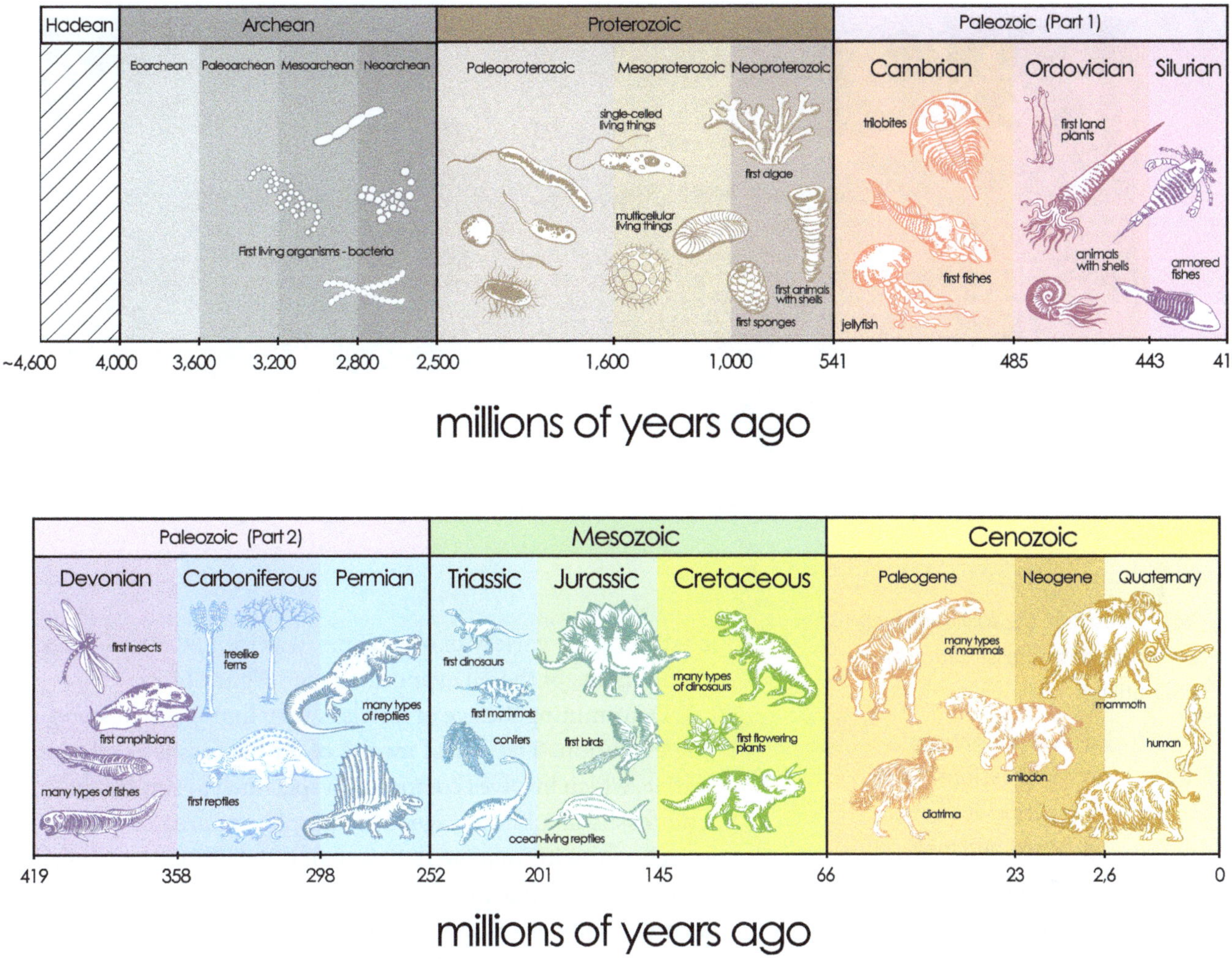

FIGURE 14.1 Geologic time

FIGURE 14.2 Fossils: **a.** Amber with ant inclusion. **b.** Ammonite. **c.** Shark's tooth

FIGURE 14.3 Bad day at La Brea Tar pits—giant mammals trapped in the tar pit ~30,000 years ago

PoI 1. Describe the following well-known means of fossilization and their utility:

a. carbonization
b. impressions
c. permineralization
d. casts and molds
e. pyritization

PoI 2. What do you think scientists can learn from studying coprolites?

coated a wide variety of organisms, was formed after lighter components of the oil evaporated (see Figure 14.3).

Determining the age of fossils is very important in studying evolution. The oldest means of dating fossils is *relative dating*, which involves comparing a specimen to similar rocks and fossils of a known age. Stratigraphy, an area of geology concerned with relative dating, can only approximate ages of fossils, not actual numbers. For example, Figure 14.4 shows a stratigraphy of a South African cave, dated (OSL technique, which determines the last date of quartz sediment's exposure to light) to be from about 68 ka (68,000 years ago) to about 85 ka, in the Middle Stone Age. *Index fossils* are representatives of certain eras or periods. For example, *trilobites* (extinct marine arthropods) are index fossils of the Paleozoic Era (540–252 million years ago), and dinosaur *T. rex* is an index fossil of the Cretaceous Period (more than 65.5 million years ago).

A more precise means of dating fossils is *absolute dating*, which can yield specific dates for rocks and fossils. The classic methods of absolute dating are based on the half-life of radioactive isotopes, such as those of carbon (*carbon dating*), which can be used for specimens less than 60,000 years old (Peppe and Deino 2013).

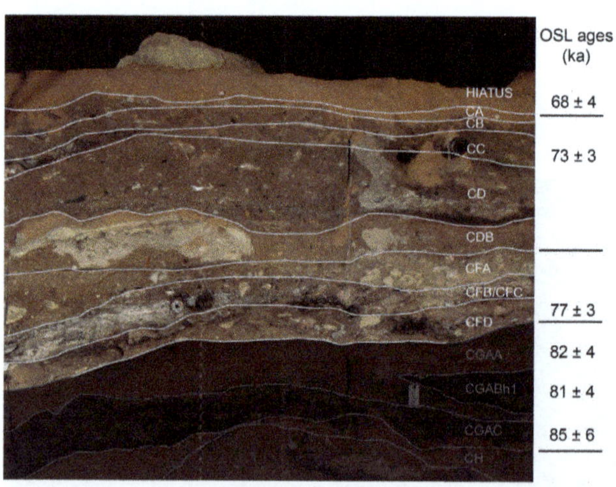

FIGURE 14.4 Stratigraphy of a cave in South Africa, Middle Stone Age

PoI 3. Discuss how each of the following absolute dating methods works and the type of objects it dates:

a. radiometric dating, such as potassium-to-argon and uranium-to-lead
b. amino acid racemization
c. paleomagnetism
d. thermoluminescence

APPLYING SCIENCE TO LIFE 2 Fossils in Myth

From the Latin, *fossil* literally means "dug up." These dug-up curiosities and giant bones intrigued people living in the ancient worlds. They were responsible for many mythical creatures, such as the cyclops, dragons, and monsters. One such myth is that of the gold-guarding griffin (see Figure 14.5). According to Greek myth, the griffin was a large creature that possessed the body of a lion and the beak of an eagle. They were thought to guard the vast source of gold near the Gobi Desert and even to prey upon humans and horses. Through the years, the legend and powers of these mythical monsters grew. Today, paleontologists speculate that there was some foundation to the legend. Fossil *Protoceratops* bones have been found in the same region as the origin of the griffin (Byrnes 2015; Mayor 2011). These large frilled and beaked dinosaur remains could be easily mistaken for a mythical monster!

FIGURE 14.5a Griffin statue

FIGURE 14.5b Protoceratops skeleton

QRQ 1. The study of life in geologic time by investigating fossils is called ______.

QRQ 2. Replace each blank with one of the following terms: centuries, eras, millennia, periods. Geologic time is divided into ______, which are subdivided into ______.

QRQ 3. Select all conditions that increase the chance of an organism becoming fossilized: ate plants, becomes completely dried, buried in ground, buried in tar, covered with ice/snow, drowned, encased in tree sap, footprints, inorganic minerals replace the organic matrix, photosynthesized, possesses hard parts, quick burial

QRQ 4. Select the best term for each description: absolute, amber, carbon, coal, fixed, fossiliferous, ichnofossil, ichnospecies, index, mummification, petrification, primary, relative, rock, sedimentary, subfossil

a. fossil representative of a certain era or period
b. fossilized tree sap
c. trace of an organism that lived in the past, such as a footprint or burrow
d. type of dating based on the half-life of radioactive isotopes
e. type of dating comparing a specimen to similar rocks/fossils of a known age
f. type of dating that can yield specific numerical dates for rocks and fossils
g. type of fossil embedded in rock formed from mud, sand, or silt
h. type of fossilization when a specimen completely dries
i. type of fossilization when inorganic minerals replace specimen's organic matrix

14.3. Extinction

Today, we are facing an extinction crisis on our fragile planet. Plant and animal species are being lost at an alarming rate. Perhaps as many as a dozen species of organisms become extinct daily, and the Center for Biological Diversity predicts that 30 to 50 percent of living species may become extinct by midcentury (Center for Biological Diversity 2020, "Extinction"). The complex web of life is threatened, and, once in motion, extinction can snowball and jeopardize the success of all life.

Three types of extinction are important in the study of evolution: background extinction, mass extinction, and extinction at the hands of humankind. *Background extinction,* or natural extinction, includes those organisms that are unable to adapt to changing environments, fall victim to disease, lose their habitat, or lose in competition with other organisms in the natural environment. Through the history of life, background extinction has occurred at a relatively steady rate and is the result of normal evolutionary processes. Background extinction only impacts a small number of species within an ecosystem and is not associated with the activities of humans. Geologic time is filled with species of plants and animals that disappeared forever as the result of background extinction. It has been estimated that two vertebrate species in 10,000 become extinct every 100 years. However, during the past 200 years, such extinctions have accelerated.

What happened to the dinosaurs? Today, we know that they, along with a multitude of life forms, were the victims of a mass extinction about 65.5 MYA. *Mass extinctions* involve the termination of many species over a relatively short period of geological time, resulting from a global catastrophe. (Greshko and National Geographic Staff 2019; National Geographic 2020).

PoI 4. Find examples of each of the following causes of mass extinctions and when they occurred:

a. asteroid strikes
b. intense volcanism (volcanoes)
c. massive methane releases
d. glaciations
e. gamma ray bursts from supernovas many light-years away

APPLYING SCIENCE TO LIFE 3 Two Critical Mass Extinctions

Scientists are slowly unraveling the mysteries of two of the greatest mass extinctions that have impacted life on this planet. The most devastating one, the *Permian-Triassic extinction event*, or the Great Dying, occurred about 251.9 MYA. This disaster was responsible for the extinction of 96 percent of marine life, such as trilobites and many fishes, and 70 percent of terrestrial life, such as early mammal-like reptiles and many swamp forest plants. Recently, scientists have determined that the cause of this mass extinction was a mega-volcano in Siberia. It is estimated that the volcano produced enough lava to cover the continental United States in a layer 1,000 feet thick, the height of some 100-story skyscrapers. In addition, the event triggered a cascade of environmental disasters, including acidification of the oceans, production of poisonous gases, and vast amounts of carbon dioxide. Thus, life itself was almost extinguished.

Another mass extinction that marks the beginning of the Mesozoic Era, 65.5 MYA, the *K-T extinction event*, is responsible for the annihilation of 70 percent of the flora and fauna on Earth, including the ammonites (a type of marine mollusk), dinosaurs, flying reptiles, and marine reptiles. Through the years, a number of explanations for the Mesozoic extinction have been proposed and are still being considered. Today, one of the leading hypotheses involves the impact of a large asteroid, about six miles wide, traveling approximately 50,000 mph, colliding with the earth. This collision had the devastating power of over a billion Hiroshima-plus-Nagasaki atomic bombs. The impact site, known as Chicxulub Crater, is on the edge of the Yucatan Peninsula and is over 110 miles wide. The event triggered a giant tsunami, firestorms, release of poisonous gases, bombardment of boulders, blockage of sunlight, and subsequent global carnage (Chu 2018; Cowen 1999; Hickey 2018; Hoffman 2020).

Although frightening and devastating, mass extinctions allow for *adaptive radiation*, an evolutionary process that occurs when a variety of new species develop from an ancestral stock. Adaptive radiation can be brought on by changes in the environment or as organisms invade new habitats. The Permian-Triassic extinction event was devastating to the majority of life forms on Earth, but it allowed for the coming of dinosaurs. The extinction event that destroyed the dinosaurs and other Cretaceous flora and fauna (65.5 MYA) opened the door for flowering plants, birds, and mammals.

There is not very much we can do about natural background extinction and mass extinction, but perhaps we can avert the third type, *extinction at the hands of humankind*, by good stewardship of our Earth and life. Over the past centuries, humankind is responsible for the extinction of such animals as the passenger pigeon, dodo bird, and Tasmanian tiger. Today, there are thousands of vulnerable, threatened, and endangered species in the world, including the African elephant, Amur leopard (see Figure 14.6), and blue whale.

FIGURE 14.6 Endangered Amur leopard, world's rarest cat

APPLYING SCIENCE TO LIFE 4 Bringing Them Back from the Brink of Extinction

When you think of the swamps of Louisiana, Florida, and other Southern states, the intimidating American alligator (*Alligator mississippiensis*) comes to mind. This impressive carnivorous reptile can reach lengths of 10–15 feet and can weigh as much as 1,000 pounds. The alligator is a *keystone species* (a species critical to the survival of the other species) in the swamp and is crucial to the survival of the system. Fortunately, now this alligator ranks of least concern (evaluated to not being near threatened) on the International Union for Conservation of Nature (IUCN) status list. The IUCN is at the forefront of saving the planet's species from extinction and maintains a global list of the extinction status of plants and animals, from deficient to extinct.

As the result of overhunting and habitat loss, alligator populations reached critically low numbers in the 1950s. The once iconic and abundant reptile was well on the way to extinction. Then, in 1973, the *Endangered Species Act* (*ESA*) was enacted, and the alligator was officially placed on the endangered species list. As the result of prohibiting alligator hunting, monitoring populations, and saving alligator habitats by the US Fish and Wildlife Service (FWS) and state wildlife agencies, the once endangered alligator began to make a dramatic comeback. In 1987, the American alligator was removed from the endangered species list and has successfully repopulated the swamps. Consequently, the swamps are recovering, and limited alligator hunting and egg collecting are multimillion-dollar industries in the southern portion of the United States. The story of the alligator is one of the first of many success stories that are the direct result of the Endangered Species Act.

The ESA is responsible for providing and maintaining a program for the conservation of threatened and endangered plants and animals, as well as the protection of their habitats. The act is maintained by the FWS and the National Oceanic and Atmospheric Administration (NOAA) Fisheries Service. The ESA, which is the strongest law of any nation designed to protect biodiversity, is responsible for protecting over 1,600 species in the United States and its territories. Many species have been saved by this act, including the gray wolf, American bald eagle, whooping crane, humpback whale, brown pelican, and the Colorado butterfly plant (Beans 2013; Center for Biological Diversity 2020, "Endangered"; USFWS 2020).

PoI 5. Write a paragraph about each of the following animals, including where and when it lived and cause(s) for its extinction:

a. Carolina parakeet
b. passenger pigeon
c. dodo bird
d. Tasmanian tiger

PoI 6. Write a paragraph about each of the following organisms, including where it lives, threats to its existence, and efforts to save it:

a. Amur leopard
b. blue whale
c. orangutan
d. Georgia aster
e. Arizona agave

There have been five major mass extinctions, and, today, many scientists are warning that Earth is on the brink of a sixth great mass extinction. Presently, our planet is experiencing the worst rate of species loss since 65 million years ago (Wong 2000). The cause of this looming mass extinction is the result of the impact of humans on the planet, such as destruction of habitat, pollution, introduction of exotic species, overfishing and overhunting, disruption of biogeographical corridors, ozone decay, high levels of carbon dioxide, ocean acidification, and global warming. Mass extinctions impact all life on Earth, but the amphibians (frogs, toads, salamanders) are in the most danger, with over one-third of all the roughly 7,500-plus known species at the risk of extinction. The destiny of life on Earth is now in our hands (E. O. Wilson Biodiversity Foundation 2019).

PoI 7. For each of the following dangers to the environment by humans, prepare a presentation:

a. destruction of habitat
b. overfishing and overhunting
c. disruption of biogeographical corridors

PoI 8. Prepare a presentation about the risks and causes of possible extinction for many amphibians.

QRQ 5. (Answer choices in parentheses) Complete the following statement with the appropriate choices: ______ (Adaptive, Background, Human-generated, Mass) extinction includes those organisms that are unable to adapt to changing environments, fall victim to disease, lose their habitat, or lose their competition with other organisms in the natural environment. This type of extinction impacts a ______ (large, small) number of species within an ecosystem and ______ (is, is not) associated with the activities of humans.

QRQ 6. ______ extinction involves the termination of many species over a relatively short period of time, geologically speaking, resulting from a global catastrophe.

QRQ 7. (T/F) The K-T extinction event, which was responsible for the annihilation of 70 percent of the flora and fauna on Earth, including the ammonites, dinosaurs, flying reptiles, and marine reptiles, enabled the rise of flowering plants, birds, and mammals.

QRQ 8. The alligator is an example of a(n) ______ species in swamps of the Southern United States, because this animal is crucial to the survival of the swamps.

QRQ 9. Select the reason(s) why our planet is experiencing the worst rate of species loss since 65 million years ago: adaptive radiation, amphibians, geopolitics, impact of humans on the planet , insects, radiation from the planets

14.4. Transitional Fossils and Living Fossils

Darwin boldly predicted that fossils would be found that link major groups of plants or animals sharing a common ancestry. These missing links, or *transitional fossils*, show change over time and descent with modification. Literally, transitional fossils serve to connect the gap between two different kinds of organisms that share a common ancestor. Although Darwin lacked fossils to provide clear evidence for his ideas, modern paleontologists have discovered transitional fossils between fishes and *tetrapods* (four-legged vertebrates), reptiles and mammals, land animals and whales, lizards and snakes, wasps and ants, and so on.

The classic example of a transitional fossil is *Archaeopteryx lithographica*. Found in limestone deposits in Germany in 1860, today this fossil serves as a transition between dinosaurs and modern birds. *Archaeopteryx* lived about 150 MYA, during the Jurassic Period, and appears to be an enigma, sharing both dinosaur and bird traits (see Figure 14.7). It was about the size of a modern raven, with broad wings and a long tail. Several dinosaur traits included teeth, three fingers with claws on the wings, a flat sternum, and belly ribs. *Archaeopteryx* also shared several characteristics with birds, including feathers, wishbone (furculum), and flexible first toe. Whether or not *Archaeopteryx* was a glider or capable of flight is still debated. Since the discovery of *Archaeopteryx*, several other fossils have been found that link dinosaurs and birds. *Archaeopterx* and the recently discovered fossil *Xiatingia zhengi* from China provided evidence that these animals were related to several types of dinosaurs (Bartfal and Kang 2009).

FIGURE 14.7 Drawing of Archaeopteryx

APPLYING SCIENCE TO LIFE 5 Dinosaur Misconceptions

A host of misconceptions exists in the general public regarding dinosaurs. Any pack of toys labeled as dinosaurs contains "marine dinosaurs," such as *Mosasaurus*, *Plesiosaurus*, and *Ichthyosaurus*, as well as "flying dinosaurs," such as *Pteranodon*, *Quetzalcoatlus*, and *Pterodactylus*. However, these extinct animals are not dinosaurs at all; they are marine and flying reptiles, lacking some of the characteristics that define dinosaurs. The average dinosaur was not the giant of movies. The vast majority of dinosaurs were small, with some being the size of a robin or smaller. In the movies, dinosaurs are represented as creatures covered in scales—no, many dinosaurs were covered with feathers (see Figure 14.7). Also, many are depicted as drab green, brown, or gray creatures. New evidence indicates that a number of color pigments were found in dinosaurs, and the colors were probably displayed in stripes and other patterns for countershading, sexual dimorphism, warnings, and camouflage. After studying the skulls of many dinosaurs, paleontologists surmised that dinosaurs were capable of communications using a number of sounds, such as hisses, bellows, squeaks, hoots, and, of course, snarls and roars. In fact, a Jurassic or Cretaceous environment may have sounded more like a concert featuring bird calls. Another major movie mistake places humans living alongside of dinosaurs. The fact is, they missed each other by millions of years, and the only humans living with dinosaurs were the cartoon Flintstones. Finally, dinosaurs live today as birds (Pickrell 2017; Service 2019)!

PoI 9. Define each of the following terms involving colors, stripes, and patterns, and give examples in living animals:

a. countershading
b. sexual dimorphism
c. warnings (patterns, colors, sounds, odors)
d. camouflage

APPLYING SCIENCE TO LIFE 6 *Tiktaalik*: To Walk or Not to Walk

Evolutionary biologists have thought for many years that a group of fishes, known as the sarcopterygians, or lobe-finned fishes, gave rise to the first tetrapods—amphibians. In 2004, the often-stingy fossil record shed light on this unique relationship. After much research and hard work, an expedition into the Canadian Arctic, led by Neil Shubin and Ted Daeschler, unearthed the fossil remains of a monumental transitional fossil.

FIGURE 14.8 Rendition of *Tiktaalik roseae*

During the Devonian Period, approximately 375 MYA, *Tiktaalik roseae* swam and hunted inlets of the once warm waters of Northern Canada (see Figure 14.8). Sometimes called a "fishopod," this unique fossil possessed fish and amphibian traits, as well as transitional traits. Fishlike traits include gills, scales, and fins, while several amphibian traits include lungs, a neck, rib bones, tetrapod pelvis, flat skull, and wrist bones. Being a transition form, *Tiktaalik* also featured transitional ear and limb bones that reflect both fish and tetrapod traits. Evolutionarily, it bridges the gap between true sarcopterygian fishes and amphibians (University of Chicago 2020).

The term *living fossil* was coined by Charles Darwin and referred to organisms that have remained virtually unchanged throughout geologic history. The best-known living fossil is a large, lobe-finned marine fish known as the coelacanth (see Figure 14.9). Thought to be extinct for 65 million years, a living coelacanth was found in 1938 by Mary Latimer off the east coast of Africa. The fish has remained unchanged for millions of years, and today two species have been catalogued. Other living fossils include dawn redwood, *Ginkgo biloba*, chambered nautilus, crocodiles, pelicans, and the horseshoe crab.

FIGURE 14.9 Coelacanth

PoI 10. Find images of the following living fossils and find out how old they are:

a. dawn redwood
b. *Ginkgo biloba*
c. chambered nautilus
d. horseshoe crab

QRQ 10. ______ fossils serve to connect the gap between two different kinds of organisms that share a common ancestor. For example, Archaeopteryx has traits of both dinosaurs and modern birds.

QRQ 11. ______ fossils, such as crocodiles, are organisms that have ancient ancestors that have changed little.

14.5. Biogeography

If Cuba originated in the Pacific Ocean, how did it get to its present location? Why does Australia have so many interesting marsupials? Why are there no penguins in the Arctic? These are a few questions that can be answered by biogeography. Substantial evidence for evolution comes from the discipline of *biogeography*, or the study of the range and distribution of organisms. Presently biogeography is divided into two subdisciplines. *Historical biogeography* involves the patterns of distribution of organisms in the past, and *ecological biogeography* addresses current patterns of distribution.

As the result of his adventures on the HMS *Beagle* and his correspondence with naturalists and scientists, Darwin became intrigued by the distribution of species on the planet. He asked such questions as: Why do islands have such unique assemblages of plants and animals? And why are the plants and animals in far-reaching continents similar yet very dissimilar? He concluded that evolution was the driving force shaping the distribution of species. His fundamental ideas were amazingly accurate, but it would take the work of others to confirm his hypotheses.

Two of the giants of biogeography were *Alfred Russel Wallace* and *Alfred Wegener*. Wallace, a contemporary of Darwin, did intense work in the Amazon Basin and the Malay Archipelago, which cumulated in his classic two-volume set *The Geographical Distribution of Animals* (1876). In this great work and others, Wallace laid down the foundations of modern biogeography. Noting the distribution of species and how many are confined to certain regions, Wallace divided the world map into six biogeographical realms (see Figure 14.10a). Today, this map, with its eight distinct *realms* and about twice as many *biomes* (biological communities sharing a particular region/climate, such as a tropical forest), provides valuable baseline information to scientists (see Figure 14.10b) (World Press 2020).

Wallace is also known for establishing an imaginary line, termed *Wallace's Line*, that runs between Australia and the Asian islands and mainland. To the west of the line, animals resemble those of Asia and include tigers, rhinoceroses, and monkeys. To the east of the line, animals resemble those of Australia and include kangaroos, wombats, and koala bears. Wallace's Line is significant because it establishes floral and faunal boundaries as they relate to species distribution and geologic features.

As early as the late 1500s, scientists speculated that Africa and South America were connected. Then, in 1915, the German meteorologist Alfred Wegener was intrigued by the observation that identical fossil plants and animals were found on opposite sides of the Atlantic. He proposed that the continents were once connected, like a giant jigsaw puzzle. His work was highly criticized, and it was not until the 1960s, after the ocean floor began being mapped, that his idea received the attention it deserved. Today, scientists readily accept that the continents reside on *tectonic plates* that are constantly moving (see Figure 14.11). The movements of these plates had great impact on the evolution and dispersal of species over time. Evidence of their activity today is found in earthquake or seismic

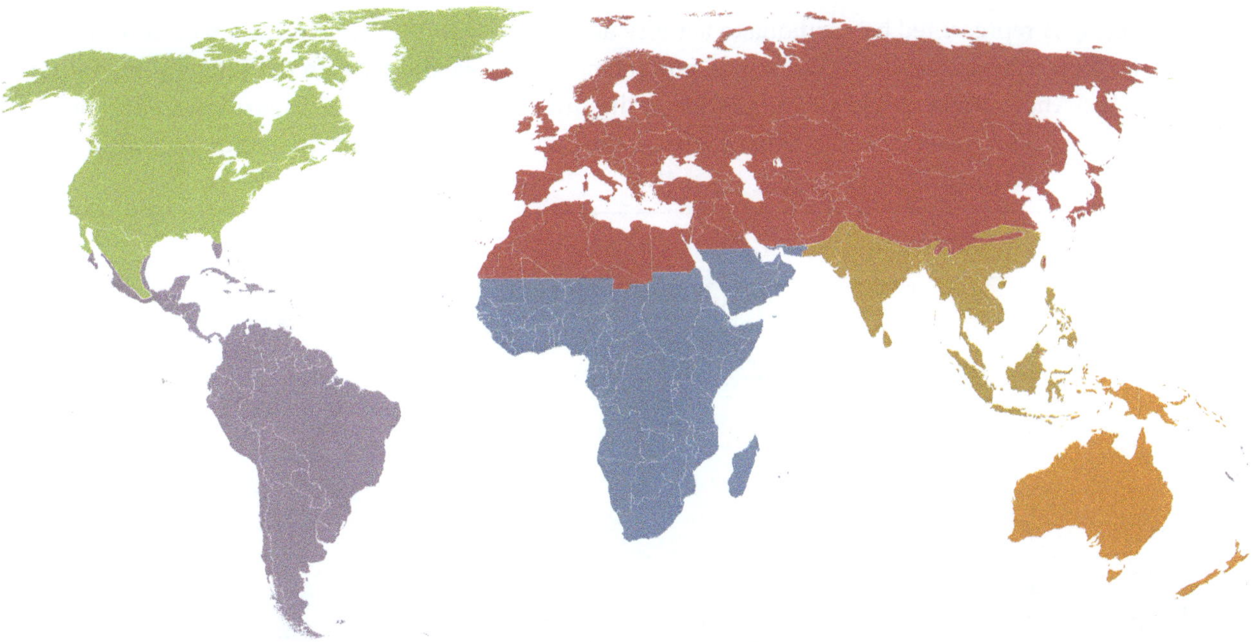

FIGURE 14.10a Main geographic realms, omitting Oceania and Antarctic ecozones

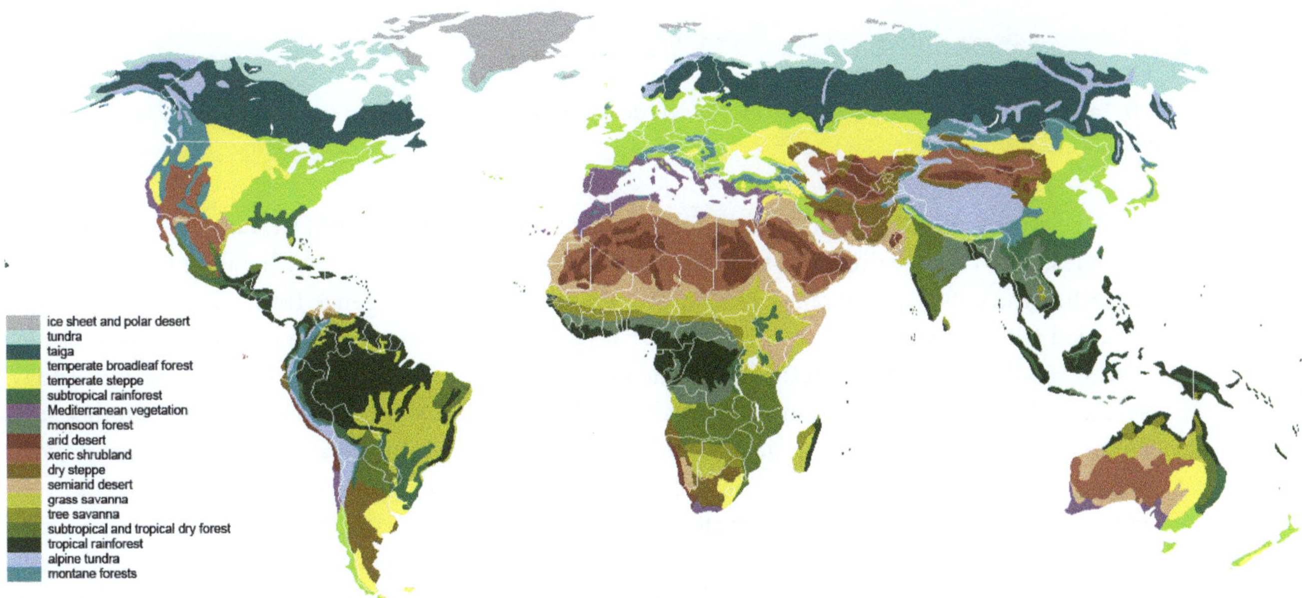

FIGURE 14.10b World's main biomes

zones, as represented by earthquake activity associated with the San Andreas fault in California, the New Madrid fault near Tennessee, and the numerous earthquakes in the Brawley seismic zone of the Andes. Interestingly, fossils of the extinct marine crocodile Mesosaurus are only found in eastern South America and southern Africa, thus reinforcing the concept of moving continents.

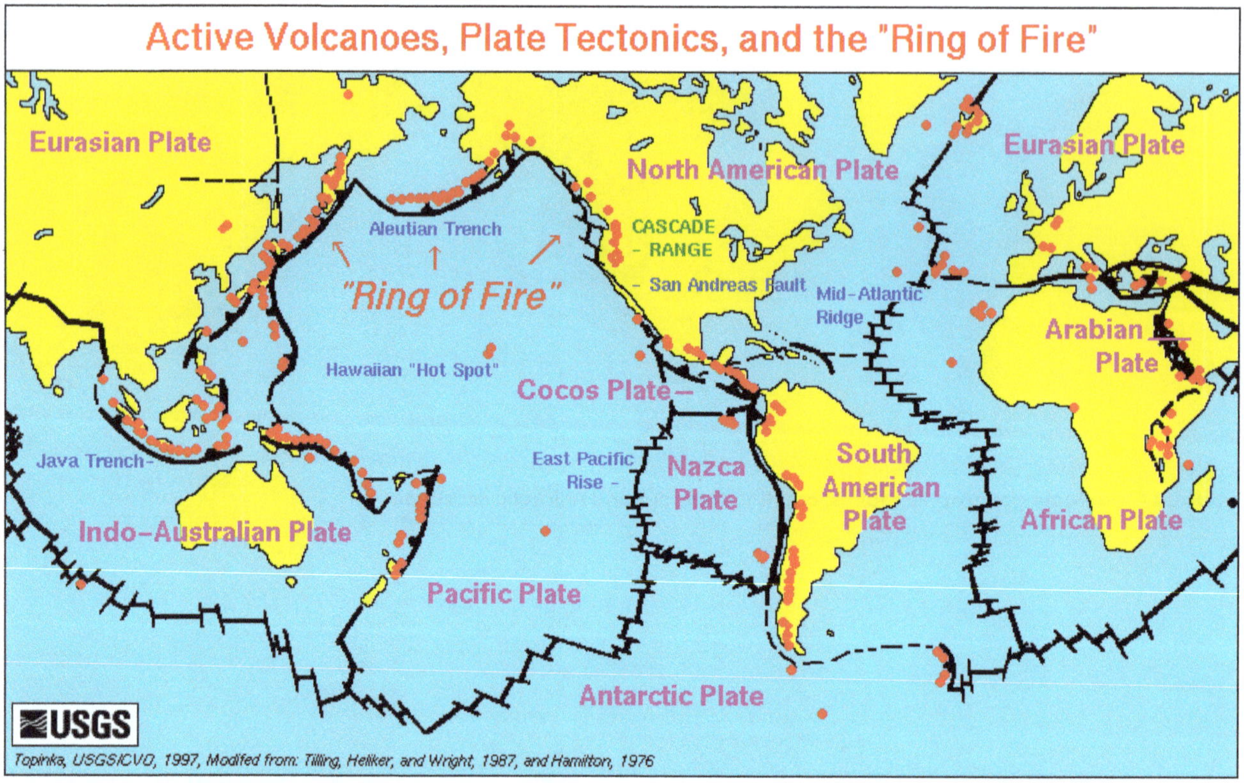

FIGURE 14.11 Tectonic plates. Check out an app that shows daily earthquake activity and relate this activity to plate tectonics.

A relatively recent and powerful tool in biogeography is the use of DNA and molecular technologies. DNA sequencing and protein sequencing provide scientists with valuable information regarding the evolutionary relationship between species and divergence from a common ancestor. Coupled with knowledge from *continental drift* (gradual movement of the continents) and *tectonic plates,* age-old questions about the origin and distribution of species are now being answered. Today, scientists have data to understand why plants and animals on different continents are similar, such as the Oriental and North American sycamore trees. Additionally, these molecular clocks provide valuable information regarding the time of divergence from a common ancestor, such as that between chimpanzees and humans approximately seven million years ago. Biogeography and genetics also answer why island populations are so unique. Being isolated in space and time, island species can be very different from their mainland relatives. After immigrating to an island, organisms adaptively radiate into new niches and can undergo speciation events. Galapagos finches and tortoises best represent this phenomenon.

QRQ 12. ______ ______ is the study of the range and distribution of organisms in the past.

QRQ 13. Select all that apply to Wallace's Line: animals to the east resemble those of Australia, animals to the east resemble those of Europe, animals to the west resemble those of America, animals to the west resemble those of Asia, establishes earthquake boundaries, establishes floral/faunal boundaries related to species distribution and geologic features, imaginary line that divides the Atlantic, imaginary line that runs between Australia and the Asian islands and mainland, line between tectonic plates

14.6. Speciation

In contrast to prevailing ideas of his time, Charles Darwin argued that species were not fixed since creation and that they were the result of evolutionary forces, such as natural selection. *Speciation* involves the formation of a new kind of organism. This new species develops unique characteristics derived from a common ancestor. *Reproductive isolation,* or the inability of organisms from related species to breed successfully, is a major force shaping the formation of new species.

Two types of reproductive *barriers* are recognized. *Prezygotic isolating mechanisms* are barriers that prevent the formation of a zygote (fertilized egg). The most obvious isolating mechanism is a *geographic barrier,* such as oceans, water bodies, deserts, and mountains. Barriers can also reflect human activities, such as building highways, walls, and fences. These physical barriers prevent organisms from finding mates. *Ecological barriers* are obstacles that include species breeding at different times or places. For example, one species of toad may prefer to mate in still puddles while another prefers moving streams. In addition, different species of flowers may produce pollen at different times in the spring. In *behavioral isolation,* mating rituals, such as courtship dances, color displays, and singing ability in birds, may greatly influence mating. Or perhaps the potentially mating couple just does not like each other. *Mechanical barriers* are also common in nature and involve the genitalia in animals or the pollinating structures in plants being incompatible. A Saint Bernard and a Chihuahua best illustrate this problem. *Gametic incompatibility* is another prezygotic isolating mechanism. Despite successful mating, these barriers impede the formation of a zygote. This is particularly true in organisms that undergo external fertilization, such as corals, sea urchins, fish, and frogs. This gametic incompatibility, driven by physical or chemical incompatibilities of the sex cells, serves to ensure that only sperm from the same species fertilizes an egg.

In *postzygotic isolation,* barriers prevent the development of an offspring or impede the offspring's ability to compete with other members of the population. *Hybrids* result from the mating of two different species or, perhaps, genera. The farther apart the mates are from a common ancestor, the less likely that the hybrid will be successful, so such legends as cabbits (cats and rabbits) and jackolopes (jackrabbit and antelope) are just myths.

APPLYING SCIENCE TO LIFE 7 Some Interesting Hybrids

Several interesting *interspecific hybrids* (different species) have been described in recent years. Mating between two species can produce offspring with new colors, shapes, and behaviors. The mule, famous for being a hard but stubborn worker, is a cross between a female horse and a male donkey; its counterpart is a hinny, or a cross between a female donkey and a male horse. Other hybrids of note include killer bees (European honeybee and African honeybee) and ligers (male lion and female tiger). Although the vast majority of hybrids are sterile, there are some exceptions. The beefalo, or cross between an American bison and a domestic cow, is fertile, as well as the wholphin and some cat, bear, and dog hybrids. Hybrids in plants are more common than animals, and many are fertile, including grapefruit, key lime, stargazer lilies, sweet corn, and many varieties of apple and cotton (O'Donnel 2018).

Biogeographic factors, such as waterbodies, deserts, and mountain ranges, have a profound influence on the formation of new species. Speciation is intimately associated with biogeography and is much more likely in a population that is geographically isolated from others. Modern evolutionary biologists have described several models of speciation.

Most speciation events fall into two categories—allopatric and sympatric. *Allopatric speciation* occurs as the result of geographical isolation, when a population is separated by an event. For example, the separation may involve the formation of a river or continents in movement, such as the separation of Madagascar and Africa by earthquake activity 200 million years ago. A geographical barrier then separates the initial population into two subpopulations. While separated, the subpopulations continue to evolve and accumulate physical, behavioral, and genetic differences. Over time, the two populations evolve into distinct populations and, eventually, species. The two populations do not have to be totally separated but need to have greatly reduced contact.

Sympatric speciation involves speciation in populations that are not geographically fragmented. The most noteworthy means of sympatric speciation involves reproductive isolation, which may involve mating seasons that occur at different times or mating based on the exploitation of a new niche. One potential example of sympatric speciation involves the hawthorn and apple maggot flies. In the 1860s, farmers described a new maggot that fed on apples, similar to the hawthorn fly maggot, which fed on hawthorns. Today, the fly's food of preference is linked to the food fed on by the maggot. Thus, flies that like hawthorns only mate with other hawthorn flies, and apple flies do the same. This host shift of fly preference may be the initial step in the formation of new kinds of flies and, ultimately, new species. In the plant kingdom, sympatric speciation is common in plants that are polyploidy (extra set of chromosomes). These plants are separated genetically from the euploid (normal number of chromosomes) plants in their range. Many useful plants today are the results of polyploidy, including strawberries, bananas, cotton, sugarcane, and potatoes.

QRQ 14. Select from the following the appropriate term for each description below: allogeneic, allopatric, behavioral isolation, breeding isolation, ecological barrier, episodic, gametic incompatibility, geographic barrier, hybrid, mating isolation, mechanical barrier, mixed, physical isolation, postfertilization isolation, prefertilization isolation, prezygotic isolation, procreative isolation, reproductive isolation, speciation, sympathetic, sympatric, terrestrial, zygote

a. despite successful mating, an obstacle to zygote formation exists
b. fertilized egg
c. formation of different species
d. inability of organisms from related species to breed
e. obstacle preventing the development of an offspring or impeding the offspring's ability to compete with other population members
f. offspring that results from mating of two different species or genera
g. reproductive isolation caused by animals having different means of finding and choosing mates
h. reproductive obstacle caused by incompatible reproductive organs
i. reproductive obstacle by a physical feature, such as a river
j. reproductive obstacle of species breeding at different places or times
k. reproductive obstacle that prevents formation of a fertilized egg
l. type of speciation in populations that are not geographically fragmented
m. type of speciation that occurs as the result of geographical isolation when a population is separated by an event

14.7. Comparative Anatomy

Comparative anatomy is the discipline of biology that investigates the structural differences and similarities among species. Important in understanding evolutionary trends, comparative anatomy has unveiled to biologists that living things are a variation of a theme.

Homologous structures are anatomical features that are shared by related organisms. Such organisms share a common anatomy as the result of having a common ancestor. The structures have similar anatomical makeup, but functions may differ. Thus, the structure has been tweaked over time as organisms adapted to different conditions or needs. The forelimbs of various mammals are good examples of homologous structures (see Figure 14.12). These structures are composed of the same bones, such as the humerus, radius, ulna, and various small bones of the lower forelimb. In whales, the forelimbs, called flippers, are used in stabilization and guidance in the world of water. However, in bats, the same bones constitute the wings, spread over elongated finger bones. In an antelope, the same bones are employed in locomotion, and in humans, in a variety of functions, including gripping. The same is true in plants—the leaves of an insectivorous sundew can capture insects, the spines of a cactus are for water conservation and protection, and the bright red or yellow leaves (bracts) of a poinsettia mimic petals of a flower. Homologous structures are examples of *divergent evolution* as species accumulate differences from a common ancestor.

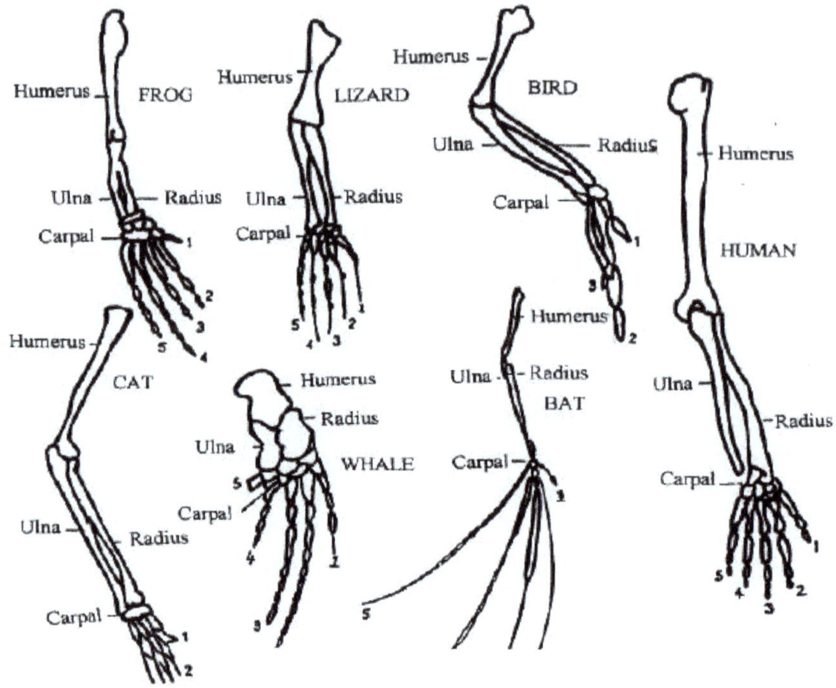

FIGURE 14.12 Anatomical drawing of forelimbs

In contrast to homologous structures, *analogous structures* involve features that have similar function but have evolved independently through time. Analogous structures are an example of *convergent evolution*, in which species have different evolutionary histories. Both birds and butterflies evolved wings for flight, but they have different origins and are different anatomically. Table 14.1 lists several analogous and homologous structures. Notice that bats and birds have structures in both categories.

TABLE 14.1. Some Homologous and Analogous Structures

Organisms	Structure	Homologous?	Analogous?
birds, butterflies	wings		yes
humans, octopi	eyes		yes
dolphins, ichthyosaurs	body shape		yes
beetles, tigers	legs		yes
bats, birds	forelimbs	yes	
bats, birds	wings		yes
various mammals	forelimbs	yes	
cacti, sundews, poinsettias	leaves	yes	

Organisms possess *structures* that have lost their original function or have been modified for new functions. Such a structure, termed *vestigial,* no longer performs the particular function for which it evolved. Some examples of vestigial structures in humans include the appendix, the coccyx, and the small muscles responsible for goose bumps. Perhaps you have seen a dog with cherry eye, or a red swollen mass in the medial corner of the eye. It is caused by the inflammation of the remnants of a third eyelid that can be easily seen in birds, reptiles, amphibians, and some mammals.

An *atavism* is a sporadic anomaly that occurs when an individual is born with an ancestral trait. Such rare anomalies result from the reactivation or expression of an ancient gene that was functional in the organism's ancestors. Examples include humans born with tails, snakes with limbs, eyes in blind cave fish, dolphins with hind limbs, and extra toes in horses.

APPLYING SCIENCE TO LIFE 8 Maybe Hen's Teeth Are Not So Scarce!

Near the Smoky Hill River in Western Kansas, the fossil remains of a curious bird was found in Cretaceous Period (145.5–65.5 MYA) deposits. *Hesperornis regalis* was a giant aquatic flightless bird that cruised the ocean searching for fish and measured just over six feet in length. Like many of the birds of its time, the loon-like *Hesperornis* possessed a beak full of reptile-like teeth. Fossil records indicate that these ancient toothed birds eventually lost their teeth between 60 and 80 million years ago. That may be good for us!

In 1821, the French naturalist Geoffroy St.-Hilaire claimed that during embryological development, some bird embryos exhibited tooth formation. Many ridiculed his observations, and his idea remained dormant for well over a century. However, in recent years, St.-Hilaire has been vindicated through research involving embryonic tooth development in chickens.

The presence of tooth development in chickens is an example of an *atavism,* or the reappearance of an ancient trait lost through the process of evolution. Genetically, atavisms result from changes in when, where, and how a particular gene is expressed in an organism. Atavisms serve as important clues to our genetic and evolutionary past.

In 2006, developmental biologists Matthew Harris and John Fallon released their research results that chickens have the ability to develop teeth like their dinosaur ancestors. In fact, these teeth resemble the teeth of baby alligators and crocodiles. Afterall, the genes regulating tooth development are still in the genome for birds.

The sonic hedgehog (SHH) gene is responsible for tooth development, number and positioning of the appendages, patterning of the internal organs, and formation of outer covering and its accessory structures. In normal chickens, the SHH gene expresses itself in the sides of the gums, but in crocodilians and mutant chickens, a mutation of the gene called *talpid 2* expresses itself in the center of the gum, causing tooth development. The gene is lethal, and the chicks, although tooth development begins, do not hatch. The exact mechanism that allows for tooth development in chickens is presently unknown. New research indicates that by engineering a virus to mimic the molecular signals of the talpid 2 mutation, developing normal chicks initiate tooth development. However, the teeth are soon absorbed by the beak.

By studying atavisms, scientists will develop a better understanding about how evolutionary changes can result from modifications in development. Perhaps one day, as a result of bird-tooth research, scientists will be able to initiate the process of regrowing teeth in individuals that have lost them as the result of disease, poor care, or accident. On a sidenote, this research vindicates the observations of Geoffroy St.-Hilaire (AAAS 2006; Everhart 2014; National Geographic 2015).

QRQ 15. Classify each structure as analogous, homologous, vestigial, or atavism:

a. eyes of humans and octopi
b. eyes in blind cave fish
c. forelimbs of various mammals
d. hind limbs in whales
e. leaves in cacti, sundews, and poinsettias
f. pelvic bones in some snakes
g. pelvis and hind limb bones in whales
h. tails in humans
i. three toes in horses that usually have one toe (a hoof)
j. tonsils in humans
k. wings in birds and butterflies
l. wings in wingless stick insects
m. wings of ostriches
n. wisdom teeth in humans

14.8. Embryology and Evolutionary Development

In the early 1840s, Darwin was made aware of the embryological studies of *Karl von Baer* and realized that they could be used in defending his ideas on evolution. Karl von Baer studied *embryology*, the investigation of the development of an embryo from fertilization through the various stages of early development (*ontogeny*).

Von Baer noted similar embryological developmental stages in the ontogeny of organisms. The more closely the organisms were related, the more traits the embryos shared. Essentially, in embryological development, it is difficult to distinguish between a fish, amphibian, reptile, bird, and mammal embryo (see Figure 14.13). For example, a human embryo is very fishlike early in development, possessing a post anal tail and "gill slits." These gill slits are openings between four pairs of embryological pharyngeal arches. The pharyngeal arches in humans and other mammals form the jaw, middle ear bones, hyoid bone, several facial nerves, thyroid gland, parathyroid glands, components of the thymus, Eustachian tubes, carotid artery, and tonsil cavity in later development. As the result of embryological anomalies, some humans are born with vestiges of "gill slits" or a post anal tail.

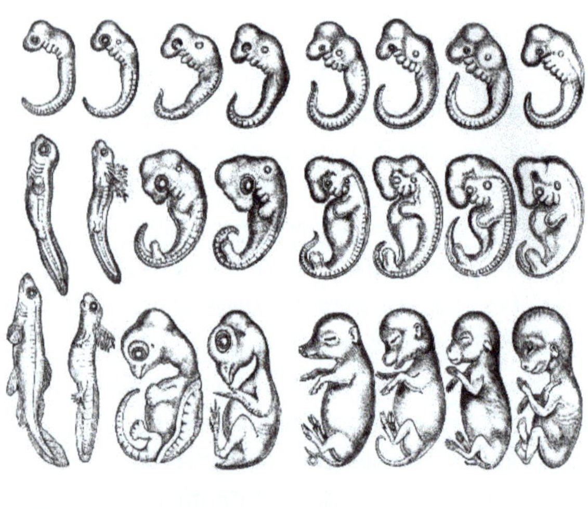

FIGURE 14.13 Drawing of embryos

Modern biologists recognize von Baer's law of embryology and its importance to evolution. Essentially, von Baer observed that in embryology, general traits appear earlier in development than specialized traits, and the development of specific embryonic characteristics progresses from general to specific through development.

In recent years, a new field of biology, known as *evolutionary development*, or *evo-devo* for short, has emerged that incorporates the sciences of evolution, developmental biology, and genetics. Through observations and research, evo-devo is answering questions such as how genes direct the formation of a species and how certain variations on a theme of these genes can form species-specific traits through evolution. For example, the same master genes that direct the development of segmentation and appendages in a fruit fly are also responsible for the development of these anatomical features in a human. Thus, animals possess a genetic toolkit that they have inherited from a common ancestor. Over time, tweaks and modifications of this toolkit have resulted in the diversity of features seen in the animal kingdom. Literally the same gene or genes are used over and over again, with modifications. Interacting with other genes has resulted in a unifying thread in the tapestry of life. Now biologists have a better understanding about how different tweaks in genes can result in the different shapes of bird beaks and how the forelimb of a bat forms a wing and of a dolphin forms a flipper.

APPLYING SCIENCE TO LIFE 9 Legs on the Head

One thing that studies in evolutionary development have revealed is that the genes involved in making a mouse, a human, or a fly are fundamentally the same. The common ancestor to animals provided them with a set of core genes, or a "genetic toolkit." Thus, studying the development of animals is a variation on a theme. This theme is very clear during development, placing the legs or organs in the right place, shaping different types of beaks, sculpting different kinds of appendages, or setting up the body axes. However, many miscues in development may result from mistakes made in this "genetic toolkit."

Studies in evo-devo have revealed that a certain *Hox gene*, known as *Antp*, is responsible for the formation and placement of appendages during the development of a fruit fly embryo. (*Hox genes* are master genes that direct development.) In some instances, the *Hox* gene sends a message to replace the antennae on the head of a fruit fly with appendages (legs), resulting in a condition known as antennapedia (see Figure 14.14). Similar errors in these genes are responsible for unusual traits in some organisms, such as in a ten-legged mutant spider.

FIGURE 14.14 Antennapedia in a fruit fly

In humans, these basic genes are responsible for establishing the left and right axes. The medical term *situs solitus* indicates that the internal organs are in the normal anatomical positions, for example with the heart tilting to the left and the gall bladder on the right side. Mix ups in these genes may result in the condition *situs inversus*, in which all of the internal organs are reversed, or mirrored, from the normal position, or another situation, *situs ambiguous*, in which the internal organs are distributed randomly in the body cavity. Could you imagine being a surgeon, before proper imaging techniques, encountering one of the two variations in organ placement?

QRQ 16. (Answer choices in parentheses) Select the appropriate phrases to complete the statement: The more closely organisms are related, the ______ (less similar the gestation period is for, more similar the gestation period is for, fewer traits are shared by, more traits are shared by) their embryos. Moreover, general traits appear ______ (about the same time, at varied times, earlier, later) in the development of an embryo as/than specialized traits.

14.9. Genetics and Evolution

Major support for evolution comes from the world of genetics, from the fundamentals of molecular genetics to new frontiers such as epigenetics. A major ingredient missing from Darwin's work on natural selection was from a lack of understanding of inheritance. Gregor Mendel established a fundamental model of inheritance in the 1860s, but its value was not realized until the early 1900s. During the 1930s and 1940s, a group of scientists formulated what Julian Huxley coined the *modern synthesis theory*, which fused the ideas of Darwin and Mendel with new concepts regarding genetics and inheritance. A *scientific theory* is a generally accepted explanation about an aspect of the natural world that has been sustained by repeated testing. It is not mere speculation. Table 14.2 lists some of the major contributors to modern synthesis theory. As a result of these studies, a new definition of evolution emerged as a change in allelic (alternative gene) frequencies within populations over time, and vast new evidence reinforced the theory of evolution.

TABLE 14.2. Major Contributors to Modern Synthesis Theory

Major Contributor	Contribution
Theodosius Dobzhansky	evolution and population genetics
R. A. Fisher	illustrated Mendelian genetics are complementary to evolution
E. B. Ford	role of ecological genetics
J. B. S. Haldane	role of population genetics in evolution
Ernst Mayr	species concept
Bernhard Rensch	biogeography
George Gaylord Simpson	paleontology
G. Ledyard Stebbins	behavior
Sewall Wright	computational work on gene frequencies

PoI 11. Discuss the work of one of the major contributors to modern synthesis in Table 14.2.

Along with natural selection, mutation, and migration, genetic drift, which we discuss in more detail in Module 15, is one of the major mechanisms of evolution. It involves the random fluctuations in the allele frequency of particular traits due to chance events. Genetic drift has a more pronounced impact on smaller populations, such as those isolated in space and time. This explains why island populations have evolved such unique flora and fauna, like those of the island of Madagascar.

QRQ 17. Select all statements true about "scientific theory": has sustained repeated testing, involves results of a few experiments by one team of scientists, is a hypothesis, is an approximation, is an explanation about an aspect of the natural world, is an explanation of one laboratory's experiments, is difficult, is not a mere speculation, is often doubted by many scientists, is provisional, is speculative, is uncertain, is unpredictable.

14.10. Macroevolution and Microevolution

A look at the history of life reflects the power of *macroevolution*, or changes that occur above the level of species. These changes occur on a grand scale, involving deep time and natural selection. They may appear incredibly gradual, as seen in the evolution of whales and dolphins from land ancestors or the evolution of flowering plants from cone-bearing stock. Although at work today, the mechanisms of macroevolution cannot be observed in the short period of a human lifetime. Thus, biologists are dependent on paleontology and molecular biology to understand the power of macroevolution.

On the other hand, *microevolution* involves small changes within a species over a short period of time. These changes reflect changes in the gene frequency within a population that is shaped by mutation, natural selection, nonrandom mating, gene flow, and genetic drift. Microevolution is much more evident in organisms that have large population sizes and short generation times, such as bacteria. Examples of microevolution abound in the everyday world and include resistance to insecticides by insect populations, herbicide resistance by certain weeds, resistance of bacteria to antibiotics, and our never-ending war against constantly changing viruses, such as HIV.

APPLYING SCIENCE TO LIFE 10 Evolutionary Wars

Today, the medical world is faced with a new kind of crisis. It is not a newly emerging disease, such as Ebola, or a new carcinogen provoking cancer, or even a new weapon, but is a problem resulting from an age-old evolutionary arms race. The new crisis involves the resistance of many pathogenic bacteria to contemporary antibiotics.

Basically, antibiotic resistance involves the ability of a microorganism, such as a bacterium, to withstand the effects of an antibiotic, such as penicillin. This resistance is the consequence of evolution by natural selection, specifically microevolution. Simplistically, microevolution involves the change in gene frequency of a particular trait over time. Microevolution can be observed in a population, such as bacteria, in a relatively short period of time, as resistance mutations increase in response to the use of antibiotics. Bacteria occur in great numbers and reproduce simply and rapidly, and as a result, genetic changes, such as antibiotic resistance, can spread quickly. The misuse and overuse of antibiotics through the years has evolutionarily selected for resistance in several species of medically important bacterial pathogens.

Bacterial populations contain individuals that vary in their genetic makeup. When the population of bacteria are exposed to a particular antibiotic, a few of the bacteria may have a new or mutated gene that will produce a mechanism to inactivate or pump the antibiotic out of the cell. Although most of the members of this population, lacking a defense mechanism, will be killed by the antibiotic, those few bacteria with the new protective adaptation will persist and produce offspring. In succeeding generations, these resistant bacteria will not be susceptible to the antibiotic. The antibiotic used is a *selection factor* (environmental feature that influences successful survival and reproduction of a particular genotype). When this factor is present, bacteria having the genotype (gene makeup) that contains the protective genes will be selected for by the antibiotic. These bacteria will survive and produce the vast majority of new bacteria as long as the same antibiotic (selection factor) is present.

The overuse and abuse of antibacterial drugs has put into motion microevolutionary forces that select for drug resistance in bacteria. These former wonder drugs have now shaped a generation of bacterial species that negatively impact humankind. This negative impact is multifold, impacting the success rate of medical procedures such as implants and transplants, impeding everyday procedures, harming immunocompromised patients, and increasing mortality rates.

Several examples of antibiotic-resistant bacteria follow:

- *Mycobacterium tuberculosis* causes tuberculosis, which mainly affects the lungs and can be fatal. New strains of *M. tuberculosis* have evolved high resistance.
- *Neisseria gonorrhea* is responsible for the sexually transmitted disease gonorrhea, causing various infections of the genitals, urethra, rectum, and throat. *N. gonorrhea* has developed medium resistance to antibiotics.
- *Escherichia coli* in the pathogenic form can cause urinary tract infections, meningitis, diarrhea, and gastrointestinal distress. Some strains of *E. coli* have developed high antibiotic resistance.
- *Clostridium difficile* is responsible for diarrhea and gastrointestinal problems, especially in hospitals. Its incidence has been linked to the overapplication of broad-spectrum drugs, which are used to treat a large variety of disease. Presently, the bacterium exhibits low resistance.
- *Klebsiella pneumoniae* and *Streptococcus pneumoniae* are both responsible for community-based pneumonia and lung infections. Both of these bacteria are highly resistant to antibiotics and dangerous.

Perhaps the classic example of the development of antibiotic resistance in bacteria involves *Staphylococcus aureus*. These bacteria can frequently be found on the skin and in the upper respiratory tract, such as inside the nose. Many times, infection from these bacteria is associated with boils, impetigo, food poisoning, toxic shock syndrome, and cellulitis. A variety of conditions increase the chances of a "*staph*" infection, including trauma to the skin, burns, diabetes, dialysis, improper food preparation, and contact sports.

In the past, methicillin was used to combat *staph* infections, but it was replaced by oxacillin because methicillin damaged the kidneys. As early as the 1960s, *MRSA* (methicillin-resistant *Staphylococcus aureus*, pronounced "mersa") was described. Today, MRSA is responsible for dangerous infections worldwide, particularly in hospitals, correctional institutions, military centers, and athletic facilities. Symptoms include but are not limited to fever, abscesses, bone or bladder infections, and endocarditis. Many times, MRSA infections are fatal. Another drug, vancomycin, was used to treat MRSA forms of *staph*, but microevolution jumped into action again, resulting in *VRSA* (vancomycin-resistant *Staphylococcus aureus*). Individuals susceptible to these infections include individuals that have had previous MRSA infections, those that have overused vancomycin, and individuals with underlying health conditions, such as diabetes and kidney disease.

Besides medical intervention, there are a few activities that can reduce your chances of developing a *staph* infection. They include thoroughly cleaning wounds and keeping them covered; wearing clean clothes and using clean bedding; washing your hands; not sharing razors, towels, and other personal items; and mostly using common sense.

What can be done to counter the antibiotic resistance war?

- Prescribe the correct drug for the correct infection.
- Only use antibiotics when they are useful—antibiotics are *not* designed for viral infections or to be wonder drugs.
- Physicians can be more conservative in prescribing antibiotics. Do not overuse antibiotics.
- The effects of specific antibiotics need to be tested diligently and monitored in patients.
- Develop risk-factor profiles to determine which patients may have antibiotic-resistant strains of bacteria.
- Be extremely careful in prescribing broad-spectrum antibiotics.
- Ensure the proper education and use of antibiotics by patients.
- Rethink the use of antibiotics in livestock.
- Promote the continuous discovery, creation, and development of new antibiotic medicines.

With these measures and others, one must keep in mind that we are fighting an evolutionary war and that bacteria, by their sheer numbers, short generation time, and prolific reproduction rates, have the advantage (Longitude Prize 2014; Mayo Clinic 2020; Palumbi 2001; Science Daily 2019).

QRQ 18. Indicate whether each situation illustrates macroevolution or microevolution:

a. evolution from antibiotic-sensitive bacteria to antibiotic-resistant bacteria
b. evolution from cone-bearers to flowering plants
c. evolution from herbicide-sensitive weeds to herbicide-resistant weeds
d. evolution from land animals to whales

QRQ 19. Select from the following the appropriate study for each description below: biochemistry, biogeography, comparative anatomy, developmental biology, ecology, embryology, evo-devo, genetics, microbiology, modern synthesis

a. combines ideas of Darwin and Mendel with new concepts regarding genetics and inheritance
b. incorporates the sciences developmental biology, genetics, and evolution
c. investigates the differences and similarities among species
d. study of the development of embryos
e. study of the range and distribution of organisms

QRQ 20. Select an appropriate person for each achievement below: Ted Daeschler, Charles Darwin, Mary Latimer, Ernst Mayr, Neil Shubin, Karl von Baer, Alfred Russel Wallace, Alfred Wegener, Sewall Wright

a. coined the term "living fossil"
b. concluded evolution was the driving force shaping distribution of species
c. helped found biogeography; noted the distribution of species and how many are confined to certain regions
d. maintained that species were not fixed since creation but were the result of evolutionary forces, such as natural selection
e. predicted existence of transitional fossils
f. proposed continents were once connected like a giant jigsaw puzzle
g. proposed natural selection
h. studied ontogeny; noted the difficulty in distinguishing between embryos of many different species

Image Credits

Fig. 14.1: Copyright © 2015 Depositphotos/alinabel.

Fig. 14.2a: Copyright © by Anders L. Damgaard (CC BY-SA 4.0) at https://commons.wikimedia.org/wiki/File:Baltic_amber_inclusions_-_Ant_(Hymenoptera,_Formicidae)8.JPG.

Fig. 14.2b: Copyright © by Tylwyth Eldar (CC BY-SA 4.0) at https://commons.wikimedia.org/wiki/File:Ammonite_-_Uptonia_jamesoni.jpg.

Fig. 14.2c: Copyright © by Ra'ike (CC BY 3.0) at https://commons.wikimedia.org/wiki/File:Shark_tooth_-_Haifischzahn.jpg.

Fig. 14.3: Source: https://commons.wikimedia.org/wiki/File:La_Brea_Tar_Pits.jpg.

Fig. 14.4: Copyright © by Turid Hillestad Nel , Christopher Stuart Henshilwood (CC BY 2.5) at https://commons.wikimedia.org/wiki/File:Blombos_Cave_stratigraphy.jpg .

Fig. 14.5a: Source: https://commons.wikimedia.org/wiki/File:Hippogriff1.jpg.

Fig. 14.5b: Copyright © by IJReid (CC BY 4.0) at https://commons.wikimedia.org/wiki/File:Protoceratops_andrewsi_ROM.jpg.

Fig. 14.7: Copyright © by Pedro José Salas Fontelles (CC BY-SA 3.0) at https://commons.wikimedia.org/wiki/File:Archaeopteryx_lithographica_-_Pedro_Jos%C3%A9_Salas_Fontelles_(flipped).jpg.

Fig. 14.8: Copyright © by Nobu Tamura (CC BY-SA 3.0) at https://commons.wikimedia.org/wiki/File:-Tiktaalik_BW_flopped.jpg.

Fig. 14.9: Copyright © by DataBase Center for Life Science (DBCLS) (CC BY 4.0) at https://commons.wikimedia.org/wiki/File:201704_Coelacanth.svg.

Fig. 14.10a: Copyright © by carol (CC BY-SA 3.0) at https://commons.wikimedia.org/wiki/File:Ecozones.svg.

Fig. 14.10b: Copyright © by Ville Koistinen (user Vzb83) (CC BY-SA 3.0) at https://commons.wikimedia.org/wiki/File:Vegetation.png.

Fig. 14.11: Source: https://commons.wikimedia.org/wiki/File:Map_plate_tectonics_world.gif.

Fig. 14.12: Copyright © by Ruth Lawson (CC BY 3.0) at https://commons.wikimedia.org/wiki/File:Anatomy_and_physiology_of_animals_Various_vertebrate_limbs.jpg.

Fig. 14.13: Source: https://commons.wikimedia.org/wiki/File:Baer_embryos.png.

Fig. 14.14: Copyright © by toony (CC BY-SA 3.0) at https://commons.wikimedia.org/wiki/File:Mutation_Antennapedia.jpg.

LAB #14

Laboratory Exercises

14.11. Laboratory on Peppered Moths

14.11.1. Preliminary Work

In this laboratory, we examine how animals adapt to changing environmental factors using a classic example, peppered moths. From the text's website, view the introductory video; download and open "*Peppered Moths.nlogo*."

1. Click the Info tab at the top of the page and read the sections "What Is It?" and "How It Works."
   a. ______ Before the Industrial Revolution, what advantage did light-colored peppered moths have? A. birds preferred dark-colored moths, B. light-colored moths blended well with the white bark of the trees so that birds could not see them as well, C. light-colored moths could find a mate easier and thus were more abundant, D. light-colored moths were healthier, E. light-colored moths were smaller
   b. ______ Give the kind of moth that began to predominate when soot darkened the bark of the trees: A. blue moths, B. dark-colored moths, C. light-colored moths, D. medium-colored moths, E. white moths, F. yellow moths
   c. **1.** ______ **2.** ______ **3.** ______ **4.** ______ A *species* includes organisms that are capable of interbreeding to produce viable offspring, while a *population* is a group of organisms belonging to the same species that reside in the same location at the same time. *Evolution* involves changes that take place in inherited physical and behavioral characteristics of populations of organisms over time. Complete the following statement: Natural selection is a principal mechanism of **1.** (A. evolution, B. genetics, C. microbiology, D. physiology, E. selectivity), whereby **2.** (F. living individuals in an area, G. members of a family, H. members of a population, I. members of a species) that are better adapted to the **3.** (M. environment, N. food, O. sunlight, P. temperature, Q. terrain) **4.** (R. are more aggressive, S. grow faster, T. grow larger, U. have more offspring, V. survive to produce more offspring). (Refer to other sources as needed.)

2. Read "How to Use It."

a. ______ Give the MUTATION value that indicates the child moth should have exactly the same coloration as the parent.

b. ______ Give the MUTATION value that indicates the child's coloration is random and has no correlation with the parent's color.

c. ______ Birds are not visualized in the simulation. What slider incorporates several factors determining how birds feast on moths? A. MUTATION, B. NUM-MOTHS, C. POLLUTE, D. SELECTION, E. SPEED

d. ________________ Select factors that SELECTION consolidates: A. growth rate of birds, B. how camouflaged moths are, C. how hungry birds are, D. how important camouflage is to escaping predation, F. number of birds, G. number of moths, H. size of the birds

e. ______ Give the SELECTION value that indicates color does not matter as far as a bird eating a moth or not.

f. ______ When we click the POLLUTE or CLEAN UP buttons, the environment darkens or lightens, respectively. What SPEED value yields the fastest change in pollution (color)?

g. ______ percent Give the maximum pollution level.

14.11.2. Laboratory Work

Default parameters: INITIAL-POLLUTION% = 0, STOP-AT-100? = OFF, CYCLE-POLLUTION? = OFF, SPEED = 10, NUM-MOTHS = 100, SELECTION = 50, MUTATION = 13.

3. **1.** ______ **2.** ______ **3.** ______ **4.** ______ **5.** ______ Except for CYCLE-POLLUTION? being ON, use the default parameters. Allow pollution to go from 0 percent to 100 percent and back to 0 percent. Complete graph descriptions of the numbers of light (LM), medium (MM), and dark (DM) moths over time as they relate to pollution level: Initially, the numbers of LM, MM, and DM are **1.** (A. approximately equal, B. very different, C. 0, D. 400). However, quickly, the number of **2.** (E. DM, F. LM, G. MM) dominates. As polution percent is near 50 percent, most moths have **3.** (H. dark, I. light, J. medium) color. With 100 percent polution, almost all moths are **4.** (H. dark, I. light, J. medium). As pollution decreases, the results **5.** (K. remain the same, L. reverse, M. start over).

4. In this question, we observe the interplay of pollution and cleanup with the color of the moths. Use the default parameters.

a. ______ Turn off the CYCLE-POLLUTION? switch. Increase simulation speed or uncheck *view updates*. In running the simulation, after there are almost all light moths, click POLLUTE and CLEAN UP until the number of LM and MM moths fluctuate in about the same range. Select the approximate pollution percent: A. 26.3 percent, B. 27.5 percent, C. 28.8 percent, D. 30 percent, E. 31.3 percent, F. 32.5 percent, G. 66.2 percent, H. 67.5 percent, I. 68.7 percent, J. 70 percent

b. **1.** ______ **2.** ______ Change SPEED to 100 so that the environment responds quickly to changes in pollution. Run the simulation at normal speed until almost all moths are light. Complete the following statement: We can "kill off" the moths by clicking POLLUTE **1.** (A. rapidly, B. slowly) several times, making the pollution **2.** (C. decrease rapidly to 0 percent, D. decrease slowly to 0 percent, E. increase rapidly to 100 percent, F. increase slowly to 100 percent).

c. **1.** ______ **2.** ______ **3.** ______ Complete the description of environment and actions to "kill off" the moths by clicking CLEAN UP: In a **1.** (A. dark, B. light, C. medium) world with almost all **2.** (A. dark, B. light, C. medium) moths, click CLEAN UP rapidly to make the pollution percent go quickly to **3.** (D. 0 percent, E. 50 percent, F. 100 percent).

d. **1.** ______ **2.** ______ Your actions from parts b and c simulate a rapid change in **1.** (A. environmental conditions, B. genetics, C. mutations, D. number of animals), in which the type of animal **2.** (E. adapts too fast, F. cannot adapt quickly enough, G. mutates).

5. In this question, we quantify the model's relationship between pollution percent and percentages of the three types of moths at 100 ticks. Therefore, except for STOP-AT-100? being ON, use default parameters. Increase the simulation speed or uncheck view updates.

a. **1.** ______ **2.** ______ **3.** ______ Complete the hypothesis. For each blank, use the following possible answers: A. decreases, B. decreases then increases, C. increases, D. increases then decreases, E. remains steady. As INITIAL-POLLUTION% increases from 0 percent to 100 percent, at time 100 ticks, the percentage of light moths **1**, the percentage of medium moths **2**, and the percentage of dark moths **3**.

b. **1.** ______ percent **2.** ______ percent **3.** ______% Start a spreadsheet named *Peppered Moths* as in Table 14.3 to record your data. Set up and run the simulation at least five times, with *view updates* unchecked or with a fast speed, and record output values under appropriate headers. Because tabs are separators, we can copy all three output values and paste them at once so that values are in different columns. Calculate and record each column average: **1.** percent light, **2.** percent medium, **3.** percent dark

TABLE 14.3. Start of Peppered Moths Lab Spreadsheet

Your Names			
Peppered Moths Lab			
	Question 5b		
	% Light	**% Medium**	**% Dark**

c. **1.** ______ **2.** ______ **3.** ______ Set INITIAL-POLLUTION% to 20 percent, for a somewhat grayer background. Repeat part b.

d. **1.** ______ **2.** ______ **3.** ______ Repeat part b with INITIAL-POLLUTION% as 40 percent.

e. **1.** ______ **2.** ______ **3.** ______ Repeat part b with INITIAL-POLLUTION% as 50 percent (medium gray).

f. **1.** ______ **2.** ______ **3.** ______ Repeat part b with INITIAL-POLLUTION% as 60 percent.

g. **1.** ______ **2.** ______ **3.** ______ Repeat part b with INITIAL-POLLUTION% as 80 percent.

h. **1.** ______ **2.** ______ **3.** ______ Repeat part b with INITIAL-POLLUTION% as 100 percent (black).

i. **1.** ______ **2.** ______ **3.** ______ Graph the percentages of light, medium, and dark moths at time 100 ticks versus INITIAL-POLLUTION%, and complete the graph descriptions using the following possible answers: A. decreases, B. decreases then increases, C. increases, D. increases then decreases, E. remains steady. As INITIAL-POLLUTION% increases from 0 percent to 100 percent, at time 100 ticks, the percentage of light moths **1**, the percentage of medium moths **2**, and the percentage of dark moths **3**.

j. ______ Do your data support your hypothesis in part a, yes (Y) or no (N)?

k. **1.** ______ **2.** ______ Consider the graphs of part i again. Complete the statement using the following possible answers: A. 0–20, B. 20–40, C. 40–50, D. 50–60, E. 60–80, F. 80–100. The biggest change in the percentage of light moths is in the INITIAL-POLLUTION% range of **1.** percent, while the biggest change in the percentage of dark moths is in the INITIAL-POLLUTION% range of **2.** percent.

l. ______ Give the INITIAL-POLLUTION% value that results in the largest percentage of medium moths.

6. Write a thoughtful paragraph about how your data and observations support the concept of natural selection.

MODULE

15

Fit for Life—Adaptation and Diversity

15.1. Introduction

Snorkeling or diving along part of Australia's Great Barrier Reef (GBR), you see an incredible variety of organisms. Large heads of hard corals surround you. More than 70 species of hard coral species here, out of 411 total species, are capable of forming the actual reefs. The reefs are beautifully festooned with an array of other plants and animals. You see soft corals, like sea fans, waving in the current and massive barrel sponges, which are some of the 1,500 species of sponges on the GBR (WWF-Australia 2020). There are plantlike anemones, a myriad of small crustaceans, colorful sea stars, sea urchins and sea feathers, various worms, clams, and colorful nudibranchs going about the business of nourishing themselves. Of course, you also see a dazzling assemblage of larger animals, like fish, shark, turtles, and perhaps a sea snake. You marvel that these organisms display significant and appropriate characteristics or adaptations that allow each species to survive and reproduce on the reef.

Although the Great Barrier Reef likely represents the most diverse ecosystem on the planet, you can find great diversity all over the globe, in every type of ecosystem. Human beings have long wondered about the diverse communities of organisms that populate the earth and pondered their origins.

15.2. Animal Diversity: The Mollusks

Imagine yourself walking down a sandy beach, enjoying the sunshine and gentle ocean breezes. As you gaze, you see that the sand is littered with various types of shells. As it happens, the creatures that produced most of these shells belong to an animal phylum (a taxonomic grouping of organisms) called the Mollusca. Experts claim that the mollusks are one of the most diverse groups of animals on earth and that there may be as many as 200,000 living species. Even though there is great variety in *morphology* (body structure), they all possess a soft body, composed of a "head," a visceral mass, and a "foot." Frequently, a calcareous cover, or shell, encloses and protects their soft bodies. Some of the shells you find there still contain the organism that produced them, but many of the shells are just remnants of former life. Familiar examples of mollusks include snails, clams, squid, oysters, and chitons (Bunje 2003). Consider for a moment how different those animals appear. How is it that such disparate creatures are all considered mollusks?

The mollusks are a large and varied animal group, and each member of that group embodies a collection of adaptations. *Adaptations* are traits or characteristics that improve an organism's chances to survive in their environment and subsequently produce more progeny. Adaptations are generally considered to be a result of *natural selection*. When Charles Darwin proposed the process of natural selection, he based his hypothesis on several principles he developed based on his data and observations (University of Michigan 2010):

1. There is some morphological, physiological, and behavioral variation within populations of organisms.
2. Many of the traits that characterize organisms are heritable. We now refer to the genes that produce these traits as *genotypes* (genetic makeup of an organism). We also know *mutations* occur that change genes and gene expression. Mutations are essential materials for natural selection. Most of these changes are detrimental or neutral, but occasionally they are advantageous.
3. Populations tend to produce numbers of offspring that exceed the capacity of the environment to support them. There is, then, a struggle for existence among the members of the population.
4. Members of a population that have traits conferring a competitive advantage (adaptations) over other members have improved chance to survive and reproduce. Hence, these members will contribute more offspring, also with similar adaptations, to the next generation.

Thus, natural selection is a process that alters the relative frequencies of different genotypes in a population, changes that result from differences in the ability of the resulting *phenotypes* to survive and reproduce. Reproductive success is synonymous with *biological fitness*. Therefore, if a mutation yields a new genotype that increases the fitness of the organism, more of this genotype successfully reproduce, and more of this genotype should appear in the next generation. If the environment remains consistent, the frequency of the genotype will continue to increase accordingly.

Let's reiterate what we mean by *genotype* and *phenotype*. *Genotype* is the set of genetic instructions that determine physical and functional characteristics of an organism. The physical expression of the genotype is termed the *phenotype*. For example, if a mouse has genetic instructions for gray fur color, the gray color is the expression of those genes—the phenotype. If another mouse has a variation of that same set of genes that code for white fur color, the phenotype would be white. (These alternative forms of the same gene are termed *alleles*.) Suppose that these animals live in an area where the soil is gray. With such a background, the gray mouse blends in with the soil color better, and, therefore, a predator would be less likely to spot the mouse. Conversely, a predator is more likely to catch and eat a white mouse, which stands out against the soil. Hence, more gray mice will survive to reproduce and pass on the genes coding for the gray phenotype.

QRQ 1. Suppose two women have the same mutation of the BRCA1 gene (BReast CAncer gene one), but one woman has breast cancer caused by the mutation while the other woman does not. We say that the women have the same ______ related to this gene but different physical expressions of this gene, or different ______.

QRQ 2. In nature, the process called ______ ______ promotes the survival of the best-adapted organisms, ensuring that they will reproduce, and their offspring, which have their genes, increase in proportions of succeeding generations.

QRQ 3. Select which of the following are principles on which Darwin based his natural selection ideas: certain animals have the ability to imitate other animals in order to survive, many of the traits that characterize organisms are heritable, members of a population that have traits that confer a competitive advantage over other members will have improved chances to survive and reproduce, populations tend to be uniform, populations tend to produce numbers of offspring that exceed the capacity of the environment to support them, species remain unchanged over time, the earth is divided into distinct biogeographic realms, there is some morphological, physiological, and behavioral variation within populations of organisms

QRQ 4. Alternative forms of the same gene are called ______.

QRQ 5. Suppose the bark of trees tends to be gray, and suppose two types of moths, with colors yellow and gray, light on the bark.

a. Which moth has the best fitness?
b. Which moth is a bird more likely to eat?
c. In the next generation, which moth is more likely to survive to reproduce?

15.3. Genetics of Populations

15.3.1. Frequencies

Evolution is simply defined as a change in the allelic frequency within a population over time. *Population genetics* is a major contributor to our understanding of the evolutionary process. In population genetics, the *gene pool* refers to the alleles for all genes available within a population (group of organisms of the same species living in the same area). For example, in your class, you have genes that influence the color of your skin, hair, and eyes; height; blood type; intelligence; and thousands of other traits. Generally speaking, the more varied the gene pool, the greater the ability of members of a natural population to adapt to changes in the environment, which also increases their survival chances.

Members of a population display similar traits because they share a common genome. However, mutations, recombination (crossover), changes in chromosome number, independent assortment, random fertilization, and some environmental factors result in variation within a population. Just look around and observe the variation in traits among your friends, even though they all may appear to be very human. In studying the genetics of populations, we consider genetic variation within populations and the variables influencing these variations. Such studies are particularly relevant in today's world for understanding evolution and disease and ensuring that life continues to flourish on Earth.

Individual organisms possess mutations in their genome that may ultimately affect their reproductive success and survival. However, the true power of the evolutionary process reveals itself at the population level. The alleles of shared genes are the working parts of variation and evolution. In studying a population, we determine the *frequencies*, or proportions, of a particular genotype, phenotype, or allelic ratio. For example, consider a hypothetical population, of 500 individuals, where we know individual genotypes; allele A is dominant, and allele a is recessive. Suppose the genotype frequency for AA is 0.56 (56 percent) and for aa is 0.15 (15 percent), making the frequency for Aa (phenotypes Aa or aA) 0.29 (29 percent). In the population, there are then 280 AA (0.56 × 500), 145 Aa (0.29 × 500), and 75 aa (0.15 × 500).

Genotype	Population	Genotype Frequency
AA	280	280/500 = 56%, or 0.56
Aa	145	145/500 = 29%, or 0.29
aa	75	75/500 = 15%, or 0.15
Total	500	100%, or 1.0

From these numbers, we can determine the *phenotypic frequency* for the population. The dominant phenotype frequency for A, or the fraction of times a genotype has A (AA or Aa), is 0.85 (0.56 + 0.29). However, there is only one recessive phenotype for this situation, with genotype aa, which has frequency 0.15.

From the genotypic frequencies, you can calculate the number of a and A alleles in the population. Computing the *allelic frequency* (frequency of alleles) is just a bit more complex. An individual displaying the A allele can be homozygous (AA) or heterozygous (Aa), and each of the 500 individuals has a pair of alleles for this gene, so there are 1,000 total alleles. Each AA contributes two As, while Aa has one A. Similarly, aa displays two *a*s, and Aa one. Thus, to calculate the allelic frequencies, we use the following equation, where the frequency for A is p, and the frequency for a is q:

$$p = freq(\mathrm{A}) = freq(\mathrm{AA}) + (1/2)\, freq(\mathrm{Aa})$$
$$q = freq(\mathrm{a}) = freq(\mathrm{aa}) + (1/2)\, freq(\mathrm{Aa})$$

In the 280 individuals with AA genotype, there are 2 × 280 = 560 As, and 145 individuals have an A from the genotype Aa. Thus, the frequency for the A allele is (560 + 145)/1000 or 705/1,000 = 0.705 (70.5 percent). For the a allele, there are 75 individuals that are aa, or 150 alleles (2 × 75), and for Aa, 145 individuals that have an a allele from the Aa genotype. The frequency for a is (150 + 145)/1,000 (295/1000) or 0.295 (29.5 percent). The sum of these frequencies is 1.0 or 100 percent (0.705 + 0.295).

In a theoretical population with frequency of A being p and of a being q, we can determine the frequencies of genotypes in the next generation using our friend from basic genetics, the Punnett Square. The following is such a square, showing the crosses with our known allelic frequencies:

	p = freq(A)	**q = freq(a)**
p = freq(A)	$p \times p = p^2$	pq
q = freq(a)	$qp = pq$	$q \times q = q^2$

The frequency, or probability, of A in the mother *and* of A in the father is the *product* of the probabilities, $p \times p = p^2$. Likewise, the probability of receiving an a allele from each parent is q x q $= q^2$. However, the child can obtain genotype Aa in one of two ways, A from the mother and a from the father, or vice versa. Thus, the frequency (probability) in A from the mother and a in the father is $p \times q = pq$, while the frequency of a in mother and A in father is $q \times p = qp = pq$. The frequency of one situation *or* the other (to obtain Aa or aA) is the *sum* of the frequencies, or $pq + pq = 2pq$. Adding all frequencies, we generate the following equation:

$p^2 + 2pq + q^2 = 1$, because the sum of the frequencies must be 1 (or 100 percent)

This equation was derived independently during the early 20th century by *G. H. Hardy* and *W. Weinberg*. The equation predicts that a population will have constant genotypic frequencies from generation to generation, as long as there are no evolutionary effects. For the most part, genotypic ratios of natural populations are rarely at equilibrium. However, the equation is still a useful tool, providing scientists with information about changes in these populations brought on by natural selection and other factors. Also, the *Hardy–Weinberg equation* has medical applications in predicting percentage of carriers of alleles associated with specific genetic diseases.

For example, let's consider the recessive genetic disorder cystic fibrosis, discussed earlier. In the United States, approximately 1 in 2,500 babies of Northern European decent are born with cystic fibrosis. Using the Hardy-Weinberg equation, we can determine approximately what percent of the Northern European population will be heterozygous (carriers) of the cystic fibrosis allele. Since 1/2,500 individuals are homozygous recessive, $q^2 = 1/2{,}500$ or 0.0004 (0.04 percent) and $q = \sqrt{0.0004}$ or 0.02. Because the sum of the alleles $p + q = 1$, $p = 1 - q = 1 - 0.02 = 0.98$, or 98%. Knowing that 1/2,500 individuals are homozygous recessive (aa), we know that the rest have a normal phenotype. Let's use our equation to calculate the number of carriers of this recessive allele: $2pq = 2(0.02)(0.98) = 0.0392$. Thus, the frequency of the heterozygotes (carriers) is 0.0392 or 3.92 percent.

15.3.2. Additional Mechanisms Underlying Evolution

Gene Flow

Throughout geologic time, organisms have been on the move. Ancestral horses emigrated out of North America, across the Bering Land Bridge, into Siberia about 10,000 years ago. From there, the horse spread throughout Asia, the Middle East, Northern Africa, and Europe, only to return to North America with the coming of the Spanish just a few centuries ago (see Figure 15.1). *Gene flow* involves the movement of alleles from one population to another. Many cases of gene flow are obvious, as in the case of the horse. In other instances, gene flow is more subtle, such as in pollen and seeds blowing in the wind, tiny invertebrates traveling on the feet of birds, pests entering a region on ornamental plants, and parasites moving from one population to another. In humans, gene flow is quite evident as people migrate from one region to another because of exploration, poverty, hunger, disease, overcrowding, and war. As a result of such movement, new alleles may be introduced into an established population. Thus, new phenotypes appear in the original population. For example, the green-eye phenotype was introduced into India by soldiers of Alexander the Great. Tay–Sach's disease, which primarily affected Ashkenazi Jews, now occurs in Louisiana Cajuns. Gene flow can act to spread from one population to another mutations that are beneficial, neutral, or detrimental. In any case, phenotypic and underlying genotypic ratios and the population's success are affected.

FIGURE 15.1 *Hydracotherium*, the first horse that appeared in North America

Genetic Drift

The equilibrium predicted mathematically in Hardy–Weinberg may occur in large populations. However, *genetic drift*, random changes in the genetic frequencies of a population *by chance alone*, is a force that particularly affects small populations. Both beneficial and harmful alleles are subject to genetic drift and potentially can be lost more easily from a small population. For example, in a small population of squirrels there exists two phenotypes, one for black fur (BB, Bb) and one for gray fur (bb). A random event, not related to fur color, that drastically reduces the squirrel population with gray fur (bb) can significantly affect the available alleles for the next generation. If the trend for eliminating the gray allele continues, the gray phenotype can be totally lost over time.

In a special case of genetic drift, the *founder effect*, genetic variation is reduced when a population descends from a small number of ancestors. An excellent example can be seen in the Afrikaners of South Africa.

APPLYING SCIENCE TO LIFE 1 Sharing Genes: No Dutch Treat

In some societies, locals thought people who showed strange behaviors, avoided light, had periodic "fits of madness," and so on to be vampires. Actually, many of these "vampires" suffered from a genetic disorder—*porphyria*. In medicine, there are several distinct types of porphyria, all of which result from a faulty mechanism responsible for building heme, a molecule essential in constructing *hemoglobin*. In blood cells, the protein hemoglobin is important in carrying oxygen throughout the body. As the result of a faulty gene, an intermediate in hemoglobin synthesis known as *porphyrin* builds to toxic levels. Several of the symptoms that are common in porphyria patients include photosensitivity, development of blisters and sores when exposed to sunlight, severe gastrointestinal pain and constipation, wine-colored urine, fits of what appears to be madness, epilepsy, convulsions, *erythrodontia* (red-colored teeth), and chronic anemia. Many individuals also exhibit an aversion to garlic. Garlic is known to stimulate blood cell production, increasing porphyrin levels in the body with the defect (APF 2020).

Variegate porphyria is a homozygous dominant metabolic genetic disorder that occurs in 1/100,000 people of European descent. The highest incidence worldwide (3/1,000) occurs in South African individuals of Dutch ancestry (*Afrikaners*). The gene for variegate porphyria (PPOX) is located on the long arm of chromosome 1, in which there are over 100 known mutations. However, in 95 percent of the variegate porphyria patients studied in South Africa, one mutation (R59W) is the most common (NIH 2020). The high number of individuals with porphyria in South Africa can be traced to 1688, when two Dutch settlers within a small Dutch community married in Cape Town, South Africa. Both of these individuals had porphyria because of the dominant nature of the disorder. The couple had eight children, four of whom had porphyria. Early Dutch settlers were a small group, and with so much inbreeding, there is an abnormally high disease incidence. The founder effect is also apparent in oceanic islands when a new population is derived from a few initial founders, like those of the Hawaiian and Galapagos Islands.

As another special case of genetic drift, a *population bottleneck* can result in a loss of genetic diversity. In a bottleneck event, a population undergoes a significant reduction in size, caused by disease, natural disasters, habitat destruction, overhunting, or overfishing. Following such an event, where some alleles might be lost, the surviving individuals may have allele frequencies different from the original population. Inbreeding within the smaller population further alters allelic frequencies and ultimately phenotypes. The African cheetah exhibits a high degree of genetic uniformity, making it difficult to respond to environmental and other changes, such as disease. Today's wild cheetahs are highly inbred. Cheetahs can attribute their current lack of genetic diversity to several population bottlenecks, the most severe one at the end of the last ice age (10,000 years ago), when there was a major extinction of various mammals

PoI 1. Relate the founder effect to the following:

a. Huntington's chorea
b. Maple syrup urine disease
c. Iguanas in the Galapagos Islands
d. Fumarase deficiency

FIGURE 15.2 Cheetah, victim of multiple population bottlenecks

(see Figure 15.2). This bottleneck decreased the genetic variety in the remaining, small population of cheetahs. Today, climate change, habitat loss, and poaching further complicate the plight of the cheetah.

PoI 2. Relate the founder effect to the following:

a. Florida panther
b. Black rhinoceros
c. Cheetah

PoI 3. *Consanguinity*, or inbreeding, is dangerous to a population because it increases the chances of potentially deleterious alleles to combine. Describe the following as related to consanguinity and its negative effects:

a. The Hapsburg dynasty, including Charles II of Spain
b. Thoroughbred horses
c. English bulldogs

Small populations may contain a *genetic load* in which certain lethal genes and deleterious allelic combinations can result in decline of the average fitness of a population, leading to fewer viable offspring.

APPLYING SCIENCE TO LIFE 2 A Devastating Disorder in Horses

American paint horses are beautiful animals that display bold and very colorful markings (see Figure 15.3). Overo lethal white foal syndrome is a genetically lethal recessive inherited disorder. An affected foal lacks melanocytes (pigment cells) and has white or nearly all white hair with blue eyes and pinkish skin. In addition, many of these foals are deaf. At birth, the horse appears normal and successfully stands and suckles its mother. About 12 hours after birth, it begins to have painful gastrointestinal problems and severe colic. If not euthanized by a veterinarian, the affected foal will suffer terribly and die from a ruptured bowel (UFAW 2020).

FIGURE 15.3 American paint horse

QRQ 6. Suppose in a population of 1,680 individuals, a particular gene has a dominant allele A and a recessive allele a. Moreover, suppose the frequencies for homogeneous dominant and heterogeneous genotypes are 0.125 and 0.625, respectively. Give the following:

a. percentage of the population with genotype AA
b. percentage of the population with genotype Aa (or aA)
c. homogeneous recessive genotype frequency
d. number of individuals with genotype aa
e. dominant phenotypic frequency
f. $p = freq(A)$
g. allelic frequency of a

QRQ 7. Suppose a particular gene in a population has dominant and recessive alleles A and a, respectively, and 49 in 10,000 individuals are homozygous recessive. Give the following:

a. percentage of the population with dominant phenotype
b. $q = freq(a)$, as a decimal number
c. $p = freq(A)$, as a decimal number
d. frequency of carriers in the population
e. percentage of carriers in the population

QRQ 8. Select the appropriate term corresponding to each description: allele flow, allele load, allele reservoir, bottleneck, founder effect, gene flow, gene pool, genetic drift, genetic load, originator, sphincter

a. certain lethal genes and deleterious allelic combinations can result in decline of the average fitness of a population, leading to fewer viable offspring
b. event in which a population undergoes a significant reduction in size, causing the loss of some alleles of a gene

c. movement of alleles from one population to another
d. random changes in genetic frequencies of a population by chance alone
e. reduction of genetic variation when a population descends from a small number of ancestors
f. set of all alleles available in a population

15.4. Classifying Diversity

Have you ever wondered about how many species of living things exist on Earth, from microscopic bacteria to the great blue whale? A *species* is classically defined as a group of organisms that can potentially or actually interbreed in the natural environment and produce fertile offspring. However, this idea does not apply to asexually reproducing organisms, such as bacteria, or hybrids that can breed and produce viable progeny, such as the beefalo, a cross between the domestic cow and the American bison. Moreover, the fossil record yields the information that modern humans (*Homo sapiens*) interbred with Neanderthals (*Homo neanderthalensis*) in Europe. Keep in mind when you tally your species count that all humans, including *Homo sapiens* and *Homo neanderthalensis*, are one species. Also, all domesticated dogs are in one species, all nonwild cats in another, all ginkgo trees in a separate one, and all bald eagles in their own species.

Calculating the number of living species (*extant*) on Earth is virtually impossible. To date, only 1.7 million species have been described by scientists. However, estimates vary widely, from 8.7 million to perhaps over a billion species. Shockingly, there are well over 350,000 described species of beetles and well over 5,000 species of mammals. Why such a range? Many organisms are too minute or elusive to describe, and new molecular technologies have shown that defining a species can be difficult. The world-renowned biologist E. O. Wilson estimates that as much as 80 percent of life on Earth remains undiscovered. It is estimated that over 99 percent of species that have lived on Earth are now *extinct*. By conservative estimates, the number of extinct species far exceeds two billion. For reference, only two species of elephantine creatures are described today, but over 150 species have been known to exist over the past 50 million years.

Taxonomy, or the discipline of biology involving the classification of organisms, is responsible for the identification, classifying, and naming of living things. Its roots can be traced back to ancient humans that needed a fundamental knowledge of the life around them in order to survive and satisfy their curiosity. Today, as in the past, taxonomy has the challenging responsibility of making sense of diversity. In addition, taxonomy is important in maintaining biodiversity, monitoring environmental changes, managing invasive species, understanding evolution, and identifying plant, animal, and human pathogens.

In 2007, biology celebrated the 300th birthday of the "Father of Taxonomy," *Carolus Linnaeus*, knighted as Carl von Linne. As a young man in Sweden, Linnaeus showed much potential in botany, despite being just an average student. In fact, Carolus was not considered university material by most of his teachers. Despite this, he attended several universities, studying botany and medicine. In 1735, he attended the University of Hardeswik, known for being a diploma mill; defended his previously written thesis on malaria; diagnosed a patient; and became a physician in just two weeks!

Through the years, Linnaeus practiced as a physician, served as a professor of botany, became the first president of the Royal Swedish Academy of Sciences, and revolutionized taxonomy.

APPLYING SCIENCE TO LIFE 3 A Little More on Carolus Linnaeus

Linnaeus was not a modest individual, as evidenced in these quotes about himself:

"No one has been a greater botanist or zoologist. No one has written more books, more correctly, more methodically, from personal experience. No one has more completely changed a whole science and started a new epoch."

"God created, but Linnaeus ordered" (Linnaeus' Philosophia Botanica 2005)

Linnaeus also modified the Celsius temperature scale, developed by Anders Celsius, by inverting the original scale to 0 degrees representing the freezing point and 100 degrees representing the boiling point of water. He is also given credit for inventing index cards to accommodate his ever-growing list of species. Dr. Linnaeus properly identified bats as mammals instead of birds, whales as mammals instead of fish, and correctly classified humans as primates. Additionally, he was one of the founders of ecology, noting the relationship of organisms to their environment.

Informally, common names are used to describe organisms, such as the Siberian tiger and the bald cypress tree. However, common names can become confusing, because a specific organism may have more than one common name. For example, a mountain lion may also be called a cougar, panther, puma, or catamount. *Binomial nomenclature* is a formalized system of uniquely naming organisms by assigning them two names. The first name is the *genus*, and the second name is the *species*. For example, the genus-species name of a mountain lion is *Puma concolor* (see Figure 15.4). The genus usually refers to a small group of closely related organisms and usually serves as a noun. The species name refers to a particular group belonging to the same genus and serves as an adjective. This two-name description is the *scientific name* of an organism. In writing scientific names, the entire scientific name has to be in italics or underlined; the genus name is always first, with the first letter capitalized; and the species name is in lower case. Originally developed in the 1500s, use of scientific names were formally adopted by biologists due to the contributions of Carolus Linnaeus.

FIGURE 15.4 The mountain lion, cougar, or puma has many common names but only one scientific name, *Puma concolor.*

One of the great contributions of Linnaeus was the development of *taxonomical ranks,* or a *hierarchical* means of organizing living things. In biology, a *taxon* (plural *taxa*) is a core unit of classification. Each taxon includes organisms that share defining characteristics in common. Taxa are organized from general characteristics to specific characteristics.

Eight mandatory ranks are in a taxa: *domain, kingdom, phylum, class, order, family, genus*, and *species* (see Figure 15.5). As an example of the highest rank, the domain Eukarya comprises organisms that possess eukaryotic cells, such as mushrooms, apple trees, dolphins, humans, and American badgers. At the second level, an example of a kingdom, Animalia, consists of multicellular heterotrophic organisms, such as mollusks, coral, squid, earthworms, grasshoppers, robins, and giraffes.

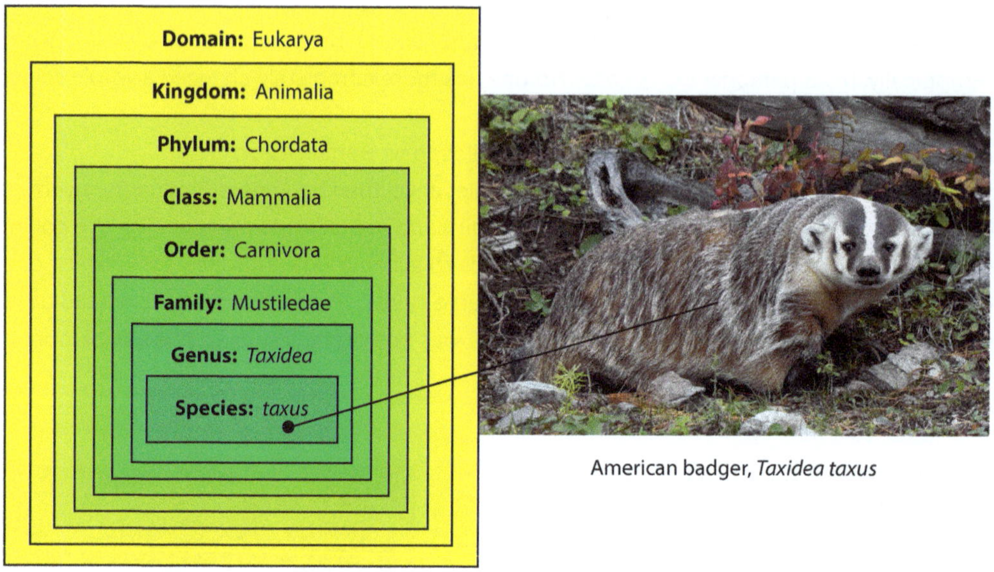

FIGURE 15.5 Taxon for the American badger, *Taxidea taxus*

Phylum (plural *phyla*), is a more specific taxonomic grouping of organisms and follows kingdom. Approximately 36 phyla of animals have been established. For example, the phylum Mollusca contains clams, oysters, snails, slugs, whelks, squid, and octopi, as well as mollusks, while sponges are in the phylum Porifera. Phyla are subdivided into *classes* according to evolutionary relationships, and, as indicated in Figure 15.5, classes are subdivided into orders, orders into families, families into *genera* (singular *genus*), and finally, genera into species.

In the human family tree, there are several genera of note, including *Sahelanthropus* (at seven million years ago, perhaps, was at a node or split between humans and chimps) and *Australopithecus* (including several species of hominids closer to the genus of modern humans, *Homo*). The fossil record yields many species of the genus *Homo*, including *Homo erectus, Homo habilis, Homo rudolphensis, Homo neanderthalensis*, and *Homo sapiens*. Modern humans are also placed in the *Subspecies* of *sapiens*. Thus, the classification of humans is as follows:

- Domain - Eukarya
- Kingdom - Animalia
- Phylum - Chordata
- Subphylum - Vertebrata
- Class - Mammalia

- Infraclass - Eutheria
- Order - Primates
- Infraorder - Catarrhini
- Family - Hominidae
- Genus - *Homo*
- Species - *sapiens*
- Subspecies - *sapiens*

The *Linnaean system* of taxonomy is a user-friendly means of making sense of diversity and adaptation. However, this taxonomy was based on characteristics other than new genetic and molecular information. Genetics and molecular biology are altering many of the classic ideas of taxonomy. *Phylogeny,* or the study of the evolutionary history of a taxonomic group of organisms, is opening new doors in understanding life on Earth. Today, *phylogenetic trees,* or *phylogenetic diagrams,* illustrate, usually with respect to evolutionary time, the relationship between species and the divergence from a common ancestor. *Cladograms,* such as in Figure 15.6, are visual tools that aid scientists in understanding how organisms are related based on derived traits, such as feathers, and their common ancestry.

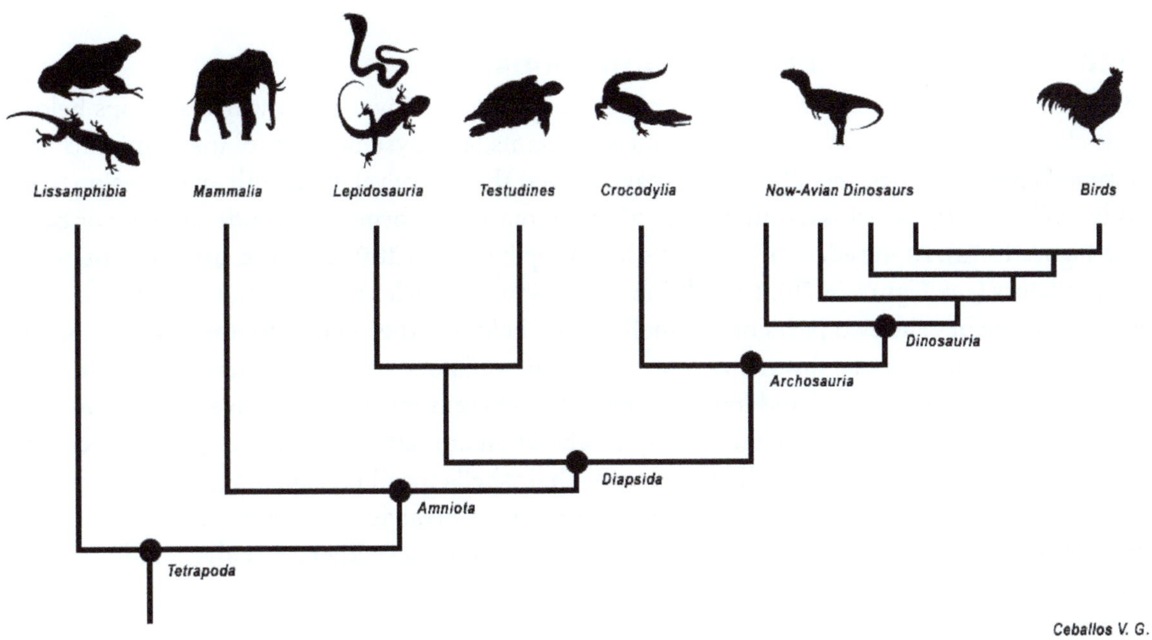

FIGURE 15.6 Simple cladogram of Tetrapods

QRQ 9. Select all that are part of the formatting for scientific names: entire scientific name italicized or underlined, first letter of first word capitalized, first letter of second word capitalized, genus first, genus second, only first word italicized, phylum first, phylum second, species first, species second

QRQ 10. Place in order, from largest (highest rank) to smallest (lowest rank), the taxa ranks: class (C), domain (D), family (F), genus (G), kingdom (K), order (O), phylum (P), species (S)

QRQ 11. Give the scientific name of humans.

QRQ 12. A visual tool that aids scientists in understanding how organisms are related based on derived traits, such as feathers, and their common ancestry is called a(n) ______.

15.5. Adaptations

All organisms are essentially collections of adaptations, and you can find excellent examples in all major phyla. Adaptations are easily seen in animal phyla, such as the Arthropoda (e.g., insects, crustaceans) or the Chordata (e.g., vertebrates). If we return to our consideration of the mollusks, we find some rather remarkable examples of adaptations. For illustrations, we consider some of the adaptations of scallops (class Bivalvia), squids (class Cephalopoda), and nudibranchs (class Gastropoda).

15.5.1. The Scallop and Its Adaptations

Many of the shells you see on the beach come from the class Bivalvia, to which scallops belong. This class includes not only scallops but mussels, clams, and oysters. Most of these animals have common characteristics, including a two-part, hinged shell (*bivalve*) lined with an extension of the body wall (*mantle*), which surrounds the soft body mass and forms the *mantle chamber* or *cavity* (see Figure 15.7a). Along edges of its mantle, a scallop has up to 200 blue eyes, used to spot potential predators (see Figure 15.7b). In addition, many short tentacles sense motion and chemicals in the environment. The two parts of the shell can be held together tightly by the contraction of a powerful adductor muscle, thwarting some potential predators. Scallops are unusual for bivalves in that adults are unattached and found in the sand, whereas most bivalves live attached to some substrate or buried in the mud. Unlike many bivalves that are rather slow moving, scallops can, for short distances, move rather rapidly by clapping their shells together, expelling jets of water. This ability to move quickly helps them avoid predation. Many bivalves, like clams, have a well-developed muscular foot for burying themselves, but scallops have a smaller muscular foot for various local movements.

Specialized to filter and trap food suspended in water, the gills function for both gas exchange and food gathering. Water entering the mantle chamber carries not only oxygen but also small planktonic organisms and bits of organic matter. The water flows over the gills, which provide a thin surface for gas exchange. Scallop gills filter the water and capture food particles. The cells of the gills produce mucus, and food particles trapped in mucus strands are transported along ciliated grooves toward the mouth. Unlike some classes of mollusks, bivalves lack a head, jaws, and rasping "tongue" (*radula*) (Kegel 2020).

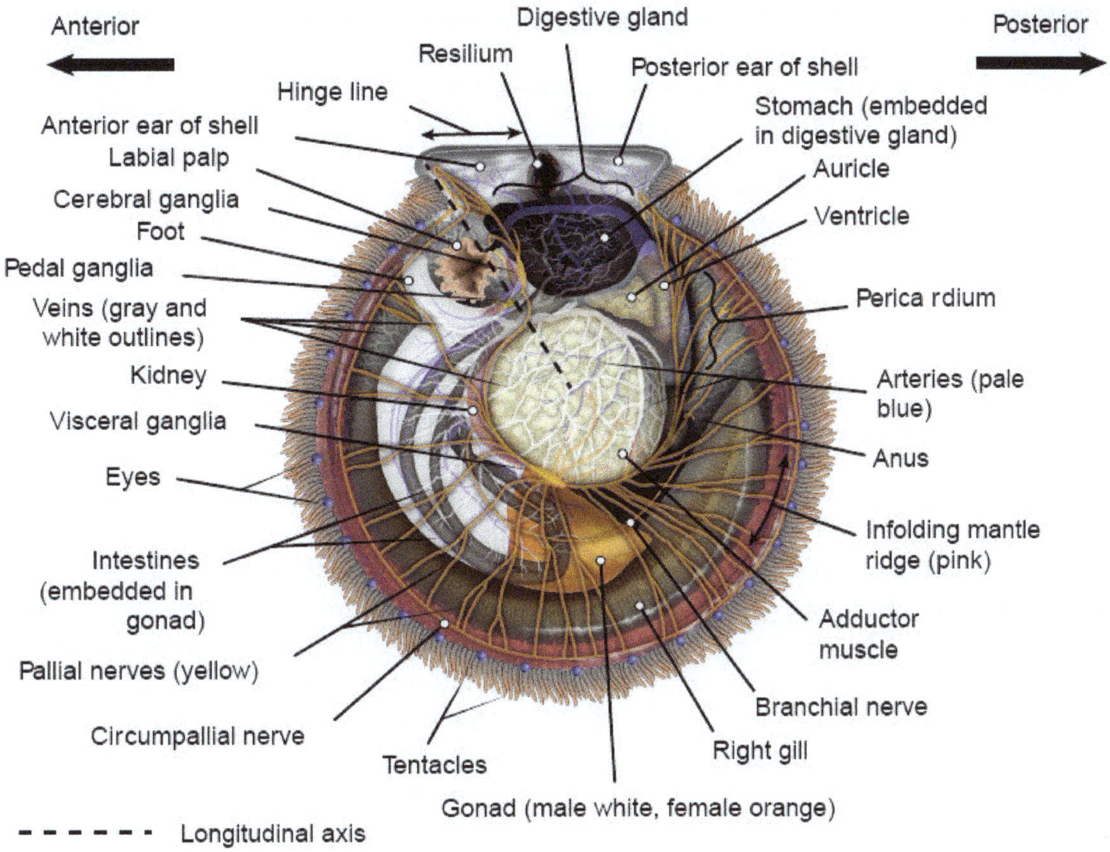

FIGURE 15.7a Scallop anatomy

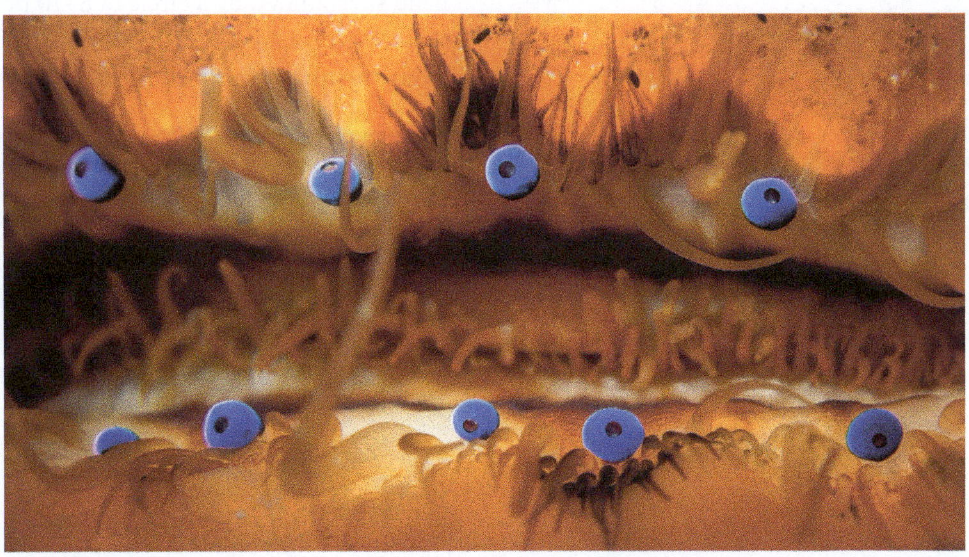

FIGURE 15.7b Edge of scallop mantle, showing short tentacles and blue eyes

QRQ 13. The extension of the body wall that lines the shell of a clam is called the ______.

QRQ 14. Select the following that are function(s) of the "foot" in bivalves: breathing, digging, eating, feeling, filtering, moving locally, smelling, tasting

QRQ 15. Scallops, clams, mussels, and oysters don't need jaws or a radula because they ______. Possible answers: absorb nutrients through bivalves, filter out food from the water, consume only soft food, have a muscular stomach, store grains of sand that grind food, swallow food whole.

PoI 4. What characteristics of bivalve mollusks would qualify as adaptations? Why?

PoI 5. What advantage might the body shape and reduction of the shell have for the cephalopods?

15.5.2. The Squid and Its Adaptations

Perhaps you have eaten calamari. In restaurants, calamari are often cut up into smaller, ring-shaped pieces and served poached or boiled atop salads or fried. What you are eating is squid. The squid belongs to the class Cephalopoda, which also includes nautilus, cuttlefish, and octopus. Most of the members of this group have streamlined bodies and a reduced and internalized shell (see Figure 15.8). Some cephalopods, like the octopus, have even lost their shell.

Squid are fascinating animals whose adaptations have made them one of the fastest-moving and smartest *invertebrates* (animals without backbones) in the ocean. Compared to other classes of Mollusca, the cephalopods have huge brains and a highly developed sensory system. At the anterior end of the elongated body is a head surrounded by a circle of 8–10 prehensile *tentacles* (or arms), which are highly modified from the ancestral foot. In the middle of the

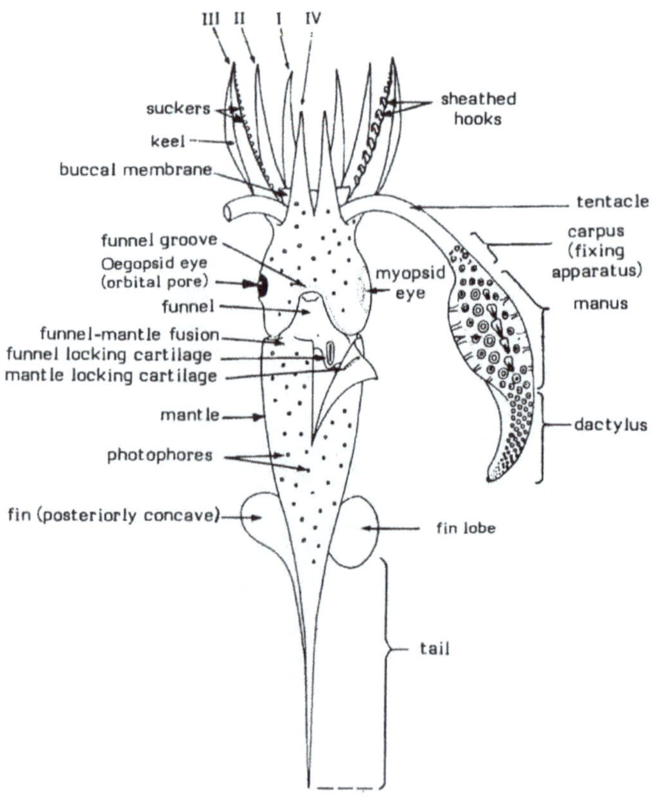

FIGURE 15.8a Squid's anatomy

FIGURE 15.8b Caribbean reef squid

tentacles is the *beak* (jaws), and inside the mouth is the *radula* (a tonguelike structure covered in tiny teeth, used for scraping off flesh of prey) . Also, beneath the head is a tubular *siphon*, through which water can enter and exit. The squid uses this siphon to move water for gas exchange and waste disposal. To move quickly, the squid can contract its body walls, forcing water as a jet through the siphon and propelling the animal in the desired direction (Wood 2020; Rutgers University 2020).

PoI 6. How do squid feed? What structural and functional features have adapted squid to this feeding method?

Unlike the bivalves, cephalopods have a recognizable head. The most prominent structures on their heads are the well-developed eyes, which look eerily like ours. Squid eyes are actually structured differently and represent an excellent example of *convergent evolution* (where distantly related organisms develop similar traits by different evolutionary steps) (Bunje 2003).

Squid and other cephalopods are also capable of very intricate and abrupt color and pattern changes of their skin. The rapid color changes are mediated partially by a collection of pigmented organs called *chromatophores,* which include a central chromatophore cell containing a sac of pigment, associated with nerve and muscle cells. There are chromatophores with red, yellow/orange, or brown/black pigments. The attached muscle cells contract under signals from the nerves, expanding the size of the sac and changing the color of the skin surface. Each chromatophore cell is under the control of its own nerve cell or cells. So the squid can expand certain chromatophores almost instantly, while leaving the others contracted. Additional cell types augment the diversity and complexity of colors and patterns, including reflective chromatophores containing silver or iridescent pigments, and fixed white chromatophores. There are even some animals that possess chromatophores that emit light. If you combine the assortment of pigmented organs and cells and place them in various ways on a surface with changeable contour, you can get skin that may vary quite considerably in color and pattern (see Figure 15.8b) (Hanlon 2007; Hanlon et al. 2009; Mäthger et al. 2009; Sutherland et al. 2008).

QRQ 16. Select which of the following are statements of how cephalopods differ from bivalves: arms/tentacles around the mouth for seizing/holding prey, beak and radula for feeding, complex colors and patterns of skin surface, contraction of mantle forms jet of water for rapid movements, diggers, highly developed brain and nervous system with sophisticated eye design, hydrodynamic shape for fast swimming, prey are particulate matter/plankton (not fish), protective shells, sessile, small size

QRQ 17. Rapid color changes in squid are mediated partially by a collection of pigmented organs called ______.

15.5.3. The Nudibranch and Its Adaptations

If you examine Figure 15.9, you will see a spectacular but little-known mollusk called the *nudibranch*. It is incredible that this animal belongs to the same phylum as clams and squid, but if you could watch its development, you would see some common features.

PoI 7. Squid are adapted to change their surface colors and patterns. How might this adaptation benefit the squid?

Nudibranchs belong to the subclass Opisthobranchia, a name that means the gills lie behind the heart. The opisthobranchs belong to a widely distributed class of Mollusca, the Gastropoda. Gastropods include organisms that we know as snails, and most of the members of this group produce

FIGURE 15.9 Nudibranch

a shell that has a spiral shape. A few opisthobranchs produce a protective external shell, but most either have only a small or internalized shell or no shell at all. The nudibranchs lack a shell and gills. Instead, the outer body surface provides an area for gas exchange. Some nudibranchs, like the one in Figure 15.9, have dorsal extensions of the body surface, called *cerata*, enclosing *evaginations* (outpocketings) of the digestive tract.

Despite their small size, nudibranchs are mostly keen predators feeding on other marine organisms. Individual species' tastes may be quite specific. Some nudibranchs feed on anemones and coral polyps, whereas others feed on sponges or tunicates. These prey animals are all *sessile* (fixed in place) and cannot evade the onslaught of hungry nudibranchs, which bear down on them with toothed tongues (radulas) and/or strong jaws. Although unable to move out of the way, many of these prey organisms are not totally helpless. For instance, the anemones and coral polyp tentacles contain cells that produce encapsulated structures called *nematocysts*. Nematocyts are small capsules having a coiled, hollow thread and toxins. When certain chemical or tactile signals are received, small, hairlike triggers are activated, and the thread is everted and penetrates the target organism, injecting toxin. Some types of nematocysts entangle prey instead. Nematocysts may help them subdue prey but may also fend off predators.

Apparently, to some nudibranchs nothing tastes better than that coral polyp, and they use mucus to take the sting out of their meal. The nudibranch is able to inhibit the release of many of the coral polyp's nematocysts and to transport these intact nematocysts into the ends of their cerata. So the nudibranch can maintain its own battery of "borrowed" nematocysts for subduing prey and for defense against predators. Besides nematocysts, some nudibranchs are often able to accumulate toxins in their skin from other prey species.

Nudibranchs are often brightly colored, frequently to match the colors of their prey. On the other hand, bright colors in nature may mean "danger, do not tangle with me." Thus, potential predators may be warned that nudibranchs contain various toxins that will not taste to their liking.

Like other gastropods, the nudibranchs have a muscular foot. Depending on where they live, (sand versus reef) and their size, the foot may work a little differently. Large nudibranchs that move across a sandy bottom typically have a broad foot that moves by waves of muscular contractions, moving posteriorly from the front. Smaller animals that inhabit coral reefs or mats of algae may move along a mucus track using cilia. If there are strong currents, the creatures may have special glands that secrete an especially sticky mucus. Trails contain chemicals that may help them communicate with other nudibranchs or potentially aid detection by predators (Rudman 2001; Wägele and Klussmann-Kolb).

QRQ 18. Select all phrases that apply to the cerata: accumulate nematocysts/toxins useful for subduing prey, dorsal extensions of body surface, for feeding, for locomotion, for reproduction, for surface respiration, outfoldings of the digestive tract, pigmented organs, prehensile arms, rasping "tongue," store stinging cells for defense

The bivalve, cephalopod, and nudibranch display a number of specialized adaptations, including those for feeding, defense, and movement. From a marine ancestral form, these and other groups of mollusks have evolved over millions of years. Under natural selection, changes that have taken place over time have allowed the mollusks to survive, diverge, and exploit different types of habitats. Examination of other phyla would reveal that they all have developed adaptations that have enabled them to do likewise. Of course, many adaptations either have no effect on fitness or reduce it. An animal with an adaptation that reduces fitness tends to die out.

PoI 8. What enables nudibranchs to tolerate the nematocysts/toxins of their prey?

PoI 9. What adaptations for locomotion, defense, and feeding characterize the following example organisms belonging to the subphylum Vertebrates: bony fish, frogs, birds, snakes, bats, and whales?

15.5.4. Alternative Sets of Adaptations

Just look around—the biological world is filled with amazing and novel adaptions. One of the most interesting animals on Earth is often overlooked or taken for granted—the ant. Fossil evidence indicates that ants evolved from wasp-like ancestors approximately 140 million years ago. A conservative estimate of the number of species of ants is 12,000. Ants are social insects, living in colonies of a few dozen to millions of individuals. Within the colony, ants exhibit a division of labor and a caste system consisting of a queen, workers, and soldiers. Generally, ants are omnivorous and eat a variety of plant and animal products. Ants exhibit complex communications by means of touch, sounds, and *pheromones* (chemical signals that are capable of influencing the behavior of animals). As you already may know, ants can employ a successful means of defense by biting, stinging, or both. Some species of ants are beneficial to humans by controlling pests, and some species can even serve as crude surgical sutures. Silver ants are one of many species of these remarkable creatures.

APPLYING SCIENCE TO LIFE 4 Reflecting on Silver Ants

Few animals can make a life under the severe environment of the Sahara Desert. It is hot (up to 70 °C [158 °F]) and dry, with little available food or water (Evarts 2015). It seems incredible that any animal could live under such inhospitable conditions. Even with summer midday sand temperatures surpassing 60 °C, Sahara silver ants emerge from their burrows to race about to forage for food. These ants have a critical thermal maximum of 53.6 ± 0.8 °C, which is one of the highest among terrestrial animals (Wehner, Marsh, and Wehner 1992). At this maximum for any animal, death will soon follow if the temperature does not decrease.

The scant food resources available include other, less temperature-tolerant species of insects that die in the oppressive heat. Maintaining their own body temperature within a range of 48–51 °C, the ant must move quickly, searching as far as they dare. (Your body temperature is around 37 °C). What are your body's ways of cooling down?

Silver ants have several adaptations and at least one advantage that favors their success. First, their main predator (a lizard) must retreat during the hottest part of the day. The ants cautiously emerge from their burrow after the lizards have left. Then the ants have maybe ten minutes to find some food and transport it back to their lair. In their scurrying about, they search by themselves, touching the sand as little as possible, running with four, or sometimes even two, of their long legs. Unlike most ants, these ants don't rely on chemical trails to find their way back to the nest; they probably don't have time. However, they have a reliable navigation system based on the position of the sun and the location of their nest, which allows them to return directly and quickly. Some of the time in the sun must be spent cooling a little by crawling up on a dead plant branch. Scientists term this a "thermal respite."

The silver color of these ants results from their upper bodies being covered in tiny hairs that in cross section are triangular. These hairs are brilliantly reflective of light rays, and much of the light that passes into the hairs is also reflected internally as heat. Reflected light gives the ants their silver color and name. Most importantly, this reflection reduces absorbed light and helps the ants remain slightly cooler. During experiments where ants with hairs were compared to those where the hairs had been removed, those with hairs were able to maintain a body temperature that was significantly lower than those without hairs (Shi et al. 2015).

One last remarkable adaptation these ants have to survive the heat are "heat-shock" proteins (HSPs). These proteins are important chaperones for newly synthesized proteins, because the HSPs help them to form a functional configuration (shape). HSPs can also help a protein regain its normal shape if it should be inactivated by some factor like high temperature. All plants and animals that have been studied produce these proteins, showing how important and what an ancient adaptation they are. It isn't unusual that silver ants have HSPs, but it is significant that they are already present before they leave their nest for the very hot desert. Most HSPs are synthesized under some condition of "shock" or environmental change. Because making the HSPs takes time in most cases, the preparative production of them without some shock gives the ants a head start in resisting destruction of needed proteins in their muscles and mitochondria, two things that definitely need to work well if you are needing to move very fast (Redmond 2018; Sokol 2015).

Although the possibilities are seemingly endless, we now consider two other remarkable examples of adaptation and diversity, one from the plant kingdom and one from the animal.

APPLYING SCIENCE TO LIFE 5 Fooled Ya!

Another, less well-known example of adaptation mimicry, mentioned in module 13, occurs in a family of orchids found only in southwestern Australia. Known as the hammer orchids, these orchids attract pollinators using fragrances that resemble those produced by female thynnid wasps (see Figure 15.10). The flowers of these orchids are highly modified to form *labellum* (lip of the orchid flower, on which pollinators land), which is attached to a hinged stem. The labellum resembles the shape and color of the wasp pollinator female, which is usually flightless. In normal mating, the male mounts the female and carries her off while copulating with her. Instead, the male, attracted by the fragrance produced by the orchid, attaches to the labellum and tries to fly the female away. At this point, the hinge moves to turn the male over until it contacts the *stigma* (upper part of the female flower parts that receive pollen). If the wasp has already visited another orchid, the flower obtains pollen deposited by the male after contact with the sticky stigma. The pollen will germinate, producing a pollen tube that will transfer the other orchid's male gametes to the *ovule*

FIGURE 15.10 Hammer orchid flower

(female structure that gives rise to the female gametes). Following fertilization, the *zygote* (fertilized egg) will develop into an embryo, which is enclosed within a seed coat and stored food (*endosperm*) to form a seed (Ogden 2015; Young 2014).

APPLYING SCIENCE TO LIFE 6 One Strange Bird: The Kakapo Parrot

Visit the isolated primeval forest of several barrier islands off the coast of New Zealand, and you may be lucky enough to get a glance at one of the strangest birds in the world. *Ornithologists* (biologists that study birds) are amazed by this highly endangered species. The kakapo, which translates from Latin as "owl-face soft-feather," goes by several common names, including owl parrot and night parrot.

This rather odd-looking bird reminds the observer of a comedic cross between an owl and a Muppet (see Figure 15.11). The kakapo is the only flightless parrot in the world and uses its oversized feet and strong legs to waddle through the terrestrial environment. Its main food source is in trees, so this oddity of the bird world uses its feet and bill to climb trees. Its vestigial wings are merely used as a parachute and for balance. In the world of parrots, the kakapo is very large, measuring over 25 inches in length and weighing up to nine pounds. The kakapo is also the only nocturnal parrot in the world, roosting in trees or hiding on the ground until nightfall. The dorsal feathers of the bird are an attractive mottled moss green, with dark brown blotches and black bars. The feathers on the ventral side, tail, and chest are a mottled yellow-green. The feathers are unusually soft, and biologists claim that they have a pleasant, sweet-musky odor. Speaking of smell, these strange birds have evolved a keen sense of smell! The face of the kakapo is adorable and features a brown facial disc similar to an owl. It also features forward facing eyes that are critical for low-light vision and fine feathers on the face for tactile purposes. Facial feathers are also important in focusing sounds to the ears. The ivory-colored beak is large and distinct. Males and females have similar coloration, but the female is slightly smaller.

FIGURE 15.11 New Zealand kakapo (*Strigops habroptilus*)

Kakapos possess a charismatic personality. Many of these interesting birds seek human company and may even preen or try to mate with an observer. When startled, they stand erect and freeze, trying to blend into the background. To "win the wing" of the female, a male kakapo emits low-frequency booms and high-frequency wheezes, called chings, for several nights, all through the night. Depending on the availability of food, kakapos only mate about every four years. The female takes on the duties of a single parent in the realm of kakapos. Kakapos are perhaps the most long-lived birds on the planet, living for up to 100 years.

Kakapos were once very common in New Zealand, but introduced species, such as rats and cats, have drastically reduced their populations. Presently, this remarkable bird is critically endangered. Perhaps less than 160 birds have been accounted for in the wild. However, conservationists are making a valiant effort to save the species. Geneticists are also involved in saving the parrot by developing family trees to ensure that selective breeding does not involve inbreeding. The kakapos are truly a marvel of evolution, but their destiny is now in our hands. Let's only hope that the species can be saved and restored before it is too late (Feldberg 2019; Lazarus 2019; Valentine 2016; Whiteway 2020).

PoI 10. How might biodiversity help human beings advance medicines and improve the ability to feed the hungry and malnourished?

QRQ 19. Select from the following the appropriate term for each description below: chemokine, endosperm, kernel, labellum, microbiologist, ornithologist, orthogist, ovule, pheromone, stigma, zygoid, zygote

a. biologist that studies birds
b. chemical signal capable of influencing behavior of animals
c. female structure that gives rise to the female gametes
d. fertilized egg
e. stored food inside a seed
f. upper part of the female flower parts that receive pollen

Image Credits

Fig. 15.1: Source: https://commons.wikimedia.org/wiki/File:Hyracotherium_Eohippus_hharder.jpg.

Fig. 15.2: Copyright © by Charles J Sharp (CC BY-SA 4.0) at https://commons.wikimedia.org/wiki/File:Cheetah_(Acinonyx_jubatus)_female_2.jpg.

Fig. 15.3: Copyright © by H2O (CC BY-SA 3.0) at https://commons.wikimedia.org/wiki/File:Am_Paint_Horse.jpg.

Fig. 15.6: Copyright © by Ceballos V. G. (CC BY-SA 4.0) at https://commons.wikimedia.org/wiki/File:Tetrapoda_Cladogram.jpg.

Fig. 15.7a: Copyright © by KDS444/KDS4444 (CC BY-SA 3.0) at https://commons.wikimedia.org/wiki/File:Scallop_Diagram2.svg.

Fig. 15.7b: Copyright © by Matthew Krummins (CC BY 2.0) at https://commons.wikimedia.org/wiki/File:Scallop_eyes2.jpg.

Fig. 15.8a: Source: https://commons.wikimedia.org/wiki/File:Composite_diagram_illustrating_basic_squid_features,_ventral_aspect.jpg.

Fig. 15.8b: Copyright © by Betty Wills (CC BY-SA 4.0) at https://commons.wikimedia.org/wiki/File:Squid_colors_2.jpg.

Fig. 15.9: Source: https://www.nsf.gov/news/mmg/mmg_disp.jsp?med_id=71085.

Fig. 15.10: Copyright © by Geoff Derrin (CC BY-SA 4.0) at https://commons.wikimedia.org/wiki/File:Drakaea_labelled.jpg.

Fig. 15.11: Copyright © by Mike Bodie (CC BY 2.0) at https://commons.wikimedia.org/wiki/File:Sirocco_full_length_portrait.jpg.

Fig. 15.12a: Copyright © by Wegmann (CC BY-SA 3.0) at http://commons.wikimedia.org/wiki/File:King_cheetah_fur.JPG.

Fig. 15.12b: Copyright © by Harlequeen (CC BY 2.0) at http://commons.wikimedia.org/wiki/File:Giraffe_markings.jpg.

Fig. 15.12c: Copyright © by Abujoy (CC BY-SA 2.5) at http://commons.wikimedia.org/wiki/File:Detail_Jaguar.JPG.

Fig. 15.12d: Copyright © by Hans Hillewaert (CC BY-SA 3.0) at http://commons.wikimedia.org/wiki/File:Equus_quagga_boehmi_(skin).jpg.

LAB #15

Laboratory Exercises

15.6. Laboratory on Fur Patterns

15.6.1. Preliminary Work

You can likely identify each animal from the pattern of its skin in Figure 15.12. These patterns vary with species and among individuals of the same species. What mechanism is responsible for converting each species' genes into patterns of such beauty and complexity? Scientists have been studying this question for decades, and although we seem to be getting closer to our understanding, there may be more than one answer. One hypothesis to explain these patterns comes from an unexpected source—Alan Turing, a 20th-century British mathematician, far better known as the "father of computer science." More recently, he has gained substantial recognition from his work in cracking the German codes during World War II. Surprisingly, in 1952, Turing wrote a paper, "The Chemical Basis of Morphogenesis," proposing a model for development of organized patterns that form during animal development (Turing 1952). He hypothesized that patterns were established from chemical gradients organized by interacting molecules (*morphogens*). Essentially, Turing suggested that some fairly complex biological organization could be explained (modeled) using a *reaction-diffusion mechanism*, or a model with two steps—chemical reactions followed by diffusion of the chemicals—that repeat over and over.

For a number of years, many experimental biologists rejected Turing's model because it was too simple, we could not isolate the morphogens, and there were no real examples where the model seemed to apply. In 2006, a paper published in *Science* presented one of the first examples of a pattern that seemed to confirm the validity of the Turing hypothesis. Sick, et al. (2006) studied the regulation of hair-follicle patterns from the skin of mice. These scientists found two underlying morphogens, *WNT* (activator) and *DKK* (inhibitor), secreted into the extracellular space, that diffuse away from the source. The larger *WNT* diffuses more slowly. Thus, we have a short-range activator and a long-range inhibitor, much like Turing's generalized model. Now, more work needs to be done to determine quantitatively whether this regulation of mouse hair follicles is a type of system described in Turing's paper.

In this laboratory, we employ Turing's model to generate a number of fur-like patterns. The model illustrates how small changes in parameters can result in a wide diversity of patterns just as, in nature, small evolutionary changes can cause an astounding diversity in life. From the text's website, view the introductory video and "Fur Patterns Expression;" download and open "*Fur Rev.nlogo.*"

FIGURE 15.12 Fur of cheetah, giraffe, jaguar, zebra

1. Click the Info tab at the top of the page and read sections "What Is It?" and "How It Works."

   a. Match each component below with the appropriate item(s) listed: A. activator, B. colorful, C. differentiated, D. inhibitor, E. morphogen, F. pigment cell, G. uncolorful, H. undifferentiated

      ________ melanocyte

      ________ states of melanocytes

      ________ types of "morphogen" that D cells secrete

      ________ morphogen that on its own causes a central cell to become colorful

      ________ morphogen that on its own causes a central cell to become not colorful

   b. Give the one-letter abbreviations for each of the following:

      ________ activator

      ________ colorful

_______ differentiated

_______ inhibitor

_______ not-colorful

_______ undifferentiated

c. _____ Each cell is a central cell for its own neighborhoods. How many neighborhoods do we consider for each cell?

d. _____ Each D cell within the inner neighborhood contributes morphogens of what type (abbreviation) to the central cell?

e. _____ Each D cell within a central cell's ring (that is, within the outer circle but not inside the inner circle), or outer neighborhood, contributes morphogens of what type (abbreviation) to the central cell?

f. _____ If the power of the activators is bigger than the power of the inhibitors, the cell will become **1.** Select from the following: A, B, C, D, E, F, G, H, I, J, U, V, W

g. _____ A colorful cell will become **1.** if the power of the inhibitors is bigger than the power of the activators. Select from the following: A, B, C, D, E, F, G, H, I, J, U, V, W

h. **1.** _____ **2.** _____ w = (**1.** power) / (**2.** power). Possible answers: A. activator, B. colorful, C. differentiated, D. inhibitor, E. morphogen, F. pigment cell, G. uncolorful, H. undifferentiated

i. _____ If $w > 1$, which is greater, the power of the activators (A) or the inhibitors (I)?

j. _____ Let's illustrate the calculation $A - wI$ and its use in the rules. Suppose the ratio, w, is 0.35 and the central cell is D. If the number of D cells in the ring (between the inner and outer circles), or outer neighborhood, of a central cell is six, then the number of inhibitors $I = 6$. Suppose the number of D cells in its inner neighborhood is eight, so the number of activators $A = 8$. Thus, we compute the difference $A - wI = 8 - (0.35)(6) = 5.9$. With this calculation being positive, according to the rules, at the next time step, what should be the central cell's value, D or U?

k. Suppose the ratio, w, is 1.5, the central cell is U, the number of D cells in the outer neighborhood is two, and the number of D cells in its inner neighborhood is two.

_______ Calculate $A - wI$.

_______ Give the central cell's value, D or U, at the next time step.

l. Suppose the ratio, w, is 1, the central cell is U, the number of D cells in the outer neighborhood of a central cell is 1, and the number of D cells in its inner neighborhood is four.

_______ Calculate $A - wI$.

_______ Give the central cell's value, D or U, at the next time step.

m. Suppose the ratio, w, is 0.5, the central cell is U, the number of D cells in the ring of central cell is ten, and the number of D cells in its inner circle is five.

_______ Calculate $A - wI$.

_______ Give the central cell's value, D or U, at the next time step.

n. Give the color for each cell type in the simulation: A. black, B. blue, C. brown, D. cyan, E. green, F. magenta, G. orange, H. pink, I. red, J. white, K. yellow

_______ D cells

_______ U cells

o. ___________ Give the shape(s) of neighborhoods in the simulation: A. circular, B. elliptical, C. pentagonal, D. semicircular, E. triangular

2. Read "How to Use It."

a. ______ percent If we set initial-density slider to 30 percent, what is the chance of a cell being white initially?

b. ______ percent If we set initial-density slider to 30 percent, what is the chance of a cell being black initially?

c. ______ If DENSITY is 50 percent and RATIO is 0.35, what is w, as employed in the section "How It Works" and in the calculation $A - wI$?

d. ______ What do we press to have the model advance a single time step? A. GO, B. RESTART, C. SETUP, D. STEP, E. none

e. ______ If we only change the INITIAL-DENSITY and RATIO sliders, or if we would like to run the model again with the same settings for speed of computation, what should we press instead of SETUP? A. GO, B. RESTART, C. SETUP, D. STEP, E. none

15.6.2. Laboratory Work

3. Read "Things to Notice."

a. ___________ Press DEFAULT-PARMS, SETUP, and GO. Several times, press RESTART and notice the similar, but different, patterns. Press GO to stop. Select all characteristics of the emergent pattern: A. all black, B. all white, C. horizontal stripes, D. lots of spots, E. some spots, F. straight lines, G. vertical stripes, H. wavy

b. ______ Click RESTART. Repeatedly pressing STEP, count the number of steps until the world reaches equilibrium, not changing any more: A. 1–12, B. 13–26, C. 27–40, D. 41–52

c. ______ Changing only RATIO and pressing buttons, not sliders, discover a very different pattern with black spots on a white background. Once you have discovered a number, press RESTART several times to make sure you consistently get a desired pattern. Give the value of RATIO.

d. ______ Repeat part c for a pattern with white spots on a black background.

e. Changing RATIO to 0.26 and INNER-RADIUS-X to 2, discover a zebra pattern, where OUTER-RADIUS-X = OUTER-RADIUS-Y. Changing the radii requires pressing SETUP before the first execution.

_______ INNER-RADIUS-Y

_______ OUTER-RADIUS-X = OUTER-RADIUS-Y

f. ______ Swap the values for INNER-RADIUS-X and INNER-RADIUS-Y. Press SETUP and GO. How does this resulting pattern differ from that for part e? A. all black, B. all white, C. horizontal instead of vertical stripes, D. spots instead of stripes, E. vertical instead of horizontal stripes, F. no change

4. In this question, we investigate why some patterns occur.

a. ___________ Press DEFAULT-PARMS and change INITIAL-DENSITY to 99 percent. Press SETUP and GO. Click RESTART several times and select all that describe the emergent pattern(s): A. all black, B. all white, C. horizontal stripes, D. lots of spots, E. some spots, F. straight lines, G. vertical stripes, H. wavy

b. ______ Vary INITIAL-DENSITY from 0 percent to 99 percent, restarting and running the simulation for each value. Does the initial density of colorful (white) patches seem to impact the general characteristics of the equilibrium pattern, no (N) or yes (Y)?

c. ______ Stop the simulation and change INITIAL-DENSITY to 100 percent. Press SETUP and GO. Click RESTART several times and select all that describe the emergent pattern(s). Use possible choices as in part a. In the next few parts, we investigate why we obtain the wide diversity of patterns, even with such a small change as 1 percent on the initial density or in the value of a radius.

d. ______ To assist us in our exploration, from the text's website, download and open *"Fur Tutor.nlogo."* Much of the Info page is the same as in *"Fur Rev.nlogo."* Under Info, read the first paragraph. This tutorial goes through the process of determining, for the next time step, the color of which patch? A. (−1, 0), B. (1, 0), C. bottom center, D. center, E. left center, F. right center, G. top center

e. ______ Read "How to Use It." What do we press to start the interactive simulation? A. GO, B. SETUP, C. STEP, D. TUTOR, E. TUTOR-FLIP-FLOP

f. Click on the Interface tab and read the notes. Match each component below with the item(s) used to highlight it: A. black, B. green dot, C. red dot, D. white, E. yellow dot

_______ inner neighborhood of Patch (0, 0)

_______ outer neighborhood of Patch (0, 0)

_______ Patch (0, 0)

g. Set INNER-RADIUS-X = INNER-RADIUS-Y = 1, OUTER-RADIUS-X = OUTER-RADIUS-Y = 2, INITIAL-DENSITY = 50 percent, RATIO = 0.30, and TUTOR-FLIP-FLOP to OFF. Start the simulation, and answer the following:

_______ number of patches in inner neighborhood of Patch (0, 0)

_______ number of patches in outer neighborhood of Patch (0, 0)

h. Complete the interactive simulation to determine and view the next value of Patch (0, 0). Press STEP or TUTOR to practice another situation. Repeat the process several times until you are confident. Then answer the same interactive questions, which follow, for Figure 15.13.

_______ Is the patch at (0, 0) of type D or U?

_______ What is the number of activators?

_______ What is the number of inhibitors?

_______ Is $(A - wI)$ positive (P), negative (N), or zero (Z)?

_______ At the next time step, will the center patch be white (D) or black (U)?

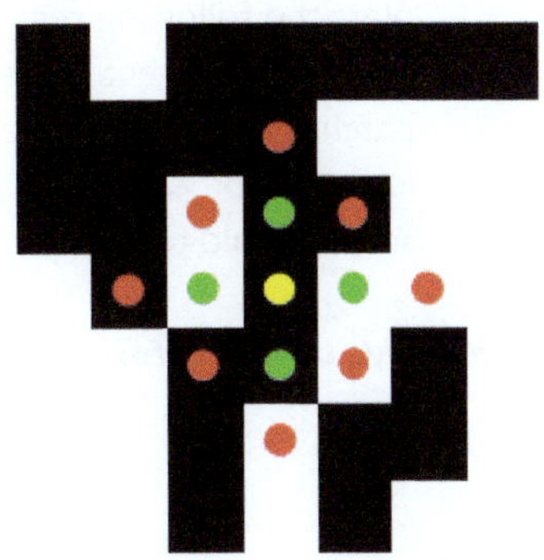

FIGURE 15.13 Circular neighborhoods of radii 1 and 2

i. Suppose the parameter settings are as in part g and the background is all white.

_______ What is the value of $(A - wI)$?

_______ At the next time step, will the center patch be white (D) or black (U)?

j. __________ $\geq w$ Suppose the parameter settings are as in part g. Give the range of w values (to three decimal places) in which a world of all white will remain all white upon execution. Recall in this case that $(A - wI)$ must be nonnegative.

k. Suppose the parameter settings are as in part g and the background is all black.

_______ What is the value of $(A - wI)$?

_______ At the next time step, will the center patch be white (D) or black (U)?

l. Set INNER-RADIUS-X = 2, INNER-RADIUS-Y = 1, OUTER-RADIUS-X = 3, OUTER-RADIUS-Y = 3, INITIAL-DENSITY = 40 percent, RATIO = 0.30, and TUTOR-FLIP-FLOP to OFF. Start the simulation, and answer the following:

_______ number of patches in inner neighborhood of Patch (0, 0)

_______ number of patches in outer neighborhood of Patch (0, 0)

m. Complete the interactive simulation to determine and view the next value of Patch (0, 0). Press STEP or TUTOR to practice another situation. Repeat the process several times until you are confident. Then answer the same interactive questions, which follow, for Figure 15.14.

_______ Is the patch at (0, 0) of type D or U?

_______ What are the number of activators?

_______ What are the number of inhibitors?

_______ Is $(A - wI)$ positive (P), negative (N), or zero (Z)?

_______ At the next time step, will the center patch be white (D) or black (U)?

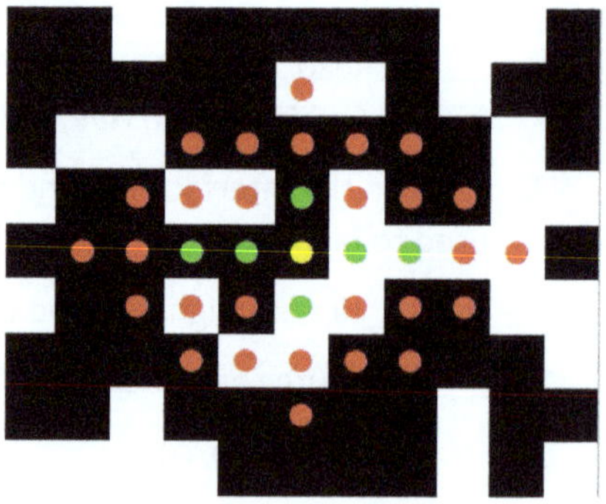

FIGURE 15.14 Inner elliptical neighborhood of x-radius 2 and y-radius 1 and outer elliptical neighborhood of x-radius 4 and y-radius 3

n. **1.** _______ **2.** _______ **3.** _______ **4.** _______ **5.** _______ Complete the following reasoning for why large outer radii and small inner radii tend to lead to an all-black emergent pattern: A large outer neighborhood has more area for **1.** (B. black, W. white) patches and thus a larger value for **2.** (A, I). Also, the inner neighborhood would be small, with less area for **3.** (B. black, W. white) patches and thus a smaller value for **4.** (A, I). Consequently, $A - wI$ would tend to be **5.** (N. negative, P. positive, Z. zero), resulting in a new color of black.

See Active Learning for an additional question on dynamic stability.

UNIT V

Ecology and Environment

MODULE

16 Population Ecology—Understanding and Caring for Earth

In nature, nothing exists alone.

—Rachel Carson, *Silent Spring*

16.1. Ecology

APPLYING SCIENCE TO LIFE 1 Unintended Consequences from Uninformed Choices

In 1958, Chairman Mao (Mao Zedong), who led China for more than 25 years, set into motion an unsuccessful plan to rid the country of four "pests": rats, flies, mosquitoes, and sparrows. The most curious choice in this group would seem to be sparrows. Apparently, Mao reasoned that the sparrows ate grain, decreasing food supply for the human population. During this campaign, millions of sparrows (and other birds) were killed. Although the birds did eat grain, they also ate insects. Without the sparrows, insects increased in numbers, decimating the crops and leaving less food for human animals. The Great Chinese Famine devastated the country between 1959 and 1961, exacerbated by the sparrow-killing program. Millions of Chinese people starved to death during this time, and the sparrows were almost exterminated.

16.1.1. Introduction

The word *ecology* comes from two Greek words: *oikos* ("house" or "home") and *logos* ("study"). Ecology, then, literally means the study of our home (Earth). Though grounded in the biological sciences, the relatively young field of ecology is interdisciplinary. Focused on interrelationships, ecologists try to understand how organisms interact with other organisms, as well as with the physical environment.

Every organism interacts with both physical and biological components of their environment. Thus, the sparrows in China had various interactions—with each other, with other organisms, with their food sources and other components of their physical environment. The sparrows did not feed on grains alone; sparrows also fed on insects. The insects also interacted with each other, with birds and other organisms, and with food sources and other components of their environment. Sparrows, insects, and other living organisms are often interconnected in complex ways. Removing a key component of any organism's environment can disrupt the interconnections, sometimes in drastic ways. When less insects were consumed by the birds, the larger population of insects ate a greater amount of the crops. The results were disastrous for the human population.

PoI 1. Do you think that Mao was thinking like an ecologist when he unleashed the assault on these pests? Explain.

PoI 2. Aside from the sparrows, most people would also classify the other targeted organisms as "pests." Ecologically, what roles do mosquitoes play?

There is a biological hierarchy going from individual cellular organelles to the *biosphere* (all regions of the earth occupied by life). Most ecologists work somewhere between the level of organisms and ecosystems, which includes *organismal*, *population*, *community*, and *ecosystem* levels.

16.1.2. Organismal Ecology

Scientists that work at this level are interested in how an organism's anatomy (structure), physiology (function), and behavior help the organism to live in the environment where it is found. To understand better this type of ecology, we can use an interesting adaptation developed by a parasite.

APPLYING SCIENCE TO LIFE 2 A Cat-and-Mouse Game

Toxoplasma gondii is a protozoan parasite that infects many different types of mammals and birds, but it needs a cat for completion of its life cycle. The cat acts as the *final host* (where a parasite reaches sexual maturity) for this parasite. The parasite also requires a bird or rodent to act as an *intermediate host* (supports nonreproductive stages of parasite life cycle), where the parasite matures. The cat feeds on the intermediate host, and the parasite continues its maturation into an infection-producing stage, which is shed into cat feces. Human beings and other animals may become infected from inadvertent intake of cat fecal material or by consuming improperly cooked pork, lamb, or venison. The ingested parasite can spread to any organ but usually causes somewhat flu-like symptoms of a disease called *toxoplasmosis*. Fortunately, the immune system in most healthy individuals limits the effects of the parasite. Pregnant women should avoid changing cat litter because intake of this parasite could seriously threaten the health of the fetus. Naturally, persons with compromised immune systems should avoid the risk-ridden sources of this infection.

Experiments using rodents have shown an interesting behavioral change in the animals infected with *T. gondii*. Normally, a rodent will avoid any type of predator. Cats are common predators of rodents, and rodents not only avoid cats but also are repelled by cat urine. Rodents infected with the parasite, however, lose their aversion to cat urine and become somewhat attracted to it. Naturally, this lack of aversion increases their chances of being eaten. Thus, it appears that the parasite, after finding its way to the rodent's brain, causes some sort of neurological change in the rat. This change is behavioral and predisposes the rodent to being caught and consumed by a cat.

Scientists are still experimenting to find out what the parasite does to alter the rodent brain in such a way that it controls the animal's behavior. One hypothesis is that it ramps up production of a neurochemical that influences this behavior. Although this is an interesting idea, the evidence is lacking to support this hypothesis. Furthermore, some of these scientists have found a possible connection between the occurrence of toxoplasmosis and the mental disorder schizophrenia. Up to now, the evidence is scant, but scientists will continue to search for more evidence of a real connection (Ingram et al. 2013; Vyas 2015).

In this example, even if we do not presently understand how the parasite influences the host's behavior, we can appreciate the benefits of such an adaptation for the parasite. By removing a rodent's natural fear of cats, the parasite ensures that the rodent is more likely to be caught, killed, and consumed by a cat. Consequently, the parasite will find the host where it can finish its life cycle, spreading infectious particles to continue future generations of these organisms.

There are a tremendous number of adaptations made by organisms to their environment, allowing them to live and reproduce under the existing conditions. For instance, many plants adapted to desert climates are succulents (plants that store water in fleshy leaves, stems, or roots). Plants in this category include cacti, aloes, and agaves. Leaves and stems are coated with a waxy cuticle to resist water loss. Desert succulents may also develop extensive, shallow root systems to absorb any available water quickly. Many of these plants also use a modified version of photosynthesis. Unlike most plants, succulents and other plants living in dry climatic conditions take in and store CO_2 during the cooler and more humid night hours. During the day, the plants close their *stomata* (gated portals for gas exchange) and utilize the stored CO_2 for producing sugar. Plants growing in moister climes take in CO_2 through open stomata during the day and use it to make sugar.

If you live in an area where it gets very cold in the winter, you are unlikely to see frogs during the coldest months. The frogs are still present during that time, but they are hibernating until warmer weather returns. In field studies of the environmental adaptations of the wood frog in Alaska, the average winter (October–May) temperature was –6.3°C, with low temperatures ranging between –8.9 and –18.1°C. In this study, scientists measured concentrations of naturally synthesized antifreeze compounds, glucose and *glycolipids* (lipids with some sugars attached). Amazingly, the frogs began producing these compounds during periodic freezes in the months preceding winter. When winter arrives, the muscles, the heart, and the liver had high levels of antifreeze chemicals, which prevents these organs and tissues from freezing. Without such protection, water in cells and the blood would freeze, generating crystals. Ice crystals can do tremendous damage to tissues, so the antifreeze compounds are critical to survival. On average, the frogs remained frozen for almost 200 days of freezing weather (Larson et al. 2014).

16.1.3. Population Ecology

APPLYING SCIENCE TO LIFE 3 Seeding Their Own Destruction/How the Mighty Fall

Long before the arrival of any Europeans, a remarkable civilization developed and collapsed in southern Mexico and northern Central America. Exploring the forests of southern Mexico, Guatemala, Belize, El Salvador, and northern Honduras, you would find the abandoned remnants of these people—the Maya.

For more than 1,200 years, the Maya were the dominant people of this region. From the ruins and artifacts they left, we know them to be quite advanced, building cities, maximizing agricultural production, creating calendars, and utilizing astronomy and advanced mathematics. The Maya population, in the millions, peaked between the 3rd and 9th centuries (AD), but at the beginning of the 10th century, their population crashed, and the Maya abandoned their great cities. Of course, ancestors of the Maya still live in various countries of the region.

The question for archeologists has been and is what caused this population crash. Recently, scientists have examined sediments below Lake Chichancanab in the Yucatan Peninsula. Analysis of these sediments revealed that between the 9th and 11th centuries AD, the region suffered a series of droughts. These droughts resulted in a 50 percent decrease in precipitation during a total of 100 years. With the peaking population and drought, the Maya cut down forests for new farming areas and for increased building in the cities. Preparation of lime plaster used in building required burning many trees. When areas are deforested, the soil tends to become drier and is destabilized. Computer models indicate that loss of these forests likely increased temperatures (3–5°C) and resulted in a 20 to 30 percent decrease in rainfall. Environmentally induced food and water shortages were undoubtedly exacerbated by the actions of the Mayans. If you combine a warmer and drier climate with inappropriate human actions, as the Mayans' story demonstrates, the problems will be accelerated, and people will die (Evans et al. 2018).

Though the Maya and other peoples have contributed to their own downfall, the worldwide trend has been to increase human population numbers. It is true that during the 14th century, the world population declined by about 200 million people due to the bubonic plague. By the 1700s, however, world population was estimated to be about 600 million. During the 19th century, world population of 1 billion was reached, and about 200 years later, the world population stands at about 7.7 billion people (Roser et al. 2019).

Normally, ecologists define *population* as a concurrent group of organisms, belonging to the same species, living in the same area. Most ecologists study populations in somewhat limited areas. The organisms in those areas likely interact and even interbreed. For example, you could study a population of largemouth bass in a lake. Or you might study a population of aspen in a national park. If you were a population ecologist, what would you want to know about these populations?

You probably would want to know how many organisms are in the population (population size) and *population density* (numbers of individuals of that species per unit area). Quantifying such populations is easier for organisms like plants that do not relocate. In migrating bird populations, for instance, populations change with the seasons. Also, if the population is sizeable, you would need to count samples and extrapolate from those numbers to get the population size.

If you know the area populated by the organisms, you can calculate density. For instance, the total population of human beings is about 7.7 billion, but the population densities vary greatly. The country of Monaco comprises less than two square kilometers (sq. km), and their population is only about 39,000 people, but the population density is over 26,000 per sq. km. Brazil, on the other hand, has a population of more than 211 million, with a land area of about 8.358 sq. km, giving it a population density of 25.25 per sq. km (CIA 2019, Statistics Times 2019). Of course, Brazil's population is not distributed evenly either. So knowing how the organisms are distributed is also important. Generally, we describe population distribution as *clumped* (distributed as bunches or groups that are close together), *random* (distributed with no discernable pattern), or *uniform* (evenly distributed).

Most commonly, organisms are distributed in a clumped pattern. This pattern may be a result of irregular or patchy distribution of resources, such as available water, shade, or food. Examples include animals whose young need food and protection from a parent, plants whose seeds fall close to the parent plant, and animals that band together in herds or schools for safety against predation. Random distribution is less common and generally occurs where there is rather uniform resource distribution. Wind-blown seeds from cottonwood trees and dandelions, as well as larval distribution of oyster larvae by water currents, are distributed in this fashion (see Figure 16.1).

FIGURE 16.1 Cottonwood seed, surrounded by "cottony" fluff that promotes wind dispersal of seeds

Uniform distribution of organisms is often found in areas where there are meager resources, even if the resources are uniformly distributed. Desert plants serve as excellent examples. For instance, in the Chihuahuan Desert, we find that the creosote bush may display a uniform distribution (see Figure 16.2). Scientists suspect that this distribution is based on inhibition through competing for scarce resources, including water (Brisson and Reynolds 1994). Also, some workers suggest that the plant produces chemicals that inhibit germination and/or growth, which hinders establishment of other plants around them. This mechanism is an example of *allelopathy* (production by a plant of chemicals that inhibit the germination or growth of another organism).

FIGURE 16.2 Creosote bush

QRQ 1. The 2010 U.S. Census counted 710,231 people in Alaska, a state with a surface area of 665,384 square miles (World Population Review 2021). Using these figures, calculate the population density of Alaska in 2010, rounded to one decimal place.

QRQ 2.

a. A population where organisms are distributed as bunches or groups that are close together has a(n) ______ distribution.
b. Select organisms that typically have this kind of distribution: animals that band together in herds or schools for safety, animals where young need food and protection from a parent, dandelions with wind-blown seeds, oysters whose larvae are distributed by water currents, plants that produce chemicals inhibiting other plants around them, plants whose seeds fall close to the parent plant, the desert plant creosote bush

QRQ 3.

a. A population where organisms are distributed with no discernable pattern has a(n) ______ distribution.
b. Using the same choices as in QRQ 2b, select organisms that typically have this kind of distribution.

QRQ 4.

a. A population where organisms are evenly distributed has a(n) ______ distribution.
b. Using the same choices as in QRQ 2b, select organisms that typically have this kind of distribution.

QRQ 5. The production of chemicals by a plant that inhibit the germination or growth of another organism is called ______.

16.2. Population Dynamics

In addition to knowing the numbers of organisms, an ecologist may want to know how these numbers change with time, termed *population dynamics*. Additionally, the scientists interested in population dynamics explore the factors that determine these changes in numbers. What factors might be examined? To answer that question, let's look at one of the tallest and rarest birds in North America.

APPLYING SCIENCE TO LIFE 4 Close Call: Whooping Cranes Pushed to the Brink of Extinction

Wildlife specialists and bird enthusiasts were whooping it up in 2018 because, for the first time since the late 1800s, the migrating population of whooping cranes exceeded 500 (see Figure 16.3). These cranes breed in the Wood Buffalo National Park in Alberta, Canada, and migrate to the Gulf Coast of Texas (Aransas National Wildlife Refuge) for the winter. An estimated 1,500 whooping cranes lived in North America around the middle of the 19th century, but the population declined for decades, until in 1938, there remained only 18. If you graphed the population numbers from 1938 to 2018, you would see a gradual increase to 500+. Although this number is encouraging, the population is still fragile and vulnerable (Butler, Harris, and Strobel 2013; Butler and Harrell 2018; Devokaitis 2018).

FIGURE 16.3 Whooping cranes

PoI 3. In Table 16.1 are population data for whooping cranes. The numbers are estimates, based on aerial or ground counts. Graph the population over time.

TABLE 16.1. Winter Whooping Crane Population Estimates from 1985–86 to 2017–18 (Butler, Harris, and Strobel 2013; Butler and Harrell 2018; Devokaitis 2018)

Winter of	Number	Winter of	Number
85–86	97	01–02	176
86–87	110	02–03	185
87–88	134	03–04	194
88–89	138	04–05	217
89–90	146	05–06	220
90–91	146	06–07	237
91–92	132	07–08	266
92–93	136	08–09	270
93–94	143	09–10	264
94–95	133	10–11	283
95–96	158	11–12	245
96–97	160	12–13	279
97–98	182	13–14	314
98–99	183	14–15	329
99–00	188	15–16	463
00–01	180	16–17	489
		17–18	505

This case leaves a number of interesting questions regarding the changes in crane populations. Numbers declined from 1,500 to 18 in less than 100 years. Why? Most research concluded that the decline was due primarily to hunting (sport, hat feathers) and habitat destruction. Also, if we look at the life history of these birds, we see that they migrate long distances, incurring many risks along the way. Although these birds have relatively long lives (~25 years), they do not become reproductive until about 4–11 years of age. The female usually lays two eggs, once per year, but only one of her chicks will fledge. Although the birds are no longer killed for hat feathers, sporadic hunting still takes place, although it is illegal. So even under rigorous management, growth of the population remains slow (Urbanek and Lewis 2015).

PoI 4. Using the data in Table 16.1, calculate the following:

a. Population growth for each year. Growth is equal to the present population – population during the previous year: $G = N_{present} - N_{previous}$.

b. Percentage growth/decline, or percentage change in population size, each year: $\Delta G = 100 \times G/N_{previous}$.

PoI 5. Suggest other limiting factors that may influence changes in population numbers of this species.

The difference between births and deaths results in change of a population. If births outnumber deaths, the population increases, while if deaths outnumber births, the population decreases. It follows that if births = deaths, the population stays the same.

population change = #births – #deaths

This relationship works if the population is isolated from all other populations of the same species. However, if organisms can move into or out of a population, we must consider gains or losses in population by *immigration* (organisms coming into the population) and *emigration* (organisms leaving the population). Where such conditions exist, the calculation is just a little more complicated. Sometimes, these movements are difficult to discern, but assuming they are observable, we must account for the changes.

population change = (#births + #immigrating) – (#deaths + #emigrating)

Mathematical models are useful in predicting the changes in populations. If you introduce an organism into an area where there are plentiful resources and count those organisms consistently over time, assuming no deaths, immigration, or emigration, the organism will reproduce without limit (at least at the beginning). If you graph those numbers of organisms versus time, you will generate what we recognize as an *exponential* or *unconstrained growth* curve. For instance, suppose you initially (time 0 hr) place bacteria onto a plate of sterile, solid culture medium. If each bacterium divides in one hour, at time one hour from each bacterium, we will have two cells, the original parent and a change of +1 offspring. With each of these two cells dividing in an hour, we have four cells at time two hours (the two original and two more bacteria). Then, after another hour, the population doubles again, so that at time three hours, we have eight cells derived from the original bacterium. Continuing in the same fashion, at hour four, the population will have 16 times as many bacteria as originally, then 32 times as many, and so on. Every cycle of division doubles the bacterial cell count. Your graph will start rising slowly but then increase faster and faster, which is typical of exponential growth (see Figure 16.4). Mathematically the relationship is expressed as follows:

$$N_{new} = N_{present} + \text{change}$$
$$N_{new} = N_{present} + rN_{present}$$

where $N_{present}$ = current population, N_{new} = population at next generation count, and r = proportionality constant. In this case $r = 1$; the change in the population is $1(N_{present}) = N_{present}$, so that $N_{new} = (N_{present}) + 1(N_{present}) = 2(N_{present})$.

In reality, no existent population is able to sustain this kind of population growth. In nature, growing populations begin to reach limits, usually as a result of reduced resources. Even bacteria in a culture dish will eventually run out of nutrients. A resource, like food, might be depleted, and/or there may be so many individuals that they compete with each other for the resource. We term the latter case *intraspecific competition*. In either case, this type of population increase is not sustainable.

For naturally occurring populations, there will be one or more limiting factors—resources or other environmental constraints. Each environment has a *carrying capacity*, which is the largest population that is supportable by that environment. Most populations, introduced into an area, begin to increase in numbers slowly, followed by a period of rapid increase. As population numbers approach the carrying capacity, growth of the population slows and then stabilizes around that carrying capacity. When we graph the numbers over time, the resulting curve takes on an S-shape (see Figure 16.5). This curve reflects *logistic growth*.

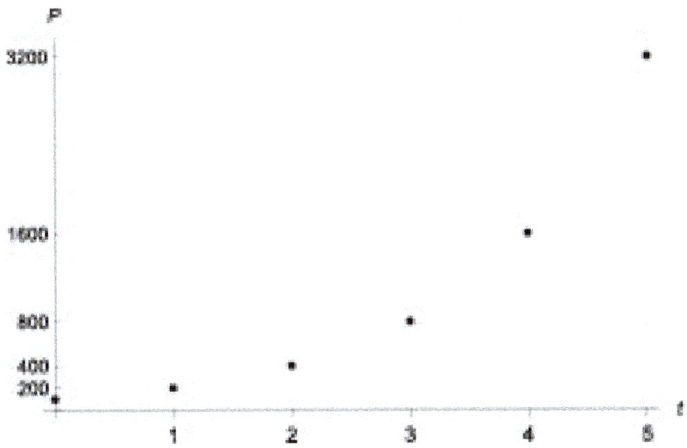

FIGURE 16.4 Graph typical of exponential growth

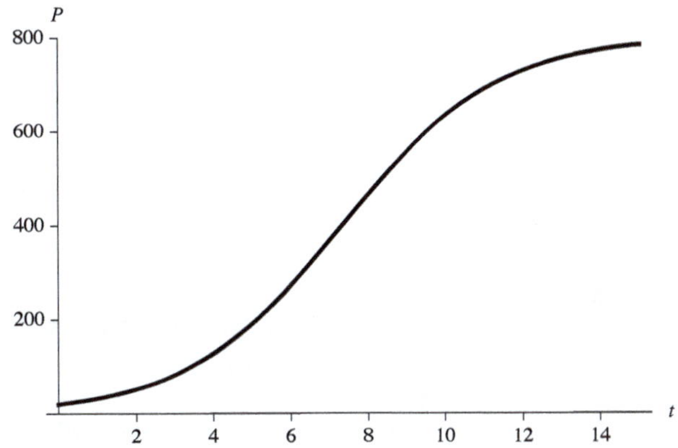

FIGURE 16.5 Graph typical of logistic growth

APPLYING SCIENCE TO LIFE 5 Rough Road to Recovery

The African country of Mozambique was torn apart by a civil war between 1977 and 1994, and it should not be surprising that one of its most beautiful national parks, Gorongosa, was abandoned. Vigorous fighting raged around and in the park, and much of the wildlife was eliminated. Elephants were killed for ivory, and other mammals were slaughtered for food or sport. Recovery efforts began in the years after a ceasefire, and some of the animal populations have recovered significantly, including magnificent antelopes called kudus (see Figure 16.6).

FIGURE 16.6 Kudu in Africa

Sample data on kudu populations, taken *intermittently* between 2000 and 2012, show a general increase in the population numbers, with number of animals leveling off (see Table 16.2). This curve indicates the population changes that seem to follow a logistic growth pattern (see Figure 16.7).

TABLE 16.2. Sample Numbers from Kudu Populations Taken between 2000 and 2012 (Stalmans et al. 2019)

Year	Numbers
2000	22
2001	39
2002	22
2004	63
2007	196
2010	259
2012	245

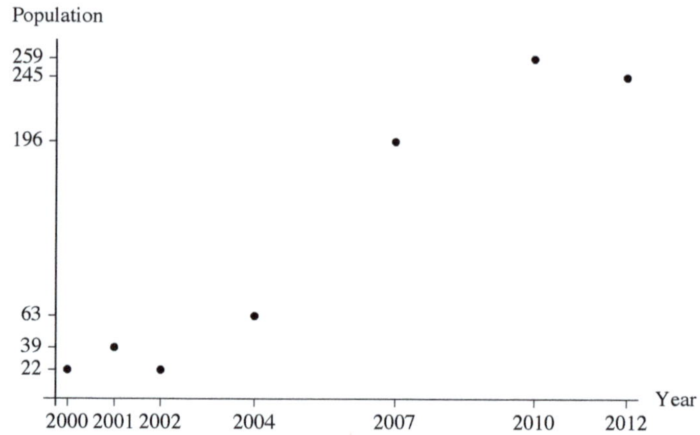

FIGURE 16.7 Graph showing changes in Kudu populations (2000–2012), using data from Table 16.2

PoI 6. Apparently, within eight months, the population reached the carrying capacity for this environment (see Figure 16.8).

a. Is this exponential or logistic growth?
b. What is the approximate carrying capacity?
c. What do you think the limiting factors were?

TABLE 16.3. Hypothetical Rodent Population Growth over a 12-Month Time Period

Months	Numbers
1	20
2	45
3	60
4	80
5	120
6	420
7	500
8	540
9	550
10	560
11	550
12	565

Many factors play roles in the dynamics of population, and these factors are classified by their dependence (*density-dependent*) or lack of dependence (*density-independent*) on population density. Regulation of population size quite often is limited by density-dependent factors. For example, suppose we release 20 adult grass-eating rodents (ten males, ten females) onto a small island. Suppose there are no predators and no other grass-feeding animals on the island, but grass covers less than half of the island's total area. Population counts, taken every 30 days, might be as shown in Table 16.3.

The island population application above is a very simple example of density-dependent limiting factors. The available food was not sufficient to support the population above the carrying capacity. Additionally, there was likely increased intraspecific competition for the available food. Food is often a limiting factor for populations, but other factors, such as number of suitable nesting sites, hiding places, number of mates, light, and so on, may also be important. Increased numbers not only increase intraspecific competition, which might include injury or death, but also lead to increased spread of disease or parasites.

Density-independent factors affect population numbers regardless of their size. For instance, natural disasters, like a fire or flood, may kill all animals that are not able to escape. Whether there are ten or 100 animals per acre, a fast-moving fire would kill all in its path. Plants can't escape; unless they are specially adapted to improve their survival in a fire, they will be burned up also. As far as we know, there is no such thing as a completely fireproof tree.

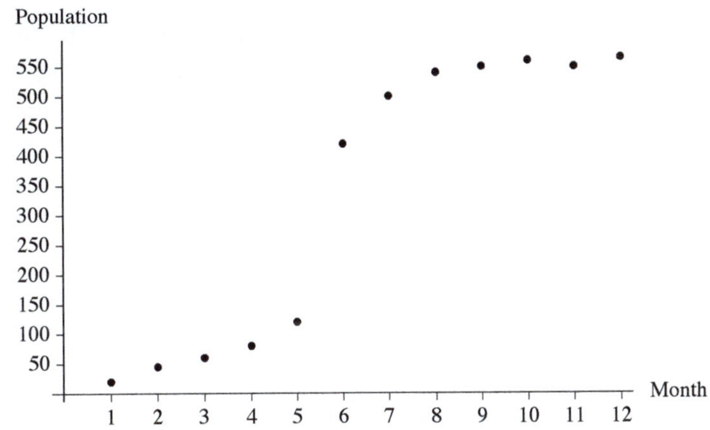

FIGURE 16.8 Graph of rodent population over 12 months, using data from Table 16.3

APPLYING SCIENCE TO LIFE 6 Living Advantages

Populations, not individuals, are the level at which *natural selection* works. Successful populations include individuals with various adaptations, but advantageous genetic changes (genetic changes that make the organism better adapted to its environment) in an individual have no effect in the next generation if those organisms don't reproduce. Even then, not all of their offspring will receive or express their advantageous genetic change. Generally, if the genetic change enhances survival, the organism is more likely to reproduce and produce more offspring carrying that genetic change. With a greater concentration of this genetic change in succeeding generations, the population gradually increases the proportion of these individuals. Overall, the population becomes better adapted to the environment.

In 2012, Spain was enduring one of the worst droughts it had experienced in the last 150 years. It is usually dry during the summers, but this one was hotter and drier than usual, which meant a terrible year for fires. Two scientists, who were studying trees in a 50,000-acre plot, were horrified that a forest fire broke out in an area that included their study site. When the fire was out, they viewed the extensive destruction of the pines, oaks, and junipers. However, there was a small portion of their plot, where just under 958 twenty-plus-foot tall Mediterranean Cypress trees had been growing. Of the original population, almost 946 survived—only 12 had been burned. Immediately, they began to determine what adaptations of the trees in this population enabled them to survive. Studies continue, but the general conclusion has been that these trees have adapted over millennia to these climates to retain water more effectively than other species, and hence, they are more fire resistant. Anatomically, their scale-like leaves are thicker and tightly compressed to the shoots, well-structured to retain water. They had a thick cuticle (outer cover), and the stomata were mostly on the inside, less exposed. Moreover, they lack some of the chemical compounds that lead to combustion. Even the litter falling from the trees holds onto water, forming a spongy matting that has few air pockets and contains significant moisture. In this case, density seems to be irrelevant for the cypress's population numbers. In fact, it might help the tree to compete with other species. Scientists are still studying the cypress for use as a firebreak for towns in these fire-prone areas (Della Rocca et al. 2014, Della Rocca et al. 2018).

There are many specializations within the field of ecology, and ecologists may work at the *organismal*, *population* (group of organisms of the same species living in the same area), *community* (group of different species occupying the same area at the same time), *ecosystem* (unit of all living organisms, as well as their physical environment), or *biome* (large community adapted to a particular environment, associated with a broad geographic area) level. In the next module, we will consider the ecology of communities.

QRQ 6. Suppose a population is 400 one year and 420 the next.

a. Calculate the population growth.
b. Calculate the percent population growth.
c. If this kind of growth occurs every year, without bound, what is the growth proportionality constant?

QRQ 7. The process of organisms leaving a population is called ______ (Answer choices: emigration, immigration).

QRQ 8. Select all characteristics that apply to a graph of the numbers in a population (versus time) that exhibits exponential growth: begins to increase slowly, concave down, concave up, decreases then increases, eventually growth of the population slows to zero, finally population numbers stabilize around a value, has a period of rapid increase, increasing, linear, S-shaped, same rate increase throughout

QRQ 9. Select all characteristics that apply to a graph of the numbers in a population (versus time) that exhibits logistic growth. Use the same choices as in QRQ 8.

QRQ 10. The largest population that is supportable by an environment is called the population's ______ ______.

QRQ 11. Classify each of the following factors in an area as density-dependent or density-independent:

a. fire
b. flood
c. food
d. suitable nesting sites

QRQ 12. Select from the following an appropriate term for each description below: biome, biometric, community, ecosystem, environmental, group, intraspecific, interspecific, organismal, population change, population dynamics, population

a. study of changes in population numbers over time
b. type of competition in which individuals of the same species compete
c. type of competition within a population
d. type of ecology that studies ecological communities adapted to a particular environment, distributed in similar environments in various parts of the biosphere

e. type of ecology that studies living and physical parts of an ecosystem and how they interact
f. type of ecology that studies responses of a population with its environment
g. type of ecology that studies the behavior, physiology, etc. of individual organisms in response to its environment and changes in that environment
h. type of ecology that studies the organization, interactions, and functions of biological communities

Image Credits

Fig. 16.1: Copyright © by G. Edward Johnson (CC BY 3.0) at https://commons.wikimedia.org/wiki/File:Cottonwood_20090521_083459_1.jpg.

Fig. 16.2: Source: https://commons.wikimedia.org/wiki/File:Creosote_bush_in_the_desert.jpg.

Fig. 16.3: Source: https://commons.wikimedia.org/wiki/File:Whooping_crane_pair.jpg.

Fig. 16.6: Copyright © by Charles J. Sharp (CC BY-SA 4.0) at https://commons.wikimedia.org/wiki/File:Greater_kudu_(Tragelaphus_strepsiceros_strepsiceros)_young_male.jpg.

LAB #16

Laboratory Exercises

16.3. Laboratory on Unconstrained Growth

16.3.1. Preliminary Work

In this laboratory, we examine unconstrained or exponential population growth (or decay), where the rate at which the number (amount) of organisms is changing is proportional to the number (amount) of organisms. From the text's website, view the introductory video for the three laboratories in this module; download and open "*Unconstrained Population Growth.nlogo*."

1. Click the Info tab at the top of the page and read the sections "What Is It?" and "How to Use It."
   a. ______ For the exponential growth in this model, the rate at which the number of turtles is changing is proportional to the number of **1**.
   b. ______ Is reproduction for the modeled organism asexual (A) or sexual (S)?
   c. ______ Calculate the growth or decay rate as a positive or negative decimal fraction, respectively, if BIRTH-RATE = 0.03 and DEATH-RATE = 0.07.

16.3.2. Laboratory Work

Default parameters are as follows: INITIAL-NUMBER-BACTERIA = 100, BIRTH-RATE = 0.10, DEATH-RATE = 0.08, STOP-AT-300 = OFF.

2. This question examines characteristics of the graphs for exponential growth/decay.
   a. ______ Using the default parameters, give the growth/decay rate, expressed as a signed (+ or –) decimal number.
   b. ______ percent By approximately what percent will the bacteria increase with each tick?
   c. ______________________ Press SETUP and GO, allowing the simulation to run for over 250 ticks. Select all characteristics that apply to the graph: A. begins to increase slowly, B. concave down, C. concave up, D. decreases to zero, E. eventually, growth/decay of the population slows to zero, F. finally, population numbers stabilize around a value, G. has a period of rapid decrease, H. has a period of rapid increase, I. increases without bound, J. same rate of change throughout

d. ______ Switch the values for birth and death rates, so that BIRTH-RATE = 0.08 and DEATH-RATE = 0.10. Also, change INITIAL-NUMBER-BACTERIA to 1,000. What is the growth/decay rate, expressed as a signed (+ or –) decimal fraction?

e. ________________________ Press SETUP and GO, allowing the simulation to run until it stops. Select all characteristics that apply to the graph. Use the same choices as in part c.

16.4. Laboratory on Constrained Growth

16.4.1. Preliminary Work

In this laboratory, we examine constrained or logistic population growth (or decay). From the text's website, download and open "*Constrained Population Growth.nlogo.*"

1. Click the Info tab at the top of the page and read the sections "What Is It?" and "How to Use It."

   a. ______ The modeled population cannot grow without bound because of competition for **1.** (A. food, B. hiding places, C. light, D. mates, E. nesting sites, F. shelter, G. space, H. water).

   b. ______ Can a turtle (bacterium) move into a neighboring occupied cell, no (N) or yes (Y)?

   c. ______ Can a turtle divide into a neighboring occupied cell, no (N) or yes (Y)?

   d. ______ The world (black area) is a 33 × 33 square of patches. Calculate the maximum number of bacteria (turtles) possible in the world if no two bacteria can occupy the same cell.

   e. ______ Initially, can two or more bacteria occupy the same patch, no (N) or yes (Y)?

16.4.2. Laboratory Work

The default parameters are as follows: INITIAL-NUMBER-BACTERIA = 100, BIRTH-RATE = 0.10, DEATH-RATE = 0.08, STOP-AT-300 = OFF.

2. This question examines characteristics of the graphs for constrained growth/decay.

   a. **1.** ______ **2.** ______ **3.** ______ **4.** ______ Use the default parameters, except turn on STOP-AT-300. Press SETUP and GO several times, observing the graph of the number of bacteria versus time. At first, the graph is **1.** (A. concave down, B. concave up, C. straight) and **2.** (D. decreasing, E. horizontal, F. increasing) before **3.** (G. approximately reaching equilibrium, H. decreasing somewhat, I. increasing somewhat), approaching **4.** (J. a positive number, K. zero).

   b. **1.** ______ **2.** ______ **3.** ______ **4.** ______ Change INITIAL-NUMBER-BACTERIA to 1,300. Press SETUP and GO several times, observing the graph. The graph is **1.** (A. concave down, B. concave up, C. straight), and at first it **2.** (D. decreases sharply, E. decreases slightly, F. increases sharply, G. increases slightly, H. is almost horizontal) before **3.** (I. approximately reaching equilibrium, J. decreasing somewhat, K. increasing somewhat), approaching **4.** (L. a positive number, M. zero).

   c. **1.** ______ **2.** ______ **3.** ______ **4.** ______ **5.** ______ Switch the values for birth and death rates (BIRTH-RATE = 0.08 and DEATH-RATE = 0.10), so that there will be more deaths than births. Also, change INITIAL-NUMBER-BACTERIA to 400. Press SETUP and GO several times, observing the graph. The graph is **1.** (A. concave down, B. concave up, C. straight), and at first it

2. (D. decreases sharply, E. decreases slightly, F. increases sharply, G. increases slightly, H. is almost horizontal) before **3.** (I. approximately reaching equilibrium, J. decreasing somewhat, K. increasing somewhat), approaching **4.** (L. a positive number, M. zero), so that the population **5.** (N. becomes, O. does not become) extinct.

3. In this question, we examine the impact of different positive growth rates on the graph. Since we are primarily interested in the graph of the number of bacteria versus time, uncheck view updates or increase the speed so that simulations run faster.

   a. **1.** ______ **2.** ______ **3.** ______ Complete the following hypothesis: In a constrained growth situation, for positive growth rates, where **1.** (A. birth rate, B. death rate) is greater than **2.** (A. birth rate, B. death rate), as the initial growth rate increases, the time until the population reaches the carrying capacity **3.** (C. decreases, D. increases, E. remains approximately the same).

   b. ______ Use default parameters, except turn on STOP-AT-300. Start a spreadsheet, named *ConstrainedGrowth*, as in Table 16.4. Setup and run the simulation at least *five* times, recording the approximate time (tick) at which the graph starts to level off, which is when the population reaches its carrying capacity. Have the spreadsheet calculate the average of these times and give this mean.

TABLE 16.4. Start of Laboratory's Spreadsheet

Names	
Constrained Growth	
	Question 3b
	BIRTH-RATE = 0.10
	DEATH-RATE = 0.08
	When level off

   c. ______ percent Give the approximate initial percentage growth rate for BIRTH-RATE = 0.10 and DEATH-RATE = 0.08.

   d. ______ Repeat part b for BIRTH-RATE = 0.12, recording the values in another part of the spreadsheet and giving the mean time here.

   e. ______ percent Give the approximate initial percentage growth rate for BIRTH-RATE = 0.12 and DEATH-RATE = 0.08.

   f. ______ Repeat part b for BIRTH-RATE = 0.14, recording the values in another part of the spreadsheet and giving the mean time here.

   g. ______ percent Give the approximate initial percentage growth rate for BIRTH-RATE = 0.14 and DEATH-RATE = 0.08.

   h. ______ Do the results of your experiments support your hypothesis in part a, no (N) or yes (Y)?

   i. ______ Considering your answers for parts c, e, and g, as the initial growth rate increased (specifically, as the initial birth rate increased), the time at which the population reached the carrying capacity **1.** (D. decreased, I. increased, R. remained approximately the same).

4. In this question, we estimate the carrying capacity of the population and consider the impact of different initial numbers of bacteria on that carrying capacity.

   a. ______ Complete the following hypothesis: As the initial number of bacteria increases, the carrying capacity **1.** (D. decreases, I. increases, R. remains the same).

   b. ______ Start with the default parameters, except turn on STOP-AT-300. In another part of your spreadsheet, indicate Question 4b, the initial number of bacteria (100), and a header for the number of bacteria at 300 ticks. At least five times, press SETUP and GO, recording the final number of bacteria. Have the spreadsheet take the average of these values and give this mean.

   c. ______ Repeat part b for INITIAL-NUMBER-BACTERIA = 500.

   d. ______ Repeat part b for INITIAL-NUMBER-BACTERIA = 900.

   e. ______ Repeat part b for INITIAL-NUMBER-BACTERIA = 1,300.

   f. ______ Do the results of your experiments support your hypothesis in part a, no (N) or yes (Y)?

   g. ______ Give your estimate of the carrying capacity to the nearest 100, such as 100, 200, 300, etc.

   h. ______ Is your estimated carrying capacity within 50 of the number of cells (see question 1d), no (N) or yes (Y)?

   i. Discuss why.

   j. ______ Set INITIAL-NUMBER-BACTERIA = 3 and run the simulation several times until the graph looks very different from the results in other parts of this question. Can the population become extinct, no (N) or yes (Y)? With small populations, surprising results can occur.

16.5. Laboratory on Competition

16.5.1. Preliminary Work

In this laboratory, we examine competition, which is the struggle between individuals of a population or between species for the same limited resource. From the text's website, download and open "*Competition.nlogo.*"

1. Click the Info tab at the top of the page and read the sections "What Is It?" and "How to Use It."

   a. **1.** ______ **2.** ______ Possible answers for each blank: A. interspecific, B. intraspecific. Competition among organisms of the same species in a population is called **1.** competition, while between two different species in a community is called **2.**

   b. ______ In the model, red and blue bacteria have **1.** (A. a combined carrying capacity, B. separate carrying capacities).

   c. ______ (T/F): Under certain circumstances, a bacterium can move to or divide into an occupied cell.

16.5.2. Laboratory Work

The default parameters are as follows: INITIAL-NUMBER-RED-BACTERIA = 50, INITIAL-NUMBER-BLUE-BACTERIA = 50, RED-BIRTH-RATE = 0.10, BLUE-BIRTH-RATE = 0.10, RED-DEATH-RATE = 0.08, BLUE-DEATH-RATE = 0.08.

2. This question examines characteristics of graphs for constrained growth/decay.

   a. **1.** ______ **2.** ______ **3.** ______ Use default parameters. At least five times, press SETUP and GO, letting the simulation run to completion and observing the graph of the numbers of bacteria versus time. After observing the world in one experiment, uncheck *view updates* or increase the speed so simulations run faster. At first, the graphs of all, red, and blue bacteria are **1.** (A. concave down, B. concave up, C. straight) and **2.** (D. decreasing, E. horizontal, F. increasing) before the total number of bacteria **3.** (G. approximately reaches equilibrium, H. decreases somewhat, I. increases somewhat).

   b. ______ (T/F): In these experiments, one or the other type of bacteria always becomes extinct.

   c. **1.** ______ **2.** ______ **3.** ______ Start the groups of red and blue bacteria with 1,000 organisms each. At least five times, press SETUP and GO, letting the simulation run to completion and observing the graphs of the numbers of bacteria versus time. At first, the graphs **1.** (A. decrease sharply, B. decrease slightly, C. increase sharply, D. increase slightly, E. are almost horizontal) before the graph of **2.** (F. blue, G. red, H. total) bacteria **3.** (I. approximately reaches equilibrium, J. decreases somewhat, K. increases somewhat), after which one of the groups of bacteria becomes extinct.

3. In this question, we examine the impact of different growth rates for the groups. Because we are primarily interested in the graphs, uncheck view updates or increase the speed so simulations run faster.

   a. **1.** ______ **2.** ______ Complete the following hypothesis: For this model, if one group (e.g., red bacteria) has a higher positive growth rate than the other group (e.g., blue bacteria), then the number of the latter group will **1.** (A. go to zero, B. sometimes approach a positive equilibrium). Moreover, the more difference there is in the growth rates, the **2.** (C. faster, D. slower) extinction will occur.

   b. ______ Give the red bacteria a competitive advantage by changing RED-BIRTH-RATE to 0.11, but otherwise, use the default parameters. We will be measuring the time at which a group of bacteria, blue or red, becomes extinct. Start a spreadsheet named *Competition*, as in Table 16.5. Set up and run the simulation at least five times, recording the time until extinction and the group (blue or red bacteria) that becomes extinct. Have the spreadsheet calculate the average of these times and give this mean.

TABLE 16.5. Start of Laboratory's Spreadsheet

Names		
Competition		
	Question 3b	
	RED- BIRTH-RATE	**0.11**
	Time to extinction	**Extinct bacteria**

c. ______ Repeat part b for RED-BIRTH-RATE = 0.13.

d. ______ Repeat part b for RED-BIRTH-RATE = 0.15.

e. ______ Repeat part b for RED-BIRTH-RATE = 0.17.

f. ______ In the set of experiments for this question, did the red bacteria ever become extinct, no (N) or yes (Y)?

g. ______ In the set of experiments for this question, did the numbers of blue bacteria increase initially, no (N) or yes (Y)?

h. **1.** ______ **2.** ______ Using the data from Parts b–e, plot the average time to extinction versus the difference in growth rates (0.01, 0.03, 0.05, and 0.07, respectively). The graph is **1.** (A. approximately straight, B. concave down, C. concave up, D. none of these) and **2.** (E. approximately horizontal, F. decreasing, G. increasing, H. none of these).

i. ______ Do the results of your experiments support your hypothesis in part a, no (N) or yes (Y)?

j. ______ With all other factors being equal, the more the birth rate for the red bacteria increases, the **1.** (F. faster, S. slower) the blue bacteria will become extinct.

4. In this question, we consider the impact of groups having different initial numbers on group survival.

a. **1.** ______ **2.** ______ Complete the following hypothesis: If one group of bacteria (e.g., red or blue) initially has more individuals, the other group (e.g., blue or red) **1.** (A. will become extinct, B. may or may not become extinct). Moreover, as the difference in the initial numbers increases, on average, the time to extinction **2.** (C. decreases, D. increases, E. remains about the same).

b. ______ Start with the default parameters, except have INITIAL-NUMBER-RED-BACTERIA = 100. In another part of your spreadsheet, indicate Question 4b, the initial number of bacteria (100), and headers for the time of extinction and which group of bacteria (red or blue) becomes extinct. At least ten times, press SETUP and GO, recording the results. Have the spreadsheet take the average of the times to extinction and give this mean.

c. ______ In these experiments, the group of blue bacteria had half the initial number of individuals as the group of red bacteria. Was it always the group of blue bacteria that became extinct, no (N) or yes (Y)?

d. ______ The results seem puzzling. Calculate the standard deviation of the times to extinction. Is the standard deviation large, say, more than 20 percent of the mean, no (N) or yes (Y)?

e. ______ Is there a lot of variation in the times to extinction, no (N) or yes (Y)?

f. ______ Let us continue the experiments and make observations whereby we might be able to refine our hypothesis. Repeat part b with INITIAL-NUMBER-RED-BACTERIA = 200 and give the mean time to extinction.

g. ______ Calculate the standard deviation of this data.

h. ______ You still are probably having experiments in which the red bacteria become extinct. Repeat part b for INITIAL-NUMBER-RED-BACTERIA = 300 and give the mean time to extinction.

i. ______ Calculate the standard deviation of this data.

j. ______ Repeat part b for INITIAL-NUMBER-RED-BACTERIA = 400 and give the mean time to extinction.

k. ______ Calculate the standard deviation of this data.

l. ______ Particularly for larger initial numbers of red bacteria, the time to extinction for what kind of bacteria (red or blue) appears to be skewed toward much larger extinction times?

m. ______ Usually, are the extinction times more spread out from the mean when red-bacteria extinctions occur in the set of experiments, no (N) or yes (Y)?

n. ______ Do the results of your experiments support your hypothesis in part a, no (N) or yes (Y)?

o. ______ Let us reconsider the data only calculating the means when the blue bacteria become extinct. Thus, calculate the mean of the time for extinction of the blue bacteria when INITIAL-NUMBER-RED-BACTERIA = 100 (part b).

p. ______ Calculate the mean of the time for extinction of the blue bacteria when INITIAL-NUMBER-RED-BACTERIA = 200 (part f).

q. ______ Calculate the mean of the time for extinction of the blue bacteria when INITIAL-NUMBER-RED-BACTERIA = 300 (part h).

r. ______ Calculate the mean of the time for extinction of the blue bacteria when INITIAL-NUMBER-RED-BACTERIA = 400 (part j).

s. Using the data from parts o–r, plot the average time to extinction for blue bacteria versus INITIAL-NUMBER-RED-BACTERIA (100, 200, 300, 400). The graph is **1.** (A. approximately straight, B. concave down, C. concave up, D. none of these) and **2.** (E. approximately horizontal, F. decreasing, G. increasing, H. none of these).

t. Develop a revision/refinement of your hypothesis in part a. Discuss other experiments that could be used to test this hypothesis.

MODULE 17

Community Ecology—In This Together

APPLYING SCIENCE TO LIFE 1 Exotic Problem

Naples, Florida, is a town on the Gulf of Mexico, in the southwestern part of the state. Naples's population is just above 20,000, and the cost of living there would be considered high. Besides the beautiful beaches and numerous golf courses, the town is located close to the Everglades and the Ten Thousand Islands areas. If you have enough income, it must be a wonderful place to live.

Let's imagine that you are a new resident of Naples, living along the fairway of one of the well-maintained golf courses. Early one evening, as you sit on your deck, you see something long and dark moving slowly along the rough at the edge of your lot. When you go to investigate what it is, you are dumbstruck to find that what you are observing is a very large snake. Upon seeing you, the snake moves more quickly into the brush. You make a frantic call to the county sheriff, who tells you they'll send someone out to investigate. About 20 minutes later, a biologist from the Conservancy of Southwest Florida shows up and quickly finds the snake. The biologist tells you that this animal is a Burmese python, and she radios for help. About an hour later, the biologists capture the snake, which is about 10 feet long. Because it is a female, they will take her in and euthanize her.

The *Burmese python* is a native of Southeast Asia and is one of the largest snakes in the world (see Figure 17.1). It is not venomous, but it is a constrictor and an ambush predator. The snake crushes, suffocates, and consumes its prey. The snake bites the head region of small prey and swallows the animal whole. For larger prey, the snake bites to hold on, coiling its body around the animal to stop the prey from breathing. Pythons eat quite a variety of small mammals, wading birds, deer, and even alligators. In Florida, they are considered an *apex predator* (top predator, with no natural predator to eat them). Alligators will infrequently attack and eat pythons, but often enough, the alligator loses, and the snake attempts to eat the attacker.

On average, a Burmese python lives about 20 years, may grow longer than 20 feet, and may weigh up to 200 pounds. The animal is solitary, but in Florida, groups may be found during

FIGURE 17.1 Burmese python

mating season, which is sometime between January and April. After 60–90 days of gestation, the females lay their eggs (20–100), coil over them, and guard them for about 60 days. Once they hatch, the young ones are on their own.

How did this exotic animal find its way to your backyard? Miami is a hotbed for the exotic pet trade, and importers have brought in vast numbers of animals from Asia to sell as pets. According to the US Fish and Wildlife Service, between 1970 and 1995, 17,000 Burmese pythons were imported, whereas between 1996 and 2006, 99,000 of these animals were brought into the United States. There are also people who breed the animals in the United States and sell them. Some of the animals undoubtedly escaped on their own, but also, these snakes grow from about 20 inches in length to 7 feet in one year, becoming less easily contained. Apparently, some owners, unable to handle such large animals, released the snakes into a convenient and vast subtropical wetland that would support them—the Everglades. The earliest sightings of pythons there, according to the Nature Conservancy, were in 1979. At that time, no one was alarmed about the effects these animals would have on the Everglade communities. Exacerbating the problem, more pythons escaped captivity from exotic wildlife holding facilities devastated in 1992 following Hurricane Andrew. Sometime during the 1990s, this python species established breeding colonies in the Everglades. Estimates of the current population vary, but it is perhaps 100,000.

These snakes have fed voraciously, especially on small mammals that are native to the Everglades. Although the snakes do not target human beings, in 2009 a couple who kept a python illegally in a home with three children had their two-year-old killed by their pet snake. Household pets may be even more vulnerable. Populations of many small mammals have declined by 90 percent or more. This area is a known breeding area for wading birds, which are also threatened by the pythons. Stomach contents from the pythons have included almost 40 species of birds, remnants of deer and bobcat, and even other snake species. On the other hand, there is only one potential predator for the snakes present in the area—the alligator.

Belatedly, as the alarming decline of native wildlife numbers and other concerns mounted, environmental, wildlife, and governmental groups began to seek solutions. The government hired contract hunters and later allowed local residents to kill the snakes. Other organizations, like the Conservancy of Southwest Florida, use internal radio transmitters in males (sentinels) during the breeding season, hoping the animals will lead them to females. After the captured snakes are euthanized with chemical injections, they are put into freezers for later examination and study. Because this organization is interested in the biology of this species, researchers also collect material to determine tissue mercury levels. Furthermore, biologists also take chemical samples from the skins for use in the development of pheromone attractants that may attract the snakes to baited areas, where they can be caught and removed. Even with the aid of various groups that kill the snakes, snake-hunting dogs, and high-tech tracking and capture methods, as well as biological studies of these invaders, officials now are resigned to containing this animal rather than eliminating it. It is worrisome that as the decline in their food accelerates, the snakes may venture more frequently into populated areas or try to extend their range. With climate changes, other states may also eventually be dealing with this exotic problem.

17.1. Introduction

As you can see from the history of the Burmese python in Florida, natural and nonnative populations do not exist alone. Each species' population is part of an interacting group of populations living within the same area, called a *community*. The introduced python is wreaking havoc on a variety of native animal populations by feeding on them. This type of disruption can ultimately alter the structure, interactions, and dynamics of the community.

PoI 1. What do you think are the major factors that promoted successful establishment of the Burmese python in the Everglades?

17.2. Community Diversity

Biodiversity describes the structure and complexity of a community. Diversity is expressed by the total number of species in the area (*species richness*) and by the proportion of each species within the entire community (*relative abundance*). Ignoring other components of the community for simplicity, let's look at an example of two hypothetical tree communities. From Table 17.1, observe that in communities A and B, there are 10 identical species of trees. So, species richness would be 10 for both. However, the relative abundance numbers show that community A has an equal percentage (10 percent) for each of the 10 species, and community B displays a community where three species make up 64 percent of the trees. Intuitively, data for these relatively simple examples point to community A displaying the most diversity, based on relative abundance. Although ecologists do use intuition, they also rely on diversity indexes when studying more complex communities.

TABLE 17.1. Comparison of Diversity of Two Hypothetical Tree Communities, A and B, Where S = Species

A's Species	S1	S2	S3	S4	S5	S6	S7	S8	S9	S10	Total
Number	5	5	5	5	5	5	5	5	5	5	50
%	10%	10%	10%	10%	10%	10%	10%	10%	10%	10%	100%

B's Species	S1	S2	S3	S4	S5	S6	S7	S8	S9	S10	Total
Number	9	1	3	12	6	11	3	3	1	1	50
%	18%	2%	6%	24%	12%	22%	6%	6%	2%	2%	100%

17.3. Keystone Species: Disproportionate Influence

Hugging the coastal areas in the western parts of North America are massive kelp forests. Kelp are very large brown algae that favor the cool, somewhat shallow ocean close to the shoreline. Incredibly important to the coastal ecosystem, they form underwater forests that provide numerous species of animals with food, nursery areas, and shelter (see Figure 17.2). Besides the kelp, there are many other species of algae that make their homes in these forests. The waters are fertile and cool, perfect for the primary producers, like the algae, which in turn support large populations of fish, marine mammals, birds, and vast numbers of invertebrates. Resident animals include sea urchins, often in large "herds," that favor the kelp as food (see Figure 17.3). These animals will even attack the structures (holdfasts) that anchor the kelp to the bottom. If unchecked, sea urchins can consume massive amounts of kelp, leaving bare huge swathes along the seafloor.

FIGURE 17.2 Kelp forest

FIGURE 17.3 Purple sea urchins

Fortunately, there is a natural solution to this natural disaster—sea otters (see Figure 17.4) and, in some cases, other predators. Sea otters often hang out in and among the kelp forests. The kelp may hide them from predators or anchor the sea otters while they munch voraciously on items harvested from a menu of abalone, crabs, clams, snails, and sea urchins. Sea otters just love sea urchins, and they eat huge quantities of them when these spiny denizens are around. The only problem is, sea otters face many threats to their survival—white shark, disease, parasites, habitat destruction, oil spills, pollution, and so on. Despite continuing conservation efforts, their population numbers are still critically low.

FIGURE 17.4 Sea otter eating urchin roe

When sea otters are absent from the kelp forests, the urchins move in and have their way with the kelp. Such destruction changes the entire ecosystem because the kelp is critical for so many organisms. Breeding and nursery grounds, food sources, and shelter for fish, marine mammals, and invertebrates are lost. The entire community structure is destroyed.

In the 1960s, a distinguished biologist, Robert Paine, was studying the tide pools along the rocky coast of the state of Washington. Each of these pools represented tiny ecosystems, with a characteristic community composed of various representatives: algae, mollusks, barnacles, and a sea star (starfish), *Pisaster ochraceus* (see Figure 17.5). The sea star was the top predator in these kingdoms by the sea, primarily feeding on the mollusks. He took two areas littered with the tide pools, one experimental and one control. He did nothing to the control pools except monitor changes in the tide pool communities. In the experimental area, he meticulously removed the sea stars from each pool, and Dr. Paine got very good at removing and hurling the animals out into the ocean.

FIGURE 17.5 Sea stars clinging to rocks with mussels (top) and greenish anemones (bottom); tentacles withdrawn at low tide

What Paine found probably surprised him, but it astounded other ecologists. At the end of the study, he found the experimental pools bereft of their former diversity. During the experiment, removal of the sea star was immediately followed by barnacles taking over most of the available surface in each pool. Within months, California mussels had overwhelmed the barnacles, snails, sponges, and anemones. In the end, each pool had only mussels and a little algae growing on the mussels. On the other hand, in the control pools, where the sea stars were still in charge, the earlier diversity was maintained.

Dr. Paine's conclusion was that not all species within an ecosystem are of the same importance. Some species have a much greater influence in maintaining the community diversity and structure than others. Apparently, the sea stars kept their diet varied and consequently prevented any one of the other species from taking over the pool. Paine called this sea star a *keystone species*. Previously, the basic tenet of ecologists was that diversity stabilizes the ecosystem. Paine's work effectively demonstrated that one species could have undue input into maintaining ecosystem stability. It was a revolutionary concept, and now we find many other examples of such keystone species (HHMI 2019).

PoI 2. Why are each of the following considered keystone species?

a. gray wolf
b. Saguaro cactus
c. showshoe hare
d. African elephant

In fact, the sea otters are considered to be such a species. In many kelp forests, loss of these keystone animals results in a collapse of the communities in that ecosystem. Some people also consider the kelp to be a keystone organism. If the kelp are lost, the community is lost, and the ecosystem is changed.

17.4. Ecological Niche and Interspecific Competition

In any community, the resources necessary for life are limited. Each species (population) in a community plays a role in that community. This ecological role is what scientist call a *niche*. The niche is determined by all the living and nonliving resources available that the organism requires or can use. For example, a cottontail rabbit is an herbivore, requiring access to appropriate food (grasses, clovers [bean family], composites [aster/sunflower family], and garden vegetables) and water. Various features of the physical environment must match with the rabbit's tolerance (e.g., temperatures, humidity range, etc.). To avoid detection by predators, cottontails also need sufficient hiding places, such as thickets or brush piles. Oh, yes, they also need other rabbits for reproduction. According to the NC Wildlife Resources Commission, adequate rainfall during summers, which stimulates plentiful vegetation growth, is needed for optimal reproduction (NC Wildlife Resources Commission 2017). Under ideal conditions, a doe may bear five litters yearly, resulting in 15–40 young. Cottontails are prey for a number of predators—foxes, badgers, hawks, owls, and the like. Because of the rapidity of rabbit reproduction, predation is important to dampen population growth. The ecological niche, or role, of rabbits or of any species in a community, is determined by the sum of the resources, environmental conditions, and interactions needed for the species to live successfully in that area.

In a community where two species with overlapping niches are trying to access and utilize a resource that is limited, there will be competition between those species. Competition may result in reduced reproductive rate and population size for one or both competitors. Ecologists describe two types of competition, *interference* and *exploitative*. When individuals of one species physically prevent competing species from accessing a *limiting resource* (resource required for population growth), the competition will likely lead to injury and/or deaths. Interference occurs directly between individuals of the two populations. In animals, such competition may include fighting, whereas a plant may use chemicals to inhibit growth of plants around it (*allelopathy*). In *exploitative* competition, one species' use of the limiting resource indirectly impacts the competing species by diminishing the availability of the resource. These interference and exploitative strategies may be applied to intraspecific competition, too.

When two or more competing species attempt to access and utilize the exact same limiting resources, the two populations will be unable to coexist over the long term. One species will outcompete the other, and the other species will die out or be forced to emigrate. Only one of the species will prevail. The *competitive exclusion principle,* alluded to during the middle of the 19th century by Charles Darwin, is expressed simply by the 20th century American ecologist Garrett Hardin: "Complete competitors cannot coexist" (Hardin 1960).

To reduce competition for the same resources, two species with overlapping resource needs, under natural selection pressure, may utilize *resource partitioning* to allow coexistence in the same area. What results is two species with more differentiated niches. This change may involve using different resources, occupying different habitat parts, or changing behavior to avoid trying to access the resource at the same time. For instance, they may feed at different times of day.

Early evidence of resource partitioning was published by R. H. MacArthur in 1958. MacArthur was studying five species of warblers in coniferous forests of New England in the United States. These small insect-eating birds, living in the same area, were assumed to fill essentially the same niche in this community, making them an exception to the competitive exclusion principle. With careful observation, however, MacArthur discovered that the birds were each foraging at different times in different parts of the trees. Furthermore, they had adjusted their mating to different times of the season. Thus, they were not competing for exactly the same resources, and therefore, they had altered their niches so that they all could coexist (MacArthur 1958).

QRQ 1. Suppose community A has three species, S1, S2, and S3, with 15, 50, and 35 individuals, respectively, and community B has 5 species with 6 individuals each.

a. Give the relative abundance (percent) of species S1 in community A.
b. Give the relative abundance (percent) of a species in community B.
c. Which community displays the most species richness, A or B?

QRQ 2. Select from the following an appropriate term for each description below: allelopathy, anentropy, area, community, competitive exclusion principle, exploitative, exploratory, interference, limiting, neighborhood, niche, nook, relative abundance, resource division, resource partitioning, species abundance, species richness

a. ecological role each species (population) plays in a community
b. interacting group of populations living within the same area
c. only one of multiple competing species will prevail
d. plant using chemicals to inhibit growth of plants around it
e. proportion of each species within the entire community
f. resource required for population growth
g. total number of species in the area
h. two species with overlapping resource needs, under natural selection pressure, evolve to have more differentiated niches
i. type of competition where individuals of one species physically prevent its competing species from accessing a limiting resource
j. type of competition where one species' use of the limiting resource indirectly impacts the competing species by diminishing the availability of the resource

17.5. Symbiotic and Other Interactions

Of special interest to ecologists are interactions between species within a biological community. In some cases, these interactions have evolved between two species that live in close association for at least part of their life cycles where one or both of the species are positively affected. This type of relationship is termed *symbiosis*. Of course, not all of interspecific reactions are symbiotic, and not all of them are positive for both species.

17.5.1. Mutualism

Mutualism is usually a symbiotic interaction where both participants benefit from the relationship. There are many examples of such interactions in nature. For instance, *zooxanthellae* (photosynthetic protists—eukaryotic, mostly unicellular) inhabit some of the soft tissues in the giant clam and in the coral species that form reefs (see Figure 17.6). In both cases, the host species provides CO_2 and nitrogen for the zooxanthellae, and in return, the zooxanthellae provide sugars from photosynthesis to the host. Both hosts can capture other sources of energy (filtering or capturing prey), but the zooxanthellae supply their host with critical sources of energy and nutrients.

Although a close relationship, mutualism does not require the partners to live in intimate associations, like zooxanthellae with clams and corals. For example, many flowering plants require pollination by animal pollinators. The pollinators are rewarded with energy-rich nectar from the flower, while the flower's pollen is dispersed to other flowers for seed production. Both participants have a net gain.

FIGURE 17.6 Giant clams

17.5.2. Commensalism

Commensalism is an interaction where one partner is benefited, while the other partner is not affected. For example, pseudoscorpions, which have pincers like a scorpion but lack the sting, usually live in the soil or under the bark of trees (see Figure 17.7). Certain pseudoscorpions are able to attach to specific kinds of beetles. Scientists have not detected effects on the beetle, but the pseudoscorpions gain from dispersal at the behest of the beetles (Zeh and Zeh 1992).

FIGURE 17.7 Pseudoscorpion

FIGURE 17.8 Cattle egret

Sometimes the effects on each partner in such a relationship can be hard to detect or observe. However, with further observation or advancing technologies and methods, scientists may uncover previously unknown effects. For example, cattle egrets are a widely distributed member of the heron family (see Figure 17.8). These birds have a close association with cattle, buffalo, and other large grazing animals. The birds are somewhat like "groupies," as they follow behind these large herbivores, which stir up insects out of the grass or other vegetation as they move along. These insects form an important food source for the birds. Consequently, this relationship has long been cited as an example of commensalism. The bird benefits, and the mammal does not seem to be affected. However, egrets will also hop on an animal's back and remove external parasites, such as ticks. Some of these cause painful tick bites that may become infected. Furthermore, the ticks carry disease organisms, such as heartwater fever, that sicken the animal. There has been some concern that the egrets could also aid in the dispersal of the ticks and, thereby, the diseases the tick harbors. Despite this concern, both species seem to benefit, making the relationship mutualistic.

Some whales, such as the humpback or gray, participate in another relationship that is classified as commensal. On humpback whales, you may observe on their heads (particularly lower jaw) or tails some mysterious bumps surrounded by calcium plates. In Figure 17.9a you can see some of these bumps on the edge of the whale's tale. It may surprise you to know that these bumps are a very specialized type of crustacean called a barnacle. We most often associate barnacles with rocky shores, piers, or the hulls of ships. Whale barnacles have specialized to attach themselves to the skin of their hosts (see Figure 17.9b). The living portion of the barnacle is the pink-tan central region, surrounded by calcium plates (white). Within the plates, the barnacle seemingly lies on its back, and a series of feathery appendages, called cirri, extend into the water to help filter out and capture food materials for the animal. The barnacles feed on *plankton* (small organisms that tend to float in bodies of water) as the whale swims through the water. At first glance, the barnacles seem like they should cause harm to the whale and therefore be parasites. However, even with heavy infestations, these hitchhikers do not seem to cause harm. So we consider this relationship a commensal one.

FIGURE 17.9a Barnacles on edge of humpback whale's tail

FIGURE 17.9b Barnacles on neck of humpback calf

17.5.3. Exploitation Interactions

There are several categories of *exploitation* interactions: *predation*, *herbivory*, and *parasitism*. In each of these categories, one member of the relationship is benefited, and the other member is harmed.

Predation

Predators are animals that feed on other animals. Obviously, the predators are benefited, but the individual prey is not. Predator-prey interactions involve the development and evolution of offensive and defensive mechanisms. Predators who develop adaptations that improve their chances of hunting and catching prey will most likely produce more offspring, many of whom may have inherited the adaptive advantage. For example, the Burmese python strikes quickly to ambush its prey and has developed impressive rows of teeth that curve backward, preventing the prey from pulling out and escaping.

On the other hand, prey that develop modifications increasing their chances of avoiding predation will also likely produce more offspring with the adaptation. Many potential prey produce toxins to inhibit predation. For instance, the cane toad (see Figure 17.10), like many other species of toads, secretes a toxin when threatened by a predator, including household pets. The poison in this species is released from the shoulder area of the toad, and a predator biting the head region will have a load of toxin absorbed by the lining of its mouth. Toxic effects include increased heartbeat, convulsions, paralysis, and, on occasion, death.

Some animals adopt *cryptic coloration*, which makes them less visible to predators. For instance, the waterfall frog, which is native to northeastern Australia, lives in and around fast-flowing streams and waterfalls. Notice in Figure 17.11a how well the frog blends with the moss, allowing it to escape the notice of predators such as snakes, birds, or, perhaps, other frogs (Wikipedia 2020). Cryptic coloration can also be used to make predators less noticeable, allowing them to ambush prey more easily. Note how the camouflaged yellow crab spider surprises its prey in Figure 17.11b.

FIGURE 17.10 Cane toad

Instead of serving for concealment, some prey coloration serves as a warning to deter predators. The poison dart frogs

FIGURE 17.11 Cryptic coloration. **a.** Camouflaged waterfall frog. **b.** Camouflaged yellow crab spider nabs prey.

are natives of the tropics of Central and South America. Predators themselves, these frogs feed on small arthropods (e.g., insects). Many of the frogs are brightly colored, with reds, blues, and yellows, often against black. Other members of this group display cryptic coloration. The brightly colored ones shown in Figure 17.12 also produce a number of toxins, which they release onto the skin. The animals may use toxins acquired from their diet instead of synthesizing their own. Toxins vary in their potency. The most toxic poisons seem to come from frogs of the genus *Phyllobates*. Predators find that the frogs are unpalatable and usually learn to avoid them, cued in by their bright colors. These bright colors represent *aposematic coloration,* which advertises to potential predators that the animal is poisonous. Indigenous people used these toxins for tips of their darts. Interestingly, the cryptically colored species of dart frogs tend to lack the poison (Smithsonian 2019).

Other potential prey, such as monarch butterflies (see Figure 17.13, left), consume milkweed foliage during their caterpillar stage. The foliage contains a substance that is toxic to most animals, and this toxin persists into the butterfly stage. Over the course of their evolution, the butterflies developed mutations that made them resistant to the toxins, allowing the butterfly to consume the milkweed foliage without harm (Karageorgi et al. 2019). The toxins can interfere with neural and cardiac function of any predator, although they may not prove fatal. However, the toxins taste bad enough that a predator will usually avoid future encounters with the monarch. Interestingly, another

FIGURE 17.12 Poison dart frogs. **a.** Golden poison dart frog. **b.** Sapphire blue poison dart frog.

FIGURE 17.13 Comparison of the monarch and viceroy butterflies

smaller butterfly, the viceroy, resembles the monarch closely enough that lizards, birds, and other potential predators avoid that butterfly, as well (see Figure 17.13, right). The viceroy's physical resemblance to the monarch is an example of *mimicry.*

Predator-prey interactions often represent good examples of *coevolution*, where each reciprocally promotes adaptations in the other. Cheetahs and antelope have roamed together on the plains and dry grasslands of Africa for a very long time. Adaptations have helped cheetahs to be successful predators. Thought to be the fastest known land animal, cheetahs are able to reach speeds of 70 mph for short times. The animal has a slight frame wherein their small collarbones do not connect to the shoulder blades. They have hips that swivel on a flexible spine, long legs, and a long tail that improves balance and agility. Their claws do not retract, helping them dig in for sharp turns. The spotted pattern of its fur helps the cheetah hide out in the tall grasses, so that this predator can approach much closer to its prey (see Figure 17.14a). Cheetahs prey on a number of animals, including smaller antelope, such as gazelles and impalas (see Figure 17.14b). Gazelles are unable to run as fast as cheetahs, but they can sustain high speeds for long periods of time. Impalas can leap up to 33 feet. Antelope also often are found in herds, which minimizes the risk of becoming dinner. Furthermore, antelope have very keen senses. Upon detecting an approaching predator, animals

FIGURE 17.14a Cheetahs

FIGURE 17.14b Herd of impala

will often produce an alarm sound to alert others. So as the cheetah adapted to improve its hunting abilities, the antelope adapted to improve their escape abilities. And so, the battle to out-adapt each other will rage into the future for the cheetah and the antelope.

Herbivory

Herbivory is an interaction wherein an animal consumes parts of plants, algae, and photosynthetic bacteria. Obviously, the animals benefit, but the plants are diminished. In terrestrial areas, herbivores include some insects, mammals, reptiles, birds, and so on (see Figure 17.15a). In the aquatic areas, many of the herbivores are small invertebrates (such as zooplankton, see Figure 17.15b), but some vertebrates (some fish, a few reptiles, and a couple of mammals, such as manatees) are also herbivores.

FIGURE 17.15a Monarch butterfly caterpillar

FIGURE 17.15b Crustacean zooplankton sample

Plants have developed defenses against the plant eaters. Structural adaptations include thorns, which physically impair herbivore damage. On the other hand, some herbivores have also evolved to produce countermeasures. Land iguanas in the Galapagos Islands (see Figure 17.16a) preferentially eat the flowers, fruit, pads (modified stems), and spines (modified leaves) of the prickly pear cactus (see Figure 17.16b). The cactus not only provides food, but the water-storing pads also provide water for the iguanas. Water is a scarce resource on these dry islands. Despite the thorns, the iguanas have very tough mouths, which counteract the spines as a defensive mechanism.

Many plants have developed bad-tasting or toxic compounds, which may be on the surface or concentrated in their tissues. For instance, raw lima beans, as well as roots and leaves of the parent plant, contain significant amounts of a chemical compound called *linamarin*, which is classified by chemists as a cyanogenic glucoside (compound that releases the poison cyanide when the proper enzyme removes the glucoside [sugars]). When cyanide is released following digestion, the cyanide inhibits a key mitochondrial enzyme that transfers electrons to oxygen to generate water at the end of the electron transport system. Of course, the electron transport system is vital to producing sufficient ATP for cellular activities (e.g., skeletal or heart muscle contraction). Without ATP, cells will be unable to function normally, ultimately leading to death. Even so, some caterpillars can consume plant parts containing linamarin because they isolate the compound and prevent contact with the enzyme that would bring about the release of cyanide (Zagrobelny et al. 2004).

FIGURE 17.16a Galapagos land iguana

FIGURE 17.16b Galapagos prickly pear cactus

Parasitism

Parasitism is a non-mutualistic symbiotic interaction where one organism benefits (*parasite*) and one organism is harmed (*host*). Parasites may be internal or external to the host. Internal parasites (*endoparasites*), which live and access nutrients from inside their hosts, include the malaria-causing protist, hookworms, and tapeworms.

APPLYING SCIENCE TO LIFE 2 Uninvited Hitchhiker

Martin and three of his friends had been planned and trained to hike the John Muir Trail for years. When they started this trek in early August last year, they were so excited. For the first 18 days, the experience was incredible, easily living up to their expectations. A black bear passing through the campsite on the sixth day just added excitement. On day 19, with only a few miles to go, Martin and one of his friends began feeling poorly. Their stomachs cramped, and they burped and had lots of gas. Then they developed diarrhea. Feeling wretched, they found enough strength to finish the hike, after which their friends took them to an emergency room. Examination of their stools confirmed the doctor's diagnosis of giardiasis, and the doctor prescribed metronidazole, an antibiotic. Furthermore, she gave them oral rehydration solutions and advised them to drink lots of fluids. Within a few days, they were back on the road home.

PoI 3.

a. What is the disease agent for giardiasis?
b. What does the disease agent get from the host?
c. How does the disease agent cause the symptoms of the disease?
d. How is the disease agent spread?
e. How did these men likely acquire giardiasis?
f. How could they have prevented this disease?
g. What is the mechanism of action of metronidazole?

Ectoparasites are those that live and feed on the external surfaces of the host, and these include ticks, dodder (vine parasite of green plants), head lice, and bedbugs. Some of these ectoparasites also carry disease organisms, such as those for Lyme disease and Rocky Mountain spotted fever. It seems that parasites exist for the majority of all organisms. Even bacteria have their own viruses (bacteriophages), which are considered a type of parasite.

APPLYING SCIENCE TO LIFE 3 Transgender Crabs?

The North American mud crab is native to the Atlantic Coast, from the Gulf of Saint Lawrence to Veracruz, on the Gulf of Mexico. They are often found in and among oyster reefs or submerged vegetation. Transported by trade shipping to Europe, they now are considered an invasive species in an expanded territory. Their diet is quite varied, consuming plant detritus (decomposing particulate organic matter), remains of dead fish, and small invertebrate animals. As they have expanded their range, they have shown themselves to be quite adaptable to varying salinities, temperatures, and other environmental conditions. Having a significant impact on native communities, the North American mud crabs have proved to be important species for study by biologists.

As we discussed in the section on commensal relationships, whale barnacles are considered to be commensals. However, for the mud crab, there is a barnacle that is definitely a parasite, *Loxothylacus texanus*, which is a member of the barnacle family Sacculinidae. Members of this family are parasitic on a variety of crustaceans, including many crabs. These parasites produce both male and female larvae. The female, which is the infectious agent, attaches and penetrates the outer covering of the host. Once inside the crab, the female barnacle develops into an internal parasite, which grows, producing a brood of sacs (containing eggs) that emerge through the abdomen. Once a male larva fertilizes the eggs, they mature in the sac. Then mature larvae are released to spread the infection.

What is so bizarre about this relationship is that the growing parasite essentially takes over the crab's body. Infected crabs, male or female, have their gonads inactivated, and the crab essentially exists to feed the parasite and protect its young. Males, effectively castrated, have their anatomy modified to resemble females. The exact mechanism is being elucidated by scientists.

Some scientists propose the use of these parasites as a biological control for invasive species of crustaceans, like the green porcelain crab, which is extending its range from the tropics to as far north as Charleston, SC (Mack et al. 2019). The scientists suggest there is little danger to other species, because these parasites are very host-specific. Before instituting this type of biological control, however, extensive research should test host specificity very carefully. Too many times, this type of control has led to disastrous ecological consequences (Glenner et al. 2000; Fowler et al. 2013).

PoI 4. The cane toad was introduced to Australia in 1935 as a biological control for beetles that were ruining the sugarcane crops.

a. Did they help the beetle problem? Why or why not?
b. What impact have the toads had in Australia?
c. What problems might ensue from using the *Loxothylacus texanus* parasite to control invasive crabs?
d. Have there been any successful uses of biological control in natural systems? If so, describe a couple of examples.

QRQ 3. Select from the following the appropriate term for each description below: aposematic, coevolution, commensalism, communal, cryptic, ectoparasite, endoparasite, exploitation, harbinger, herbivory, host, inner parasite, interdependence, mimicry, mutualism, outer parasite, parasite, parasitism, predation, reciprocity, symbiosis

a. ability of an animal to imitate another in order to survive
b. coloration that advertises to potential predators the animal is poisonous
c. coloration that makes an animal less visible to predators
d. each of two organisms reciprocally promotes adaptations in the other
e. interaction where an animal consumes parts of plants
f. interaction where an offensive predator hunts a defensive prey
g. interaction where both participants benefit from the relationship
h. interaction where one organism is benefited and one organism is harmed
i. interaction where one partner is benefited and the other is not affected
j. internal parasite
k. organism benefiting from a parasitism relationship
l. organism harmed from a parasitism relationship
m. parasite that lives and feeds on external surfaces of another organism
n. predation, herbivory, and parasitism interactions
o. relationship between two species living in close association for at least part of their life cycles where one or both species are positively affected

QRQ 4. Classify each relationship as commensalism, herbivory, mutualism, parasitism, or predation:

a. Burmese python and rodent
b. cat and mouse
c. cattle and grass
d. deer tick and deer
e. Egyptian plover eating meat from a crocodile's teeth, thereby cleaning the teeth
f. fish and seaweed
g. flowering plants and bees
h. hookworms and humans
i. humpback whales and barnacles
j. pseudoscorpion attached to and dispersed by beetle
k. zooxanthellae living inside a giant clam, each providing nutrients for the other

QRQ 5. The nontoxic viceroy butterfly physically resembles the toxic monarch butterfly, which is an example of ______.

QRQ 6. The following is an example of ______: Cheetahs and antelope have evolved together, with the cheetah becoming a better predator as the antelope becomes a more elusive prey, and vice versa.

17.6. Trophic Structure

Ecological communities have a *trophic* (feeding) structure, which embraces all of the interactions we have discussed. Trophic structure is based on levels of nutritional relationships. Typically, connections between trophic levels can be illustrated using what ecologists call a *food web*. More simply, we sometimes describe basic trophic relationships using a linear *food chain*. The trophic levels usually include primary producer, primary consumer, secondary consumer, tertiary consumer, and perhaps quaternary consumer. In each case, a lower level provides food and energy to the level above. Most communities display more complex, interconnected feeding relationships, in the form of a web where you can highlight individual food chains within these webs. Additionally, a web interconnects with another important food chain that is based on the utilization of dead organic material, called *detritus*.

17.6.1. Producers and Consumers

Let's dissect the trophic structure of a community. In both terrestrial and aquatic communities, *primary producers* are made up of photosynthetic or, in some extreme conditions, chemosynthetic organisms. Terrestrial primary producers include green plants and photosynthetic algae or bacteria. In aquatic communities, this level is dominated by algae or photosynthetic bacteria and protists, although there are aquatic plants in shallow zones. Chemosynthetic organisms are primary producers around deep water thermal vents and in geothermal springs. Thus, primary producers use light or chemical oxidations to produce the organic molecules and the energy to support the entire community. For simplicity, we are going to use photosynthetic primary producers as the basis for our discussion.

Primary consumers are herbivorous animals that are adapted for eating leaves, fruit, seeds, etc. In a terrestrial setting, these animals might be insects; grazing, browsing, or seed- or fruit-eating mammals; or seed- or fruit-eating birds. In most aquatic systems, algae and other photosynthetic microorganisms feed protists, zooplankton, larger invertebrate animals, some fish, and a number of birds.

The animals at the next level will eat the primary consumers. *Secondary consumers* include carnivores, insectivores, and perhaps some animals that eat some parts of primary producers. Animals that consume both animal and vegetable materials are called *omnivores*, and these organisms can feed at both levels lower in the food web. At this level, we would find some insects, arachnids (e.g., spiders), various mammals (e.g., racoons), reptiles, fish, and some birds.

Although *tertiary consumers* are *carnivores*, eating meat, many are also omnivores. In some communities, this level may be the topmost level. Animals in this category include tuna and other fish, dolphins and toothed whales, sharks, snakes, cats, foxes, birds, and coyotes.

Quaternary consumers are truly the "top dog" in some communities, primarily feeding on other carnivores at the secondary and tertiary levels. Usually these animals lack natural predators, other than human predators. Quaternary consumers include raptors, such as hawks and eagles, and killer whales. Roughly 10 percent of the energy available in the *biomass* (organic material containing stored energy) of one level is available for the next higher level. Consequently, some communities lack this trophic level, because there is not enough energy available from the tertiary consumers to support another level. In those cases, tertiary consumers are at the top of the food web.

Envision these trophic levels as a pyramid, with the base formed by the primary producers and the top consumer at the apex of the pyramid (see Figure 17.17). At every level, energy and nutrients are lost through respiration (heat).

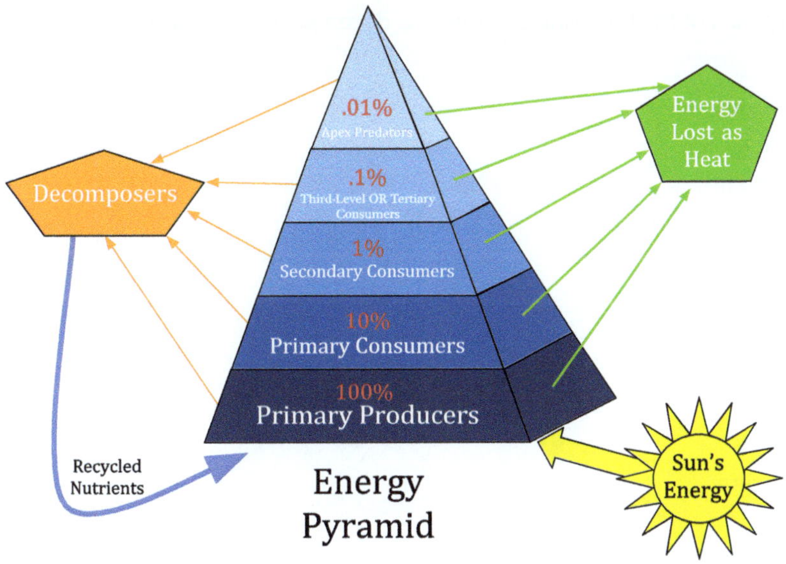

FIGURE 17.17 Ecological (energy) pyramid

17.6.2. Food Webs

The above food chain describes what might be called the *grazing food web,* because the first consumers are grazers, feeding on energy-capturing organisms. Interwoven with these types of feeding relationships is the *detritus food web.* Detritus is dead organic matter, which is formed from dead organisms. There are three groups of organisms involved in this food web, *scavengers, detritivores,* and *decomposers. Scavengers* are animals, such as vultures, that consume the flesh of dead animals (*carrion*). Other animals, such as the hyena, are primarily carnivores but also eat carrion. Insects such as the cockroach are also considered scavengers, but they will scavenge materials other than that derived from dead animals. Scavengers reduce the number of dead animals and help to produce smaller bits of detritus that is further digested by detritivores and decomposers. *Detritivores* are animals that consume organic material, as well as some decomposer organisms, further breaking down the organic matter, thereby absorbing and excreting nutrients. Detritivores include animals such as millipedes, earthworms, sea cucumbers, and crustaceans. In turn, the detritivores are eaten by various consumers, including arthropods and birds. *Decomposers* (bacteria, fungi, and other microbes) also use detritus for nutrients and energy, usually secreting enzymes to digest the detritus, which produces molecular nutrients the decomposers can absorb. In all of the categories in this web, nutrients are released into the soil. These nutrients are useful to support plants and other life in the grazing food webs, which employ the nutrients to build new organic materials for growth and reproduction.

QRQ 7. Complex, interconnected ______ (feeding) relationships, including utilization of detritus, is called a(n) ______ ______. Basic linear feeding relationships is called a(n) ______ ______.

QRQ 8. Leaves, fruit, seeds, and algae are on the lowest level of the complex interconnection of feeding relationships and so are called ______ ______. Deer, fruit bats, finches, and zooplankton, which eat these, respectively, are on the next higher level of feeding relationships and, so, are called ______ ______.

QRQ 9. Classify each of the following as carnivore, decomposer, detritivore, detritus, omnivore, or scavenger:

a. bacterium
b. cockroach
c. dead fish
d. decaying leaves
e. earthworm
f. hawk
g. human
h. mushroom
i. rat
j. shark
k. vulture

17.7. Community Dynamics: Succession

One of the characteristics of any community is that it changes. Whether short, seasonal, or yearly, the community will change. Longer term, there are changes in species composition and relative abundance within communities, and these bring about a set of alterations in community structure that is both directional (a certain trajectory of events) and predictable. We term such a series of changes *ecological succession*. Most of the focus of past research has been with changes in plant communities. For that reason, descriptions of successional stages usually emphasize the predominating plant species.

After some sort of disturbance, such as a volcanic eruption, succession commences. For example, there are ongoing volcanic disturbances on the Big Island of Hawaii. This island is actually made up of five volcanoes, three of which are still active. The most active is Kilauea, which has had frequent eruptions during the past 35 years. In 2018, a dramatic eruption produced enough lava to cover almost 14 sq. mi, adding new surfaces on which succession could begin.

Ecologists term succession that starts on a biologically barren site, such as a fresh lava flow, *primary succession*. The colonizing organisms are called *pioneer species*, and the first pioneers arrive quickly. Generally, there are islands of intact vegetation within the flows, or unaffected vegetative areas are nearby to supply spores and seeds for propagation on the lava surface. As rain and wind begin to wear away at the lava, pioneer plants may add some acidic secretions that further the breakdown of the rock (lava), releasing nutrients such as iron and potassium. Early phases of rainforest succession on Hawaiian lava include *Stereocaulon* lichen stage, *Dicranopteris* fern stage, and *Cibotum* tree fern stage (Mueller-Dombois and Boehmer 2013)

17.7.1. Early Stages

Most of these pioneer organisms are very hardy, low to the ground, and often growing on bare lava rock. Pioneers augment the effects of weather in breaking down the rock, trapping particulates, and adding to the organic matter, building soil. Some ferns (*Nephrolepis*) may join this early community, growing with mosses in the cracks or other irregularities of the lava flow. This stage of succession is referred to as the *Stereocaulon lichen stage*. The lichen is quite apparent within about four years. Pioneer organisms include the following:

1. **Cyanobacteria** (blue-green algae), which fix nitrogen (convert atmospheric nitrogen, unavailable to plants, into an available form of nitrogen)
2. **Mosses**
3. **Lichens**, organisms made up of various species of fungi and algae or cyanobacteria, considered a mutualistic relationship

As soil continues to develop, particularly in crevices, larger ferns and seed-producing plants can be successful. The deeper soil helps to hold some moisture, and shrubs such as heather and tree seedlings of a species of flowering evergreen trees (*Metrosideros polymorpha*) establish themselves. Shortly, additional shrubs also become established.

After 50 years or more, the tree saplings grow to form a stand of young trees. During this time, a native fern, *Dicranopteris*, has taken its place under the rather open canopy of the saplings. Branches on the younger trees tend to be short, and the light-loving fern grows with abandon. This fern spreads from underground stems that branch out, producing new vertical shoots, which can branch in an indeterminate fashion. Soon, the fern covers the ground, forming mats and some dense thickets. With plenty of sunlight, the fern will thrive, which is a reason for calling this stage the *Dicranopteris fern stage* (see Figure 17.18). During that time, few if any new seeds germinate, but the very young trees (seedlings) are favored by the sunlit forest and continue to grow. During this stage, fresh *humus* (dead organic matter), produced from decaying plant material, begins to accumulate to form a thin layer of organic matter called *peat*. The peat layer helps the area hold onto even more moisture.

FIGURE 17.18 Dense patch of *Dicranopteris* ferns

The community continues to grow, accumulating a matted layer of organic matter. When the trees mature, their branches begin to overlap, reducing the sunlight to the ground significantly and stimulating a decline in the *Dicranopteris* ferns. Meanwhile, a new, larger fern begins to become more prominent, the *Cibotium* tree fern. Consequently, this phase of succession is referred to as the *Cibotium tree fern stage.*

At this point in succession, the established forest, with *Metrosideros* as the only *canopy tree* (tree species forming the topmost cover or crown of a plant community), is self-perpetuating, with younger *Metrosideros* trees taking over as the older trees fade. Additionally, the *Cibotium* tree ferns remain to dominate the *understory* (the layer underneath the canopy). Diversity is limited because the changes take place in relative isolation, meaning that few new species, other than a few subcanopy trees, are introduced (see Figure 17.19).

FIGURE 17.19 Small tree ferns (*Cibotium*) under a canopy of *Metrosideros* near the Kilauea Visitor Center

17.7.2. Secondary Succession

Secondary succession occurs when an area that has already been colonized by living organisms is recolonized. Such succession follows various disturbances, such as forest fires or clearing of an established group of organisms, often for an agricultural purpose, or abandonment of agricultural lands. The classic studies of Keever (1950) in old-field succession in the Piedmont region of North Carolina, which Table 17.2 summarizes, are still illustrative and relevant.

TABLE 17.2. Dominant Plants Present at Different Times in Old-Field Succession (Keever 1950)

Years After Abandonment	Plants
0^{+}	crabgrass
1	horseweed, crabgrass
2	white aster, crabgrass, horseweed
3	broomsedge, mixed herbaceous
4	broomsedge, pine seedlings, mixed herbaceous
5	young pines, broomsedge, mixed herbaceous
18^{+}	closed pine stands, young hardwoods

Actually, hardwood species, such as oak and hickory, begin to grow under the pines, eventually, replacing the pines after 70^{+} years. This community is sometimes referred to as the *climax forest community*, dominated by species of oak and hickory with a mixture of understory hardwood tree species. Well-adapted to the environment (climate, topography, etc.), numerous species interactions stabilize the community.

Keever's focus was on the herbaceous plants, and she drew a number of conclusions from data, from both field and greenhouse studies, with regard to the major causes of this rather certain set of changes. She determined the following underlying processes to be most important:

1. **Seed dispersal.** Horseweed has early success because of its prolific production of windblown seeds. Crabgrass, often present early, may be obscured by taller horseweed plants. Horseweed produces sufficient numbers of seeds, which do not germinate until the spring, and displays spreading and vegetative growth that slows as temperatures cool in the fall/winter.
2. **Interspecific (and a little intraspecific) competition.** During the second year, horseweed plants lose out in competing for water and minerals with other horseweed plants, as well as young aster plants that begin to grow in the area. This competition accounts for the primacy of the white aster during the second year after cultivation ceases. During the third year, the asters, which are not drought-resistant, lose the battle for water with incoming drought-resistant broomsedge, which also reduces the water available in the soil. Also, mature asters like bright sun, and the broomsedge reduces the amount of sunlight available for the aster plants. Aster seedlings are also shaded out by taller plants, including mature broomsedge and asters.
3. **Life history characteristics.** Horseweed is dominant first because its seeds are mature about the time that many of the crops grown in the area are last cultivated. The seeds produced do not require a dormant period, so they germinate right away. The plant overwinters, matures, blooms, produces seeds, and dies the following year. Other competing species have spring-germinating seeds (white aster) or are slow-growth perennials.

17.7.3. Animals

During the changes that take place in the plant community, there are also changes in and contributions by animals and microbes. Plant ecologists were the primary investigators and advocates of succession for many years. Therefore, not as much literature reflects the changes in and roles of other organisms, although scientists at least mentioned in passing the contributions of algae and fungi.

Animals are scant in the pioneer stages of plant succession. Little food is available, and the living space is nonexistent for anything other than very small critters, especially in primary succession. The animals that might be able to tolerate fairly rigorous conditions of newly formed habitats might include a few arthropods, such as ants, mites, and small spiders. As larger and a greater diversity of plants become established, more soil forms; moisture and shade increase; more insects, their larvae, and nematodes (roundworms) arrive; and there may be changes in species of pioneer animals. As taller plants, including seed-producers, appear and reproduce, the potential niches for animals increase for various invertebrates (more insects, worms, snails, etc.) and small vertebrates, such as lizards, mice, squirrels, moles, amphibians, and some birds. As the *climax plant community* approaches, the presence of a more enticing habitat plus the availability of prey attract more predators, such as snakes, foxes, hawks, and so on.

It is difficult to describe the successional changes and roles microbes play during plant succession. Pioneer species often include bacteria, cyanobacteria, lichens, and fungi. Bacteria and fungi, in particular, play a large role in decomposition of organic material, making nutrients available, which pass first to plants but which help to support the rest of the community as well.

17.7.4. Conclusions

What kind of conclusions should we draw based on ecological succession research? Nature isn't simple to understand. Yes, nature is complex, and scientists are working hard to understand this complexity. At this point, studies of succession would lead us to conclude that there are a variety of factors that influence succession—both *biotic* (living) and *abiotic* (nonliving). Untangling the underlying interactions will certainly take more work, but we do have and continue to develop more sophisticated tools, and powerful modeling capabilities should bring us to a closer understanding.

QRQ 10. List in order the early phases of rainforest succession on Hawaiian lava (omit inappropriate selections): *Cibotum* tree fern, *Dicranopteris* fern, *Opuntia* prickly pear cactus, *Pisaster ochraceus*, *Stereocaulon* lichen

QRQ 11. List in order the changes in animals during plant succession: animals are scarce; more insects and roundworms arrive; more predators such as snakes, foxes, hawks, etc. arrive; small spiders, ants, mites arrive; various invertebrates and small vertebrates arrive

QRQ 12. Select from the following the appropriate term for each description below: biological succession, bushes, climax community, *Dicranopteris* fern, ecological succession, first, forerunner, peat, pioneer, primary, secondary, understory

a. colonizing species in primary succession
b. dead organic matter
c. forest dominated by hardwoods
d. layer underneath the tree canopy
e. series of long-term changes in a community
f. species that break down rock, trap particulates, build soil
g. stage of succession when humus begins to accumulate
h. steady state community in ecological succession
i. succession following disturbances or abandonment of agricultural lands

j. succession starting on a biologically barren site, such as a fresh lava flow
k. thin layer of organic matter

Image Credits

Fig. 17.1: Source: https://commons.wikimedia.org/wiki/File:Burmese_python_(6887388927).jpg.
Fig. 17.2: Copyright © by NOAA's National Ocean Service (CC BY 2.0) at https://commons.wikimedia.org/wiki/File:Kelp_Forest_(12801115735).jpg.
Fig. 17.3: Copyright © by Ed Bierman (CC BY 2.0) at https://commons.wikimedia.org/wiki/File:Purple_Sea_Urchins_(7622488604).jpg.
Fig. 17.4: Copyright © by matt knoth (CC BY 2.0) at https://commons.wikimedia.org/wiki/File:Sea_otter_with_sea_urchin.jpg.
Fig. 17.5: Copyright © by Steven Pavlov (CC BY-SA 4.0) at https://commons.wikimedia.org/wiki/File:Starfishes.jpg.
Fig. 17.6: Source: https://commons.wikimedia.org/wiki/File:Giant_clams_(12197415505).jpg.
Fig. 17.7: Source: https://commons.wikimedia.org/wiki/File:Pseudoscorpion_-_A._Wild.jpg.
Fig. 17.8: Source: https://commons.wikimedia.org/wiki/File:Cattle_egret_Niger.jpg.
Fig. 17.9a: Source: https://commons.wikimedia.org/wiki/File:Humpback_tail_Fallarones.jpg.
Fig. 17.9b: Source: https://commons.wikimedia.org/wiki/File:Anim1032_-_Flickr_-_NOAA_Photo_Library.jpg.
Fig. 17.10: Copyright © by Froggydarb (CC BY-SA 3.0) at https://commons.wikimedia.org/wiki/File:Bufo_marinus_from_Australia.JPG.
Fig. 17.11a: Copyright © by Sztraka38 (CC BY-SA 2.0) at https://commons.wikimedia.org/wiki/File:Litoria_nannotis_camouflage.jpg.
Fig. 17.11b: Copyright © by Zeynel Cebeci (CC BY-SA 4.0) at https://commons.wikimedia.org/wiki/File:Thomisus_onustus_-_Yellow_crab_spider_04.jpg.
Fig. 17.12a: Copyright © by Brian Gratwicke (CC BY 2.0) at https://commons.wikimedia.org/wiki/File:-Golden_Poison_dart_frog_Phyllobates_terribilis.jpg.
Fig. 17.12b: Source: https://commons.wikimedia.org/wiki/File:Poison_dart_frog_229.JPG.
Fig. 17.13: Copyright © by PiccoloNamek (GNU 1.2) at https://commons.wikimedia.org/wiki/File:Monarch_Viceroy_Mimicry_Comparison.jpg. A copy of the license can be found here: http://www.gnu.org/licenses/gpl.html
Fig. 17.14a: Source: https://commons.wikimedia.org/wiki/File:Young_Cheetahs.jpg.
Fig. 17.14b: Source: https://commons.wikimedia.org/wiki/File:Aepyceros_melampus_(South_Africa).jpg.
Fig. 17.15a: Source: https://commons.wikimedia.org/wiki/File:Monarch_caterpillar_eating_swan_plant_leaf.jpg.
Fig. 17.15b: Source: https://commons.wikimedia.org/wiki/File:Zooplankton.jpg.
Fig. 17.16b: Copyright © by Jason Hollinger (CC BY 2.0) at https://commons.wikimedia.org/wiki/File:Opuntia_echios_var._echios_(12663871645).jpg.
Fig. 17.17: Copyright © by Swiggity.Swag.YOLO.Bro (CC BY-SA 4.0) at https://commons.wikimedia.org/wiki/File:Ecological_Pyramid.svg.
Fig. 17.18: Copyright © by Filo gèn' (CC BY 4.0) at https://commons.wikimedia.org/wiki/File:Dicranopteris_linearis_(Gleicheniaceae).jpg.
Fig. 17.19: Source: https://commons.wikimedia.org/wiki/File:Hiking_through_the_forest_on_the_Halema‘uma‘u_trail.jpg.

LAB #17

Laboratory Exercises

17.8. Laboratory on Community

17.8.1. Preliminary Work

We will use a community model to investigate various predator-prey-invasive species scenarios. From the text's website, view the introductory video; download and open "*Community Rev.nlogo*."

1. Click the Info tab at the top of the page and read the sections "What Is It?" and "How to Use It."
   - **a.** ______ Give the maximum number of offspring of an eagle when giving birth.
   - **b.** ______ Give all that each of the following eats from the following list: eagles (E), rabbits (R), vegetation (V), wallabies (W)

     _______ eagles

     _______ rabbits

     _______ wallabies

2. Read "How to Use It."
   - **a.** Select the model color for each component: A. black, B. blue, C. brown, D. gray, E. green, F. orange, G. red, H. rose, I. white, J. yellow

     _______ eagle

     _______ rabbit

     _______ wallaby
   - **b.** Select from the following the appropriate term for each description below: A. %-VEGETATION-COVERAGE, B. CHANCE-EAGLE-CATCHES-RABBIT, C. ENERGY-FROM-WALLABY, D. SIMULATION-LENGTH, E. TIME-FOR-INVASION, F. VEGETATION-INTENSITY, G. WALLABY-VEGETATION-ENERGY

     _______ amount of energy a wallaby gains from eating patch's vegetation

     _______ amount of energy an eagle gains from eating a wallaby

_______ for the model not to automatically stop, this value should be 0

_______ in a patch with no wallabies, eagle will eat a rabbit with this probability

_______ minimum vegetation value in a cell with vegetation

_______ percentage of patches with a random vegetation value

c. _______ percent Suppose the current amount of vegetation is 47,870, and the original amount of vegetation is 155,750. Calculate vegetation level, rounded to one decimal place.

d. _______ percent Suppose the current amount of vegetation is 112,052, and the original amount of vegetation is 86,733. Calculate vegetation level, rounded to one decimal place.

e. _______ The MEAN EAGLES monitor displays the average number of eagles from tick **1.** to the end of the simulation.

f. _______ After execution of an experiment that stops with a SIMULATION-LENGTH greater than 1,000, what is the easiest way to transfer values for the various means from the model to a spreadsheet?

  a. copy and paste individual values
  b. copy and paste the line of values
  c. read values from monitors and type into spreadsheet
  d. read values from graph and type into spreadsheet

17.8.2. Laboratory Work

Give numeric values rounded to one decimal place. Default parameters: %-VEGETATION-COVERAGE = 95, VEGETATION-INTENSITY = 50, WALLABY-VEGETATION-ENERGY = 4.0, RABBIT-VEGETATION-ENERGY = 4.0, SIMULATION-LENGTH = 2,000, INITIAL-EAGLES = 20, CHANCE-EAGLE-CATCHES-RABBIT = 1.00, INITIAL-WALLABIES = 100, ENERGY-FROM-WALLABY = 25, MAX-WALLABY-OFFSPRING = 1, RABBIT-INVASION? = OFF, TIME-FOR-INVASION = 500, RABBITS-TO-ADD = 25, ENERGY-FROM-RABBIT = 25, MAX-RABBIT-OFFSPRING = 10

3. In this question, we consider the simplest of all predator-prey situations, one predator population and one prey population. To do so, we will only have wallabies and vegetation, where the former is the predator and the latter is the prey. Thus, set INITIAL-EAGLES to 0 and RABBIT-INVASION? to OFF. Only graph the wallabies and the vegetation.

   a. **1.** _______ **2.** _______ **3.** _______ **4.** _______ **5.** _______ Set WALLABY-VEGETATION-ENERGY to 1.5 and run the simulation for 1,000 ticks. Otherwise, use parameter default values. Initially, as the wallaby population increases, the vegetation **1.** (A. decreases, B. increases, C. stays approximately the same). With less food, the wallaby population **2.** (A. decreases, B. increases, C. stays approximately the same). With fewer wallabies, the vegetation **3.** (D. can, E. cannot) recover somewhat. Then, with more vegetation, the number of wallabies **4.** (A. decreases, B. increases, C. stays approximately the same). The system cycles through these results, although the maximum and minimum numbers **5.** (F. do, G. do not) dampen.

   b. In the next few questions, we will investigate carrying capacity and the impact of changes in the amount of vegetation a wallaby eats. Change SIMULATION-LENGTH to 2,000 and uncheck *view updates*. Run the simulation at least five times, recording MEAN WALLABIES and MEAN

VEGETATION LEVEL in a spreadsheet. Label the data with the question number and indicate WALLABY-VEGETATION-ENERGY = 1.5. Over all experiments, calculate and give the averages of the following:

_______ MEAN WALLABIES

_______ MEAN VEGETATION LEVEL

c. ______ Do you think these values will change if we increase the initial number of wallabies, no (N) or yes (Y)?

d. ______ Increase INITIAL-WALLABIES from 100 to 300 and repeat part b. Does this increase change the long-term number of wallabies or the amount of vegetation, no (N) or yes (Y)?

e. ______ Thus, what is the approximate carrying capacity (midpoint of the eventual stable cycle) of the wallabies?

f. ______ We now investigate the impact of reducing how much a wallaby eats even further. Change WALLABY-VEGETATION-ENERGY to 1.0 and INITIAL-WALLABIES back to 100. Run the simulation for 500 ticks at least three times, recording the *final* number of wallabies and vegetation level with appropriate labels in your spreadsheet. Calculate and give the average *final* vegetation level.

g. Discuss why the predator becomes extinct.

h. **1.** ______ **2.** ______ Complete the hypothesis, using the following choices for each blank: D. decrease, I. increase, R. remain the same. As we increase WALLABY-VEGETATION-ENERGY, when extinction does not occur and the system stabilizes, the mean midpoint of the wallaby population size will **1.** and the mean midpoint of the vegetation level will **2.**

i. Repeat part b with WALLABY-VEGETATION-ENERGY = 4.0.

_______ MEAN WALLABIES

_______ MEAN VEGETATION LEVEL

j. Repeat part b with WALLABY-VEGETATION-ENERGY = 8.0.

_______ MEAN WALLABIES

_______ average MEAN VEGETATION LEVEL

k. ______ Do your results support your hypothesis in part h, no (N) or yes (Y)?

l. Discuss why you think increasing WALLABY-VEGETATION-ENERGY causes these results.

4. This question explores a community with eagles, wallabies, and vegetation with no rabbit invaders. Thus, we have an apex predator, or a predator at the top of the food chain, and a prey, which, in a sense, is also predator to the vegetation, the lowest level of prey.

a. **1.** ______ **2.** ______ **3.** ______ Use the default values (with WALLABY-VEGETATION-ENERGY = 4.0 and INITIAL-EAGLES = 20), except have SIMULATION-LENGTH = 1,000. Graph eagles, wallabies, and vegetation. Check *view updates* and run the simulation. For each blank, use decreases (D), increases (I), or stays the same (S): Initially, as the wallaby population increases, the amount of vegetation **1.** and then the number of eagles **2.** Shortly after the number of wallabies starts decreasing, the amount of vegetation **3.**

b. Discuss why the eagle population continues decreasing for a while after the wallaby population starts increasing. Also, consider why the vegetation level responds faster to changes in the number of wallabies.

c. **1.** ______ **2.** ______ Complete the following hypothesis, using less (L), more (M), or about the same (S) for each blank: Long term, the number of wallabies will be **1.** and the amount of vegetation will be **2.** when eagles are present than when this apex predator is not in the area.

d. Change SIMULATION-LENGTH to 2,000 and uncheck *view updates*. Run the simulation at least five times, recording MEAN EAGLES, MEAN WALLABIES, and MEAN VEGETATION LEVEL in your spreadsheet. In a separate column, indicate experiments where wallaby extinction occurs. Label the data with the question number and indicate WALLABY-VEGETATION-ENERGY = 4.0. Have your spreadsheet calculate the averages and standard deviations of the values below. Give the averages (not standard deviations):

_______ MEAN EAGLES

_______ MEAN WALLABIES

_______ MEAN VEGETATION LEVEL

e. ______ Compare your wallaby and vegetation answers from part d with those from question 3i. Do the results support your hypothesis from part c?

f. Discuss how and why the presence of eagles impacts the number of wallabies and the amount of vegetation in the long term.

5. This question considers the interrelationship of all species in the community, including the European rabbit, an invasive species. Start with the default parameters, except turn on RABBIT-INVASION. Also, to have an invader that is in many ways comparable to the current eagle prey (wallabies), set MAX-RABBIT-OFFSPRING to 1. Display all graphs.

   a. Suppose the rabbits have an advantage because the eagles are not as adept at catching the invaders. Maybe the rabbits are faster or hide better than the wallabies. Thus, change CHANCE-EAGLE-CATCHES-RABBIT to 0.80. Run the experiment for at least five times when the eagles do not become extinct before the rabbit invasion (tick 500). Record MEAN EAGLES, MEAN WALLABIES, MEAN RABBITS, and MEAN VEGETATION LEVEL in your spreadsheet. In a separate column, indicate experiments where wallaby extinction occurs by tick 2,000. Be sure to include in your spreadsheet the question number and the changed parameters. Have your spreadsheet calculate the averages and standard deviations of the values below. Give the averages (not standard deviations):

   _______ MEAN EAGLES

   _______ MEAN WALLABIES

   _______ MEAN RABBITS

   _______ MEAN VEGETATION LEVEL

   b. **1.** ______ **2.** ______ For each blank, use the corresponding letter for less (L), more (M), about the same (S), or varied, with some more and some less (V). Comparing the results for question 4d and part a, the averages for MEAN EAGLES, MEAN WALLABIES, and MEAN VEGETATION LEVEL are **1.** and the standard deviations are generally **2.** when there is an invasion with CHANCE-EAGLE-CATCHES-RABBIT = 0.80 than when there was no invasion.

c. Write a thoughtful discussion about why each of the changes might occur.

d. Let us now consider the situation where eagles catch rabbits as readily as wallabies, but where the rabbits have more voracious appetites and, consequently, gain more energy from eating. Thus, for this experiment, return CHANCE-EAGLE-CATCHES-RABBIT to its default value of 1.00, but change RABBIT-VEGETATION-ENERGY to 4.2, a slightly larger value than the wallaby vegetation energy. Run the experiment for at least five times when the eagles do not become extinct before the rabbit invasion at time 500. Record MEAN EAGLES, MEAN WALLABIES, MEAN RABBITS, and MEAN VEGETATION LEVEL in your spreadsheet. In a separate column, indicate experiments where wallaby extinction occurs. Be sure to include in your spreadsheet the question number and the changed parameters. Over all experiments, have your spreadsheet calculate the averages and standard deviations of the values below. Give the averages (not standard deviations) of each of the following:

_______ MEAN EAGLES

_______ MEAN WALLABIES

_______ MEAN RABBITS

_______ MEAN VEGETATION LEVEL

e. **1.** _______ **2.** _______ **3.** _______ For each blank, use less (L), more (M), or about the same (S): Comparing the results for part d to those of question 4d, the average for MEAN EAGLES was **1.** when there was an invasion with RABBIT-VEGETATION-ENERGY = 4.2 than when no rabbits appeared. Moreover, the average for MEAN WALLABIES was **2.** and the standard deviation was **3.** when we had the invader.

f. Write a thoughtful discussion about why each of the changes in mean eagle and wallaby numbers and mean vegetation levels might occur.

g. Suppose the invasive rabbits are harder to catch and eat more vegetation. Thus, set CHANCE-EAGLE-CATCHES-RABBIT to 0.80 and RABBIT-VEGETATION-ENERGY to 4.2. Run the experiment for at least five times when the eagles do not become extinct before the rabbit invasion, recording output in your spreadsheet. Calculate and give the following averages:

_______ MEAN EAGLES

_______ MEAN WALLABIES

_______ MEAN RABBITS

_______ MEAN VEGETATION LEVEL

h. Looking at your data for parts a, d, and g, indicate which change(s) of parameter—A. CHANCE-EAGLE-CATCHES-RABBIT = 0.80, D. RABBIT-VEGETATION-ENERGY = 4.2, G. both—has the biggest impact on each of the following situations:

_______ hurting the number of eagles

_______ increasing the chance of collapse of the wallaby population

_______ hurting the amount of vegetation

MODULE

18 Ecosystems—Natural Networks

18.1. Introduction

Studies of *communities* investigate interrelationships between populations of different species sharing a given geographic area. These populations represent *biotic* (living) components—plants, animals, microorganisms. Studies of *ecosystems* include both biotic and *abiotic* (nonliving) components. Thus, ecosystem ecologists study the living communities as they interact with their physical environment. You may think an ecosystem includes a fairly large area, and indeed many ecosystems, like tropical rainforests, deserts, oceans, and rivers, may be quite extensive. However, a small pond or a decaying trunk of a tree could also be considered an ecosystem.

18.2. An Estuary

APPLYING SCIENCE TO LIFE 1 Bountiful Waters

Indian River Lagoon (IRL) is a narrow *estuary* (coastal body of water, partly enclosed, where freshwater from land meets the saltwater of the ocean) that extends for more than 150 miles along the eastern coast of Florida (see Figure 18.1). IRL can be better described as a *lagoon* (small body of water separated from a large body of water by some natural barrier [e.g., reefs, barrier islands]) system. In this case, the estuary is formed and protected by a chain of *barrier islands* (islands that lie parallel to the coastal shore).

During the 1830s, a US Army surgeon, Jacob Rhett Motte, was posted in Florida during a phase of the Seminole Wars. While there, Motte explored the Indian River Lagoon, and we have his journal that describes some of what he observed (Motte 1953). If you had been with him, you would have experienced a natural ecosystem saturated with thousands of species of plants and animals. In these waters, you would have encountered extensive open water, seagrass beds, salt

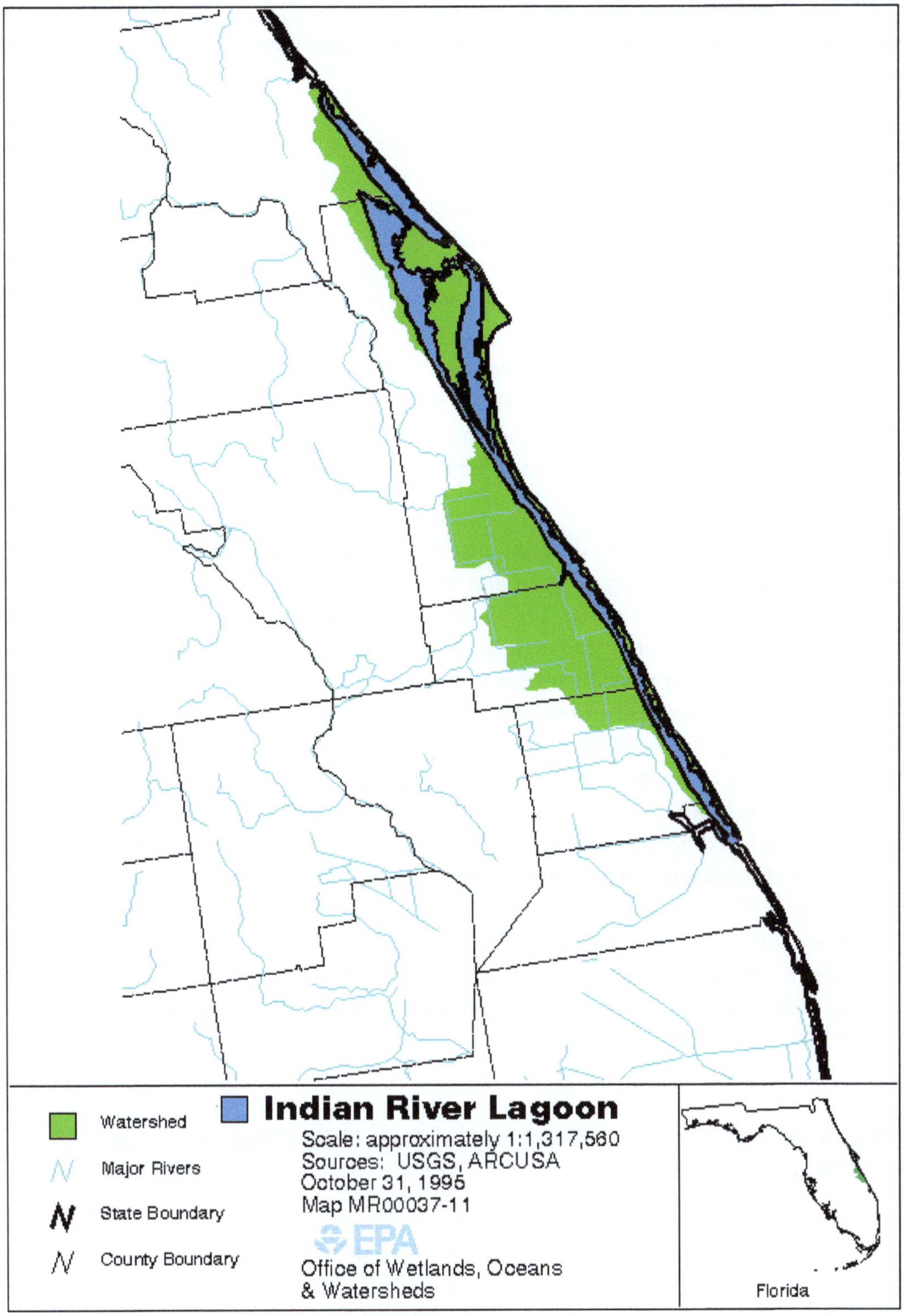

FIGURE 18.1 Map of Indian River Lagoon

marshes, and mangroves. Animal diversity, particularly birds, would have surpassed in number and variety most people's imagination. You can get some idea of his observations of the animal life by reading words from Motte's edited journal:

> Vast numbers of aquatic birds covered the waters, those unfrequented lagoons being the resort of millions of all the varieties of waterfowl that were generally found in that latitude. Innumerable species of wild-ducks, pelicans, gulls, etc. floated around us, whilst in all directions as far as the eye could see, the dark and leafless limbs of the closely growing mangrove were dotted with thousands of white cranes (probably egrets), as with flakes of feathery snow, giving to the scene the peculiar appearance of extensive cotton fields after a frost. ...
>
> Inadequate are the words to express the quantity and quality of the fish that abounded in these waters. In one hour one may take as many Bonetas, Redfish, Groupers, Bass, Trout, and myriads of other kinds....As for the oysters; six are a comfortable meal for one person, indeed without exaggeration, the greatest abundance may be easily obtained, which after removal from the shell measure six or seven inches in length, and two or three inches in breadth.

As you can conclude from Motte's descriptions of the variety and abundance of life in the IRL, estuaries are extremely rich biologically. In and around the mangrove, seagrass beds, and salt marshes, a wide variety of animals feed, breed, and nest. Besides numerous species of birds and fish, estuaries often are inhabited by young turtles, mammals, and a vast array of invertebrate animals. Moreover, estuaries still provide at least temporary homes to almost all recreationally and commercially caught species of fish and shellfish in the state of Florida.

On the roots of mangroves live a variety of animals, like anemones, snails, echinoderms (brittle stars, etc.), sponges, and sea urchins. Shielded in hiding places within the roots are many species of small or immature fish. An assortment of wildlife, including birds, may use mangrove stands for breeding, nesting, and feeding. The presence of mangroves helps to stabilize the shoreline, helping it resist the impact of storms. By absorbing some of the excess nutrients that enter the estuary, these trees help to deter the development of an overwhelming growth of algae, called a *bloom*. Such blooms may kill aquatic animals by removing oxygen for their own respiration and/or by releasing toxins. Because mangroves are most prevalent in tropical and subtropical areas, they occur primarily in the southern areas of IRL. In fact, the southern IRL represents the northernmost limit for the growth of mangroves (Florida Museum 2019, "Mangroves"; Smithsonian 2018).

APPLYING SCIENCE TO LIFE 2 Poisoned Waters: Little Organisms but Enormous Effects

Algal blooms like those in the IRL are not restricted to the east coast of Florida, of course, and the west coast of Florida has had its share of outbreaks. At the root of these blooms is a dinoflagellate, *Karenia brevis*, or *K. brevis*. Dinoflagellates are flagellated photosynthetic *protists* (one-celled organisms with nuclei) found in the *plankton* (very small floating creatures) of marine and freshwater habitats. These creatures produce a red pigment, peridinin, and when they occur in very large numbers, they lend that color to the water. Therefore, although not a tide, such an outbreak has been termed a "red tide." Other blooms may be associated with other causative organisms that produce different colors, like brown or green, depending on the predominating pigments present. Such blooms are natural occurrences in many areas and have been recorded regularly along the coastal zones of Florida for centuries. Usually developing during the summer and fall, for the past few years, the red tides seem to have lingered or at least prevailed for extended periods of time. In 2017, a red tide welcomed the new year, and after a brief respite, it returned in the fall of 2019.

In the Gulf of Mexico, the bloom begins offshore from a combination of conditions that lead to a preponderance of this dinoflagellate. *K. brevis* is a relatively slow-growing organism and is often outcompeted by faster-growing algae, like *diatoms* (single-celled algae with cell walls made from silica). However, if there are nitrogen-fixing blue-green algae present, they provide *K. brevis* with the nitrogen needed for proliferation (Weisberg et al. 2016). This species of dinoflagellate produces a neurotoxin that can cause the death of fish and invertebrates in the surroundings. With the demise of these organisms, decomposers get to work, releasing even more nutrients, which promote further growth of the *K. brevis* population. Currents and winds carry the floating hordes of death into the shores, and as it is fed further by human input of nutrients, the concentration of toxin becomes high enough to kill not only fish but sea turtles and marine mammals. As the water breaks along the shoreline, small, toxin-filled droplets are released that can be inhaled by human beings, causing respiratory irritation. Furthermore, the toxins can stimulate asthma attacks and aggravate emphysema in chronic sufferers (Perkins 2019).

Why the little creatures produce such a toxin is being debated and researched by scientists, but we don't yet have a satisfying justification. Likely, it initially evolved to weed out predators, which would normally consume large quantities of the dinoflagellate population. One suggested alternate function has been the toxin's participation in regulating cellular salt concentrations. The evidence supporting this hypothesis has not been completely convincing (Errera and Campbell 2011, 2013; Sunda et al. 2013) .

As the dead animals wash up on the attractive Florida beaches, they do not make walking on the beach very enjoyable. With the stench and necessary avoidance of the dead, putrid animal carcasses, tourists tend to go elsewhere. Tourism in Florida is big business, and the economic impact of these killer tides is significant. Furthermore, some of the other animals caught or collected for food during a red tide might sicken consumers. Thus, Floridians have both economic and public health crises on the horizon.

Of course, scientists are trying to determine what the critical factors are that lead to such toxic blooms, and they are making progress. Because of the scale of the event, it is likely impossible to prevent them, but scientists are considering some novel ideas to mitigate existing red tides. The Chinese and South Koreans, for instance, have used clay particles coated with polyaluminum chloride, which gives the particles a charge that attracts oppositely charged phytoplankton to them. These clay particles, attached to red tide organisms, aggregate with other particles and sink. As they sink, they pick up more phytoplankton cells. Some of the algae break open, killing the organism, and others die buried in sand. Interestingly, the clay particles yield an additional benefit by removing some of the free toxins from the water (Anderson Lab 2020).

Seagrasses are often found in coastal areas, including estuaries. Extensive prairies of various grass species, like turtle grass (*Thalassia testudinum*), live submerged in the waters of estuaries. The extent of seagrass spread is limited by water clarity, but there are some seagrass species found in deeper, very clear waters. The roots of these plants stabilize the sediments, which also helps to make the water clearer. Seagrass beds are also home to a variety of animals, some of which feed on the grasses (sea urchins, turtles, manatees), and provide shelter for many fish and invertebrate animals. Secondarily, many predaceous birds (egrets, pelicans, osprey) and marine animals (dolphins) feed on the *fauna* (animals) inhabiting seagrass beds (Florida Museum 2019, "Seagrasses").

Periodically inundated by tidal flows, salt marshes occur in many estuaries. These marshes are generally dominated by species of rushes and grasses. In Florida, the black needle rush (*Juncus roemerianus*) occupies the highest (and driest) portions of the marsh, and various species of cord grass (*Spartina*) live in the lower, wetter zones. The salt marshes may provide food from flowers, fruits for some animals (insects, small mammals), and dead organic matter for detritivores (fiddler crabs). Oysters, crabs, shrimp, and some juvenile fish may live within salt marshes, feeding on organic matter and algae. Filter feeders, like oysters and mussels, feed on the high levels of particulate organic matter of the marsh, whereas carnivores, like birds, mammals, and crabs, visit the marsh to feed on the smaller animals (FDEP 2019; Dybas 2002; Florida State Parks 2019).

PoI 1. Match the following organisms to their tropic levels (primary producer, primary consumer, secondary consumer, tertiary consumer, decomposer, scavenger/detritivore): birds, algae, fish, turtle young, bacteria, manatees, cord grass, oysters, sea urchins, snails, dolphins, zooplankton, crabs, mangroves, fungi, conchs, seagrass

Estuaries are ecosystems that display rich biodiversity. From the above descriptions, what kind of food web or pyramid could we construct for an estuary? Of course, with the limited numbers of specific examples, the pyramid may not be complete, but let's see what organisms fall into each of our main categories.

At this point, we have a picture of the kinds of plants and animals inhabiting estuaries, but keep in mind that we haven't discussed in any detail the microbial communities. However, with our general picture of the plant and animal communities, we can move on to examine a few of the abiotic factors of this ecosystem.

18.2.1. Salinity, Anaerobic Sediments, and Tidal Fluctuations

Estuaries are typically filled with *brackish* water, which is water that may have a variety of salinities greater than that of freshwater (~0). The salinity of estuary water varies by depth or location. Many of the flowering plants are exposed to fluctuating depths (tidal) and salinities. Seawater contains various salts, primarily sodium chloride (NaCl), in concentrations of 35 grams dissolved in 1,000 mL of water. Concentrations may also be expressed as 35 *parts salt per thousand parts water*, or *per liter (‰ or ppt)*. The salinity of freshwater is virtually zero. Depending on the proximity to sources of freshwater and the amounts of freshwater entering an estuary, the salinity can vary somewhere between 0 and 35‰.

Organisms that live in such environments must be able to withstand the changes that take place in salinity as it varies with tidal changes and amounts of precipitation or discharge. Plants of the salt marshes and some mangroves experience varying water levels and salinities. *Spartina* (cordgrass) can withstand almost a full range of salinity variations and high temperatures that may stress other plants. While taking up water, *Spartina* filters out salt through its roots and can secrete excess salt through its leaf surfaces (NOAA 2019). Red mangroves that live along the edges of the water produce arched prop roots that suspend the trees over the water, in addition to adventitious roots that drop from branches. Both root types help to stabilize the tree in the sediments. Red mangroves also experience very high-salinity water (60–65 ppt), but their roots filter out the salt, so that most of the water that enters the plant is freshwater. Like the marsh grass, some mangrove species can also secrete excess salt from their leaf surfaces. Furthermore, air can pass through the root parts that are not submerged, which allows extra oxygen for the parts of the roots in the rather anaerobic sediments below (Florida Museum 2019, "Seagrasses"; FDEP 2019).

Many estuarine animals can also adjust to changes in salinity as they move to and from the salinity of the ocean. Invertebrates like the blue crab have special ion transporters in the membranes of the cells of their posterior gills that help to regulate the salt concentration in their *hemolymph* (equivalent to blood, found in many invertebrates). In the ocean, the crab's internal fluid conforms to the salt concentration of its surroundings. As it moves toward the coast, where freshwater input occurs, the salinity of the water declines. To maintain a functional concentration of salts, the transporters actively move Na^+ into the hemolymph. This ability to adjust the internal concentrations with those in the surrounding water helps the animal live successfully in an environment that is characterized by salinity fluctuations (Henry et al. 2012; Towle and Weihrauch 2001).

PoI 2. Salinity is only one variable that is important to life in the estuary.

a. Name five other abiotic factors that affect the plant and animal communities in the estuary.
b. Briefly explain how each might be important.

Ecosystems encompass all organisms in the community, plus factoring in the physical environment. Thus, we will consider organisms interacting with each other as well as with the environment. The biotic components of the ecosystem (organisms) are interconnected by food webs. Primary producers absorb energy from the sun and CO_2 from the environment and transform that energy and CO_2 through photosynthesis into organic molecules. When the producers are eaten by the primary consumers (herbivores), those organic molecules are used for energy and synthetic reactions needed for growth, reproduction, movement, and maintenance. Subsequently, the remaining energy within the flesh of primary consumers passes through the secondary and tertiary consumers, with the most energy released as heat. Energy, therefore, flows *unidirectionally* through the food webs of the ecosystem's community.

In the Indian River Lagoon estuarine ecosystem, the primary producers gather the sun's energy for their own purposes, losing some energy as heat during respiration. Primary consumers consume these algae and seagrass. As primary producers die or defoliate, they contribute much organic matter to the estuary. This organic matter is utilized by detritus feeders, and then the remainders are broken down by decomposers. Considerable energy from the primary producers flows through the decomposers. However, some of the primary consumers are eaten by the secondary and/or tertiary consumers. Again, at each consumer level, as animals die, their energy is utilized by the scavengers and decomposers, so much less energy remains available for the secondary and tertiary consumers. Even the scavengers and decomposers die, providing energy for other scavengers and decomposers.

Besides energy flow, there are chemical elements (nutrients) that are needed for synthesis, reproduction, repair, maintenance, and other necessary activities. The chemical elements needed in greatest quantities include carbon (C), hydrogen (H), oxygen (O), nitrogen (N), phosphorous (P), and sulfur (S). Although these elements may be part of living organisms (serving as a *reservoir*, or long-term storage location, of biotic components), they also are transferred to water, air, or soil/sediment (in reservoirs of abiotic components). Ecosystems are characterized by a dynamic, cyclic circulation of these nutrients, sometimes called *biogeochemical cycles*, where the nutrients move from one component (reservoir) of the ecosystem to another. Because water is essential for all of life, let's explore the hydrologic cycle.

QRQ 1. Select the following that are reasons an overwhelming growth of algae may kill aquatic animals: preventing reproduction, releasing toxins, removing oxygen for respiration, stopping development of young, using water

QRQ 2. Select the following that are advantages of mangroves to the ecosystem: a toxin in their roots prevents establishment of invasive species; a variety of animals live on their roots; a wide variety of animals feed, breed, and nest there; their roots filter out the salt, so most water entering the plant is freshwater; they absorb some of the excess nutrients that enter the estuary; they do not need many nutrients to flourish; they help ensure stabilization of the shoreline against storms; those along water edges produce prop roots that suspend the trees over the water

QRQ 3. Select from the following the appropriate term for each description below: abiologic, abiotic, biologic, biotic, bloom, blossom, brackish, community, ecosystem, estuary, fauna, flora, hemolymph, lagoon, lake, mangrove, population, reservoir

a. animals
b. coastal body of water, partly enclosed, where freshwater and saltwater meet
c. set of all living things in an area that interact with each other and with nonliving elements
d. conforms to the salt concentration of the crab's surroundings
e. fluid of the internal cavity of many invertebrates with similar functions as blood
f. interacting group of populations of different species living within the same area
g. living
h. long-term storage location
i. nonliving
j. overwhelming growth of algae resulting from excess nutrients
k. small body of water separated from a large body of water by a natural barrier
l. water that is saltier than freshwater but less salty than sea water

18.3. Hydrologic Cycle

APPLYING SCIENCE TO LIFE 3 Water Rights

If you live in the United States, one of the wealthiest countries in the world, you might be surprised to find that, according to a recent report released by nonprofits DigDeep and the U.S. Water Alliance, more than two million Americans live without access to running water and indoor plumbing. Even more of our citizens apparently live without proper sanitation. In other nations, principally developing nations, millions of people lack access to clean water. With the dependence we all have on safe drinking water, it is little wonder than many organizations and public health agencies consider access to such water a "basic human right." In the United States, one of the populations in greatest need are Native Americans, where they are 19 times more likely to lack running water in their homes as white Americans (DigDeep 2019). Many of the Americans that lack such access obtain water from wells or windmills, which pump water that is often polluted. The windmills on the lands of the Navajo Nation, for instance, can provide water that is contaminated with arsenic and/or uranium from old mining operations. There are over 500 abandoned uranium mines in the area that have polluted the groundwater in the area.

In this section, water, which contains hydrogen and oxygen, is a "nutrient." We'll come back to the issue raised in Applying Science to Life 3 later, but we begin by focusing on the *hydrologic cycle*. Like other nutrients, water is stored, at least temporarily, in interconnected abiotic and biotic *reservoirs* (long-term storage locations). When you examine the literature on this cycle, the number of natural reservoirs may differ, but we will consider seven of them. The numbers in Table 18.1 are estimates (rounded off) of water held in each reservoir. These numbers are meant to provide you with some general values that will allow you to compare the relative proportions of water stores in each reservoir. For most organisms, only freshwater sources are usable, so the largest reservoir, the ocean, is not easily utilized. Also, freshwater in deep groundwater or frozen in ice mass is not readily available for use.

TABLE 18.1. Major Reservoirs of Water in the Hydrologic Cycle. Broadly Based on Henshaw, Charlson, and Burges (2000).

Reservoir	Volume (km³)	% of Total Water
Atmosphere	0.0129×10^6	0.0009
Soil	0.0165×10^6	0.0012
Groundwater	23.4×10^6	1.6887
Oceans	1338×10^6	96.5563
Rivers, Lakes	0.1896×10^6	0.0137
Ice	24.1×10^6	1.7392
Biosphere	0.00112×10^6	0.0001
Total Water	**1385.72×10^6**	**~100.0000**

Table 18.1 describes the water reservoirs, but it doesn't inform us about the dynamics of the hydrologic cycle. However, Figure 18.2 does. The water cycle is powered by the sun's energy, which warms surfaces of bodies of water, the land, and plant surfaces. As heating occurs, liquid water is converted into water vapor. From most surfaces, this process is termed *evaporation*, but water is also drawn upward in plants toward the upper surfaces and evaporates. Most of this kind of evaporation, called *transpiration*, occurs through the stomata in leaves. Also, some water vapor is formed from ice and snow by a process called *sublimation* (solid water converted directly to water vapor). Evaporated

water, which is less dense than other major components of the atmosphere (N_2 and O_2), moves upward into the atmosphere. As the humid air rises, the atmospheric pressure decreases, and the temperature drops, eventually causing the water vapor to condense, forming clouds, and if the droplets become heavy enough, they fall as precipitation. Liquid or frozen precipitation falls mostly into the ocean. Frozen precipitation may accumulate as ice caps on high mountains or glaciers. Liquid precipitation and melted ice and snow on land either form runoff that enters streams and rivers or percolates into the soil. Some of the water that enters the soil eventually finds its way into the *groundwater* (water held below ground in soil, pores, or crevices in rock). Some groundwater moves from *aquifers* (underground water supply found in porous rock, sand, and gravel (not shown in Figure 18.2)) to the surface (springs) or into the ocean. Other water can be taken up and utilized by plants.

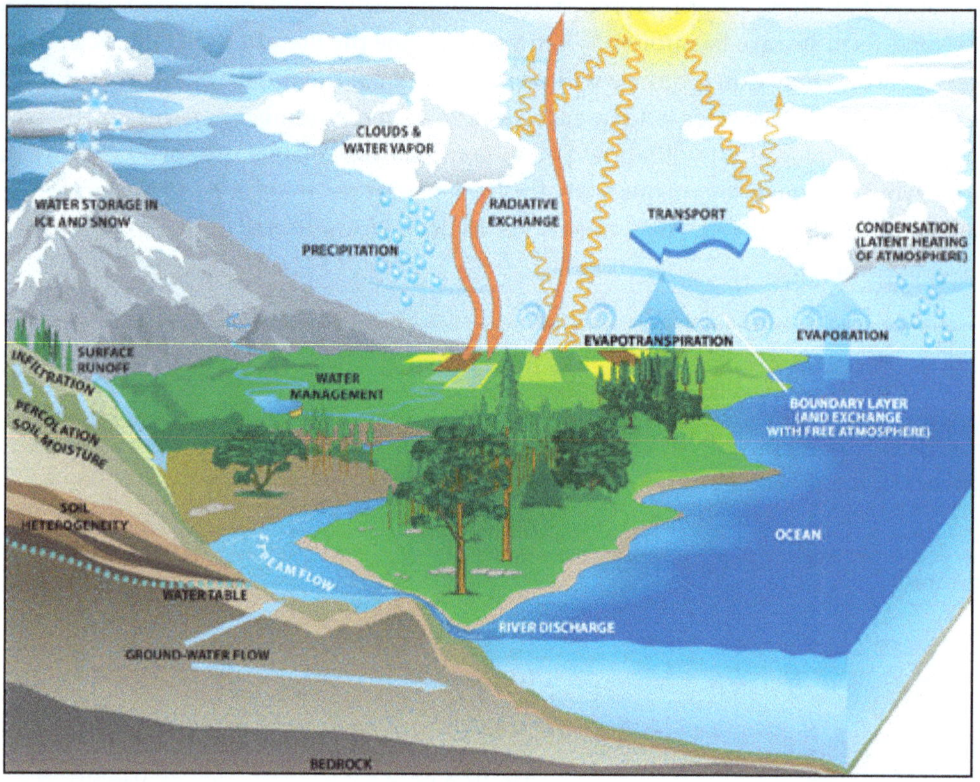

FIGURE 18.2 Hydrologic cycle

The hydrologic cycle, like other nutrient cycles, is made up of reservoirs and *fluxes* (movement of nutrients from one reservoir to another). In this particular cycle, the major abiotic reservoirs are the ocean (all saline), surface water (lakes, streams, rivers), ice, soil, groundwater in aquifers, and the atmosphere. Plants, animals, and other organisms are also components of the *biosphere* (parts of the earth supporting life), which is, of course, a biotic reservoir. Notice in Figure 18.3 that the largest reservoir (~96+ percent) for water is the ocean (including all saline water), with freshwater reservoirs (including biotic) making up just over 1 percent. Because this small portion of freshwater is not universally available for consumption, organisms must adapt to access and conserve it to survive.

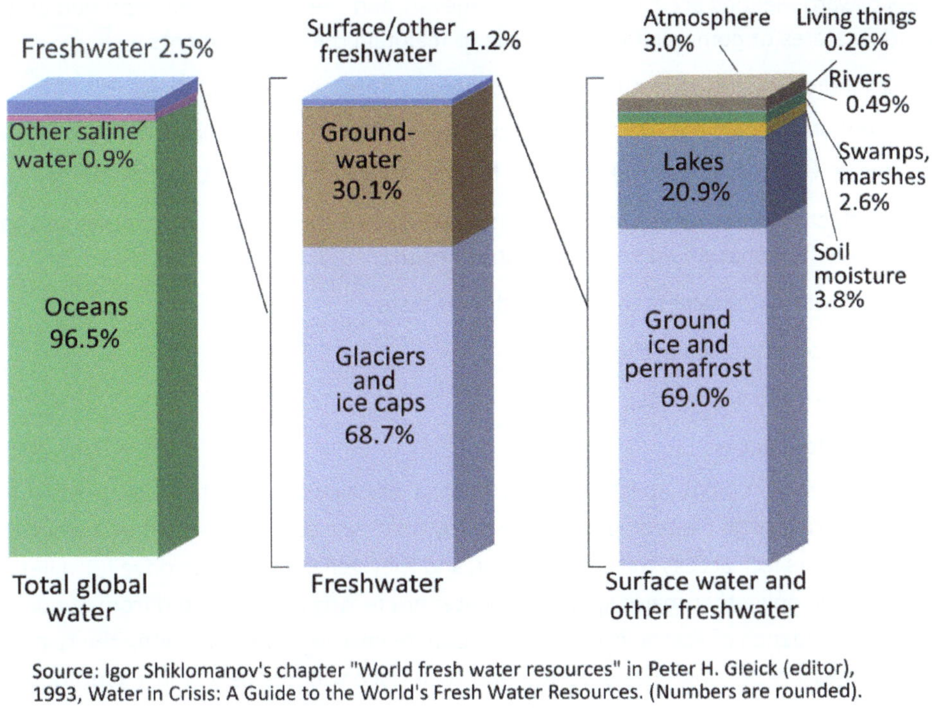

FIGURE 18.3 Earth's reservoirs for water

APPLYING SCIENCE TO LIFE 4 Water Rights, Continued

Researchers at Michigan State University have studied the "water access gap," or the disparity between those who have and those who do not have access to clean and reliable drinking water and safe sanitation, in the United States and reported five major areas important to this issue (DigDeep 2019):

1. Federal data does not accurately describe the water access gap.
   a. Data collected is inconsistently collected and undercounts underserved communities.
   b. It doesn't include wastewater treatment.
2. Race is the strongest indicator of water and sanitation access.
   a. 0.3 percent of white households lack access to complete plumbing.
   b. 0.5 percent of major minority households lack such access.
   c. 5.8 percent of Native American households lack such access.
3. Poverty is a key obstacle to water access.
   a. Higher income households were less likely to lack access to complete plumbing.
   b. Households with higher educational attainment were less likely to lack such.
4. Water access challenges impact entire communities.
   a. The lack of sufficient water infrastructure can affect entire communities.
   b. Such lack can also be much higher in pockets within areas with generally higher access.

5. Progress toward meeting access challenges is uneven, and even when some improvements are made, some states or communities are unable to maintain the improvements.
    a. The backsliding tends to occur in less prosperous states and communities.

Advocates for equitable access to water and wastewater services strive to achieve the following:

- Access to safe water that meets EPA and WHO (World Health Organization) standards
- Access to effective wastewater systems, treating sewage that avoids human contact and any type of leakage or overflow that would endanger public health
- Water supply without unacceptable color, odor, or taste
- Sink, tub, or shower located in the home
- Affordable water and wastewater services
- Nondiscriminatory access

As an example of the problem, Apache County, Arizona, has a population that is 90 percent Native American, a poverty rate of 36 percent, and an unemployment rate above 10 percent. An estimated 30 percent of the residents lack access to running water, and they generally have to drive for miles to haul water back home. The only other major sources of water are mostly contaminated from hundreds of abandoned mines. Incidence of gastric cancers in the county have skyrocketed during the last 20 years. With climate changes, precipitation has dropped, and so surface water is very scarce. Outhouses abound in the area. Consumption of sugar-rich drinks often is high because the drinks are more accessible than clean water. Navajo have diabetes II rates two to four times that of whites.

The Native American community often struggles for water rights, and there are virtually no funds to construct the necessary water infrastructure. The population is of low density in mountainous areas and extending water and sewer lines to reach them is very costly. Estimates to provide access to clean running water and proper sanitation are around $200 million.

It is obvious that there are no simple solutions, but one nonprofit that partners with DigDeep in that area is COPE (Community Outreach and Patient Empowerment). This organization tries to encourage local convenience stores and trading posts to foster clean water choices and to stock adequate fresh fruits, vegetables, and traditional foods. COPE also works with early childhood education centers to ensure students' access to filtered water systems. Obviously, there is still much need (DigDeep 2019; Navajo Nation Department of Water Resources 2011).

PoI 3. Below are organizations that work on various water problems. Summarize their major projects related to clean water, access to clean water, and so on:

a. Alliance for Water Efficiency
b. Alaska Native Tribal Health Consortium
c. Water.org

PoI 4. Access to clean water is a global problem. Various agencies and organizations, using the ideas of creative people, have devised some innovative solutions for this problem. Find out about the following solutions, including a description of the product or approach and applications of the solution. Also, can the solution be applied successfully more broadly?

a. hippo roller
b. drinkable book
c. SunSpring

QRQ 4. The conversion of liquid to gas is called ______.

QRQ 5. The conversion of ice directly to water vapor is called ______.

QRQ 6. Loss of water vapor from the stomata of a leaf is called ______.

QRQ 7. The largest reservoir for water is the **a.** ______ (Answer choices: atmosphere, biosphere, groundwater, ice, oceans, rivers and lakes, soil), which has about **b.** ______ of the water. The total of percentage of water in freshwater reservoirs is about **c.** ______. Choices for b and c: 1 percent, 13 percent, 24 percent, 47 percent, 58 percent, 62 percent, 73 percent, 80 percent, 96 percent.

QRQ 8. The major reservoir to which water flows from the ocean reservoir is the ______ (Answer choices: atmosphere, biosphere, groundwater, ice, oceans, rivers and lakes, soil).

18.4. Nitrogen Cycle

APPLYING SCIENCE TO LIFE (OR DEATH) 5 Giving Back to Nature

If you like solving mysteries and enjoy watching forensics experts reveal how they solve mysteries about death, natural or not, you might be interested in participating (before or after death) in an anthropological program at the University of Tennessee Anthropological Research Facility, or one of the six other related facilities in the United States. It was established in the early 1980s, and researchers at the "Body Farm" place human cadavers in various situations (e.g., in water, buried in a shallow grave in the woods, in a car trunk) and follow the decomposition process. At first, the founder, Dr. William Bass, used bodies to determine cause and time of death. As the center developed, it assisted law enforcement and forensics investigators, helped make advancements in forensic techniques, and assisted in solving a number of crimes (Roach 2003).

An advantage to having corpses decompose "out of the box" (coffin) is that the decomposition products can reenter the ecosystem to support the life of other organisms—natural recycling. Furthermore, the bodies allowed to decompose in the soil are not embalmed, which means toxic preservation chemicals are not released into the burial site. Most people also have a coffin, surrounded by a vault. The vault is employed to prevent damage to the coffin and sinking of the gravesite, but your body will still decompose, even if more slowly.

Detritivores (invertebrate animals that ingest and break up organic matter; scavengers) and *decomposers* (mostly bacteria and fungi that absorb nutrients they have digested externally) are found at the base of food webs and are very important for returning to any ecosystem nutrients for reuse by living organisms. In doing so, the decomposers and detritivores gain energy and nutrients for their own structure and function.

Most of the literature represents studies of plant decomposition, but animals also die and produce waste in the wild. By following the decomposition process of carrion (flesh of dead animals), scientists can study the quantities and rates of nutrients returned. Researchers can also use decomposition to reveal the succession of scavengers and decomposers over time. Sequential breakdown of animals in terrestrial areas generally involves three major groups of animals: invertebrates, particularly flies and beetles; scavenger vertebrates, such as vultures and racoons; and microbes, such as bacteria and fungi.

Dead vertebrates (field mice, rabbits, etc.) may be scavenged by vertebrate animals, which may carry significant amounts of carrion from the site of death. Generally, *necrophagous* invertebrates (those feeding on carrion), such as certain flies and beetles, are responsible for much of the carcass breakdown. The succession of insects involved in decomposition may vary seasonally in temperate regions, but quickly after death, insects, such as blow flies, detect the carcass. Perhaps within minutes, female blow flies lay several hundred eggs near the mouth, eyes, or other openings. These eggs hatch into larvae (maggots), which grow while feeding on the flesh. Also, some of the flesh is converted to CO_2 (respiration) and various excretory wastes, including carbon- and nitrogen-containing molecules. The larvae eventually leave the carcass, individually or *en masse*, and bury themselves in the soil, perhaps meters from their former source of food, and begin pupation (transformational change into an adult fly). The adults mate and, after finding a new carrion source, begin a new cycle. Overlapping with the flies, other insects, such as beetles, may congregate to feed on the carcass or the fly eggs and larvae (Lewis 2011; Farwig et al. 2014; Tomberlin and Adler 1998).

Meanwhile, from the beginning, an unseen set of decomposers have been at work—*microbes* (microscopic organisms, such as bacteria and some fungi). Many of these organisms are found within the gut of the animal, and others are found on external surfaces. Some soil organisms can contribute to their efforts, as well. The microbial community includes many species, and species composition changes as decomposition proceeds. Long held back by the defenses of the living animal, these ungrateful bacteria and fungi, once comfortably lodged in and on the animal, use the tissues of the recently departed as a food source. As the animal cells die, their enzymes are released to help promote the food bonanza. The body becomes liquified and gaseous. Microbial metabolisms convert the remains into some pretty smelly stuff—ammonia, hydrogen sulfide (think rotten eggs), cadaverine, and putrescene (think dead fish), as well as many other chemicals.

With the efforts of decomposer communities, the stuff that once made up the dead animal becomes part of the stuff of living organisms. Carbon dioxide and other gases, produced by respiration or fermentation of decomposer organisms, enter the atmosphere, and other decomposition byproducts are released into the surrounding soil. Nutrients entering the soil may change the soil chemistry, including the pH and concentrations of nutrients (e.g., nitrogen, phosphorous). These changes become part of the local environment for plants and soil organisms (Pechal et al. 2013; Parmenter and MacMahon 2009).

Most human beings in the United States forestall entry of their bodies' nutrients to the environment, prolonging the process through embalming and boxing up the cadaver in a casket and vault. Most funeral home personnel take great care to make the dead body resemble the living form of that person as closely as possible. Depending on lots of environmental factors, the body may decompose in weeks or may not decompose for decades.

PoI 5. What are some examples of detritivores in a forest of the temperate zone?

PoI 6. What is cremation? What are the products of cremation?

Although significant amounts of the element *nitrogen* (N) enter the soils and sediments, waters, and organisms of any ecosystem, most of that nitrogen is in the atmosphere (see Figure 18.4). In fact, *nitrogen gas* (N_2) is the most common gas (almost 80 percent) in the atmosphere, making the atmosphere the most important reservoir for that nutrient. Major reservoirs for nitrogen are found in Table 18.2. The other principal reservoir for nitrogen is sedimentary rock. Note that atmospheric and sedimentary nitrogen pools contain 99.5 percent of the earth's nitrogen, and neither is directly available for organismal use. Ironically, all living organisms need nitrogen, but very few of them can use it as nitrogen gas in the atmosphere. Sedimentary sources represent essentially unavailable nitrogen. Before we consider the workaround for the relative scarcity of nitrogen available to support life, let's consider a couple of fundamental questions: Why is nitrogen so important? Can you think of any chemical compounds in your body that have nitrogen as part of their structure and function? In answering these questions, at least a couple of molecules should come to mind—proteins (amino acids), nucleic acids (nucleotides) like DNA and RNA, ATP, and chlorophyll. Even viruses utilize proteins and nucleic acids. Another question to ponder is the following: With most of the earth's nitrogen being stored in an unusable form, how do organisms get their nitrogen?

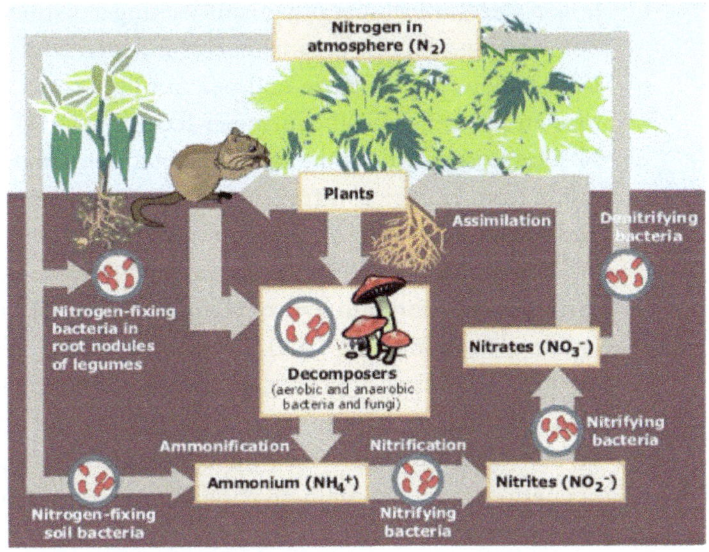

FIGURE 18.4 Nitrogen cycle

TABLE 18.2. Major Reservoirs and Estimated Quantities of Nitrogen They Contain (*Tg = 1 trillion [10^{12}] grams)

Reservoir	Quantity (Tg N/yr)*	% Total
Atmosphere (N_2)[1]	3.95×10^9	79.5
Sedimentary Rocks[1] (unavailable)	1.0×10^9	20.1
Ocean (mostly N_2)[2]	2×10^6	0.4
Soil Organics (detritus)[1]	1.9×10^5	0.0
Land and Marine Organisms[1]	10.5×10^4	0.0
Total	**~ 5×10^9**	**100**

Sources for data: [1]Mackenzie (1998), [2]Schlesinger and Bernhardt (2013)

Nitrogen is often a limiting factor for photosynthesis and growth in green plants. For agriculturally grown crops, farmers often add a usable form of nitrogen, such as nitrate (NO_3^-) or ammonium (NH_4^+). Synthetic fertilizers depend on the *Haber–Bosch process,* which combines N_2 with H_2 under high temperatures and pressures, producing ammonia (NH_3). The ammonia can be used to generate fertilizers, such as ammonium nitrate and urea. These synthetic fertilizers have been the underlying reasons that world agriculture can provide enough food to feed growing populations. On the other hand, this important process consumes incredible amounts of the world's energy resources.

For millennia, before the Haber–Bosch process was ever envisioned, the earth had biological *nitrogen fixation,* wherein certain microorganisms, primarily certain types of bacteria and other *prokaryotes* (unicellular organisms lacking a distinct nucleus), can convert N_2 into ammonia (NH_3). Some nitrogen-fixing organisms live independently, and others live in association with host plants. Leguminous plants (bean and pea family) attract nitrogen-fixing bacteria to form *nodules* (swellings on the roots housing nitrogen-fixing bacteria) where N-fixation can occur. What unifies these N-fixing organisms is the presence of an enzyme complex that includes the enzyme *nitrogenase*. Aside from biological nitrogen fixation, small amounts of N_2 are converted to nitrate by lightning and are added to the soil through rainfall.

The ammonia (NH_3) produced by N-fixation can be converted by certain microorganisms through oxidation into hydroxylamine (NH_2OH) and then into *nitrite* (NO_2). Nitrite can then be oxidized to form *nitrate* (NO_3) by another group of microbes. This overall process is referred to as *nitrification*. Primary producers can readily utilize this nitrate for their nitrogen source. Specific types of microorganisms (ammonia-oxidizers and nitrite-oxidizers) perform these reactions in many aerobic environments. These microorganisms utilize the reactions to generate very small amounts of energy. If you maintain an aquarium, these same types of organisms can detoxify the ammonia contained in the urine of your fish.

Nitrate absorbed into and assimilated by primary producers is used to produce new proteins and other nitrogen-containing molecules for growth, maintenance, and reproduction. When the primary producers die or drop their leaves, the nitrogen-containing molecules can be metabolized by detritivores and decomposers for their own growth, maintenance, and reproduction or broken down and released into the surroundings (soil, sediment, water). Through death and waste production, consumer organisms also contribute to the nutrients available to other organisms. Living animals produce *ammonia* during the metabolism of nitrogen-containing molecules like amino acids. The ammonia is excreted quickly by most aquatic organisms, but it is converted into some less toxic products in terrestrial animals. In birds, most reptiles, and terrestrial invertebrates, ammonia is converted using lots of energy to *uric acid*, which is less toxic and less water-soluble than ammonia. Uric acid is released as a white paste. Mammals and amphibians convert ammonia into *urea*, which is also not as toxic and requires a smaller amount of water, for excretion into their urine. Once released, these excretory waste molecules may be converted by decomposers back to ammonia, which can enter the nitrification pathway to provide nutrients for other primary producers.

Some of the nitrogen-containing molecules can be converted by various microorganisms into N_2 through a process called *denitrification*. In this process, nitrates are reduced to form nitrites, which may then be reduced to form various reaction products, including *NO* (nitric oxide) and *N_2O* (nitrous oxide). The reaction products can be further reduced to yield N_2 for the atmosphere. It should be noted that nitrous oxide is considered an important greenhouse gas contributing to global warming. Furthermore, N_2O can interact with ozone (O_3) to damage the protective layer in the stratosphere (layer of the atmosphere just above the troposphere, which is the lowest layer of the atmosphere).

PoI 7. How might excessive denitrification affect farmers?

PoI 8. Primary production is often reduced by a shortage of a critical limiting factor. What is a limiting factor (with regard to primary productivity)? Give some examples.

Choices for a and b: atmosphere, biosphere, groundwater, ice, oceans, rivers and lakes, sedimentary rock, soil.

QRQ 9. Choices for a and b: atmosphere, biosphere, groundwater, ice, oceans, rivers and lakes, sedimentary rock, soil. The largest reservoir for nitrogen is the **a.** ____, and the other principal reservoir for nitrogen is **b.** ____. Most of the nitrogen is **c.** ____ (Answer choices: available, unavailable) to support life.

QRQ 10. (Answer choices in parentheses) The most common gas in the atmosphere is ____ (carbon dioxide, chlorine, hydrogen, methane, nitrogen, oxygen, water vapor), making up about ____ (1 percent, 13 percent, 24 percent, 47 percent, 58 percent, 62 percent, 73 percent, 80 percent, 96 percent) of the atmosphere.

QRQ 11. T/F: Proteins, nucleic acids, ATP, and chlorophyll have nitrogen as part of their structure and function.

Choices for b and c: ammonia, chlorine, methane, nitrate, nitrite, nitrogenase, nitrogen, urea.

QRQ 12. Choices for b and c: ammonia, chlorine, methane, nitrate, nitrite, nitrogenase, nitrogen, urea. The **a.** ____ (Answer choices: Bausch & Lomb, Decomposition, Habeas-Corpus, Haber-Bosch, Nitrification) process combines **b.** ____ gas and hydrogen gas under high temperature and pressure to produce **c.** ____, which can be used to generate fertilizers.

Choices for a, c, e, and g: ammonia, chlorine, methane, nitrate, nitrite, nitrogenase, nitrogen, urea.

QRQ 13. Other choices in parentheses. Certain microorganisms that have the enzyme **a.** ____ use **b.** ____ (denitrification, nitrogen fixation, nitrification) to convert nitrogen gas to **c.** ____. Other microbes can **d.** ____ (oxidize, reduce) the latter into hydroxylamine and then into **e.** ____, which other microorganisms can **f.** ____ (oxidize, reduce) to form **g.** ____. We call this overall process **h.** ____ (denitrification, nitrogen fixation, nitrification).

QRQ 14. Select the way(s) primary producer organisms contribute to the nitrogen available to other organisms: death, drop their leaves, fertilizers, photosynthesis, waste production

QRQ 15. Using the choices in QRQ 14, select the way(s) consumer organisms contribute to the nitrogen available to other organisms.

QRQ 16. Birds, most reptiles, and terrestrial invertebrates convert ammonia to ____, while mammals and amphibians convert ammonia into ____. Decomposers can convert these waste products to ____.

QRQ 17. (Answer choices in parentheses) In a process called ____ (denitrification, fixation, nitrification), various microorganisms can convert some of the nitrogen-containing molecules to nitric oxide and ____ (ammonia, carbon dioxide, nitrate, nitric oxide, nitrite, nitrous oxide, ozone), an important greenhouse gas contributing to global warming.

18.5. Carbon Cycle

18.5.1. Reducing Corporate Footprints

APPLYING SCIENCE TO LIFE 6 Credit Where Credit Is Due

Google Corporation announced its intention to reduce its *carbon footprint* with a three-pronged effort: using solar power (on site), buying "green power," and buying "carbon offsets." *Green power*, as defined by the Environmental Protection Agency (EPA), is a type of renewable energy where electricity is generated by wind, solar, geothermal, biogas, approved biomass, or small hydroelectric sources. But what are "carbon offsets"? *Carbon offsets* are investments that are used to fund projects and activities that reduce carbon emissions. They are generally expressed in terms of *carbon credits*, which are the metric tons of *carbon dioxide* (CO_2) or CO_2*-equivalent gases* (CO_2e) reduced or eliminated from the atmosphere by an emission-reducing project (Google Corporation 2011; Novey 2014). Accordingly, companies like Google, government entities, and individuals can purchase these offsets to counterbalance the emissions they may be releasing. There are a variety of qualified emission-reducing projects, examples of which include the following: construction of a wind farm, as opposed to a coal-powered power plant; reforestation and *afforestation* (planting trees on land barren of any trees); capturing methane from animal waste. Once certified to be emission-reducing, these activities can be designated for use as carbon offsets.

Other corporations/industries have also become interested in procuring these carbon offsets. For instance, Enbridge, a major transporter of gas and oil in North America, has purchased offsets through the Carbon Farmer (Carbon Farmer 2014). An enterprising couple in Alberta, Canada, have a business where they plant native trees for industries, corporations, and individuals, which pay them for certified carbon credits. Each tree planted sequesters an amount of carbon, at least for the life of the tree. Moreover, the Chevrolet division of General Motors has invested in carbon offsets that will help fund the development of a geothermal energy system for Ball State University, replacing coal-fired boilers that are about 70 years old. Ball State gets the investment for a project that will reduce their carbon emissions by 85,000 tons per year, saving the university approximately $2 million in yearly utility costs, and Chevrolet gets carbon credits to offset the carbon emissions it generates (Ransford 2014).

Before we explore the reasons these companies have taken steps to reduce carbon emissions or to balance out their carbon emissions through investing in emission-reducing projects, we focus on the element involved—*carbon*. Carbon is a naturally occurring element on Earth, which composes only a little (0.02 percent) of the earth's crust. However, it is ubiquitous in the life inhabiting the earth. Plants and some microbes use the energy of the sun to convert (chemically) molecules of carbon dioxide in the atmosphere or water to produce sugars and other important organic molecules. These organisms are termed autotrophs, but actually, they are photoautotrophs, because they utilize the sun's energy and inorganic carbon (CO_2) to produce their own nutrients. Autotrophs are vital, directly or indirectly, for the growth, reproduction, and survival of animals and other heterotrophs. Carbon is incorporated into and accumulates in living organisms as they grow and is released back into the air or water through *cellular respiration*. When an organism dies, its body is consumed by heterotrophic scavenger organisms (e.g., insects), which utilize the intake for food and release nutrient-containing wastes as smaller bits. These small bits are broken down further by decomposer organisms (e.g., fungi, bacteria), which are also heterotrophs. Regardless of the trophic levels, living organisms (both auto- and heterotrophs) release CO_2 from cellular respiration. Other significant ways CO_2 is released from organisms into the environment include the combustion of organic matter (e.g., forest fires and the burning of fossil fuels).

QRQ 18. A common type of carbon offset where the metric tons of carbon dioxide or carbon dioxide-equivalent gases reduced or eliminated from the atmosphere by an emission-reducing project is a(n) ______. Select answers from the following: afforestation, carbon credit, carbon footprint, carbon offset, carbon reduction, green power.

QRQ 19. Reforestation is ______, while afforestation is ______. Select answers from the following: planting of trees in an area previously forested, planting of trees in an area not previously forested, removing trees from a forested area

QRQ 20. Organisms that utilize the sun's energy and inorganic carbon (CO_2) to produce their own nutrients are termed ______, while human beings would be considered ______, because they use organic molecules for energy and carbon sources. Select answers from the following: autotrophs, heterotrophs, photoautos, phototropic

18.5.2. The Cycle

Figure 18.5 displays an artist's depiction of the carbon cycle with data as of 2006. The values given in the figure may vary somewhat from those discussed in the text, because the diagram was based on estimates from a different source.

Carbon Pools

The uptake and release of carbon dioxide is but one part of a global movement of carbon through various *reservoirs* (*pools* or *sinks*). This movement is commonly known as the *carbon cycle*. Major carbon pools include the following:

1. earth's crust
2. atmosphere
3. oceans (organisms and water)
4. terrestrial ecosystems (organisms and soil)

Most of the earth's carbon lies in the earth's crust. The crust contains about 100 million *gigatons (Gt)* of carbon, which is an enormous amount. One Gt equals 10^{15} grams, or one billion metric tons (metric ton = 1,000 kg or 2,205 pounds) (GLOBE 2014).

The carbon in the *crust* (outermost solid layer of Earth) is found in deposits of sedimentary rocks—*shale* and *limestone*. Besides crustal carbon, there is also about 4,000 Gt of hydrocarbons formed from the deposits of organisms that were exposed to high temperature and pressure for millions of years (GLOBE 2008). We commonly refer to these hydrocarbon deposits as *fossil fuels*.

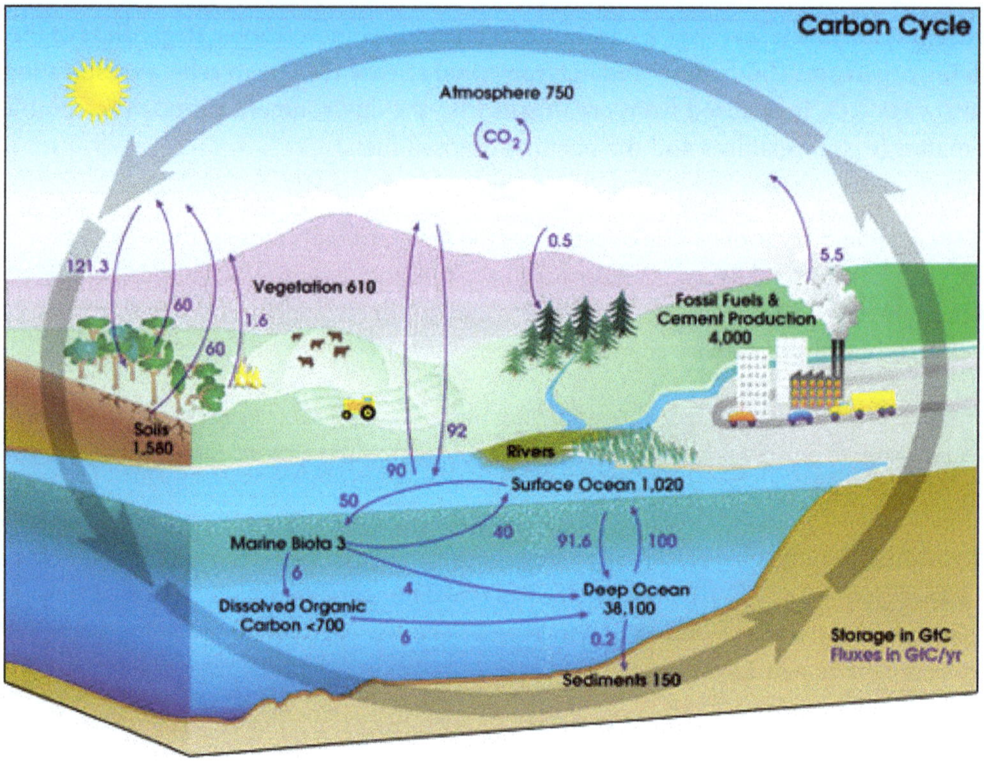

FIGURE 18.5 Carbon cycle with amounts stored carbon (gigatons (GtC)) in black and annual carbon fluxes (GtC/yr) in purple; human contribution = ~5.5 GtC/yr in the form of CO_2 to the atmosphere in 2006

The *atmosphere* contains a relatively small amount of carbon when compared with the crust—750 GtC, or gigatons of carbon. However small, the carbon gases in the atmosphere form a crucial pool of carbon. Most of this carbon is in CO_2, and a much smaller amount is in CH_4 (*methane*) and other gases. This number is relatively high compared to earlier periods of the earth's history (GLOBE 2008).

Although *terrestrial ecosystems* contain tremendous amounts of life (growing in and from the soils), for global considerations, we often focus on the plants and soils. Soils contain various animals and microbes, in addition to large amounts of *organic material*. Other life forms compose only small pools of carbon relative to the plants and soil. Plants, mostly in woody stems, encompass about 560 GtC, and the first meter of soil is estimated to include about 1,500 GtC (GLOBE 2008).

Most carbon (~38,000 GtC) in the *ocean* is found at great depths (dissolved inorganic carbon) and is generally slow to join the carbon cycle. In shallower waters, we find a much smaller concentration of dissolved carbon (~1,000 GtC). Some of this carbon originates from CO_2 dissolving, and the rest comes mostly from planktonic organisms. As these small organisms die, some of the carbon is lost to the ocean depths. The carbon in this shallower layer tends to cycle more rapidly (GLOBE 2008).

PoI 9. Why is the relatively small amount of carbon in the atmosphere especially significant?

PoI 10. What are the primary reasons that carbon dioxide levels have increased during recent times?

QRQ 21. Select which of the following are the major pools/sinks on the planet for carbon: asthenosphere, atmosphere, Earth's core, Earth's crust, large lakes, mesosphere, mountains, oceans, terrestrial ecosystems (organisms, soil)

QRQ 22.

a. One gigaton (Gt) is 10 raised to the power of ______ grams.
b. To one decimal place, one kilogram is approximately ______ pounds.
c. Approximately ______ (10 billion metric tons, 10 million metric tons, 100 billion Gt, 100 million Gt, 100 Gt) carbon is found in the earth's crust.

Carbon Fluxes

Although much carbon stays in a particular pool for long periods of time, we are especially interested in the portion that moves more frequently from one pool to another. We call these movements between sinks *carbon fluxes,* and we express them as the quantity of carbon moving between pools during a specific amount of time. Figure 18.5 depicts these fluxes as arrows with quantities in *gigatons of carbon per year (GtC/yr)*, as of 2006. If we look at terrestrial ecosystems, organisms and soil release carbon most frequently as CO_2 from respiration or combustion into the atmospheric pool. Simultaneously, terrestrial plants and some microbes remove CO_2 from the atmosphere through photosynthesis. So carbon fluxes represent a two-way street for carbon movements (Riebeek 2011; GLOBE 2008; Google Corporation 2011).

In examining the flux between the terrestrial ecosystems and the atmosphere, about half of the CO_2 released into the atmosphere each year is from plant respiration (60 GtC), and the other half is from organismal respiration and decomposition that takes place in the *litter* (undecomposed or partially decomposed organic matter) of the soil. To balance this release, green plants remove about 121 GtC/yr. Overall, however, carbon stored by plants, particularly in woody plant stems, amounts to over 600 GtC (GLOBE 2008).

QRQ 23. Movements of carbon between sinks are termed ______.

QRQ 24. Select the following things that produce the most CO_2 in terrestrial ecosystems: afforestation, combustion, decomposition, photosynthesis, reforestation, respiration.

QRQ 25. CO_2 from the atmosphere is primarily removed through the process of ______ by plants and some microbes.

Dissolved Inorganic Carbon and Ocean Acidification

Carbon dioxide is removed from the ocean through the activity of photosynthetic organisms, formation of *calcium carbonate* ($CaCO_3$), and loss to the atmosphere (sun). Input of carbon dioxide to the ocean occurs from the *solubilization* of calcium carbonate, respiration (*oxidation*) of organic carbon, and dissolution of CO_2 from the atmosphere. For many years, the oceans have been able to absorb the ever-increasing carbon dioxide released into the atmosphere from our burning fossil fuels.

Much of the carbon dioxide from the atmosphere dissolves in ocean water, where it forms *carbonic acid* (H_2CO_3). Some of the carbonic acid then dissociates into *bicarbonate ions* (HCO_3–) and *hydrogen ions* (H+), as the following reactions indicate:

$$CO_{2\,gas} \leftrightarrow CO_{2\,aqueous}$$
$$H_2O + CO_{2\,aqueous} \leftrightarrow H_2CO_3 \leftrightarrow HCO_3^- + H^+$$

From the above reactions, we can see that increasing amounts of dissolved carbon dioxide lead to an increase in hydrogen ion concentration. More hydrogen ions means that the pH will become more acidic (pH decreases). According to the Royal Society in England, the average pH of the ocean's surface water has declined by 0.1 unit of pH since the Industrial Revolution (Royal Society 2005). A change of one pH unit represents a tenfold change, and although this change seems small, it represents a 30 percent increase in H^+. Therefore, a decline of this magnitude in pH is significant. If CO_2 concentrations continue to increase, we are likely to see an even more dramatic decrease in pH. In the past, we have depended on the oceans to buffer our increased CO_2 emissions, but it seems that even our vast oceans have their limits (Doney et al. 2009; PMEL n.d.).

More alarming are the effects on ocean-dwelling organisms, many of which produce external skeletons of *calcium carbonate* ($CaCO_3$). Some of the bicarbonate, under the right conditions of pH, ionizes into hydrogen ions and *carbonate ions* (CO_3^{2-}). Marine organisms can then combine the carbonate and calcium to form calcium carbonate. The reactions are as follows:

$$HCO_3^- + H^+ \leftrightarrow 2H^+ + CO_3^{2-}$$
$$Ca^{2+} + CO_3^{2-} \leftrightarrow CaCO_3$$

These equilibrium reactions involving ionized forms of carbon, beginning with dissolving atmospheric CO_2, are sometimes termed the *marine carbonate system.*

With increasing hydrogen ion concentration, the equilibrium shifts from right to left, from free CO_3^{2-} to form HCO_3^- (Doney et al. 2009; PMEL 2014). Effectively, the calcium carbonate shells, and other exoskeletal structures will dissolve. These structures are integral to the survival of the organisms that produce them and weakening the shells will decrease the chances that the organisms will survive to reproduce. Some of these shelled animals, like crabs or clams, you may have seen at the beach, but there are others that are much smaller but quite important to marine food webs. For example, there are small swimming snails called pteropods, which are important members of the *zooplankton* system. They feed on other *plankton* and also serve as important food sources for fish, other zooplankton, and birds (Center for Biological Diversity n.d.). If unable to secrete their shells, zooplankton may not survive. Loss of such an important member of the marine food web would likely be devastating. If you eat seafood, you would be affected directly also.

Coral reefs, which support some of the most diverse communities in the world, are under tremendous environmental stress from warming ocean waters, coastal development, pollution, overfishing, and now, acidification. Growth of the reefs is based on the production of calcium carbonate ($CaCO_3$) skeletons, which support the soft tissues of the individual coral polyps. Acidification, therefore, hurts coral reefs in at least two ways—lack of sufficient calcium carbonate slows the growth process and weakens the skeletons, making those skeletons more susceptible to physical damage by waves and storms.

The dissolved, inorganic forms of carbon present in the ocean surface waters (collectively referred to as *dissolved inorganic carbon*) can be released again into the atmosphere, taken up by *phytoplankton*, or moved to deeper levels of the ocean. Carbon in dead plankton is converted to various organic carbon compounds, some of which are soluble. A small portion of the carbon may move to deeper parts of the ocean floor. As you might surmise, upward movement in the water column may take a very long time. Variation in surface water temperature at different latitudes and seasons influences greatly the movement of these carbon entities. In high latitudes, colder waters with higher solubility for CO_2 and thereby enriched sinks carry the carbon deeper. Carbon can be brought up from the depths at various upwellings, which often happen along the coastlines. Prevailing winds blow surface waters away from the coast, and this action helps to draw deeper water up from the ocean depths (GLOBE 2008; IPCC 2007).

QRQ 26. Carbon dioxide is removed from the ocean by which of the following process(es)? activity of photosynthetic organisms, dissolution of CO_2 from the atmosphere, formation of calcium carbonate, freshwater moving from streams and rivers, loss to the atmosphere, respiration of organic carbon, solubilization of calcium carbonate, storm activity

QRQ 27. Carbon dioxide is added to the ocean by which of the process(es)? Use the same choices as for QRQ 26.

QRQ 28. Carbon dioxide from the atmosphere dissolves in ocean water to produce ______ ______.

QRQ 29. In the ocean, carbonic acid dissociates to produce what kind(s) of ion(s)? (bicarbonate, calcium carbonate, carbon, carbon dioxide, carbonate, dioxide, hydrogen)

QRQ 30. Some of the bicarbonate in the ocean, under the right conditions of pH, ionizes into hydrogen ions and ______ ion. Use the same choices as for QRQ 29.

QRQ 31. With increasing concentrations of CO_2 in seawater, the marine carbonate system reactions are shifted to the right, resulting in ______ being converted to ______. Use the same choices as for QRQ 29.

QRQ 32. Marine organisms combine ______ and calcium to form their external skeletal structures.

QRQ 33. Select the chemical notation for each molecule below from the following: $CaCO_2$, $CaCO_3$, CO_2, CO_3^{2-}, H^+, H_2CO_2, H_2CO_3, HCO_2, HCO_3^-

a. bicarbonate ion
b. calcium carbonate
c. carbon dioxide
d. carbonate ion
e. carbonic acid
f. hydrogen ion

QRQ 34. (T/F) When pH decreases, ocean water is more acidic.

QRQ 35. Balance the equation using the following possible answers: CO_2, CO_3, H_3CO_2, H_2CO_3, H_2O, HCO_3

$$H_2O + CO_{2\ aqueous} \leftrightarrow \_\_\_\_\_\_ \leftrightarrow \_\_\_\_\_\_ + H^+$$

PoI 11. Review the following references on ocean acidification and summarize pertinent results:

a. http://www.pmel.noaa.gov/co2/story/What+is+Ocean+Acidification%3F
b. http://www.nap.edu/download.php?record_id=12904#

PoI 12. Find at least three recent news stories and other references about one of the following topics. As indicated by your professor, make a presentation or write a paper on the topic. Be sure to attribute proper credits and to have a list of references at the end.

a. Ocean acidification and its impacts on life.
b. What your hometown (or another city close to where you live) can do/has done to reduce carbon emissions.
c. What farmers can do to reduce carbon emissions.
d. What the federal government can do/has done to reduce carbon emissions.
e. What state government can do/has done to reduce carbon emissions.
f. What can be done in constructing and operating buildings to reduce carbon emissions.
g. Pros and cons of carbon credits.
h. Pros and cons of biofuels, such as corn ethanol.
i. Pros and cons of nuclear power, considering its impact on carbon emissions as well.
j. Deforestation of rainforests in developing countries, considering conflicting interests, such as farmers, loggers, environmentalists, etc.
k. Pros and cons of natural gas.
l. Pros and cons of solar panels.

PoI 13. What can we do to encourage solutions to increased carbon emissions?

We have focused on three of the biogeochemical cycles that characterize all ecosystems. Three cycles should be sufficient to give you an idea of how such cycles work and how important they are to supporting life. Energy flows through ecosystems from primary producers to the top consumers, and nutrients are recycled by way of fluxes that connect all of the nutrient reservoirs.

18.6. Valuating Ecosystems: Ecosystem Services

Ecosystems are quite often undervalued when calculations are done to support human activities, such as development, resource utilization, and agriculture. Disrupting the structure and function of ecosystems also disrupts what are called *ecosystem services* (contributions by ecosystems to sustain human survival and ways of life). Human civilization has utilized advances in technology to supplant services once provided by overly exploited ecosystems.

TABLE 18.3. Services That Exemplify Each Category of Ecosystem Services

Provisioning	Supporting	Regulating	Cultural
Products obtained from the ecosystem	Services sustaining all other categories Indirect long-term benefits	Services that result in benefits from regulating ecosystem functions	Nonphysical benefits obtained from ecosystems
food	primary production	water quality	recreation
raw materials	oxygen production	air quality	reflection
water	appropriate habitats	pollination	spiritual
biochemicals	maintenance of genetic and species diversity	buffer weather, climate	psychological
medicines	erosion control	biological control	cognitive development
building materials	soil formation	sequester carbon	
genetic resources	biogeochemical cycles	recycle wastes	

Ecosystem services are typically grouped into four categories: provisioning, supporting, regulating, and cultural. Table 18.3 lists some of the services that fall under each category.

What has been overlooked and underestimated by politicians, economists, and the general public is the values of these services. In 2014, a paper by Costanza et al. estimated the total worldwide value for ecosystem services at $125 trillion/yr. The last valuation had been done in 1997, and during the intervening years, about 25 percent of the earth's surface area performing these essential services had decreased through changes in land use. Most of the changes occurred in the tropics (rainforest destruction), wetlands, and tundra (treeless plains in regions where there are very low temperatures, scant precipitation, nutrient-poor soils, and short growing seasons). Over half the area of coral reefs was lost during this same time period. With the additional stresses of increasing population demands and changes in the world's climate, ecosystems are facing even more stress. We may find that technology will not be able to make up for the damage we are doing to the sources of essential services performed by healthy ecosystems. To be certain, the human species is a significant part of the biosphere (all regions of the earth where life exists), and our actions are making enormous demands on our environment. In the next module, we will examine some important ways we have harmed the earth and will explore some ways we might mitigate the damage.

QRQ 36. Contributions by ecosystems to sustain human survival and ways of life are called ______ ______.

Image Credits

Fig. 18.1: Source: https://commons.wikimedia.org/wiki/File:Indianriverlagoon.GIF.

Fig. 18.2: Source: https://science.nasa.gov/earth-science/oceanography/ocean-earth-system/ocean-water-cycle.

Fig. 18.3: Source: https://www.usgs.gov/media/images/distribution-water-and-above-earth-0.

Fig. 18.4: Source: https://commons.wikimedia.org/wiki/File:Nitrogen_Cycle.jpg.

Fig. 18.5: Source: https://www.nasa.gov/centers/langley/news/researchernews/rn_carboncycle.html.

LAB #18

Laboratory Exercises

18.7. Laboratory on the Carbon Cycle

An additional assignment on "Your Carbon Footprint" is available in Active Learning.

18.7.1. Preliminary Work

This laboratory investigates the fast carbon cycle, examining the effect of human activities on the amount of carbon in the atmosphere and the upper oceans and the pH of the ocean. From the text's website, view the introductory video; download and open "*Climate Change Rev.nlogo.*"

1. Click the Info tab at the top of the page and read sections "What Is It?" and "How It Works."
   a. ____________ Give the options for the year in which to start the simulation: A. 1610, B. 1710, C. 1810, D. 1910, E. 2010
   b. ______ Select a definition of "terrestrial biomass": A. cumulative mass over time of all living organisms in a certain area or ecosystem, B. mass of all organisms, living or dead, in a certain area or ecosystem at a certain time, C. mass of all living organisms in a certain area or ecosystem at a certain time
   c. ______ Why did the terrestrial biomass decrease in the earlier scenario? A. deforestation for fuel and agriculture, B. dinosaur extinction, C. evaporation, D. factories, E. Ice Age
   d. ____________ Give the pools of carbon that the simulation monitors: A. atmosphere, B. deposition of calcium-shelled organisms, C. forests, D. lakes, E. mountains, F. upper ocean, G. volcanos
   e. ____________ Give the sources of carbon in the model: A. emissions, B. factories, C. fish death, D. land plant (forest) photosynthesis, E. ocean photosynthesis, F. soil respiration, G. volcanos
   f. ____________ Give the model's sinks, or destinations, of carbon: A. land plant (forest) photosynthesis, B. land plant respiration, C. ocean photosynthesis, D. ocean respiration, E. plant death, F. soil respiration, G. tree growth
   g. ______ Which type of respiration increases with the decrease in forests? A. atmospheric, B. land plant, C. plant, D. soil

2. Read "How to Use It" and the NASA article, "The Carbon Cycle," referenced in the model's Info section, (Riebeek 2011).

a. ____________ How many metric tons are in 1 gigaton (Gt)? Select all that apply: A. 1 billion, B. 1 million, C. 1,000, D. 1 trillion, E. 10^3, F. 10^6, G. 10^9, H. 10^{12}

b. ______ Through a series of chemical reactions, the slow carbon cycle is the movement of carbon **1.** (A. between rocks, soil, ocean, and atmosphere, B. in rocks, C. through life forms on Earth).

c. ______ How long does it last? A. 100–200 billion years, B. 100–200 million years, C. 500–600 billion years, D. 500–600 million years

d. ______ What is the fast carbon cycle, which this simulation models? A. movement of carbon through life forms on Earth, B. movement of carbon between rocks, soil, ocean, and atmosphere, C. movement of carbon in rocks

e. __________________ Referring to Figure 18.5 of the fast carbon cycle in the section "The Cycle," select all the ways carbon, in the form of carbon dioxide, gets into the atmosphere: A. human emissions, B. microbial respiration and decomposition, C. photosynthesis in the sea, D. photosynthesis on land, E. plant respiration, F. sea-to-air gas exchange

f. ____________ How is carbon dioxide taken out of the atmosphere? Use the same choices as in part e.

g. ____________ How does carbon get into the soil? A. decomposition of plants, B. microbial respiration and decomposition, C. photosynthesis, D. plant respiration

h. ____________ How does carbon get into the deep ocean? A. microbial respiration and decomposition, B. ocean sediments, C. photosynthesis, D. sea-to-air gas exchange

i. ______ Carbon dioxide in the atmosphere is **1.** (G. greater, L. less, S. the same) in the summer than during the winter. (See "The Fast Carbon Cycle" on the NASA site.)

j. ______ Why? A. growing plants release carbon dioxide, B. growing plants take up carbon dioxide, C. warmer temperatures negate the impact of growing plants

k. ____________ Give two major ways in which people are impacting the carbon cycle today: A. burning fossil fuels, B. clearing land, C. eating plant foods, D. fishing, E. not recycling (See "Changes in the Carbon Cycle" on the NASA site.)

l. ______ According to the caption for "Global Carbon Dioxide Emissions" in that same section, humanity has been emitting a steadily increasing amount of carbon dioxide each year since the beginning of the Industrial Revolution. About what fraction of these emissions are removed each year by the fast carbon cycle? A. one-fourth, B. half, C. three-fourths

m. ______ Carbon dioxide that is not removed from the atmosphere each year **1.** (A. evaporates, B. remains in atmosphere, C. rises to beyond the atmosphere).

n. **1.** ______ **2.** ______ Excess carbon being in the atmosphere **1.** (A. cools, B. warms) the planet and **2.** (C. helps, D. inhibits) plant growth. (See "Effects of Changing the Carbon Cycle" on the NASA site.)

o. ______ Excess carbon in the ocean makes the water more **1.** (A. acidic, B. basic, C. dark), putting marine life in danger.

p. ______ The most important gas controlling the temperature of the earth is **1.** (A. carbon dioxide, B. chlorofluorocarbons, C. hydrogen, D. methane, E. nitrogen, F. oxygen, G. ozone).

q. Give the percentage breakdown (A. 5 percent, B. 10 percent, C. 15 percent, D. 20 percent, E. 25 percent, F. 30 percent, G. 35 percent, H. 40 percent, I. 45 percent, J. 50 percent, K. 55 percent, L. 60 percent) of the causes of the earth's greenhouse effect. (See "Effects of Changing the Carbon Cycle" on the NASA site.)

_______ carbon dioxide

_______ water vapor

_______ clouds

r. ______ Warmer Earth temperatures lead to **1.** (H. higher, L. lower) water vapor concentrations in the air.

s. ______ If carbon dioxide concentration rises, the air temperature will **1.** (F. fall, R. rise).

t. ______ If the air temperature rises, the amount of water vapor in the atmosphere will **1.** (F. fall, R. rise).

u. ______ If there is more water vapor in the air, greenhouse heating will be **1.** (L. less, M. more).

v. ______ By about how much have average global temperatures increased since 1880? A. 0.2°C (0.4°F), B. 0.4°C (0.7°F), C. 0.6°C (1.1°F), D. 0.8°C (1.4°F), E. 1.0°C (1.8°F), F. 1.2°C (2.2°F)

w. ______ Greenhouse warming does not happen immediately, because **1.** (A. the atmosphere soaks up heat, B. of a delay in carbon reaching the atmosphere, C. the earth soaks up heat, D. the ocean soaks up heat)

x. ______ Dissolving carbon dioxide in the ocean makes the water **1.** (L. less, M. more) acidic.

y. ______ A lower pH means the water is **1.** (L. less, M. more) acidic.

z. ___________ Give two results of more acidic ocean water: A. fish cannot absorb nutrients as well, B. tends to dissolve shells, C. thicker shells, D. thinner and more fragile shells, E. water tastes bad

aa. ___________ Plants take up carbon dioxide. Availability of which two things limits the ability of carbon dioxide fertilization to increase plant growth? A. hydrogen, B. methane, C. nitrogen, D. oxygen, E. water

bb. ______ In 2008, deforestation accounted for **1.** (A. 2, B. 5, C. 8, D. 12, E. 15, F. 18) percent of all human carbon dioxide emissions.

cc.______ Because of increased permafrost thawing, the earth's temperature will **1.** (F. fall, R. rise).

dd. ______ "Net primary productivity" is the measure of the amount of **1.** (A. carbon, B. hydrogen, C. nitrogen, D. oxygen, E. water) plants and phytoplankton turn into matter as they grow. (See "Studying the Carbon Cycle" on the NASA site.)

18.7.2. Laboratory Work

Default parameters: EMISSIONS = 1.5, YEARS = 100, %_OF_FOREST_CLEARED_YEARLY = 0.2, CALCIUM-SHELLED_ORGANISMS = 2.25, Add_Forest = 0, Add_Ocean_Photosynthesizers = 0, Add_Photosynthesizers_Every_Year = OFF.

3. Read "Things to Notice." The model in this simulation is substantially deterministic, with only the placement of turtles being random. Thus, only run each experiment once.

a. On the interface level, click 1710 LEVELS and record the following:

_______ GtC in plants

_______ GtC in the upper ocean

_______ GtC in the atmosphere

_______ pH of the ocean

b. Click 2010 LEVELS and record the following:

_______ GtC in plants

_______ GtC in the upper ocean

_______ GtC in the atmosphere

_______ pH of the ocean

c. _______ The ocean was more **1.** (A. acidic, B. basic) in 2010 than in 1710.

d. Use default parameters. Starting with 2010 levels, run the simulation. Give the four final reported values and indicate for each whether the simulated value is greater (G), less (L), or the same (S) in 2110 as the value in 2010.

_______ GtC in the plants, _______ (G, L, S)

_______ GtC in the upper ocean, _______ (G, L, S)

_______ GtC in the atmosphere, _______ (G, L, S)

_______ pH of the ocean, _______ (G, L, S)

e. **1.** _______ **2.** _______ **3.** _______ **4.** _______ Complete the following possible hypothesis using choices (N. cause no change in, D. decrease, I. increase): Higher emissions will **1.** the amount of carbon in plants, **2.** the amount of carbon in the upper ocean, **3.** the amount of carbon in the atmosphere, and **4.** the pH of the ocean.

f. Use default parameters, except EMISSIONS = 10.0. Start with 2010 levels and repeat the experiment. Give the four final simulated values in 2110, and compare your results to those of part d, where EMISSIONS = 1.5, using (N. cause no change in, D. decrease, I. increase). (Observe how the results support or refute your hypothesis in part e.)

_______ GtC in the plants, _______ (N, D, I)

_______ GtC in the upper ocean, _______ (N, D, I)

_______ GtC in the atmosphere, _______ (N, D, I)

_______ pH of the ocean, _______ (N, D, I)

g. **1.** _______ **2.** _______ **3.** _______ **4.** _______ Complete the following possible hypothesis using choices (N. cause no change in, D. decrease, I. increase): A higher percent of forest cleared yearly will **1.** the amount of carbon in plants, **2.** the amount of carbon in the upper ocean, **3.** the amount of carbon in the atmosphere, and **4.** the pH of the ocean.

h. Use default parameters, except %_OF_FOREST_CLEARED_YEARLY = 0.3. Start with 2010 levels and repeat the experiment. Give the four final simulated values in 2110, and compare your results to those of part d, where %_OF_FOREST_CLEARED_YEARLY = 0.2 percent, using (N. cause no change in, D. decrease, I. increase). (Observe how the results support or refute your hypothesis in part g.)

_______ GtC in the plants, _______ (N, D, I)

_______ GtC in the upper ocean, _______ (N, D, I)

_______ GtC in the atmosphere, _______ (N, D, I)

_______ pH of the ocean, _______ (N, D, I)

Human Impact—Over Our Heads? Atmospheric Carbon and Climate Change

MODULE

19

APPLYING SCIENCE TO LIFE 1 Drying Up: Confronting Effects of a Warming Climate

Santiago, in the Maipo River Basin of Chile, is home to 40 percent of the country's 18 million people . . . about 7 million. For a relaxing break from the urban life, residents that could get away might travel only 45 miles north of the city to the shores of a lake called Laguna de Aculeo. There, visitors could enjoy fishing and various water sports on the lake, birds and other wildlife, and the peaceful and scenic surroundings of the valley. Many even built or bought cottages and villas on the shores of the lake. In 2010, the lake covered about 3,000 acres, and villas on the lake sold for up to $500,000 (Reeves 2017).

Laguna de Aculeo is a lagoon, or shallow lake, lacking tributaries and dependent on rainfall to replenish its water. Over the past seven years or so, the lagoon has shrunk drastically, leaving many of the homes of vacationers, farmers, and tour companies with only mud and grass. Surely, the receding lake has adversely affected those vacationing and those who provide services for the tourists. Fewer tourists visit, and those who own housing formerly on the lake have seen prices for their property plummet (Reeves 2017).

Tourists may find other destinations, but for Ramón Muñoz, who has farmed melons and vegetables and raised chickens around the lake for most of his 55 years, the loss of the lake and its irrigation water is devastating. Mr. Muñoz says, "My two sons have had to leave the farm for Santiago to find work to support the family. Even with their help, we barely get by." Likewise, Roberta and Carlos Silva, who have operated a nature adventure tour company, have many fewer clients. "I was lucky to find a job at the mine, which is one hour away. Otherwise, we'd have to move," says Carlos.

The changes that have occurred for Laguna de Aculeo serve as a model for the increasingly dry conditions in other parts of Chile. Garreaud et al. (2017) indicate that from 2010 to 2017, Central Chile, which includes Santiago, has endured a continual succession of years with low precipitation levels. These researchers considered this ongoing event remarkable because it has been so persistent and extensive, with yearly rainfall deficits varying between 25 and 45 percent. The consistently lower rain levels have led to prolonged drought conditions in many areas, which have taken tremendous tolls on the economy and workers. Sufficient supplies of water are vital to maintain agricultural production for internal consumption and export (WTO 2019).

Drought not only affects human populations but also is devastating to the ecology of the region, altering native plant and animal communities and "ecosystem services." Moreover, drier conditions also create conditions favoring wildfires, which can destroy habitat, cause economic damage through direct and indirect effects, and sicken or kill plant and animal (including human) populations.

When we consider what brings on drought conditions, we generally think in terms of reduction in precipitation. In central Chile, there are natural causative factors, characteristic of the South Pacific (El Niño-Southern Oscillation [ENSO] and Pacific Decadal Oscillation [PDO]) that strongly influence the precipitation deficit. However, Boisier et al. (2016) found in their analysis that the natural oscillations were not sufficient to explain the decline in precipitation during the past 30 to 40 years. Only when the effects of human activities (*anthropogenic forcing*) are considered is the precipitation decline explained.

There are many human activities that may promote drought conditions. Primarily, the decline in precipitation can be brought on by warming that results from the production of greenhouse gases, like carbon dioxide (CO_2). Such gases trap heat reflected from the earth's surface, and this warmer air can accommodate more moisture that evaporates from the soil, lake surfaces, and so on. As precipitation decreases, human populations draw more water from reservoirs, rivers, and aquifers.

In the Maipo River Basin, which includes so much of the country's population, water is critical for personal consumption, agriculture, power generation, and mining. Researchers from Pontifical Catholic University's Centre for Global Change predict that the temperature will increase by 2.5 to 3.5 °C and precipitation will decline by 40 percent by the end of the century for this basin (Vicuña, Garreaud, and McPhee 2011).

It is difficult to find a segment of Chile's society that is not affected by the increased warming and drought. In response, the government is investing in desalinization plants, reservoir construction, and access to underground water supplies. Additionally, the government is encouraging improved water conservation, fire prevention, and development of green spaces in urban communities. Recently, they have also instituted climate change instruction in public schools (Cooke 2015; Reeves 2017). Chileans, including their political leaders, are convinced that something has shifted in the climate patterns for the country, which is resulting in a shortage of water, so essential to life.

PoI 1. What are droughts/drought conditions?

PoI 2. What is ENSO? How does ENSO influence drought conditions in central Chile?

PoI 3. What is groundwater? What is an aquifer? Why are they important?

PoI 4. Besides rainfall, much of the water in Chile comes from glacial melt. How is this water used by human populations in Chile?

PoI 5. How might warming climate affect water supplies for Chile?

PoI 6. How do desalination plants work?

PoI 7. Why would the government build reservoirs in response to a water crisis?

PoI 8. What are the positive and negative effects of reservoirs?

PoI 9. What human activities, besides the production of greenhouse gases, might bring on drought?

19.1. Introduction

The disturbing story of Jahanara Khatun was recounted in a *New York Times* article in March 2014. Five years earlier, Ms. Khatun, like many others living on the lowlands of river deltas in Bangladesh, lost her modest home to Cyclone Aila. In the aftermath, her husband became ill and died in 2013. To pay for the construction of a new hut and for her husband's treatment, the family borrowed the money in exchange for labor in a local brickmaking facility. With continuing severe financial obligations, she also sold two of her children to a brickmaking factory owner in Dhaka. She still lives in the hut situated below sea level, protected by a single, shoddy berm, trying to grow a little food (Harris 2014).

Storms in this area are frequent and accompanied by storm surges that often overrun the inadequate sea walls to inundate the delta and its inhabitants. Flooding not only destroys property and lives, it also deposits salt on the land, making the land inhospitable for the growth of a meager set of crops (Harris 2014). Scientists predict that with global warming, the inhabitants of low-lying areas like deltas will be at an even greater risk for flooding. Even what we might consider to be small increases in the earth's temperature will cause seawater to expand and ice stocks (e.g., ice caps, glaciers) to melt, resulting in a higher sea level. We know from tide gauges and satellite data that the sea level has risen about eight inches since 1870 (NRC 2013). Rising sea levels have the potential to displace millions of people, mostly poor, from coastal and island communities.

19.2. Climate and Weather

Before we consider further *climate change* and *global warming*, we should clarify what the terms *climate* and *weather* mean. Have you read or heard someone ask a question similar to the following? "This winter we had record snowfall, so how can there be global warming/climate change?" Questions such as the one above indicate that many people, even political figures, do not understand the distinction between *climate* and *weather*. The National Aeronautics and Space Administration (NASA) defines the terms as follows:

> **weather:** "atmospheric conditions that occur locally *over short periods* of time—from minutes to hours or days. Rain, snow, clouds, winds, floods and thunderstorms are good examples"
> **climate:** "*long-term* regional or even global average of temperature, humidity and rainfall patterns over seasons, years or decades" (NASA 2020)

Thus, it is possible for the *weather*, as reported by a local television station or on the internet, to include record snowfalls this winter, in your town, and yet the *climate* may still be changing overall for the northeast and the entire United States. *Global warming* refers to a trend in the data, which indicates that the average temperature for the entire earth has increased. Data collected over the entire 20th century indicates a warming, particularly during the last quarter. *Climate change* includes an array of global phenomena besides global temperature increases. For instance, increased occurrence of extreme storms, droughts, fires, loss of glacial ice mass, and rise of sea level are all part of a global trend from which scientists have concluded that the global climate is indeed changing.

QRQ 1. Indicate whether each of the following statements/phrases refers to weather or climate:

a. "It sure is cold today!"
b. The 10-day forecast
c. Average high temperature in June for Paris
d. Partly cloudy with a 20 percent chance of rain
e. Hurricane warning
f. The frequency and violence of hurricanes
g. Number and severity of droughts last century
h. A drought this summer

19.3. Warming to the Reality of Climate Change

There are some people who deny that the earth has warmed, but denial is futile and potentially disastrous. We have measured surface (land and ocean) temperatures since the late 19th century (see Figure 19.1). Many independent groups of scientists have studied and analyzed the data, allowing for local conditions and improving instrumentation. They all agree that the earth has warmed by 0.8°C (1.4°F) during the past 100 years, with most of the increase during the last 35 years. Although that seems like a small increase, it is a significant rise for a short span of geological time. Moreover, climate scientists predict that over the next 100 years, the temperature may rise between 2 and 11.5°F (NRC 2010). Predicting the sea-level rise is a bit more complicated, but a commonplace estimate is a rise of two feet (Karl, Melillo, and Peterson 2009). Depending on the topography of the coastal area, this kind of rise will drastically change the shapes of many coastlines.

The question does not seem to be *if* the earth has warmed, but whether this warming is part of a natural fluctuation, or cycle, or whether our species has fueled the rise in temperature. There is general agreement from the scientific evidence that we have amassed that human activity is responsible for the largest portion of the change in climate. Our impact has been achieved primarily through altering one of the most important nutrient cycles on our planet—the *carbon cycle*.

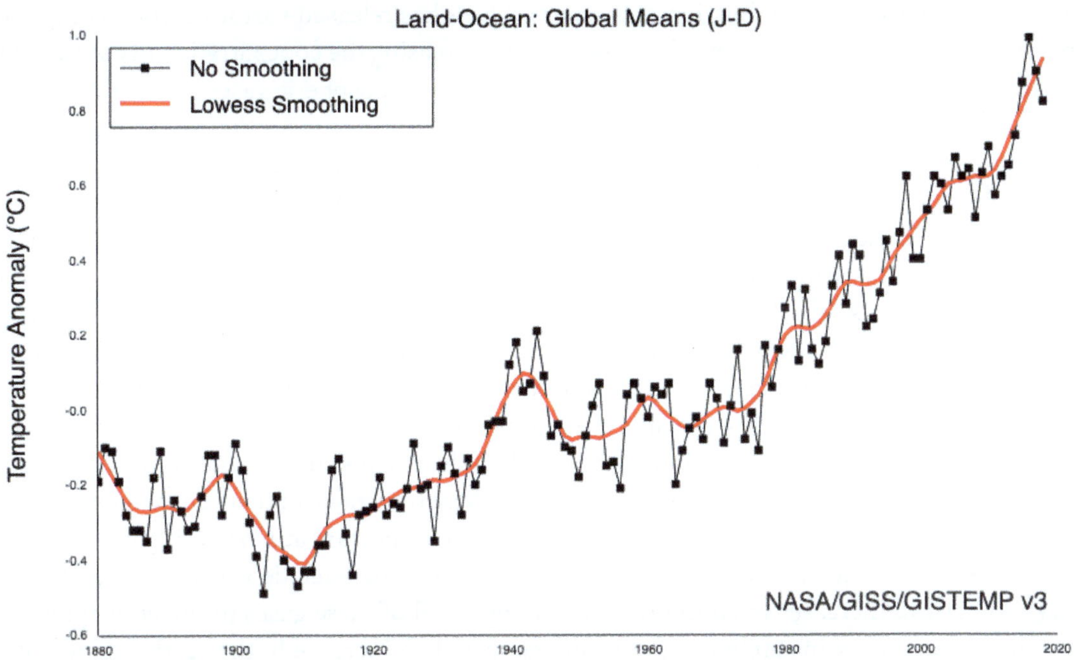

FIGURE 19.1 Global annual mean surface air temperature change (black), with five-year running mean (red)

QRQ 2. Select the ways in which it was suggested that global warming would make living on the delta in Bangladesh more problematic: decrease in animal life, increase in heat stroke, increased flooding from storms and/or ice melt, loss of property and life from flooding, loss of vegetation, salt deposits on farmland from flooding

QRQ 3. The primary way in which human beings have affected a warming of the earth is altering the ______ (Answer choices: carbon cycle, climate cycle, nitrogen cycle, oxygen cycle, water cycle).

19.4. Hijacking the Carbon Cycle

We discussed the basics of the carbon cycle in the module "Ecosystems—Natural Networks," and now we will apply our knowledge of that cycle to help understand how human beings are implicated in global warming.

The Industrial Revolution, which began during the middle of the 18th century, brought the human family many good things, making lighter the work that had previously been accomplished only through physical labor and sweat. There is no denying that new mechanisms and methods for manufacturing, transportation, and agriculture improved the living standards of many people. On the other hand, these changes were made possible by human exploitation of carbon stores formed millions of years ago from dead organisms covered by sediments and exposed to high temperature and pressure (fossilized). These stores generally take the form of coal, natural gas, and oil (DOE

2021). Over the years, our combustion of these *fossil fuels* has released various carbon compounds from a relatively stable reservoir into the atmosphere, increasing the concentration of carbon gases, especially carbon dioxide (CO_2) and methane (CH_4). These atmospheric gases belong to a category of chemical compounds that generate a *"greenhouse effect."*

If you have ever walked into a greenhouse during a warm, sunny day, you may have noticed that it can be hot, much hotter than the outside temperature. A similar situation occurs when you get into your car in the summertime after you have parked the vehicle in the sun with the windows up. Some of the light passing through the glass of the greenhouse or your car windows is absorbed, and some of the light is reflected as *infrared radiation*. This radiation does not pass readily through the glass, but the heat energy of the radiation raises the internal temperature of the greenhouse or your car. Most greenhouses have ventilation and cooling systems to prevent temperatures from rising too much so that the enclosed plants are not damaged or killed. Nevertheless, light from the sun passes through the atmosphere and is absorbed by the earth's surface, but some of the light is reflected toward space. Greenhouse gases in the atmosphere trap much of that heat, which maintains the earth at a temperature that is habitable by living things. Important atmospheric gases associated with the greenhouse effect, besides carbon dioxide and methane, include water vapor and nitrous oxide (NRC 2013). Increases in concentrations of any or all of these gases in the atmosphere will add to the infrared trapping ability of the atmosphere and an increase in the earth's temperature. Compare the situation to the effect of adding another blanket, or a heavier one, when you get into bed. The physicist Joseph Fourier first suggested in 1824 that the earth's atmosphere might act to hold in heat, and in 1896 Svante Arrhenius, a founder of physical chemistry, predicted a warming trend, now observed, resulting from human activity increasing CO_2 levels in the atmosphere (NRC 2013).

QRQ 4. Fossil fuels are ______ (Answer choices: ancient rocks that can be burned, dead organisms, covered by sediments and exposed to high temperature and pressure, methane buried deep in the ground, natural gas, oil).

QRQ 5. The use of fossil fuels changes the carbon cycle by releasing various ______ compounds from a relatively stable ______ into the atmosphere, increasing the concentration of carbon gases, especially ______ (CO_2) and ______ (CH_4). Possible answers for each blank: carbon, carbon dioxide, CFCs, methane, nitrous oxide, ozone, reservoir, sink

QRQ 6. ______ gases in the atmosphere trap much of the heat that maintains the earth at a temperature habitable by living things.

19.5. Human Activities and Climate Change

PoI 10. What is the Industrial Revolution, and how did it change the human condition?

Measurements of CO_2 levels in the atmosphere have been collected at Mauna Loa Observatory in Hawaii since the late 1950s (see Figure 19.2). Carbon dioxide concentrations prior to that time have been determined from air bubbles from ice cores taken in Greenland and

Antarctica. From these data, we know that CO_2 concentrations varied only slightly prior to the beginning of the Industrial Revolution (NRC 2013). For about 2,000 years before 1750, CO_2 levels were stable around 280±10 ppm. By the latter part of the 19th century, CO_2 levels began to rise, and in 2015 the annual mean was 399.47 ppm (Scripps CO_2 Program 2019). Thus, we had little or no change in CO_2 concentrations for at least 20 centuries, and then we have a 41 percent increase in only 2.6 centuries. These numbers provide good circumstantial evidence that the warming of the planet is being promoted by our own activities. What further evidence exists to support such a hypothesis?

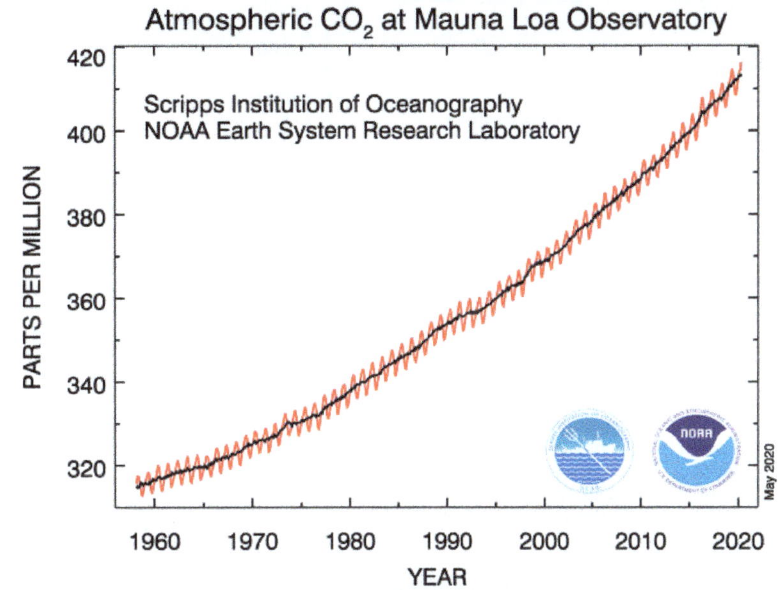

FIGURE 19.2 Monthly mean atmospheric CO_2 at Mauna Loa Observatory, Hawaii. Black curve indicates seasonally corrected data.

A report from the National Research Council (NRC) in the United States contains compelling arguments that human activities have led to climate change. Of course, the NRC cites the evidence represented in Figures 19.1 and 19.2 relating temperature rises with the increase in atmospheric carbon dioxide levels. Beyond that association, they have analyzed the isotopic carbon concentrations in the atmosphere and compared these concentrations with the carbon forms in living organisms. Isotopes of an element have nuclei with the same number of protons but a different number of neutrons (same atomic number, but different mass number). The most commonly found carbon isotope is Carbon-12 (six protons, six neutrons), which is stable. Carbon-14 (C-14, six protons, eight neutrons) is unstable (*radioactive*), decomposing over long periods of time. There is little C-14 in the atmosphere when compared with the levels in organisms living today. The low level of C-14 in the atmosphere tells scientists that much of the carbon is *not* coming from currently living sources but from old sources, where the C-14 has had millions of years to deteriorate. Their conclusion is that the CO_2 in the atmosphere is from the fossil fuels that we burn (NRC 2013).

Scientists estimate that about 75 percent of carbon dioxide emissions are from the burning of fossil fuels, including a small subset from cement production (Ding et al. 2001). Carbon dioxide, because of its great abundance and longevity in the atmosphere, is the most important greenhouse gas or *forcing agent* (forcing climate change). However, there are other important contributors, such as methane and nitrous oxide (see Figure 19.3). Methane generally comes from wetlands, but we have destroyed many of these, and human agricultural activities now discharge significant amounts of methane. Ruminant farm animal digestion, rice growth in paddies, leaking natural gas, and landfill emissions all contribute methane to the atmosphere. Nitrous oxide is produced from agricultural fertilizers, vehicle exhaust, and power plant emissions.

There are two natural influences that must be considered in any discussion of climate change, including volcanic activity and solar energy output. Between 1970 and 2000, there were two major

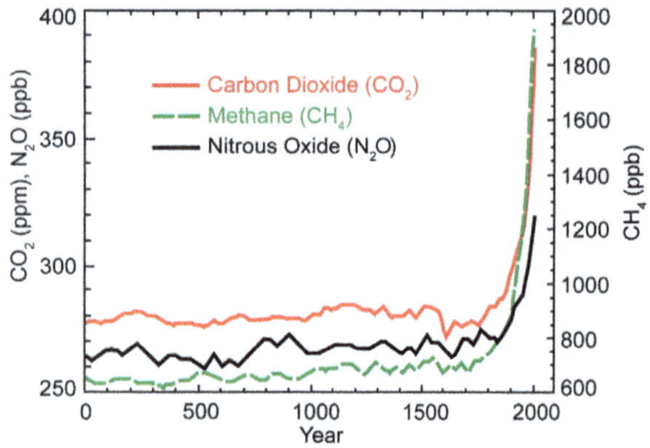

Forster *et al.*[3];Blasing[7]

Increases in concentrations of these gases since 1750 are due to human activities in the industrial era. Concentration units are parts per million (ppm) or parts per billion (ppb), indicating the number of molecules of the greenhouse gas per million or billion molecules of air.

FIGURE 19.3 Increase in greenhouse gas concentrations (parts per million [ppm] or parts per billion [ppb]) in the atmosphere over the last 2,000 years (USGCRP 2009)

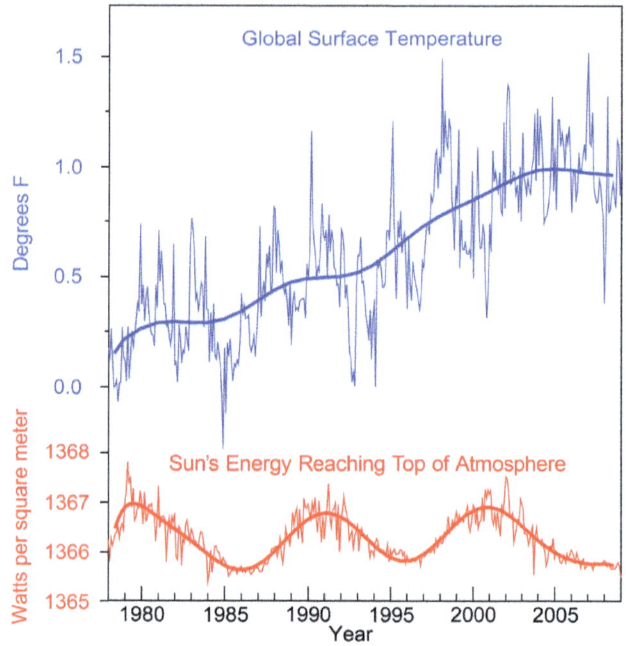

NOAA/NCDC; Frölich and Lean; Willson and Mordvinov; Dewitte *et al.*[59]

The Sun's energy received at the top of Earth's atmosphere has been measured by satellites since 1978. It has followed its natural 11-year cycle of small ups and downs, but with no net increase (bottom). Over the same period, global temperature has risen markedly (top).[60]

FIGURE 19.4 Eleven-year cycle of sun's energy received at the top of Earth's atmosphere, with no net increase (bottom), while global temperature has risen markedly (top) (USGCRP 2009)

volcanic eruptions, but they had only transient cooling impact (Hansen et al. 2000). Satellites have monitored the sun's energy output, and at least for the period of 1979–2009, it has followed a typical 11-year cycle (see Figure 19.4) (USGCRP 2009). While the energy reaching the outer atmosphere has been consistent during that interval, the earth's surface temperature has risen. Scientists have also reviewed other natural factors that might impact climate, and none of them has proved to be as important as human impact through our use of fossil fuels (USGCRP 2009).

QRQ 7. Select the following that are scientific evidence that global warming is largely a result of human activity: CO_2 concentrations in ice core air bubbles have increased since the Industrial Revolution; CO_2 concentrations in fossil fuels have increased since the Industrial Revolution; CO_2 concentrations in the ocean have increased since the Industrial Revolution; there is little isotopic carbon in the atmosphere compared with levels in living organisms, thus much of this carbon is not from living sources but from fossil fuels we burn; there is little moisture in the atmosphere compared to levels in living organisms

QRQ 8. Select the four major greenhouse gases: carbon dioxide, carbon monoxide, hydrogen, hydrogen chloride, methane, nitrogen, nitrous oxide, ozone, propane, water vapor

PoI 11. What evidence does each of the following references provide for climate change? Refer to Module 19 references at the end of the book for links and citations.

a. (Church and White 2006)
b1. (Levitus et al. 2001)
b2. (Levitus et al. 2009)
c. (Kwok and Rothrock 2009)
d1. (Gleason et al. 2008)
d2. (Frich et al. 2002)

PoI 12. Discuss the impact of global warming on the following:

a. health
b. agriculture
c. oceans
d. fishing
e. ecotourism
f. coastal environments
g. the poor

Image Credits

Fig. 19.1: Source: https://data.giss.nasa.gov/gistemp/graphs_v3/customize.html.
Fig. 19.2: Source: https://www.esrl.noaa.gov/gmd/ccgg/trends/.
Fig. 19.3: Source: https://nca2009.globalchange.gov/2000-years-greenhouse-gas-concentrations/index.html.
Fig. 19.4: Source: https://nca2009.globalchange.gov/measurements-surface-temperature-and-sun%E2%80%99s-energy/index.html.

LAB #19

Laboratory Exercises

19.6. Laboratory on Climate Change

Another assignment, "Gapminder Assignment on Climate Change," is available in Active Learning.

19.6.1. Preliminary Work

This laboratory investigates some of the impact of clouds and greenhouse gases on the temperature of the planet. From the text's website, view the introductory video; download and open "*Climate Change Rev. nlogo.*"

1. Click the Info tab at the top of the page and read sections "What Is It?" and "How It Works." Note that the surface of the earth is represented by a strip whose color (from black to shades of green to white) depends on the value of ALBEDO (from 0.0 to 1.0).

   a. Match each component with its color in the model: A. black, B. blue, C. brown, D. cyan, E. gray, F. magenta, G. orange, H. red, I. rose, J. white, K. yellow

   _______ earth

   _______ atmosphere

   _______ space

   _______ sunlight energy

   _______ heat energy in the earth

   _______ infrared (IR) energy

   b. ______ When the earth is warm, energy in the earth is **1.** (L. less, M. more) likely to produce infrared (IR) energy.

   c. ______ Does IR go through clouds, yes (Y) or no (N)?

   d. ______ Does IR go through carbon dioxide molecules, yes (Y) or no (N)?

   e. ____________ What two types of energy add thermal energy to the earth? A. chemical, B. geothermal, C. infrared (IR) light reflected down to the earth, D. kinetic, E. nuclear, F. sunlight, G. volcanic

f. ______ How is energy in the earth related to temperature? A. cubic, B. directly proportional, C. exponential, D. indirectly proportional, E. quadratic, F. square

g. ________________________ In formulating a model, computational scientists attempt to be as simple as reasonably possible. Most problems are entirely too complex to consider all details, and doing so would only make the model impossible to solve or to run on a computer in a reasonable amount of time. Thus, computational scientists make *simplifying assumptions* for the model, streamlining some of the factors and eliminating others that do not seem as important. Give simplifying assumptions for this model on climate change: A. IR light does not bounce off clouds, B. temperature is constant, C. the earth has a single albedo, D. the earth has a single heat capacity, E. the earth is a single temperature, F. visible light is constant, G. visible light is not absorbed by CO_2

2. Read "How to Use It."

   a. ______ What value of the SUN-BRIGHTNESS slider (to one decimal place) corresponds to our sun?

   b. ______ Give the value of ALBEDO (to one decimal place) that indicates that the surface reflects all sunlight.

   c. ______ Give the color of the model's surface that indicates that the surface reflects all sunlight: A. black, B. blue, C. gray, D. green, E. pink, F. white

   d. ______ In the model, do clouds block sunlight, yes (Y) or no (N)?

   e. ______ In the model, do clouds block IR, yes (Y) or no (N)?

   f. ______ In the model, do carbon dioxide molecules block sunlight, yes (Y) or no (N)?

   g. ______ In the model, do carbon dioxide molecules block IR, yes (Y) or no (N)?

   h. ______ Buttons add/subtract carbon dioxide molecules in groups of how many?

   i. ______ Give the maximum number of carbon dioxide molecules we should use.

19.6.2. Laboratory Work

For each configuration (i.e., group of settings) of an experiment, run a simulation for at least 6,000 ticks at least four times, unless otherwise instructed. Use a spreadsheet to record results and calculate averages.

3. This question involves the impact of carbon dioxide on global temperature.

   a. ________________________ For the default settings of SUN-BRIGHTNESS = 1.0, ALBEDO = 0.60, no clouds, and no additional CO_2, run the model. Slow the speed of the simulation. Run the simulation several times and watch a ray. Select all statements that describe several ray paths: A. Heat energy sometimes gives off infrared light, which goes in a straight line. B. In the earth, heat energy moves randomly. C. Often, sun energy strikes the earth and becomes heat energy. D. Sometimes a sunray is immediately reflected as infrared light into the atmosphere. E. Sometimes a sunray is immediately reflected from the earth into the atmosphere. F. Sometimes infrared light is reflected back to the earth to become heat energy.

   b. low: ______ high: ______ Uncheck *view updates* or increase the speed of the simulation. Run the simulation at least four times. Using the spreadsheet design in Table 19.1, run the simulation record the high and low temperatures that occur between 0 and 6,000 ticks. Give the mean temperatures.

TABLE 19.1. Template for Lab Spreadsheet

Names		
	3b Temperatures	
	low	high

c. ______ Click the ADD CO_2 button six times. Check *view updates* or decrease the speed of the simulation. Run the simulation and watch the IR rays. What additional behavior(s) beyond those answered in part a do you observe? A. Sometimes a new IR ray emanates from a CO_2 molecule. B. Sometimes the IR ray bounces off CO_2 molecules. C. Sometimes the IR ray just disappears. D. Sometimes the IR ray starts following the path of a CO_2 molecule.

d. ______ Complete the following hypothesis: Additional CO_2 results in a(n) **1.** (L. lower, H. higher, U. unchanged) maximum temperature.

e. ______ One experiment to test this hypothesis might be the following: Run the simulation several times for each of the parameters **1.** (A. 50, 100, and 150 CO_2 molecules; B. 0, 5, and 10 clouds; C. 1.0, 3.0, and 5.0 sun-brightness; D. 0.9, 0.5, and 0.1 albedo), recording the resulting high temperatures and calculating the averages.

f. ______ Repeat part b for 50 CO_2 molecules at least four times, recording only the *high* temperatures between 0 and 6,000 ticks on your spreadsheet (indicating Question 3g and 50 CO_2). Calculate the mean high temperature.

g. ______ Repeat part f with 100 CO_2 molecules.

h. ______ Repeat part f with 150 CO_2 molecules.

i. ______ Do the results of your experiment support your hypothesis in part d, yes (Y) or no (N)?

j. Write a thoughtful paragraph about the impact of increasing carbon dioxide emissions from various sources on climate change, using your experiments as evidence.

4. This question involves the impact of albedo on global temperature.

a. **1.** ______ **2.** ______ **3.** ______ Refer to other sources as needed to complete the following statement: Albedo is the portion of **1.** (A. CO_2, B. heat, C. solar energy, D. visible light) reflected from **2.** (E. clouds, F. Earth, G. space, H. the sun) back into **3.** (E. clouds, F. Earth, G. space, H. the sun).

b. **1.** ______ **2.** ______ Refer to other sources as needed to complete the following statement: Charcoal has an albedo of about 0.04, meaning that this solid reflects approximately **1.** (A. 0.04, B. 0.4, C. 4, D. 40, E. 0.96, F. 9.6, G. 96) percent of the incoming solar radiation, while absorbing the remaining **2.** (A. 0.04, B. 0.4, C. 4, D. 40, E. 0.96, F. 9.6, G. 96) percent of energy.

c. Refer to other sources as needed to match each item with its approximate albedo: A. 0.04, B. 0.06, C. 0.17, D. 0.50–0.70, E. 0.80–0.90

______ bare sea ice

______ bare soil

______ fresh asphalt

______ fresh snow

______ typical ocean

d. ______ Check *view updates* or decrease the speed of the simulation. Change ALBEDO to 1.0 with SUN-BRIGHTNESS = 1.0, no clouds, and no additional CO_2. Run the simulation several times and watch a ray. What initial behavior of a sunray striking the earth do you observe? A. Sometimes sun energy is immediately reflected as infrared light into the atmosphere. B. Sun energy striking the earth is always absorbed, becoming heat energy. C. Sun energy striking the earth is always reflected. D. Sun energy striking the earth is sometimes reflected and sometimes absorbed.

e. ______ Change the ALBEDO slider bar to 0.0. Run the simulation several times and watch a ray. What initial behavior of a sunray striking the earth do you observe? A. Sometimes sun energy is immediately reflected as infrared light into the atmosphere. B. Sun energy striking the earth is always absorbed, becoming heat energy. C. Sun energy striking the earth is always reflected. D. Sun energy striking the earth is sometimes reflected and sometimes absorbed.

f. ______ Complete the following hypothesis: Lower albedo results in a(n) **1.** (L. lower, H. higher, U. unchanged) maximum temperature.

g. ______ One experiment to test this hypothesis might be the following: Run the simulation several times for each of the parameters **1.** (A. 0, 5, and 10 clouds; B. 0.9, 0.5, and 0.1 albedo; C. 1.0, 3.0, and 5.0 sun-brightness; D. 50, 100, and 150 CO_2 molecules), recording the resulting high temperatures and calculating the averages.

h. ______ Uncheck *view updates* or increase the speed of the simulation. Run this experiment for ALBEDO = 0.9 at least four times. Record the *high* temperatures between 0 and 6,000 ticks on your spreadsheet (indicating Question 4h and 0.9 albedo) and calculate the average high temperature.

i. ______ Repeat part h for ALBEDO = 0.5.

j. ______ Repeat part h for ALBEDO = 0.1.

k. ______ Do the results of your experiment support your hypothesis in part f, yes (Y) or no (N)?

l. Write a thoughtful paragraph about the impact of the melting of glaciers and the polar ice caps on climate change, using your experiments as evidence. Also discuss the problems resulting from such melting.

5. This question involves the impact of clouds on global temperature.

a. ____________________ Change the ALBEDO slider bar to 0.6 with SUN-BRIGHTNESS = 1.0, no clouds, and no additional CO_2. After setup, click ADD CLOUD ten times. Check *view updates* or decrease the speed of the simulation. Running the simulation several times, watch a ray. What additional behavior(s), besides those you answered for question 3a, do you observe? A. IR rays go through clouds. B. IR rays that hit clouds are reflected. C. Sometimes a sunray is immediately reflected from a cloud as infrared light into the atmosphere. D. Sunrays go through clouds. E. Sunrays that hit clouds are reflected.

b. ______ Complete the following hypothesis: Increased cloud cover results in a(n) **1.** (L. lower, H. higher, U. unchanged) maximum temperature.

c. ______ One experiment to test this hypothesis might be the following: Run the simulation several times for each of the parameters **1.** (A. 0, 5, and 10 clouds; B. 0.9, 0.5, and 0.1 albedo; C. 1.0, 3.0, and 5.0 sun-brightness; D. 50, 100, and 150 CO_2 molecules), recording the resulting high temperatures and calculating the averages.

d. ______ In question 3 parts b and c, you ran an experiment for 0 clouds. Copy the high temperature column from that experiment to another area of the spreadsheet, indicating question 5d and 0 clouds. Give the average high temperature.

e. ______ Uncheck *view updates* or increase the speed of the simulation. Repeat part d with five clouds at least four times. Record the high temperatures between 0 and 6,000 ticks on your spreadsheet (indicating question 5e and 5 clouds) and calculate the average high temperature.

f. ______ Repeat part e with ten clouds.

g. ______ Based on these results, increased cloud cover results in a(n) **1.** (L. lower, H. higher, U. unchanged) maximum temperature.

MODULE 20

Commandeering the Nitrogen Cycle—Human Transformation of the Biosphere

20.1. Introduction

Nitrogen gas (N_2) is an inert molecule and of no use to living creatures. Prior to the early 20th century, organisms, including our own species, relied largely on *biological nitrogen fixation* (*BNF*) to convert N_2 into a form of nitrogen that they could use for life support. As you remember from our discussion of the nitrogen cycle in the module on ecosystems, plants that are unable to utilize N_2 depend on certain microbial organisms to fix nitrogen into ammonia (BNF). Subsequently, ammonia is converted into nitrite (NO_2^-) and then into nitrate (NO_3). These forms are usable by plants to synthesize biomolecules like amino acids and nucleotides that are necessary for growth, metabolism, and reproduction. We and other animals depend directly or indirectly on this chain of production for growth, metabolism, and reproduction as well. The nitrogen cycle functioned normally, supplying enough usable forms of nitrogen to provide for our ancestors' needs. Wastes and decomposed materials were either recycled or converted back into N_2.

As human populations expanded over the earth, they always had at least local impact on the soil, the air, and the water, but BNF still sufficed to supply their needs for nitrogen. However, with the Industrial Revolution (which began during the 18th century in England and spread) and the invention and application of the internal combustion engine (during the 19th century), the human species began to accelerate the extent of its influence over the earth. Afterward, human activities, particularly those in industry, transportation, and agriculture, inflicted changes in the nitrogen cycle that have affected almost every ecological system. Examples of human-induced effects include air pollution, greenhouse gas emissions (and therefore climate change), water pollution, acid rain, and loss of biodiversity. Some of the changes that developed were also positive, at least for the health and well-being of human beings. On the other hand, many changes have been detrimental to human food production and health.

With the development of the *Haber–Bosch Process* (*HBP*) in the early 20th century, human beings have had the ability to increase the productivity of crops through the addition of nitrogen-containing fertilizers. Over 50 years ago, the expanded availability and use of fertilizers began what some describe as a "Green Revolution," which dramatically stimulated farm production, crucial for people in developing nations. This revolution also was catalyzed by the widespread use of pesticides and crop (especially grain) varieties that were specifically bred to be high yielding. More food meant that more people accessed sufficient amounts of nourishment to survive and reproduce, which led to an increase in the size of the human population.

Reactive nitrogen species (N_r) constitute the types of nitrogen that are biologically, radioactively, or photochemically active, and the biologically active forms are requisite for all living organisms. We usually express amounts of these N_r in *teragrams* (*Tg*) (1 Tg = 1 trillion [10^{12}] grams). Today, increasing amounts of these reactive nitrogen compounds, which include *nitrogen oxides* (*NO_x*) (such as NO and NO_2), *nitrous oxide* (N_2O), and *ammonia* (*NH_3*), are outcomes of the way we produce and consume energy and food. We highlight these compounds because unusually large inputs from anthropogenic sources disturb the nitrogen cycle and affect the environment in various ways (see Table 20.1). In 1860, yearly N_r production is estimated to have been ~15 Tg, but by 2005, yearly N_r production had mushroomed to 187 Tg (FAO 2006). Much of the increase was fed by the development of the HBP, which supplied enough fertilizer to meet the swelling demand for agricultural products (Galloway et al. 2008). According to Battye, Aneja, and Schlesinger (2017), the rate of fertilizer use has surged 100-fold since the HBP was developed. Increases in N_r for fertilizers has bolstered the quantity and nutritional quality of food to meet the needs and improve the wellness of many more people. Generally, a healthier population is a more productive one. However, how does the human hijacking of the nitrogen cycle affect the cycle, the biosphere, and human health?

TABLE 20.1. Reactive Nitrogen Compounds, Primary Natural (Italics) and Anthropogenic (Regular Font) Sources, and Major Associated Environmental Effects (Vitousek et al. 1997; Solomon et al. 2007)

N_r	Name	Primary Sources	Environmental Effects
NO_x	nitric oxide (NO), nitrogen dioxide (NO_2)	*tropical soils, lightning*, fossil fuels, biomass burning, agriculture	acid rain, photochemical smog, ozone production (troposphere), respiratory irritation and damage
N_2O	nitrous oxide	*oceans, tropical soils*; agricultural soils (fertilizers, manures), fossil fuels, industry	powerful greenhouse gas, damages ozone layer (stratosphere); higher UV exposure can lead to sunburn, cancer, eye damage
NH_3	ammonia	*oceans, soils; fertilizers*, waste disposal, industry	damages plant leaves, acidification of soils, toxic to aquatic organisms, respiratory irritation and damage

Note the influence agricultural practices have on levels of all forms of N_r. *Nitrate* (*NO_3^-*)- and ammonium-containing fertilizers provide forms of nitrogen that can be utilized by plants. Overfertilization can exceed the nitrogen needed by plants and microbes. Nitrate is quite easily leached from the soil, followed by a number of important cationic (positively charged) nutrients (calcium, potassium, magnesium). Leaching leaves the soil nutrient-poor and eventually affects the growth of plants. Also, ammonia (from ammonium fertilizers) may be nitrified to nitrate. Some nitrate can be denitrified to form N_2. Nitrification and denitrification can lead to the production of nitrous oxide (N_2O), an important greenhouse gas that is significant for climate change. Also, N_2O interacts to destroy ozone in the stratosphere, which can lead to increased UV radiation reaching the earth's surface.

Fossil fuel combustion, depending on the fuel type, can produce a number of nitrogen oxides. NO_x are important pollutants because they can cause acid rain, ozone production (troposphere), and photochemical smog. Almaraz et al. (2018) found that fertilizer-treated soils in the Central Valley of California were a major source of NO_x pollutants. At least in certain areas of California, the soils represented the major contributor to NO_x emissions.

Ammonia is found in some agricultural fertilizers and is released from animal wastes. High levels of ammonia may damage plants directly, contribute to acidification of soils, and be toxic, especially to aquatic organisms. Nitrifying bacteria may convert some of the ammonia to nitrate, which can be used by plants and microbes or leached from the soil and enter bodies of water. *Eutrophication* (dense growth of aquatic plants, which is followed by oxygen depletion and subsequent animal death) can result in aquatic systems with too much nitrogen.

APPLYING SCIENCE TO LIFE 1 Returning to the Indian River Lagoon (IRL): Too Much of a Good Thing

From the earliest descriptions of the Indian River Lagoon (IRL) estuary, we imagine an idyllic body of clean water where a diverse and healthy set of wildlife flourishes (see Figure 20.1). Human beings have lived in the surrounding areas of the IRL for thousands of years. However, in recent years, the human population has increased dramatically. In 1960, about 250,000 people lived in the region, but by 2014, that number had grown to about 1.7 million (US Census Bureau 2019). Along with the population increase came significant urban and agricultural developments. In 1920, 95 percent of the land was designated as forest, grass, or pasture (Kim et al. 2002). However, currently the IRL region is only 4.5 percent natural forest and 12.1 percent wetland, with the remainder (83.4 percent) classified as urban, agricultural, and rangeland (Bricker et al. 2008).

With increased development along many coastal areas, higher levels of various pollutants can affect estuaries, greatly altering estuarine communities and their related food webs. Common forms of water pollution for estuaries are nutrients released from sources such as sewage and septic systems, agricultural lands, and home lawns. In the IRL, the sewage treatment plants no longer discharge directly into surface waters. However, there are hundreds of thousands of septic systems in the IRL, which have turned out to be significant sources of nitrogen contamination, originating from human wastes. Stormwater (precipitation runoff) flows upon and through soils or in urban areas over water-impervious surfaces (e.g., roads, parking lots). Stormwater that collects beneath the soil surface joins groundwater on its way to rivers, streams, estuaries, and the ocean. As the water travels, it can pick up nutrients, such as nitrogen (N) and phosphorous (P). If excessive amounts of these nutrients find their way to a lake or an estuary, they can cause drastic changes in that

FIGURE 20.1 Aerial view of Indian River Lagoon

body of water. In the shallow IRL estuary, nitrogen and phosphorous interact to form a tar-like concoction called "muck." This muck can accumulate on the bottom of the estuary, forming thick mats that smother life living on the bottom.

Nutrient enrichment can also stimulate eutrophication, which leads to other complications. With a glut of nutrients like nitrogen, an estuary may suffer a harmful *algal bloom* (rapid proliferation of planktonic algae, and sometimes seaweeds), which can lead to the devastation of seagrass beds. The loss of these seagrasses leads to the loss of wildlife associated with these plants (Burkholder, Tomasko, and Touchette 2007; Lapointe, Herren, and Bedford 2012). Algal blooms are accompanied by low concentrations of oxygen and sometimes by the production of toxins, which can also bring about fish kills and death of other wildlife. Such has been the case with the IRL estuary, and many scientists think the excessive nutrient enrichment may lead to a collapse of this precious ecosystem.

PoI 1. What leads to the low levels of oxygen during an algal bloom?

PoI 2. If sewage runoff is the major source of nutrient pollution in the IRL, what other risks might there be for the estuary communities and the human inhabitants of the area?

PoI 3. Besides low oxygen and toxins, what other effects of algal blooms might destroy seagrass?

PoI 4. What forms of nitrogen are in total dissolved nitrogen (TDN)? What would be the primary form of nitrogen found in runoff from septic systems?

PoI 5. What is wastewater? What happens to wastewater that is not directly released into surface waters?

PoI 6. What is a *dead zone* in a large aquatic system, and what causes it?

QRQ 1. The process of converting nitrogen gas into a form of nitrogen that can be used for life support is called biological ______ ______.

QRQ 2. The ______ ______ Process combines nitrogen gas with hydrogen gas under high temperature and pressure to produce ammonia, which can be used to generate fertilizers.

QRQ 3. Select the formula (NH, NH_3, NH_2, NO_3^-, NX, RN, N, N_2, N_3, NO_x, NO, N_2O, N_r) representing each chemical.

a. ammonia
b. nitrate
c. nitric oxide
d. nitrogen
e. nitrogen dioxide

f. nitrogen gas
g. nitrogen oxide
h. nitrous oxide
i. reactive nitrogen species

QRQ 4. The dense growth of aquatic plants from excessive amounts of nutrients is called ______.

QRQ 5. Select the following that are major problems with overfertilization: algal blooms with resulting loss of biodiversity, leaching of nutrients from the soil, leaving the soil nutrient-poor, major source of nitrogen oxides, production of nitrous oxide

QRQ 6. Select the following that are major problems associated with fossil fuel combustion: acid rain, acidification of soils, algal blooms, damage plants directly, major source of nitrogen oxides, ozone production in the troposphere, photochemical smog, production of nitrous oxide, respiratory irritation and damage, toxic to aquatic organisms

QRQ 7. Select the following that are major problems associated with high levels of ammonia: acid rain, acidification of soils, algal blooms, damages ozone layer in the stratosphere, damages ozone layer in the troposphere, damage plants directly, major source of nitrogen oxides, ozone production in the stratosphere, ozone production in the troposphere, photochemical smog, powerful greenhouse gas, contributing to climate change, production of nitrous oxide, respiratory irritation and damage, toxic to aquatic organisms

QRQ 8. Using the same possible answers as in QRQ 7, select the major problems associated with nitrous oxide.

QRQ 9. Using the same possible answers as in QRQ 7, select the major problems associated with nitrogen oxides.

20.2. Impact of Reactive Nitrogen Species on the Environment

20.2.1. The Two Faces of Ozone

As you can see from Table 20.1, not all of our wishes have come true with regard to the effects of increasing amounts of reactive nitrogen compounds in the biosphere. For instance, nitrous oxide (N_2O) is long-lived, and it has two particularly bad effects on the atmosphere: it is a powerful greenhouse gas, and it damages the ozone layer in the stratosphere. As a greenhouse gas, it absorbs some of the heat reradiated from the earth's surface, instead of letting it escape into space. Overall, greenhouse gases contribute to a warming of the earth's surface. Besides contributing to warming, N_2O in the *stratosphere* (layered region of our atmosphere lying above the troposphere, see Figure 20.2) catalyzes the destruction of the *ozone layer*, a stratum in the stratosphere containing *ozone* (O_3), which functions to absorb most of the damaging ultraviolet (UV) radiation from the sun. Ultraviolet radiation is the major cause of skin cancer, but it also may cause sunburns, aging of the skin, and eye damage. In these cases, ozone is a positive force, and destroying the ozone layer would, at the very least, be bad for our health.

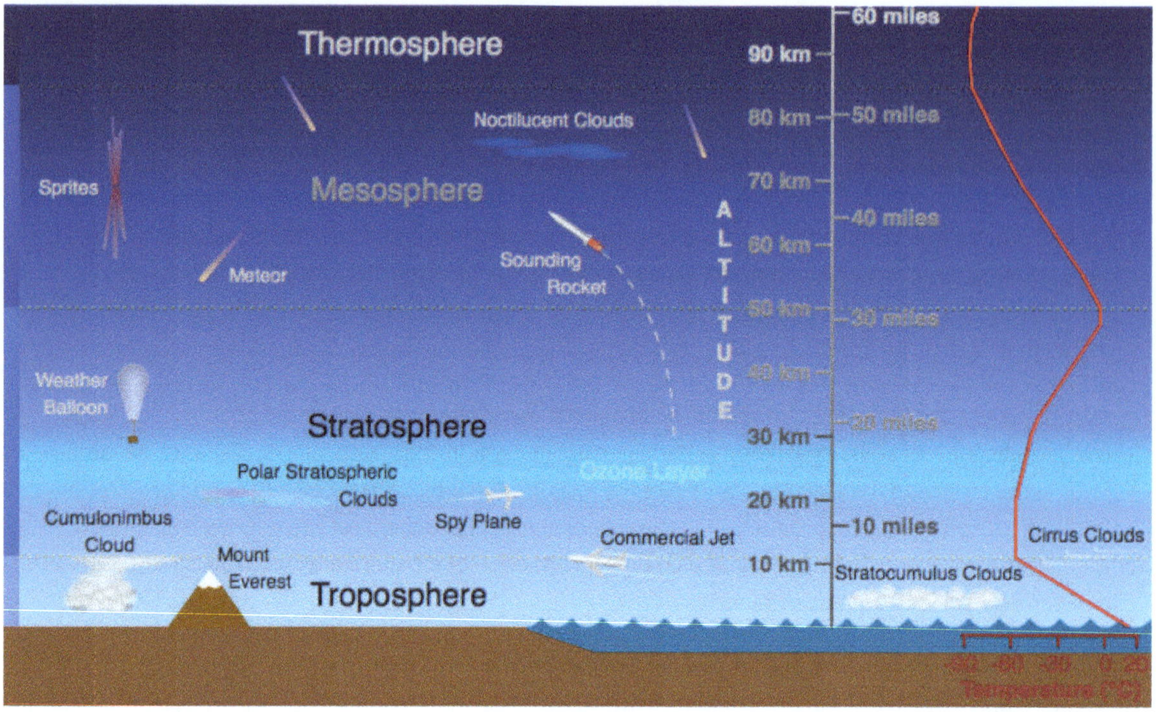

FIGURE 20.2 Layered structure of Earth's atmosphere: troposphere from ground to ~20 km altitude; stratosphere, containing ozone layer, above troposphere to 30 km

NO_x is shorthand for a number of nitrogen oxides. One of these oxides is *nitrogen dioxide* (NO_2). In the stratosphere, ozone (O_3) is formed when sunlight splits an oxygen *molecule* (O_2) to form two oxygen *atoms* (2O). When an oxygen *atom* (O) interacts with another oxygen *molecule* (O_2), they bond to form O_3, which may join other ozone molecules to form a layer that protects us from UV radiation. In the *troposphere* (lowest layer of the earth's atmosphere, extending from the earth's surface to about 20 km in altitude), also using sunlight, free oxygen (O) is formed from NO_2 instead of an oxygen molecule. Then, as in the stratosphere, the oxygen atom bonds with oxygen molecules to form ozone molecules. Thus, NO_2 is a precursor to ozone, helping to increase O_3 concentration near the ground. However, tropospheric ozone is not a welcome addition. Ozone itself is a powerful *oxidizing agent* (removes electrons from other chemical compounds) and can cause inflammation and damage to your respiratory tract. People with chronic respiratory issues often have to avoid the outdoors during times of high levels of ozone. Scientists are now working to determine if ozone is also *carcinogenic* (cancer-causing).

Ozone near the ground can also damage plants. Most of what we know about ozone effects on plants comes from studies of crops. The interactions between ozone and the plant's metabolism and cellular structure is quite complex, which means we may see effects, but we may not completely understand the underlying mechanisms that cause those effects. Ozone may damage the outsides of the plants, and on plant leaves we can observe alterations of shape, color, and color patterns. Entering the leaves through the stomata, ozone can interact with organic molecules produced by the plant. Scientists suspect that the interactions form other oxidizing chemicals. Such oxidizing agents could interact with plasma membrane components, like proteins, which can weaken the cells and cause tissue injuries. Furthermore, membrane damage may lead to abnormal cellular functions like the regulation of ion concentrations, which negatively affects chloroplast activity. If chloroplast function is reduced, photosynthesis rates lag, and less carbon dioxide will be assimilated to produce sugars and other organic molecules (NASA 2002, WHO 2000)

CO_2 assimilation may be slowed further by ozone exposure directly to the guard cells of the stomata. Researchers at Pennsylvania State University have discovered that ozone decreases the uptake of potassium by the guard cells, which inhibits the opening of the cells. If the guard cells do not open, the stomata are not open. Closed stomata will mean less CO_2 can enter the leaves for assimilation during the Calvin cycle, reducing the productivity of the plant (Torsethaugen, Pell, and Assmann 1999).

20.2.2. Acid Rain

Most people either have not heard of *acid rain* (a form of acid deposition where precipitation has a pH of 5.2 or less), or think it is a problem that has been solved, part of history. Although the problem had been recognized in the 19th century, acid rain became more prominent, especially to scientists in western Europe and eastern North America, during the 1960s and 1970s. What scientists determined was that emissions from the combustion of fossil fuels was the underlying cause of the problem.

The effects of acid rain were manifesting themselves in the trees, soil, and surface waters of regions downwind of emissions produced by industry, power plants fueled by coal, and large amounts of exhaust pollution from transportation. Lakes in the region had declines in animal populations, who were unable to survive in pH ranges of 3.0–4.5 or because their food sources were reduced by the acidic waters. Soils in many of these areas lacked the buffering capacities to neutralize the increases in acidity. Changes in the soil pH can cause *leaching* (washing away by precipitation) of nutrients, such as calcium and magnesium, that are necessary for plant growth. Furthermore, the acidity makes aluminum more soluble, and aluminum is toxic to many living organisms.

Some tree species are directly injured by acid precipitation (see Figure 20.3). For instance, in Red Spruce, acid deposition causes leaching of calcium from cell membranes in the leaves (needles). These injured leaves are more susceptible to freezing temperatures, which is really bad news for the trees that live at higher altitudes. Furthermore, red spruce trees are sensitive to aluminum in the soil, and spruce trees growing in high altitude habitats may periodically experience very acidic clouds and fogs.

FIGURE 20.3 Dead Fraser firs, killed by acid rain, at Clingmans dome, Great Smoky Mountains National Park

In combination, these stresses can make plants more vulnerable to insects and disease (Jiang and Jagels 1999; Driscoll et al. 2003)

Fossil fuel emissions contain a number of potential threats to living things. Two components closely associated with acid precipitation are *sulfur dioxide* (SO_2) and nitrogen oxides (NO_x), NO and NO_2. Normally, rainwater is slightly acidic, but when sulfur dioxide and nitrogen oxide molecules interact with water, we get sulfuric acid (H_2SO_4) and nitric acid (HNO_3), respectively. Rain, fog, snow, and other precipitation that contain these acids are much more acidic than normal rainwater. Fossil fuel combustion produces both sulfur dioxide and nitrogen oxide molecules. Electric utilities produce most of the SO_2 and a smaller amount of NO_x. Most of the NO_x originates from gasoline-powered vehicles.

APPLYING SCIENCE TO LIFE 2 Reining in Acid Rain

Your parents often tell you about their memorable summer vacation in Killarney Provincial Park (KPP) in Ontario, Canada, in the mid-1980s. Your older sister was born less than a year later. They camped, hiked, canoed, and rode bikes, but they always tell you about the beautiful aquamarine water of one of the lakes they canoed. You could clearly see the bottom of the lake. Park officials told them that the lake was about 50 feet deep at its deepest part. Although the lake was incredibly beautiful, they thought it was odd that they didn't see any living creatures, including fish.

Later they learned that the lakes in that area were very clear because of acidification of the water from acid rain. The acid conditions wiped out most, if not all, of the regular community. Approximately 60 km away from the park, nickel and copper smelters produce prolific amounts of sulfur dioxide, one of the two major components of acid rain. In studies of an experimentally acidified lake, Schindler et al. (1985) found that many prey animals for lake trout were reduced in numbers or completely disappeared. Without their primary food sources, trout were so hungry, they responded by not reproducing. During the lake acidification experiments, scientists made an interesting observation. To reduce the pH of the lake took much more acid than was calculated. When microbiologists investigated, they found that there were bacteria that could increase *alkalinity* (capability of water to reduce acid) of the water. Most of this alkalinity came from microbial denitrification of nitric acid, but some of the alkalinity was contributed also by reducing sulfuric acid (Rudd et al. 1986).

The problem was not limited to Canada, because acid rain was also having serious effects along the Eastern Seaboard of the United States. Most of the emissions came from the industrial regions of the Midwest (Ohio Valley), Pennsylvania, and New England and were responsible for acidification of precipitation in Canada and the United States. Despite much evidence, there were vested interests in the United States and Canada that denied culpability in causing acid rain. It took 27 years (1963–1990) for scientists and environmental groups to convince the US government to pass legislation requiring more rigorous air quality standards, the *Clean Air Act*. Canada took similar legislative action. Since 1990, levels of atmospheric sulfates have fallen greatly, and although the nitrogen oxide levels have also declined, NO_x levels should be further reduced. In addition to the NO_x pollution, ammonia from fertilizers and livestock feed contribute to nitric acid formation.

Today, the lakes of KPP have recovered somewhat, or at least their pH is closer to normal. Some of the lakes, however, have little ability to buffer pH, because much calcium has leached out of the soils. Animals that have calcareous exoskeletons, like crustaceans, depend on calcium for exoskeleton secretion.

Therefore, these animals are stressed to the point that they die, or at least don't reproduce. Many of these animals are key parts of the lake food web. Some recolonization by pre-acidification flora and fauna has occurred, but reestablishing the original biological community is complicated (Queens University 2011). Studies of the lake recovery continue.

In upstate New York, hundreds of lakes and streams in Adirondack Park have been recovering, albeit very slowly, from the effects of acid rain that occurred from the 1960s into the 1990s. With the enactment of additional state air quality regulations, the lakes' water quality has gradually improved. In fact, the New York State Department of Environmental Conservation (DEC) announced recently that a self-sustaining population of brook trout had been discovered in Lake Colden, which is a small lake at an elevation of more than 2,700 ft. (DEC 2019; Wikipedia 2019). This lake had been without such a fish population for more than 30 years.

PoI 7. Investigate why it took so long to get the Clean Air Act passed in the United States.

PoI 8. What parallels are evident between deniers of impacts of acid rain and deniers of climate change?

PoI 9. Why is it so difficult to restore lakes that are damaged following an environmental catastrophe like acid rain?

20.2.3. Photochemical Smog

NO_x and *volatile organic compounds* (*VOCs*) are important components of *photochemical smog*, the air pollution resulting from reaction of solar radiation with airborne mixtures of nitrogen oxides (NO_x) and VOCs. VOCs are produced as emissions from combustion of gasoline and from various home products like furniture polish, house paint, and laundry detergent. To simplify our understanding of the process, let's start with nitrogen dioxide (NO_2). We have already learned that NO_2 provides the atomic oxygen for the formation of ozone in the troposphere. Remember that NO_2 may also combine with water to form nitric acid (HNO_3), associated with acid fog and rain. The acid, VOC, and nitrogen oxides form a variety of reactive molecules that are incorporated into smog. Think of smog as a hazy soup of irritating and toxic chemicals (see Figures 20.4 and 20.5). This soup also contains ozone. Smog irritates your

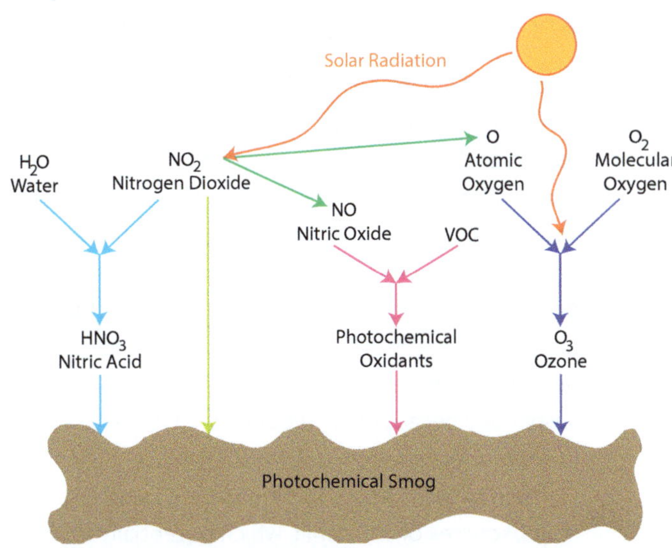

FIGURE 20.4 Formation of photochemical smog

FIGURE 20.5a NASA photograph of smog over China

FIGURE 20.5b Smog in Beijing

eyes and your respiratory system, exacerbating chronic diseases such as asthma and emphysema. Smog does direct damage to plants as well.

20.2.4. Increasing Ammonia Emissions: Not for the Faint of Heart

APPLYING SCIENCE TO LIFE 3 Big Stink

If you are a lover of good cheese, the Dutch Cheese Museum, located in Alkmaar, Netherlands, is well worth the easy travel from Amsterdam. Dairy products like cheese are the top agricultural (consumable) export of the Netherlands (Statistics Netherlands 1 n.d.). With more than 17 million people and almost four million cattle (1.5 million dairy cattle), the Netherlands has a high density of people and cows for its 33,900 km^2 of land (CIA n.d.; Statistics Netherlands 2 n.d.).

During November 2019, drivers headed toward The Hague from Amsterdam had some difficult travels because they were trapped behind an army of unhappy cattle farmers on their tractors. The convoy was slowly moving toward the Dutch seat of government to protest a spring court decision that suspended permits for construction projects to expand cattle, pig, and chicken farms. The construction projects were suspended because they would increase the already high concentrations of ammonia and other N_r compounds released into the atmosphere.

Animal farms produce mountains of wastes from which ammonia vapors emanate. Such wastes are major sources of ammonia, which potentially harms natural areas. With large numbers of animals, the amount of wastes exceeds the amount farmers can use for fertilizing fields, so they have to do something with the surplus. They can pay someone to pick up and haul away the manure, or in some cases, they say they did but didn't. The farmers were already having problems with excess phosphate wastes.

Ammonia can be deposited in an ecosystem by gas absorption, dry deposition, or dissolution in rain, fog, and so on. Animal waste is only one—although major—source of ammonia. Dutch agriculture is thought to produce about half of the nitrogen pollution falling in the Netherlands (Stokstad 2019). Significant amounts of ammonia come from fertilizers and biomass burning (combustion of organic matter, natural or human-made), so an overwhelming percentage of ammonia emissions are *anthropogenic* (from human activity) (Fenn and Hossner 1985; Schlesinger and Hartley 1992).

Natural areas in the Netherlands are very alarmed by the rapidly increasing inputs of nitrogen into the region. Many of the natural areas in the Netherlands are characterized by dunes, bogs, and heaths, where many of the native plants are adapted to low concentrations of nitrogen. Unlike many plants, grasses are stimulated to grow rapidly with the higher levels of nitrogen. Consequently, the grasses outcompete the native plants, with subsequent loss of plant biodiversity, as well as the animals and microbes associated with them.

The farmers left The Hague after the protest but got no assurances of how the government would respond to this complex issue.

PoI 10. Generally, what is the value of biodiversity? How do we value biodiversity economically?

Ammonia from the atmosphere can be taken up through the stomata, deposited on plant surfaces, or washed off into the soil as ammonium (NH_4^+). Ammonia in the soil can also volatilize, adhere to a particle or be converted to nitrite or nitrate by nitrifying bacteria. At some point, the concentration of nitrogen, from nitrite or nitrate, in a system may become saturated. Nitrification, which occurs at a high rate when nitrogen deposition is high, releases hydrogen ions, lowering the soil's pH. Soluble nitrate can be washed out of the soil and may enter streams and lakes. As NO_3^- is leached out of the soil, positively charged ions like calcium, potassium, and magnesium are lost as well. With the loss of nutrients, trees may become stunted or die, and the addition of nitrogen to bodies of fresh or saltwater can promote eutrophication. The acidification of the soil and the loss of calcium can also mobilize aluminum in the soil, which may have toxic effects on plants and aquatic animals (Vitousek et al. 1997, Galloway et al. 2003).

QRQ 10. (Answer choices in parentheses) Ozone (____ (CO, CO_2, CO_3, O, O_2, O_3, OZ, O_2Z, O_3Z)) in the atmosphere's second layer, the ____ (mesosphere, stratosphere, thermosphere, troposphere) is ____ (bad, good) because the ozone layer there ____ (enhances, protects us from) UV radiation.

QRQ 11. (Answer choices in parentheses) Ozone in the lowest layer of the earth's atmosphere, the ____ (mesosphere, stratosphere, thermosphere, troposphere), is ____ (bad, good).

QRQ 12. Possible answers: nitrogen dioxide, nitric oxide, nitrous oxide, oxygen. In the lowest layer of the earth's atmosphere, ozone is formed, using sunlight, from splitting an oxygen atom off of a(n) ____ molecule and bonding that free oxygen to a(n) ____ molecule.

Possible answers: nitrogen dioxide, nitric oxide, nitrous oxide, oxygen.

QRQ 13. (Answer choices in parentheses) Ozone is a powerful ____ (oxidizing, reducing) agent, meaning that it ____ (adds, removes) electrons from other chemical compounds.

QRQ 14. (T/F) Ozone near the earth's surface can cause respiratory tract inflammation and can damage plants.

QRQ 15. Select the following that are major problems caused by acid rain: algal blooms, can cause decline of animal populations in lakes, can cause smog, can injure leaves, changes the soil's pH (which can cause leaching), respiratory tract inflammation

QRQ 16. (Answer choices in parentheses) Sulfur dioxide (____ (SO, SO_2, PO, PO_2)) combines with rainwater to form ____ (citric, dioxide, dioxidic, oxidic, sulfur, sulfuric) acid, while ____ (nitrogen, nitrogen gas, nitrogen oxide, nitrous oxide) molecules interact with rainwater to form nitric acid.

QRQ 17. Select the following that are major sources for the molecules that interact with rainwater to cause acid rain: animal wastes, combustion of organic matter, electric vehicles, electric utilities, fertilizers, fossil fuel combustion, gasoline-powered vehicles, nuclear plants, service industry, volcanoes

QRQ 18. Photochemical smog is air pollution resulting from the reaction of solar radiation with airborne mixtures of ____ ____ and volatile organic compounds.

QRQ 19. Select the following that are major sources for ammonia: animal wastes, combustion of organic matter, electric vehicles, electric utilities, fertilizers, fossil fuel combustion, gasoline-powered vehicles, nuclear plants, service industry, volcanoes

Image Credits

Fig. 20.1: Source: https://commons.wikimedia.org/wiki/File:IndianRiverLagoon.jpg.

Fig. 20.2: Copyright © by eo.ucar.edu (CC BY-SA 4.0) at https://commons.wikimedia.org/wiki/File:Atmosphere_layers.jpg.

Fig. 20.3: Copyright © by Brian Stansberry (CC BY 3.0) at https://commons.wikimedia.org/wiki/File:-Clingmans-dome-newfound-gap-tnncl.jpg.

Fig. 20.5a: Source: https://commons.wikimedia.org/wiki/File:Smog_over_China.jpg.

Fig. 20.5b: Copyright © by Daugilas (CC BY-SA 3.0) at https://commons.wikimedia.org/wiki/File:Smog_in_Beijing_-_panoramio.jpg.

LAB #20

Laboratory Exercises

20.3. Laboratory on Smog

20.3.1. Preliminary Work

In this laboratory, we examine formation of photochemical smog. From the text's website, view the introductory video and "Smog Equations"; download and open "*Smog Rev.nlogo*."

1. Click the Info tab at the top of the page and read the sections "What Is It?" and "How It Works."
   a. ____________ Select the primary pollutants that are released into the atmosphere as a result of fossil fuel combustion: A. nitric oxide, B. nitrogen dioxide, C. nitrous oxide, D. oxygen, E. ozone, F. photochemical oxidants, G. sunlight, H. volatile organic compounds
   b. ____________ Select the principal components of smog. Use the same possible answers as in part a.
   c. **1.** ________________________ **2.** ______ In the presence of sunlight, NO_2 breaks down into an atom of oxygen and a molecule of **1.** (write out the word[s]), whose chemical formula is **2**.
   d. **1.** ____________ **2.** ______ In the presence of sunlight, an atom of oxygen and a molecule of oxygen combine to form **1.** (write out the word[s]), whose chemical symbol is **2**.
   e. ______ Ozone is destroyed when combined with what? Use the same possible answers as in part a.
   f. ______ Nitric oxide and VOC combine to form what? Use the same possible answers as in part a.
   g. Match each model component with the color of its dot: A. black, B. blue, C. brown, D. cyan, E. gray, F. magenta, G. orange, H. red, I. violet, J. white, K. yellow

      ________ NO

      ________ NO_2

      ________ O

      ________ O_3

      ________ O_x

      ________ VOC

2. Read "How to Use It" and "Things to Notice."

   a. ______ The sliders NITROGEN-DIOXIDE and VOLATILE-ORGANIC-COMPOUNDS indicate how many of each type of molecule is created **1.** (A. each hour, B. each minute, C. each second, D. initially).

   b. ______ Which slider do we turn ON to simulate a situation where normal airflow does *not* cause dispersion of pollutants away from their source? A. ATMOSPHERE-INVERSION, B. DAY-NIGHT-CYCLE, C. MOLECULE COUNTS, D. NITROGEN-DIOXIDE, E. POLLUTANTS-CYCLE, F. SUN-LIGHT-INTENSITY, G. TOTAL SMOG, H. VOLATILE-ORGANIC-COMPOUNDS

   c. ____________ If DAY-NIGHT-CYCLE is turned ON, during the simulated night hours, which reaction(s) do(es) *not* occur? A. nitrogen dioxide breakdown, B. ozone creation, C. ozone destruction, D. photochemical oxidants creation

   d. ____________ If POLLUTANTS-CYCLE is turned ON, during the simulated night hours, which pollutants(s) is/are *not* released into the atmosphere as a result of fossil fuel combustion? A. NO, B. NO_2, C. O, D. O_3, E. O_x, F. VOC

   e. ____________ In the model, the amount of primary pollutants is interpreted as the number of what kind of molecule(s)? Use the same possible answers as part d.

   f. ____________ In the model, the amount of smog is interpreted as the number of what kind of molecule(s)? Use the same possible answers as part d.

   g. ______ (T/F) The model assumes that the amount of oxygen is finite.

20.3.2. Laboratory Work

The default parameters are as follows: NITROGEN-DIOXIDE = 1, VOLATILE-ORGANIC-COMPOUNDS = 1, SUN-LIGHT-INTENSITY = 90, DAY-NIGHT-CYCLE = ON, POLLUTANTS-CYCLE = ON, ATMOSPHERE-INVERSION = ON, STOP-AFTER-3-DAYS = ON.

3. This question considers the situation when no atmospheric inversion exists, so that normal airflow causes dispersion of pollutants away from their source. Thus, use the default parameters, except turn off ATMOSPHERE-INVERSION.

   a. ______ Slow down the simulation a little so that you can observe the molecules more closely. Are there any molecules involved in smog (ozone and photochemical oxidants) immediately after nightfall, no (N) or yes (Y)?

   b. ______ Do the smog molecules dissipate in the night, no (N) or yes (Y)?

   c. ______ Are the numbers of each type of molecule periodic, no (N) or yes (Y)?

   d. To record your data, create a spreadsheet, called *Smog*, as in Table 20.2. Type parameter values and have appropriate headers for the output values, which are the maxima (plural of "maximum") of the numbers of NO_2, VOC, primary pollutants, O_3, O_x, and smog molecules. Set up and run the simulation at least five times with *view updates* unchecked or with a fast speed and record the output values under the appropriate headers. Because tabs are separators, we can copy the six output values at once and paste them into the spreadsheet so that the values are in different columns. Calculate the average of each column and give these means.

      ________ NO_2 maximum

      ________ VOC maximum

________ primary pollutants maximum

________ O_3 maximum

________ O_x maximum

________ smog maximum

TABLE 20.2. Start of Smog Lab Spreadsheet

Name(s)						
Smog Lab						
	Question 3d					
	NO_2					1
	VOC					1
	Sunlight					90
	DAY-NIGHT-CYCLE					ON
	POLLUTANTS-CYCLE					ON
	ATMOSPHERE-INVERSION					**OFF**
	Maxima for					
	NO_2	**VOC**	**pollutants**	**O_3**	**O_x**	**smog**

e. **1.** ______ **2.** ______ **3.** ______ **4.** ______ **5.** ______ **6.** ______ At night, if factories are not working and there is normal airflow, the factories stop producing primary pollutants nitrogen dioxide (NO_2) and VOC. Under these circumstances, complete the following observations using equation numbers from "How It Works," under the Info tab: Intermediates (nitric oxide, NO, and atomic oxygen, O) cannot form without **1.** (A. more atomic oxygen, B. more nitrogen dioxide, C. more NO, D. more ozone, E. additional VOC, F. sunlight) and additional NO_2 (Equation **2.** [1, 2, 3, 4]); Equation **3.** (1, 2, 3, 4) cannot continue to generate ozone without sunlight and **4.** (A. more atomic oxygen, B. more nitrogen dioxide, C. more NO, D. more ozone, E. additional VOC, F. sunlight); photochemical oxidants (O_x) cannot form without the intermediate NO and **5.** (A. more atomic oxygen, B. more nitrogen dioxide, C. more NO, D. more ozone, E. additional VOC, F. sunlight) from fossil fuel combustion (Equation **6.** [1, 2, 3, 4]); and smog and any remaining model molecules/atoms blow away.

4. In this question, we compare the previous results with a situation where atmospheric inversion exists, so that pollutants do not disperse. Thus, use the default parameters.

a. **1.** ______ **2.** ______ Complete the following hypothesis: Suppose there is an atmospheric inversion and no fossil fuel combustion at night. Then, numbers of the considered molecules **1.** (A. decrease, B. increase, C. remain constant) during the day and **2.** (A. decrease, B. increase, C. remain constant) at night.

b. ______ Do the smog molecules dissipate in the night, no (N) or yes (Y)?

c. ______ Are the numbers of each type of molecule periodic, no (N) or yes (Y)?

d. __________ Notice on the graphs that as night begins, NO_2 rises, while NO and O_3 fall. This phenomenon is explained by which equation(s) **1.** (1, 2, 3, 4) listed in "How It Works"? Thus, this observation would probably cause you to revise or refine your hypothesis.

e. ____________ In contrast, after daybreak, with sunlight available, NO_2 falls, while NO and O_3 rise quickly. This phenomenon is explained by which equation(s) **1.** (1, 2, 3, 4) listed in "How It Works"? Again, the original hypothesis needs revision.

f. **1.** ______ **2.** ______ **3.** ______ **4.** ______ Moreover, early in the daylight hours, NO **1.** (A. decreases, B. increases, C. remains constant) because of the action of Equation **2.** (1, 2, 3, 4), while as the day goes on and Equation **3.** (1, 2, 3, 4) comes into play, NO **4.** (A. decreases, B. increases, C. remains constant). Once more, the original hypothesis needs revision.

g. In another part of your *Smog* spreadsheet, indicate "Question 4g," list the parameter values, and have appropriate headers for the output values. Set up and run the simulation at least five times with *view updates* unchecked or with a fast speed, and record the output values under the appropriate headers. Calculate the average of each column and record the means below.

________ NO_2 maximum

________ VOC maximum

________ primary pollutants maximum

________ O_3 maximum

________ O_x maximum

________ smog maximum

h. **1.** ______ **2.** ______ **3.** ______ **4.** ______ Comparing your results and observations from this and question 3.2.1, complete the following statement: Suppose no fossil fuel combustion occurs at night over a three-day period. With normal airflow, smog **1.** (A. continues to increase, B. dissipates, C. remains virtually constant) at night. However, when an atmospheric inversion occurs, the amount of smog **2.** (A. continues to increase, B. dissipates, C. remains virtually constant) during the night and **3.** (A. continues to increase, B. dissipates, C. remains virtually constant) during the day. Moreover, on average, the maximum amount of smog is about **4.** (D. 2, E. 5, F. 7, G. 10, H. 12, I. 15, J. 17, K. 20, L. 22, M. 150, N. 160, O. 170) times larger with an atmospheric inversion than with normal airflow.

5. What happens if factories run through the night too? In this question, we consider the impact of that decision on the amount of smog.

a. ______ Complete the following hypothesis: Over a three-day period, when atmospheric inversions exist, the maximum amount of smog when factories operate all the time will be about **1.** times as much as when the factories close at night, on average.

b. Turn the POLLUTANTS-CYCLE switch off and repeat the experiments of question 4g, labeling the spreadsheet appropriately.

________ NO_2 maximum

________ VOC maximum

________ primary pollutants maximum

________ O_3 maximum

________ O_x maximum

________ smog maximum

c. Discuss a comparison of your results from question 4g, where factories close for the 12 hours at night, and part b, where factories operate 24 hours a day. Refer to the appropriate equations as needed. Why is the nitrogen dioxide maximum much more than doubled for the latter? Similarly, for VOC? Why is the maximum of ozone not quite double when the factories are operational twice as long? Similarly, for photochemical oxidants?

6. In this question, we examine the impact of the amount of sunlight on the amount of smog.

   a. ______ Complete the following hypothesis: As the amount of sunlight increases, when atmospheric inversions exist, over a three-day period, the maximum amount of smog **1.** (A. decreases, B. increases, C. remains the same) on average.

   b. ______ Use the default parameters, except set SUNLIGHT-INTENSITY to be 30. Repeat question 4g, recording only the smog maximum.

   c. ______ Set SUNLIGHT-INTENSITY to 50 and repeat part b.

   d. ______ Set SUNLIGHT-INTENSITY to 70 and repeat part b.

   e. **1.** ______ **2.** ______ Using the results of parts b, c, and d and question 4g, have your spreadsheet graph the (mean) maximum amount of smog versus the amount of sunlight intensity. Complete the following statement about the graph: The graph is **1.** (A. decreasing, B. horizontal, C. increasing, D. sometimes goes up and sometimes down) and is **2.** (E. concave down, F. concave up, G. straight, H. of varying concavity).

   f. ______ percent Using your answers from part b and question 4g, calculate the percent increase in the (mean) maximum amount of smog with a 60-point increase in the level of sunlight intensity (from 30 to 90). Your answer will probably show that increasing sunlight intensity, at least in the range from 30 to 90, has some, but limited, impact on the maximum amount of smog.

7. In this question, we consider the impact on the amount of smog from various amounts of nitrogen dioxide and volatile organic compounds.

   a. **1.** ______ **2.** ______ **3.** ______ Complete the following hypotheses: Suppose factories are only operating during daylight hours, and there is an atmospheric inversion over a three-day time period.

   1) Increasing NO_2 without increasing VOC will **1.** (A. decrease significantly, B. decrease somewhat, C. increase significantly, D. increase somewhat, E. approximately maintain) the maximum amount of smog.

   2) Increasing VOC without increasing NO_2 will **2.** (A. decrease significantly, B. decrease somewhat, C. increase significantly, D. increase somewhat, E. approximately maintain) the maximum amount of smog.

   3) Doubling both NO_2 and VOC will approximately **3.** (F. double, G. triple, H. quadruple, I. maintain) the maximum amount of smog.

   b. To consider the first part of the hypothesis, use the default parameters, except change NITROGEN-DIOXIDE to 2, and repeat question 4g.

   ______ NO_2 maximum

   ______ VOC maximum

   ______ primary pollutants maximum

_______ O_3 maximum

_______ O_x maximum

_______ smog maximum

c. **1.** _____ **2.** _____ **3.** _____ **4.** _____ **5.** _____ **6.** _____ **7.** _____ **8.** _____ **9.** _____

10. _____ **11.** _____ **12.** _____ **13.** _____ Complete the justification for why NO_2 is so much greater, VOC is so much smaller, O_3 is smaller, and O_x is somewhat larger than the corresponding results in question 4g. Use the following choices, except where numbers are indicated: A. NO, B. NO_2, C. O, D. O_2, E. O_3, F. O_x, G. VOC

By Equation **1.** (1, 2, 3, 4), more NO_2 results in more of the following: **2.**

As indicated in Equation **3.** (1, 2, 3, 4), the additional **4.** can combine with VOC to form O_x (part of smog), but there is not enough VOC to use these additional molecules completely. Thus, most of the **5.** molecules are used up, and more **6.** is formed than in question 4g.

As indicated in Equation **7.** (1, 2, 3, 4), the additional **8**, indicated in the first sentence of the argument, can combine with O_2 to form more O_3 (part of smog). However, once VOC runs out, by Equation **9.** (1, 2, 3, 4), the excess NO helps to destroy **10.** and create more **11**. Thus, we have less **12.** and more **13.** than in question 4g.

d. _____ Do your results support your hypothesis 1 from part a?

e. To consider the second part of the hypothesis, use the default parameters (including NITROGEN-DIOXIDE = 1), except change VOLATILE-ORGANIC-COMPOUNDS to 2, and repeat question 4g.

_______ NO_2 maximum

_______ VOC maximum

_______ primary pollutants maximum

_______ O_3 maximum

_______ O_x maximum

_______ smog maximum

f. _____ Do your results support your hypothesis 2 from part a?

g. Using an argument similar to that in part c, explain why NO_2 is so much less, VOC is so much more, and O_3 and O_x are somewhat more than the corresponding results in question 4g.

h. To test the third part of the hypothesis, use the default parameters, except change NITROGEN-DIOXIDE to 2 and VOLATILE-ORGANIC-COMPOUNDS to 2, and repeat question 4g.

_______ NO_2 maximum

_______ VOC maximum

_______ primary pollutants maximum

_______ O_3 maximum

_______ O_x maximum

_______ smog maximum

i. _____ Do your results support your hypothesis 3 from part a?

MODULE

21

Human Activity—The Future of Water Is Not So Clear

21.1. Introduction

In this text, we have considered various aspects of water, focusing early on its special chemical and physical characteristics, as well as some of its crucial functions in the cell. Later, we studied the water cycle and accessibility of clean water. Water is fundamentally important to life on this planet. Considering its indispensable nature, we will examine three more issues involving water. In each case, there are human behaviors that have had negative impacts on this essential component of our world. We begin with how we have accelerated the decline of one of the most beautiful and ecologically important ecosystems on earth—coral reefs. Coral reefs are not just pleasing to view, they are the cradle for much of the biodiversity of the ocean.

Second, consider how most of us take for granted our access to affordable, clean water. You may already be aware that we face a crisis in the infrastructure that provides us with that water. Many existing water systems date to the end of World War II. Not only old, the infrastructure is poorly maintained (usually because of costs). Many of the materials that make up the supply pipes and in-home pipes are made out of lead, which is fearfully toxic to our nervous systems. In many situations, the monitoring and reporting of problems with our water have been inefficient or even negligent, also often because of cuts in funding. Clean water affects the health of every consumer, and there are many risks besides lead, including the spread of disease-causing organisms. Experts estimate that the costs of replacing and upgrading the infrastructure are in excess of $1 trillion over the next 20–25 years. The costs of water will increase, and there will be more people who are unable to pay their water bills. To cover the expenses, we all will be paying more. The case we consider involves the water supply of Flint, Michigan. Unfortunately, there will very likely be more cases like Flint.

Lastly, we consider an issue that will begin to affect more people throughout the world—inaccessibility to sufficient amounts of water. Water is used for many more things than drinking. Can you think about what human beings use water for, besides quenching their own bodily needs? We will catalyze your consideration as we take up the last case—the ongoing battle over water supplies affecting three southeastern states—Georgia, Florida, and Alabama. Such disputes have been common in the western states, as well as between countries in Asia.

21.2. Inadvertent or Intentional: How Human Behavior Is Punishing Coral Reefs

Coastal zones in tropical climates are a prime habitat for the growth of coral reefs. Such coastal areas are also an attractive habitat for human populations and tourists, which makes reefs extremely vulnerable to damage from human activity. We include here five general categories of human activity that are harming one of the most diverse ecosystems on Earth. Those categories are sedimentation, pollution, careless recreation, overfishing and coral harvesting, and anthropogenic global warming (EPA 2018; Coral Reef Alliance 2020).

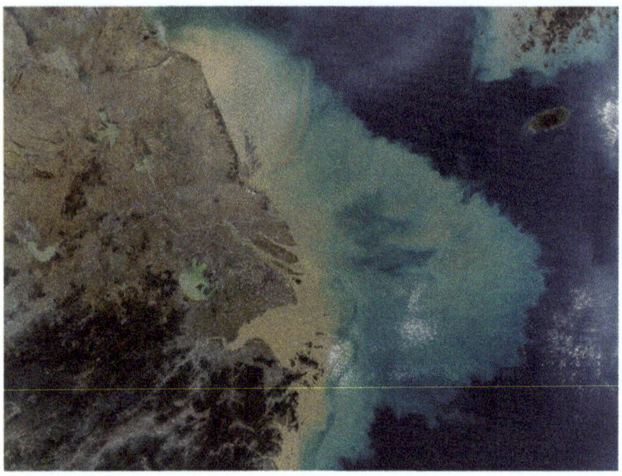

FIGURE 21.1 Suspended sediment flowing from the Yangtze, Yellow, and other rivers into coastal waters of eastern China

Sedimentation is the movement of soil, sand, and silt from terrestrial areas by wind or drainage into the oceans, which clouds the water (see Figure 21.1). Cloudy water reduces the amount of light reaching the corals, which normally supports the photosynthetic activities of the zooxanthellae that provide food essential for the health and growth of the coral animals. Furthermore, too much sediment can foul and smother the coral animals themselves. Sediments also normally contain many nutrients, which can also stimulate the overgrowth of algae. Such overgrowth can also decrease growth or smother the corals. Increased sedimentation is promoted by such activities as clearing of land (construction), agriculture, dredging, deforestation, road construction, and so on (Carlson, Foo, and Asner 2019).

Pollution can take many forms, and all of it can harm the reefs.

- Stormwater runoff may carry toxins, nutrients, and pathogens.
- Oil and chemical spills from tankers or from industrial plants and refineries can poison the coral reef ecosystem (see Figure 21.2).
- Pesticides, chemicals, and fertilizers used in agriculture can pollute rivers or coastal areas directly. Pesticides may poison many of the reef inhabitants, and fertilizers and other nutrients may stimulate the overgrowth of algae (see Figure 21.3).
- Poorly operating septic systems or fecal material washed out of farm animal operations following storms flood the reefs with nutrients and pathogens.

FIGURE 21.2 BP oil spill in Gulf of Mexico, May 2010

FIGURE 21.3 Aerial view of toxic chemicals flowing into the coastal waters near Kuala Lumpur, Malaysia

Despite this list of pollution threats, we haven't even considered the amount of trash, fishing gear, and so on that finds its way into the ocean and then onto the reefs. Such pollution, particularly that made of plastic, has now become a major agent of damage (see Figures 21.4 and 21.5). Marine mammals like whales can become entangled in lost commercial fishing equipment. Seabirds and sea turtles may starve to death if they eat too much plastic, which they can't process and which blocks normal digestive activities.

A group of researchers studied the health of corals from 159 reefs in the Asia-Pacific region. They found that coral diseases increase from 4 percent to 89 percent when corals are in contact with plastic. Furthermore, coral species with more complex structures are more likely to be affected. These species have more nooks and crannies, or microhabitats, and these small spaces are crucial to reef-associated organisms and fisheries. Plastic can have a shading effect, but its influence extends to promoting disease. For example, white syndrome is an assemblage of deadly coral diseases for which the causative agents remain elusive. However, their incidence in reefs is markedly increased when plastics are present (Lamb et al. 2018; Pollock et al. 2017).

FIGURE 21.4 Plastic bag and lines on coral reef

FIGURE 21.5 Plastic collected along black sand beach on the island of Hawaii

If we include sedimentation as a form of pollution, we can summarize the damage done by all forms of pollution. Pollution can inhibit coral reproduction and growth, weaken and increase susceptibility of corals to disease, alter food chains, cause disease and mortality, and disturb normal ecosystem functions (providing habitats, protecting coastlines, etc.).

Coral reefs are so beautiful and diverse, it is little wonder that they are popular for tourists to visit. The beauty of reefs may tempt diving or snorkeling tourists to touch or break off a souvenir. However, touching or standing on coral can do severe damage to some very delicate animals, and such behavior, done frequently enough, may cause a deterioration of the reef. Driving your boat over or dropping an anchor on the reef can also cause damage to a fragile system. The coral may seem like rocks, but the thousands of coral animals only form a thin layer over the calcareous skeleton. It doesn't take much physical stress to kill the animals that are building new reef. Carelessness may be part of human behavior, and *careless tourism* can do tremendous damage to the reefs (Coral Reef Alliance 2020).

Overfishing is a big problem for coral reefs. People catch fish for food, for sport, or to stock an aquarium. In fact, millions of people depend on getting their income and/or their dinner from reef fish. Fishing may be part of the cultural and social life of many coastal communities. These fishing pressures not only reduce the populations of various marine species, but also overfishing reduces the biodiversity of the ecosystem. The reef ecosystem has complex sets of biological interactions, and reducing or eliminating any species may have disastrous effects on the ecological functions of the reef. It doesn't take long for a reef and associated waters to be overfished by situations such as the following:

- Nets that capture young fish found in nurseries in nearshore habitats removes them before they can reproduce and replenish the population.
- People fishing always enjoy getting the biggest fish. However, the larger fish produce the most offspring and are more likely to survive naturally and continue reproducing.
- Fish species that aggregate to spawn are greatly harmed by removal in large numbers.
- The balance of the reef ecosystem depends on algae-eating fish, which keep impinging algae from overgrowing the reef. Nets and traps remove fish of all types, including the algae eaters (NOAA 2018).

Besides fish themselves, people harvest coral for jewelry. People also use coral and reef organisms for aquaria. A relatively small piece of coral for your tank represents about ten years of growth for the coral. Efforts are being made to develop policies and practices to sustainably harvest reef species, but much of the trade remains unregulated (Bruckner 2000; Domínguez, Ferrer, and Lopez 2006).

Anthropogenic (from human activity) *warming* of the planet affects coral reefs in a number of ways. Changes in climate caused by warming have the following effects on reefs:

- Warming oceans
    - Warm-water coral reefs are very sensitive to temperature changes in the ocean. One of the outcomes is coral bleaching, which we will discuss in a later section.

- Rise of sea levels
  - As the earth warms, more ice melts, causing sea levels to rise. These levels will result in increased shoreline erosion, which may have a smothering effect on the reefs, weakening the health of the reef and inhibiting growth (Field, Ogston, and Storlazzi 2011; Blanchon et al. 2009; Nature Conservancy 2020, "Sea-Level Rise").
- Alteration of ocean currents
  - Surface currents are often wind-driven. Trade winds in the Pacific tend to blow from east to west, generating movements in that direction by the surface waters. Changes to ocean currents have the potential to affect entire marine food webs. Warmer surface water does not mix as well with deeper, cooler water, and nutrients are unable to reach the surface if ocean circulation is disturbed. Phytoplankton, the principle primary producers in the open ocean, depend on nutrients brought up from rich waters deep in the ocean, and if these upwellings decline, there will be fewer nutrients, and phytoplankton populations may decline. A reduced base of primary producers will reduce the amount of food that is available for important marine food webs. In addition, many organisms depend on currents for transport. For instance, normal currents transport many small bits of life, like eggs and larvae, from one reef to another, which helps to invigorate reef communities (Alpert 2014; GBRMPA 2020).
- Changes in precipitation patterns
  - Precipitation patterns are likely to be altered by various factors from warming of the planet. Some coastal areas will receive greater average precipitation, and some will receive less, producing more or less runoff. Changes in precipitation and runoff may affect salinity, temperature, and clarity of coastal waters. These factors are crucially important to the vibrancy of coral reefs and other coastal communities (Nature Conservancy 2020, "Sea-Level Rise").
- Changes in storm patterns
  - Climate models predict that tropical storms are likely to increase in intensity. Reefs form protective barriers for coastal areas, but they are vulnerable to violent effects of powerful storms (Nature Conservancy 2020, "Sea-Level Rise").
- Ocean acidification
  - With increased CO_2, the ocean lowers its pH and makes the water more acidic. Acidic waters inhibit calcium deposition for reef growth. Hence, the reef grows more slowly, and its structure is weakened (Nature Conservancy 2020, "Sea-Level Rise"; Madin 2009; Doney et al. 2009).

QRQ 1. Categorize the cause (anthropogenic global warming, careless recreation, overfishing/coral harvesting, sedimentation/pollution, none of these) of each of the following instances of damage to coral reefs from human activity:

a. changing precipitation patterns affecting salinity of coastal waters
b. cloudy water reducing photosynthetic activities of the zooxanthellae
c. coral bleaching
d. damage to reefs by powerful storms
e. decreasing upwellings of nutrients
f. discarded plastics making corals more vulnerable to diseases
g. discarded plastics shading corals
h. driving a boat over or dropping an anchor on a reef
i. excess nutrients stimulating overgrowth of algae
j. farm fecal material flooding reefs with nutrients and pathogens
k. fouling and smothering coral animals
l. increasing acidity of water inhibiting calcium deposition for reef growth
m. increasing sunlight causing loss of nutrients
n. oil and chemical spills poisoning the coral reef ecosystem
o. pesticides poisoning inhabitants
p. reducing the biodiversity of the ecosystem
q. removing algae-eating fish, which keep algae from overgrowing the reef
r. removing coral
s. shading of reefs by large boats
t. smothering effect on the reefs caused by shore erosion
u. spread of disease from swimmers
v. sunspots causing whitening of the coral
w. touching or standing on coral

21.2.1. A Thorny Problem for the Reef

As if the coral reefs aren't under enough stress, we can add one more. There is a coral-eating predator that appears periodically on the Great Barrier Reef (GBR) in "outbreak" numbers. Widely distributed in the Indo-Pacific region, it is not an invasive species, but it is surely a very hungry one. In normal numbers, they have their role in the ecosystem, which appears to be promoting biodiversity. By consuming the fastest-growing food sources first, this predator's attacks give slower growers a chance to expand. Healthy reefs may sustain up to 20 of these creatures per hectare (10,000 sq. meters), but once the numbers surpass 30, scientists (and the corals?) begin to worry. Much higher numbers, up to 1,000/hectare, have been observed (Great Barrier Reef Foundation 2020; Nature Conservancy 2020, "Crown of Thorns Starfish"). So what is this predator, and why does it have these outbreaks?

The villain in this story is a sea star (starfish) called the "crown-of-thorns" sea star, often referred to as COTS (see Figure 21.6). Sea stars like the COTS belong to the phylum Echinodermata, a group of entirely marine invertebrates, usually displaying a five-pointed, radial symmetry. COTS are formidable-looking creatures, getting their name from the numerous spines on their body surface.

The COTS is large (larger than one foot in diameter), has fairly long spines (see Figure 21.7), and has lost the sea star's typical five-armed body plan. In fact, COTS can have from seven to 23 arms radiating from a central disk. On the underside of these arms are hundreds of tube feet (see Figure 21.8), which move the animal along. These arms are quite flexible, allowing the animal to conform and attach to the surface of the coral using its tube feet. After attaching, this predator extrudes its gut through its mouth to encompass a large area of the coral. Then glandular cells secrete a host of digestive enzymes to kill and digest the delicate coral polyps. From this digestive soup, COTS absorbs nutrients while decimating the colony (see Figure 21.9). One large specimen can feed on and kill about ten square meters of coral polyps in a year. Australian scientists say that during the past 30 years, the two most important causes of reef deterioration are the increased numbers of violent tropical storms and the crown-of-thorns sea star (Great Barrier Reef Foundation 2020; Nature Conservancy 2020, "Crown of Thorns Starfish").

FIGURE 21.6 Crown-of-Thorns sea star

FIGURE 21.7 Close-Up view of spines from COTS

FIGURE 21.8 Underside of sea star, showing tube feet (red tips) along arms, surrounding mouth

FIGURE 21.9 Surface of hard coral, showing living coral tissue (right) and dead remains (left), covered in algae, where COTS fed

When the density of COTS is high enough, the predators can wipe out large swaths of coral polyps. If the "crown-of-thorns" sea star population declines significantly, the reef might eventually recover, but it may take years outside the range of the relentless march of tube-footed menaces (Done 1988).

Although populations of COTS have fluctuated for centuries, we have identified several factors that *may* promote these outbreaks. Bear in mind that the science for these causes is unresolved, and it is likely a combination of these factors and possibly other unidentified factors that catalyze these outbreaks.

1. Many of the top COTS predators' populations have been drastically reduced. For instance, Triton's trumpet (see Figure 21.10), a very large marine snail, is an important predator, but the animals are favorites for collectors. There are a number of reef fish that also feed on COTS, such as the humphead wrasse (see Figure 21.11) and the titan triggerfish. However, these reef fish have been overfished, and many of their habitats have been damaged or lost as the reefs die back (Babcock et al. 2016; Nature Conservancy 2020, "Crown of Thorns Starfish").

FIGURE 21.10 Triton's trumpet, withdrawn into shell, with sea cucumber (another echinoderm living on seafloor) nestled by aperture

FIGURE 21.11 Humpheaded wrass

2. Warming ocean waters promote the growth and development of larval COTS. Female COTS can produce up to 65 million eggs, which after fertilization become larvae. With more larvae, more adult sea stars are produced (Uthicke et al. 2015).
3. Pollution in the form of agricultural runoff contaminates coastal areas, overwhelming the habitat with nutrients. Extra nutrients promote the growth of algae, which means more food for the larval sea star and, consequently, more adult sea stars (Brodie et al. 2005).

We now have considered the biology and the destructiveness of outbreaks on the health of coral reefs. Furthermore, we have explored the underlying causes of these COTS outbreaks. With outbreaks now affecting already weakened reefs, what can be done? On the GBR, various groups have tried removing them by hand, but the process is labor-intensive and is slow to diminish such a huge population. You can kill them by injection, using a mix that includes bile salts, but that solution is also very labor-intensive (Rivera-Posada et al. 2014). Recently, Matthew Dunbabin, a scientist specializing in robotics at the Queensland University of Technology, developed over a 10-year period a robotic sea star assassin. He calls the device COTSbot. With its very sophisticated visual system and artificial intelligence, COTSbot can successfully identify a crown-of-thorns sea star. The robot can then "zero in" on the target and inject it with the bile salt mix. The bile salts destroy many tissues, while the sea star's immune response does even more damage. The animal is dead within 20 hours. With the robot, many more can be eliminated each day, and COTSbot doesn't require a team of divers with injection guns to get the job done. Hopefully, it will help to control the COTS, and the coral reefs can face the other environmental challenges with one fewer source of stress (Braun 2018).

QRQ 2. Select which of the following are reason(s) the "crown-of-thorns" sea star in normal numbers can be beneficial to a coral reef: corals can eat larval sea star; they attack large animals that can be destructive to corals; they consume faster-growing food sources first, allowing slower growers to expand and thus promoting biodiversity; they consume dead coral, acting like debris cleaners for the ocean floor; they exude nitrogen and other useful nutrients for the coral

QRQ 3. Select which of the following are reason(s) the "crown-of-thorns" sea stars in large numbers are detrimental to a coral reef: their spines can lodge in the coral polyps, killing them; they consume too many nutrients, leaving fewer for the coral; they destroy the coral habitat; they eat too much coral; they poison the coral

QRQ 4. Select which of the following are possible reason(s) for outbreaks of COTS: agricultural runoff promoting the growth of algae, a food for larval sea stars; changing precipitation patterns affecting salinity of coastal waters; decreasing upwellings of nutrients; overcollecting of an important COTS predator with an attractive shell; overfishing reef fish that eat the "crown-of-thorns" sea star; reduced fishing of COTS; stronger storms that favor the growth of COTS; warmer ocean waters, favorable to larval COTS

21.2.2. Coral Bleaching: Pale Remnants and Stark Warning

Off the coast of northeastern Australia is one of the marvels of the natural world—the *Great Barrier Reef* (*GBR*). The GBR grows on the edge of the continental shelf, separated from the coastal plain by water. During the past 6,000–8,000 years, the sea level has risen, flooding the continental shelf between the edge of the reef and the mainland.

Visible from space, the reef, which is a collection of about 3,000 coral reefs, hundreds of coral cays, and mangrove islands, stretches along the Australian mainland for about 2,300 km (over 1,400 miles). The GBR covers about 344,000 square kilometers (about half the area of Texas). There are some 600+ species of coral, more than 1,600 species of fish, 30 species of whales and dolphins, and thousands of species of macroinvertebrate animals besides the corals, including mollusks, sponges, worms, jellyfish, and so on. There are also countless species of macroalgae (seaweeds) and microscopic life (GBRMPA 2020). Besides the animals that are established residents, the reef nourishes the early developmental stages of many transient residents. The reef domain

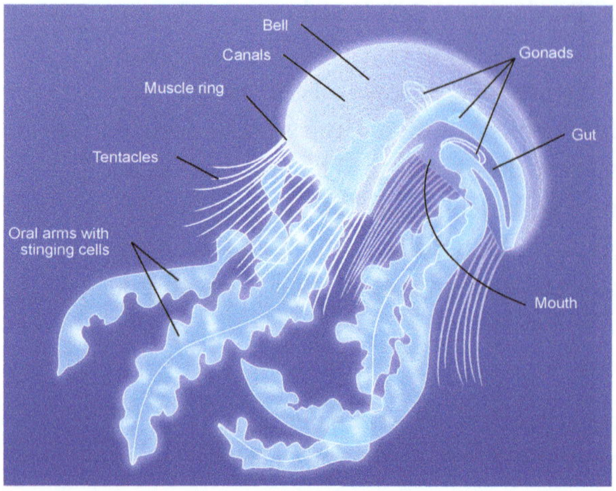

FIGURE 21.12a Structure of medusa stage

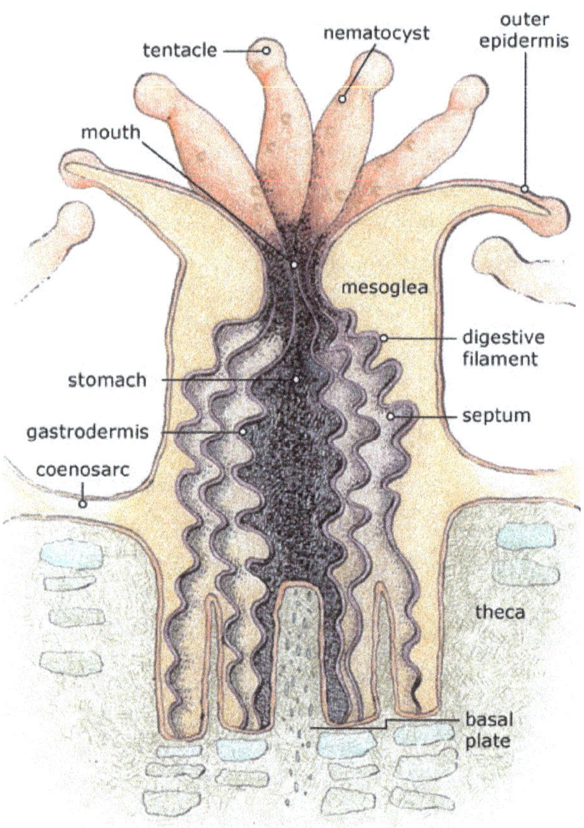

FIGURE 21.12b Structure of polyp

supports nesting of sea turtles, breeding/birthing of some whales, and breeding colonies of birds, as well.

Remarkably, coral reefs, even one as extensive as the GBR, are formed by *colonies* of individual coral animals, each one only a few millimeters in diameter. These colonies, made up of *stony corals*, secrete calcium carbonate exoskeletons for support and protection. The mass of each reef is composed of layer upon layer of calcium carbonate, deposited over millennia by successive colonies of these reef-building (*hermatypic*) corals. The living coral forms the outer surface (Castro and Huber 2016).

Stony corals belong to the phylum Cnidaria, which also includes many species of jellyfish, anemones, and soft corals (e.g., sea fans). Like other Cnidaria, stony coral individuals are constructed with a fairly simple body plan—two layers of cellular tissue, usually separated by a noncellular layer (*mesoglea*). The cellular layers include the interior *gastrodermis* lining the gastric cavity (gut), and the *outer epidermis*. Two body forms are typical for members of this phylum—a swimming form, called the *medusa*, and an attached form, called the *polyp* (see Figure 21.12). In both cases, there is an opening (mouth) into a blind pouch (open at only one end), referred to as the *gastric cavity* (gut or stomach), surrounded by a ring of *tentacles* and sometimes oral arms. The corals only have the polyp body form, living attached by way of a *basal plate* to something solid (see Figure 21.13) (Castro and Huber 2016).

FIGURE 21.13 Closeup of a star coral, showing extended polyps

The reef-building corals find nourishment from the surrounding seawater. The tentacles are armed with specialized cells called *nematocysts,* which contain barbed, sticky, and venomous threads. Contact with a small organism floating in seawater triggers the release of the threads from the nematocysts, which paralyze, entangle, or attach the prey (see Figure 21.14). The prey is then drawn into the mouth and into the gastric cavity. The gastrodermis contains various types of glandular cells. Some of these cells secrete digestive enzymes, which help to digest the prey. Other cells absorb soluble nutrients or phagocytize small particles of food. The gastric cavity has vertical folds, which may extend as filaments from the mouth. These filaments have nematocysts and glandular cells for capture and digestion of prey, and they can also help to prevent competing organisms from invading the coral's territory (Castro and Huber 2016).

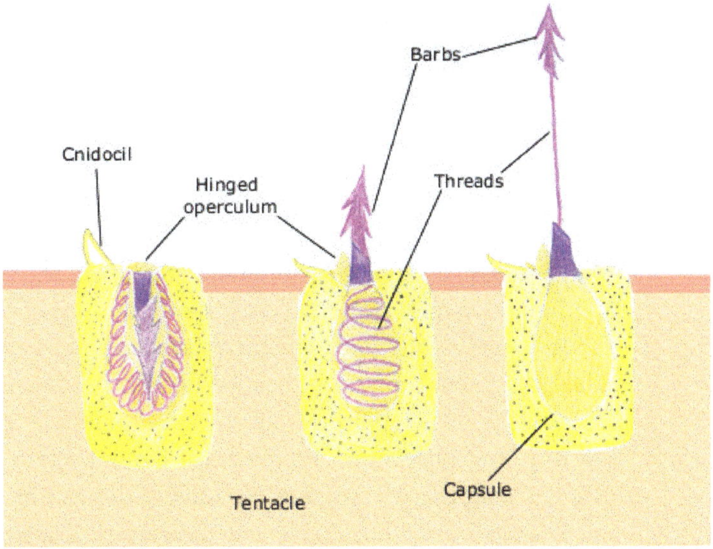

FIGURE 21.14 Nematocyst with its firing sequence

Coral polyps also have an internal source of nourishment. Within the upper reaches of the gastrodermis, cells house small, unicellular photosynthetic organisms—*dinoflagellates.* These dinoflagellates have two life forms—a free-living, flagellated form and a form called *zooxanthellae,* which lives symbiotically in vacuoles of the gastrodermis of a coral. The pigments of these zooxanthellae are responsible for the coloration of living coral colonies. Although, zooxanthellae may also be associated with other marine animals, such animals do not build reefs. To biologists, a symbiotic relationship called *mutualism* represents interactions between two species of organisms where both organisms benefit from those interactions. In this case, besides CO_2 produced by respiration of the coral tissues, the dinoflagellates utilize other nutrients (e.g., nitrogen and phosphorous) accumulated by the coral. In exchange, the zooxanthellae produce organic products from photosynthesis that the coral can use for energy or synthesis of other organic compounds (Henkel 2010; Smithsonian Ocean Portal Team 2018).

Stony corals may reproduce sexually. In many cases, the eggs and sperm are released synchronously into the sea water, where fertilization takes place (see Figure 21.15, steps 1–3). Development proceeds to produce a small larval form, called the *planula* (step 4). The planula settles on a suitable substrate and continues its metamorphosis to become a polyp (steps 5–7). The polyp will begin producing its exoskeleton, grow, and then divide asexually (*budding*) to produce another polyp. Undisturbed, each polyp will grow and divide, forming a colony. All of the polyps remain

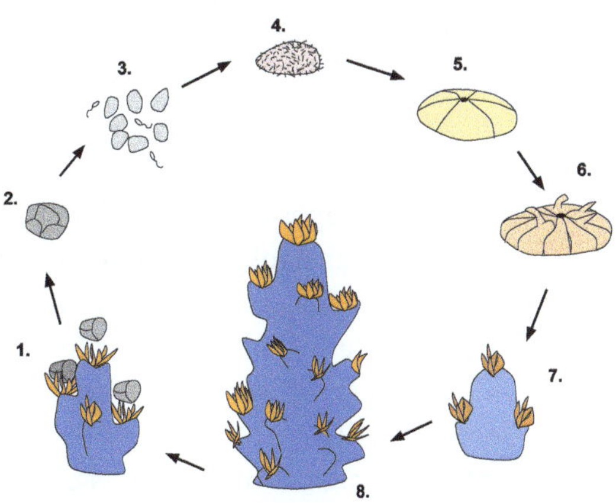

FIGURE 21.15 Coral life cycle

interconnected by an extension of the body wall, called the *coenosarc*. These connective extensions also contain canals that interconnect the gastric cavities. Through these connections, the polyps can share such things as digested materials and nerve impulses. Reefs are composed of countless numbers of coral colonies (Castro and Huber 2016).

For a coral reef to develop, a number of requirements must be satisfied. Active coral growth normally takes place in fairly shallow (< 25 M), warm (> 23°C, but < 29°C), and rather clear water, where sufficient light penetrates to promote the photosynthetic activity of the zooxanthellae. Consequently, reefs are associated with shallower, warmer zones surrounding continents or islands, or perhaps atop underwater volcanic mountains (seamounts). Temperature tolerance varies for different types of zooxanthellae, which partially accounts for the reefs developing below 25 m and outside the normal temperature range. We know that corals can adapt somewhat to varying temperatures. However, in very shallow waters, where there is abundant sunlight, or in coastal areas, where runoff from power plants occurs, temperatures may exceed tolerance and damage the reefs. Salinity, as well as temperature, influences calcium deposition rates, so most reefs are found within salinities of 32 to 40°/oo (°/oo = parts per thousand) (Castro and Huber 2016; Henkel 2010).

Because of the occurrence of reefs near coastal zones, sediments often interfere with the penetration of light to the zooxanthellae but also may foul the polyps themselves. Consequently, reefs associated with coastal areas with considerable human activity may have their growth and development impaired.

During the early part of 2016, Australia was conducting flyover surveys of the Great Barrier Reef. Professor Terry Hughes, director of the Australian Research Council Centre of Excellence for Coral Reef Studies, confirmed that about 99 percent of the reefs making up the GBR showed evidence of coral bleaching. After personally spending four days participating in the survey, Dr. Hughes said, "It was the saddest reef trip of my career" (Normille 2016). Much of the bleaching in the northern reaches of the reef were quite severe, with some of the reef dead.

This story prompts several questions: What is coral bleaching? What causes it? Why is this episode of bleaching of so much concern?

As we learned in the introduction, the zooxanthellae sequestered within the gastrodermal cells of the coral are important suppliers of energy compounds. The pigments of these symbionts also color the flesh of the coral polyps. Bleaching is the result of the expulsion or destruction of the zooxanthellae by the gastrodermal cells. If the pigments are not present, the calcium carbonate skeleton shows through the translucent polyps, giving them a bleached appearance.

Why would the polyps make such a drastic move? Bleaching of coral reefs generally reveals that the reefs are under stress. Although many stressing factors may influence the lives of the coral animals, temperature appears to be the most common one. Corals have reasonably narrow ranges of temperature tolerance. Glynn (1991) demonstrated that only a 1–2 °C increase over the highest mean temperature for a given location may lead to coral bleaching. Temperature may also work synergistically with increased levels of UV radiation to promote stress.

Lesser (1997) found that zooxanthellae cultured under elevated temperature and UV exposure produced higher concentrations of excess O_2 (byproduct of photosynthesis), which was converted into very high concentrations of *reactive oxygen species* (*ROS*), including superoxide radicals and hydrogen peroxide. Such reactive compounds can interact with myriad cellular components (e.g., proteins, DNA, etc.), causing great damage. Sufficiently damaged cellular molecules can inhibit normal cellular functions that may stymie growth or even kill the cell. Normally, aerobic cells have enzymes that interact with ROS to detoxify them, but these enzymes become overwhelmed if the ROS levels become too high. So it seems that the coral polyps, in an effort to prevent damage to their own cells, get rid of the zooxanthellae producing the toxins.

With changes in climate that yield higher water temperatures, coral reefs are coming under a great deal of stress. Strong El Niño events, which result in warmer surface waters in the Pacific, only aggravate the problem. Scientists are alarmed at the current health status of the world's coral reefs, because under prolonged stress, the reefs may die. Widespread loss of coral reefs and the communities they support would be catastrophic, causing fantastic losses of biodiversity, altering countless animal and plant populations, dislocating marine food chains, ravaging local human communities reliant on the reefs for food, and devastating tourism-based economies.

QRQ 5. Select from the following the appropriate term for each description below: basal plate, budding, coenosarc, colony, commensalism, dinoflagellate, dinophotocite, gastrodermis, gut, hermatypic, medusa, mesoglea, mutualism, nematocyst, nematode, outer epidermis, planula, polyp, ROS, sedimentation, society, stony, tentacle, zooplankton, zooxanthella

a. asexual division of a polyp, producing another polyp
b. attached form of an animal in the phylum Cnidaria
c. blind pouch that is the stomach in a coral
d. collection of individual coral animals
e. compounds that can interact with cellular components, causing much cell damage
f. extension of the body wall of a polyp, connecting it to other polyps
g. exterior layer of an animal in the phylum Cnidaria
h. interior layer lining the gastric cavity of an animal in phylum Cnidaria
i. its pigments are responsible for the coloration of living coral colonies
j. movement of soil, sand, and silt from land by wind or drainage into oceans, clouding the water
k. part of a polyp that attaches to a solid surface
l. reef building
m. small larval form, produced sexually, of a stony coral
n. specialized cell on a coral's tentacle containing barbed, sticky, and venomous threads
o. structure extending from a coral's mouth that has nematocysts and glandular cells for capturing and digesting prey
p. structure that can help prevent competing organisms from invading the coral's territory
q. swimming form of an animal in the phylum Cnidaria
r. symbiotic relationship where both organisms benefit from their interactions
s. type of coral that secretes calcium carbonate exoskeleton for support and protection
t. unicellular photosynthetic organism with a free-living, flagellated form and a form that lives symbiotically in vacuoles of a coral's gastrodermis

QRQ 6. Carbon dioxide, produced by respiration, and other nutrients accumulated by _______ are used by _______, while organic products are produced from photosynthesis by _______ and used by _______ for energy or the synthesis of other organic compounds. Possible answers: coral, fish, mesoglea, nematocyst, zooxanthellae

QRQ 7. (T/F) Polyps in a colony share digested materials and nerve impulses.

QRQ 8. Select the following that are requirements that must be satisfied for a coral reef to develop and survive: fairly deep, fairly hot, fairly shallow, low salinity, salinities of 32 to 40°/oo, somewhat cool, sufficient light for photosynthetic activity, warm

QRQ 9. If a coral reef is stressed by too much heat and UV radiation, an excess of _______ gas is produced by photosynthetic organisms within the gastrodermal cells of the coral and is subsequently converted into very high concentrations of a reactive compound that can damage the coral cells. To protect themselves from these toxins, the coral _______ expel or destroy the photosynthetic organisms, which also provide pigments responsible for coral coloration, causing _______ _______ and, ultimately, perhaps, death of the coral.

QRQ 10. Select which of the following are things that would be severely hurt by the widespread loss of coral reefs: animal and plant populations in the ocean, biodiversity, calcium levels, local human communities reliant on the reefs for food, marine food chains, salinity, tourism-based economies, water temperature

21.3. Troubles in the River City

By almost any measure, Flint, Michigan is a difficult place to make a life. According to US Census estimates, the population stands near 96,000, over half of which (53.7 percent) is African American. About a third is white (39.7 percent), and the remainder is made up mostly of Hispanics and individuals of mixed racial heritage. The unemployment rate was 12 percent in 2019, compared to ~5.0 percent for the whole of Michigan. The mean household income for Flint was just about half of the Michigan mean ($25,650 vs. $50,803). An astounding 41.9 percent of people living in Flint are living in poverty (14.1 percent for the state). Flint used to be a thriving manufacturing center for the automobile industry, but most of those jobs were eliminated there by 2010 (US Census Bureau 2020, "Flint City"; AreaVibes "Flint, Michigan").

For 50 years, Flint, Michigan, utilized water from Lake Huron, serviced by the Detroit Water and Sewerage Department (DWSD). With the closing of the GM plants, the city, half the size of its glory days, was now impoverished, perpetually teetering on the edge of financial crisis. From 2011 until 2015, the governor of Michigan appointed a succession of emergency managers to help administer the city's budget.

Beginning in June 2012, the city council discussed various ways to save money, including alternatives to the water service agreement with DWSD. A new service provider, the Karegnondi Water Authority, was a future option, but a connector would be needed to get the water to Flint. This connector would take more than two years to construct, and DWSD would up their rates for two years during construction. In 2013, the emergency manager for the city informed DWSD that they would not be renewing the contract, so that water supplied through DWSD would end in April 2014. In the meantime, the city, desperate to cut costs, determined they could use water from the Flint River as a temporary measure before completion of the connector. Subsequently, the city hired an engineering team to help the water utility make the switch. Although DWSD agreed to continue supplying water to the city, the Flint manager vetoed that idea, and switching over to the water from the Flint River took place on April 24, 2014. The state and local governments released statements assuring residents that the Flint River water was safe to drink, but by June, residents of Flint began complaining about the water. It had a brown-red color, and it smelled and tasted bad (Smith 2015; Denchak 2018; Ruckart et al. 2019; Wikipedia 2020).

APPLYING SCIENCE TO LIFE 1 Troubled Waters

LeeAnne Walters lived with her husband and four children in Flint, and during the summer she always set up a swimming pool in the side yard. During the summer of 2014, her twin boys (four years old) developed skin rashes. She took them to the doctor, who thought it might be a contact allergy, and later *eczema* (chronic skin condition of uncertain cause but probably a combination of genetic, environmental, and immune factors). Treatment for eczema involves keeping the skin moisturized and using topical steroid creams. When the rash returned, the doctor suggested it might be *scabies* (skin inflammation caused by an infestation of mites). The doctor prescribed a topical cream that contained chemicals (also contained in some pesticides) to kill the mites. Additionally, clothing and bed linen must be washed in hot, soapy water and dried at high heat. At first the rash diminished, but it returned, and the treatment stopped working. So she took the boys to a dermatologist, who could not identify a cause but eliminated scabies or eczema as possibilities.

With no diagnosis, LeeAnne noticed that the rash reappeared on one of the twins every time he took a bath or went into the pool. Eight months after the switch, she and her family stopped drinking tap water. Despite a whole home water filtration system, the water came out an orangish color. So she phoned the city to complain. The city representative took a water sample, sent it to the lab, and waited. When the results came back, he called LeeAnne and told her that her water tested very high for lead (Pb)—104 ppb (parts per billion). Anything above 15 ppb was a serious problem. He told her not to let anyone drink the water and to flush the line out by running the water for 10–15 minutes before showering, doing dishes, and so on. The representative returned to do a follow-up test on her water, and those results were even worse—397 ppb. She was already hauling in bottled water for drinking and cooking, but now she used even more water, boiling it first, for bathing the twins. With this disturbing news, however, she took her kids to the doctor, who revealed that one of the twins had lead poisoning (Smith 2015).

21.3.1. More Troubles for Flint

In August 2014, after some residents said the water was making them sick, microbiological surveys found the bacterium *E. coli*. Some strains of this bacterium can cause serious gastrointestinal infections. So the city sent out boil-water advisories and began adding chlorine disinfectant to the water. Then LeeAnne Walters and her family began losing hair, and her eyelashes fell out.

In October, discovering that the Flint water was corroding some of the truck parts, the GM plant in Flint declared it would no longer use the Flint city water. As it turns out, the large facility was partially in Flint Township, next door to but independent of Flint. They would now take their water from the Township, which was still using water from the Detroit system. GM had been a major water consumer for the city of Flint, and their disconnection cost the city half a million dollars.

In January 2015, the Michigan Department of Environmental Quality (MDEQ) said that Flint was in violation of the Safe Drinking Water Act for compounds called *trihalomethanes,* a class of known carcinogens. The large quantities of chlorine used for disinfection of water were combining with organic matter to make the most common type of trihalomethane, chloroform ($CHCl_3$). Thus, the chlorination of the water supply led to increased production of this byproduct. Citizens gathered to complained at City Hall, but officials essentially dismissed them and their concerns. The mayor and a new emergency manager declared that the water was safe to drink. Some generous citizens began distributing free bottled water because the city was ignoring them (Smith 2015; Denchak 2018; Ruckart et al. 2019; Wikipedia 2020).

Water pipes in older cities like Flint may be made of lead, especially pipes running from the water main toward a house water meter (service lines). Water often carries oxidants (e.g., dissolved oxygen, chlorine) that interact with metal, like lead, causing corrosion. The Flint River water contained naturally high levels of chlorine. As pipes corrode, some of the erosion products form a thin, passive layer along the inside of the pipe. Many water treatment facilities treat the water to promote the formation and maintenance of this layer, which protects the pipes from interacting with the oxidants in the water. This layer-forming process is calls *passivation.* Unfortunately, advised by the MDEQ, local officials opted to wait on treating the water with chemicals that would inhibit a type of corrosion (*pitted corrosion*) that destroys a pipe's passive layer. It would also save money. Introduction of water from this new source, untreated to inhibit pitted corrosion, changed the environment for the pipes that had previously carried treated water from Detroit for many years, causing *depassivation,* or breaking down of the protective layer. As this layer was broken down, the oxidants and acids in the water could interact with the iron or lead, causing the release of iron (iron pipes) and lead (lead pipes) into the water. Although iron causes the discoloration of the water, invisible lead is far more dangerous to human beings, particularly children (Torrice 2016; Pieper et al. 2018).

With the high levels of chloroform and a call from Walters, in February 2015, the EPA inquired of MDEQ about the water treatment. Initially, the MDEQ responded that Flint was treating the Flint River water properly, but eventually they recanted and admitted that Flint was not doing so. Flint should have checked the water to homes with lead service lines to determine if they all had high levels of lead. However, the city officials really did not know which houses had lead service lines, or they considered it too difficult to determine the information from their filing cabinets.

During all of this confusion, another serious problem reared its head. Within weeks of the water switchover, there was a spike in the number of cases of another waterborne disease—Legionnaires' disease (LD), a serious type of bacterial pneumonia caused by the bacterial species *Legionella pneumophila.* This increased number of cases extended into 2015 but was diminished by the city reconnecting to Detroit water in late 2015. The release of iron from the pipes combined with some of the chlorine, reducing the free chlorine concentration in the water. A lower free chlorine concentration meant less disinfectant, which likely allowed the *Legionella* to grow and spread through the water system. Residents may have acquired the bacterium directly from the water or from shower or faucet sprays. At least 87 people were sickened, and 12–14 people died from Legionnaires' disease. Some of those who survived had life-changing damage to their kidneys and other organ systems. The general consensus is that the organisms in the water pipes were responsible for many or all of the cases. There is still some concern that a few of the cases were acquired in one of the hospitals (Zahran et al. 2020; Schultz 2018).

PoI 1. What are the toxic effects that lead has on the human body?

PoI 2. Why is there really no safe level of lead for the human body?

APPLYING SCIENCE TO LIFE 2 Troubled Waters 2

In April of 2015, LeeAnne Walters was told that her son, Gavin, had lead poisoning. The city concluded that the high lead in her family's water was from the service line, which they temporarily shut off, and then they connected them to a neighbor's water with a garden hose. The Walters were only using bottled water by then, anyway. Then, in June, an EPA official who had been checking her tap water sent Walters a draft report telling her of the extraordinarily high lead levels in her water. LeeAnne forwarded the draft to a reporter, who published it, stirring further fear. For months the EPA had been warning state officials about the poor quality of Flint's water. In July, MDEQ officials told the citizens of Flint that the EPA report was wrong and there was no widespread problem with the water and to relax. By this time the EPA report was all over the media. In July, LeeAnne and other Flint residents drove to Lansing, but a high-ranking MDEQ official essentially "blew them off."

In August, the state took a set of new water samples but eliminated from consideration two of the samples. One of these samples was from the Walters' home. Without those two samples, the lead concentration fell just below the 15 ppb "action level" (not safe). They claimed that the Walters' sample, with their home filtration unit, would have artificially lowered the numbers. Oddly, the Walters' samples were the highest of those measured.

In September, after a call from Walters, an engineering professor, Marc Edwards, traveled 15 hours from VPI (Virginia Polytechnic Institute and State University) to Flint with four of his graduate students. Professor Edwards specialized in studying water-pipe corrosion. He and his team collected samples throughout the city, returning to Virginia to analyze the samples for lead. He found that there were high lead levels city-wide but that ones from the Walters' home were the highest. One of the samples had a concentration of 13,200 ppb. Any lead concentration above 5,000 ppb, the EPA considers "hazardous wastes." In September, Edwards addressed reporters, where he was especially critical of the fact that Flint had no corrosion control policy in place. He advised Flint residents to stop using and consuming water because of the high concentration of lead. Furthermore, he advised them to buy a certified lead filter or to use bottled water if they could afford to do so. EPA also warned the state that the water in Flint may not be safe to drink. It was the first time any *agency* had made such a statement, and of course, it was denied by the MDEQ and city officials. Both groups claimed the water was safe to drink (Smith 2015; Denchak 2018; Ruckart et al. 2019; Wikipedia 2020).

21.3.2. The Last Drop Falls

A pediatrician, Dr. Mona Hanna-Attisha, was alerted by a friend seeing the crisis in Flint that she should check on the lead levels of the blood from children less than five years of age. Dr. Hanna-Attisha found that the percentage of kids with elevated blood levels of lead had virtually doubled following the water source switch. She knew, and both the EPA and the CDC (Centers for Disease Control and Prevention) agreed, that there is *no* safe blood level for lead. She rechecked her data and consulted other doctors. Only then did this group of physicians go public about her findings. In late September, the physicians asked the city for the sake of public health to switch back to Detroit water. At first, the state tried to discredit her and debunk her findings. However, then they admitted mistakes had been made in their analysis. On October 1, a public health crisis was declared for Genessee County, which includes Flint. Later, the state, the city, and the Mott Foundation found $12 million to help the city reconnect to Detroit water. Reconnection took place on October 16, 2015 (Ruckart et al. 2019; Gross 2018).

APPLYING SCIENCE TO LIFE 3 What If It Were Your Child?

LeeAnne Walters attended the press conference held by Dr. Hanna-Attisha and colleagues. Part of that news conference detailed the life perils that a child with lead poisoning might deal with as they got older. The child might need special consideration for education and would have increased risk for various behavioral problems and even criminal activities. All she could think about was her son Gavin. It was a much more dreadful scenario than just losing a few IQ points, as the state nurse she consulted had told her (Smith 2015; Gross 2018). How would you feel if it were your own child?

PoI 3. Of the characters mentioned in this case, who or what group do you think failed the people of Flint, and why do you think so? Who do you think were the heroes, and why do you think so?

PoI 4. What are some of the other pollutants that can make drinking water a public health issue?

PoI 5. Access to clean water is becoming a critical issue, even in the United States. Why do you think this is true in the United States? Why is this a critical issue for people in developing areas of the world?

QRQ 11. As pipes corrode, some of the erosion products form a layer along the inside of the pipe, which protects the pipes from interacting with the oxidants in the water. This layer-forming process is called ______, while the process of breaking down of the protective layer is called ______.

QRQ 12. As the protective layer in Flint pipes was broken down, the oxidants and acids in the water could interact with the materials of the pipes, causing the release of which of the following kinds of disturbing or serious particle(s) from some pipes into the water? aluminum, carbon, chlorine, copper, *E. coli*, iron, lead, plastic

QRQ 13. In Flint's water, the metal ______ caused discoloration. However, the metal ______ could cause neurological damage, particularly in children. Use possible answers from QRQ 12.

QRQ 14. After switching to the Flint River for Flint's water supply, large quantities of ______ used for disinfection of water were combining with organic matter to make ______, a known carcinogen. Possible answers: aflatoxin, alcohol, ammonium, arsenic, chlorambucil, chlorine, chlornaphazine, chloroform

QRQ 15. The general consensus is that organisms in Flint's water pipes were responsible for many or all of the cases of ______ (Answer choices: anthrax, cholera, Legionnaires' disease, meningitis, tuberculosis, typhoid).

21.4. Rivers for Life—Conflicting Needs for Water

If you enjoy eating oysters from the eastern United States, then you have likely consumed arguably the best-tasting specimens—those from Apalachicola Bay, Florida. Apalachicola Bay is located along the warm, humid Florida panhandle, about halfway between Panama City and Tallahassee. Oysters growing in this estuary have been harvested since the middle of the 19th century (Elliott 2020), and some oystermen represent multiple generations. The long-traditional tong methods to collect the oysters are still used, supplying about 10 percent of the US supply of oysters (Ward 2015). How can one bay supply so many oysters?

Apalachicola Bay (the Bay) is an estuary covering an area of approximately 630 km^2, separated and protected by barrier islands from the Gulf of Mexico (see Figure 21.16) (Pine et al. 2015). Averaging less than three meters in depth, this bay has historically been characterized by a perfect balance of saline water from the Gulf and freshwater, largely from the Apalachicola River. This river's inflow affects not only the salinity of the Bay but also the nutrient levels, especially those of nitrogen and phosphorous (Florida Sea Grant 2013; Frick, Buell, and Hopkins 1996). The floor of the Bay also provides a set of shallow bars (shoals) upon which oyster beds develop. Consequently, the oysters normally thrive with ideal seafloor geology and ranges of temperatures and salinities (5–25 ppt) (Twichell 2007; Twichell et al. 2010). Given the favorable conditions, oysters in the Bay grow continuously and spawn from April through October (Oczkowski et al. 2011). According to De Villegas (2016), Apalachicola Bay oysters mature to marketable size in just 18 months.

PoI 6. What kind of animal is an oyster? In what habitats do oysters live?

PoI 7. Describe the life cycle of an oyster.

PoI 8. What is an "ecological niche"? What is the oyster's niche?

PoI 9. What is an estuary? Where are they found? Why are they important?

PoI 10. What is salinity? How is it measured?

PoI 11. What is the range of salinities found in the following?

a. open ocean
b. estuary
c. freshwater river

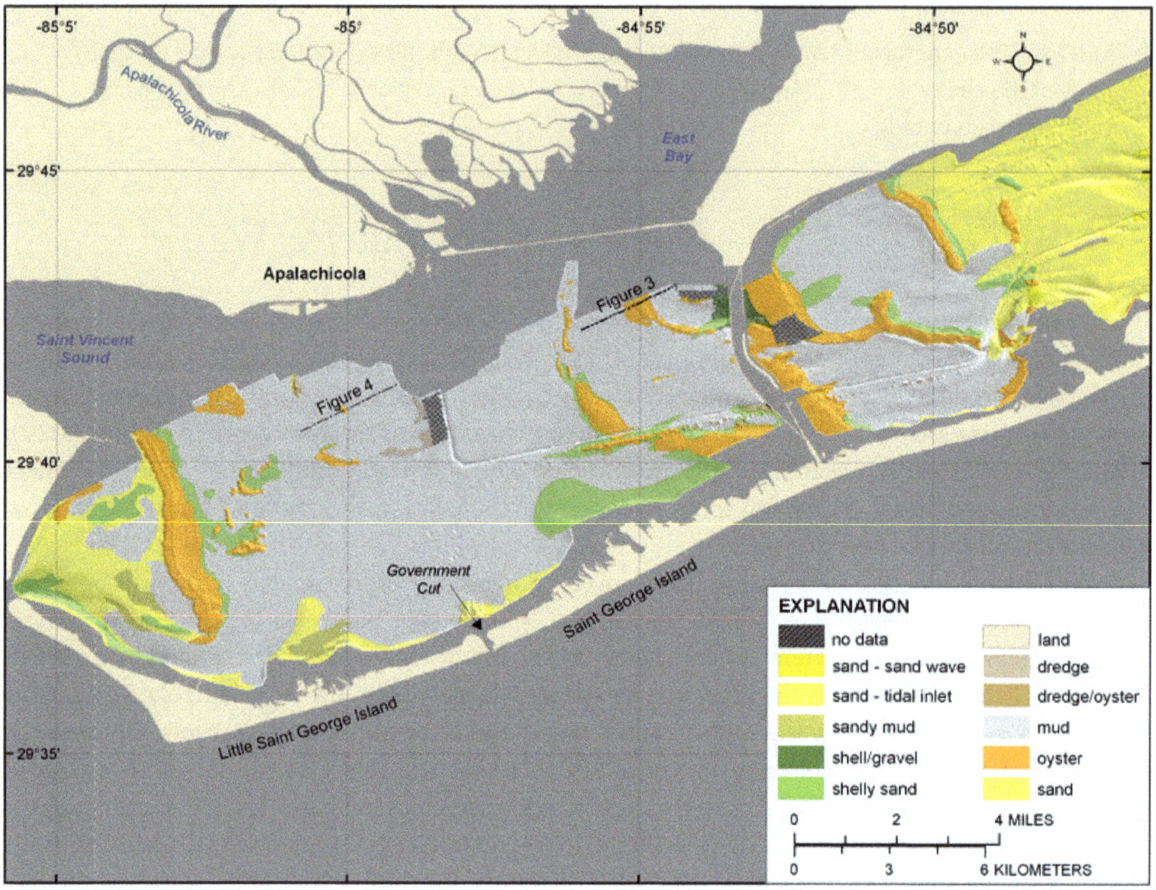

FIGURE 21.16 Surficial geology of Apalachicola Bay, emphasizing the distribution of oyster growth (orange) within the bay

21.4.1. Water Allocation Storm

In 2012, following a severe drought in the state of Georgia, the Apalachicola Oyster Fishery failed (Havens et al. 2013), prompting the National Oceanographic and Atmospheric Administration to declare a fisheries disaster (Fisch and Pine 2016). Besides the economic and biological losses in the Bay, the collapse was "the latest straw" to an almost 30-year legal battle between the states of Florida and Georgia, as well as Alabama, over water allocation from the Apalachicola-Chattahoochee-Flint (ACF) River Basin (Hawkins 2016). In 2014, the US Supreme Court agreed to hear Florida's complaint. In the suit, the State of Florida cited the economic damage done to the oyster industry from water allocation policies carried out in the State of Georgia. In short, Georgia was using too much of the river water, and that overuse was causing ecological and economic damage to the oyster industry in Florida (Manganiello 2019).

The 2012 drought was not the only drought endured recently by the ACF River Basin. Drought conditions also occurred between 1999–2001 and 2006–2008 (Livingston 2015; Leitman, Pine, and Kiker 2016). Reduced rainfall, combined with the tremendous increase in Georgia's water demand during the late 20th century, resulted in substantially reduced water flow from the Apalachicola River (Fisch and Pine 2016).

21.4.2. ACF River Basin

From the mountains of north Georgia, the Chattahoochee River flows southwest, through Metro Atlanta and southcentral and southwestern parts of the state, joining the Flint River at Lake Seminole. The river runs for over 400 miles and drains over 8,700 square miles. From Lake Seminole, the flow continues as the Apalachicola River, which empties into Apalachicola Bay. Together, these rivers comprise the ACF (Apalachicola-Chattahoochee-Flint) River Basin, which covers almost 20,000 square miles of Georgia, southeastern Alabama, and northeastern Florida (see Figure 21.17) (Couch, Hopkins, and Hardy 1996).

The rivers in this region are no longer free-flowing but highly interrupted and regulated. During the 1950s, the US Army Corps of Engineers completed construction of the Buford Dam at the headwaters of the Chattahoochee River in northeastern Georgia. This dam, one of 13 on this river, converted the upper reaches of this free-flowing river into a reservoir, Lake Lanier (Georgia Department of Natural Resources 1997). Initially, the dam was commissioned to control flooding, produce hydroelectric power, and provide for navigation. During those days, metro Atlanta, which is located roughly 50 miles southwest of the lake, was much smaller (1,398,465), but the city has grown considerably in area and population since the construction of the dam. In 2018, the estimated population of Metro Atlanta was 5,950,828 (Census Reporter 2018). Accordingly, as Atlanta grew, the lake became a very popular recreational destination, and the city and its suburbs began using increasing amounts of water released from Lake Lanier as one of the primary sources of water for its population and industries.

21.4.3. US Supreme Court Case

In 2014, the US Supreme Court appointed a Special Master to take evidence and make a recommendation to them in the case. The trial was held by the Special Master in late 2016, and he announced his recommendations on February 14, 2017. The judge recommended the court deny Florida's request, because he was unconvinced that a consumptive water-use cap for the State of Georgia could be guaranteed to result in sufficient additional freshwater input needed by the Apalachicola Bay. Florida had failed to include the US Corps of Engineers in their suit, which controls water flow in the Basin. Consequently, the judge could not induce the Corps to take the appropriate actions to alter the water flow into Florida. However, he also indicated that the State of Georgia had not regulated adequately agricultural water usage, which was "largely unconstrained" (Manganiello 2019; Reilly 2017).

The recommendations of the Special Master were not binding but only an advisory to the Supreme Court. Florida has filed exceptions to the Special Master's Report, and Georgia may respond to

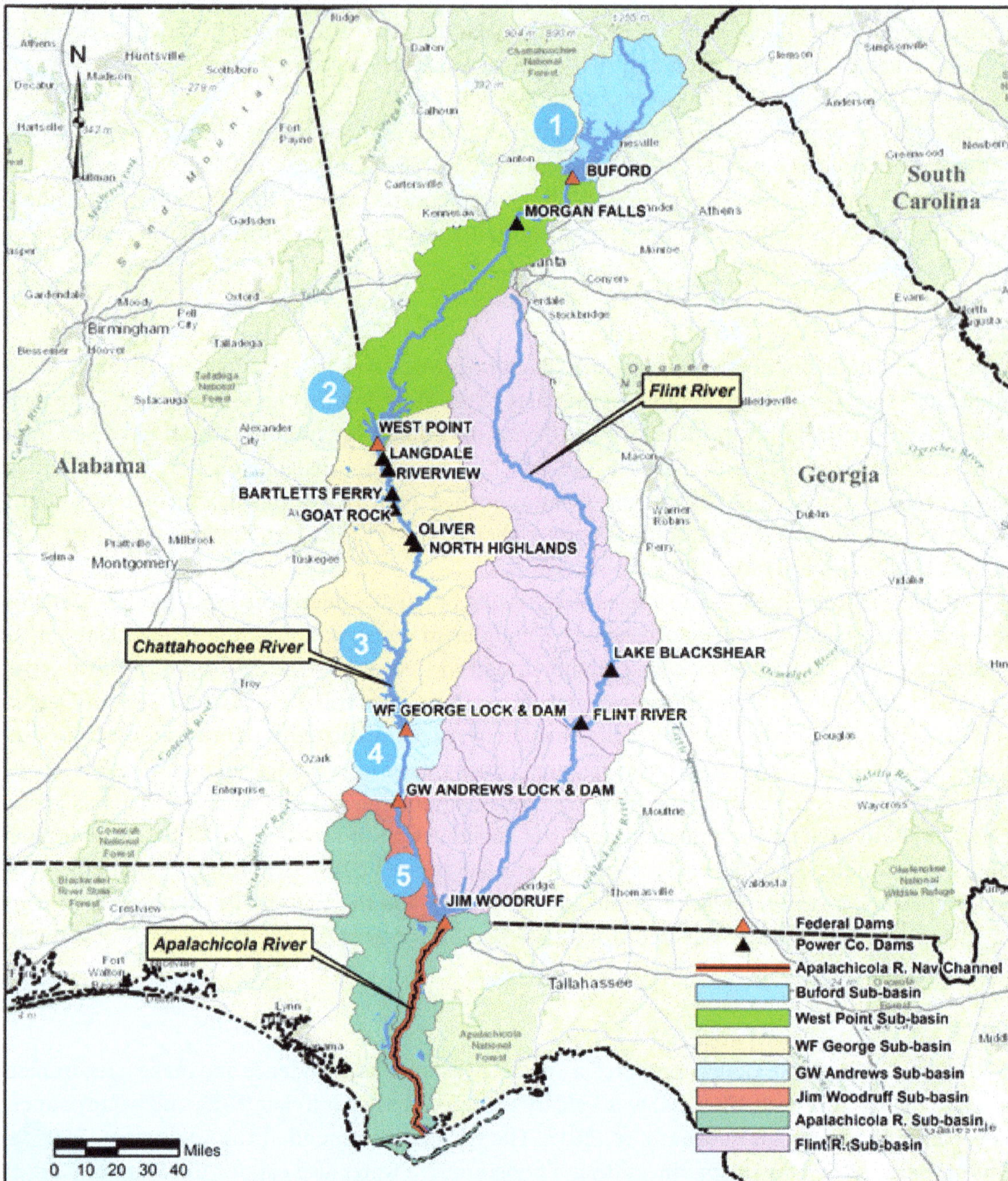

FIGURE 21.17 Diagram of the ACF Water Basin

these exceptions. The Supreme Court was originally expected to issue their decision later in 2017 or in 2018 (Manganiello 2019). In 2018, the Court appointed a new Special Master, who in 2019 sent his report back to the Supreme Court. Exceptions to the report can be submitted, and the Court should finally rule on the dispute during 2020. Water allocation issues are likely to intensify in the southeast, particularly with the recurring droughts.

PoI 12. In the court case, Florida accused Georgia of not properly allocating the freshwater from the ACF River Basin so that the lack of sufficient freshwater caused alterations in the Bay's salinity, to the detriment of healthy oyster populations. Although oysters are generally quite tolerant of salinity fluctuations, there may be a correlation between high salinity and higher mortality of the oysters. In estuarine habitats, what other factors might also influence the survival of a healthy oyster population in Apalachicola Bay?

PoI 13. Let us assume that bottom salinities for various oyster reefs in the bay may vary between 2 ppt (parts per thousand) and 30 ppt. *Perkinsus marinus*, a protozoal parasite of oysters and other shellfish, may kill up to 80 percent of oysters on a reef (Andrews 1988). Given the following data (generalized from various sources), how might periodic influxes of freshwater into Apalachicola Bay influence oyster production?

Animal	Salinity Tolerance Range
oyster	5–35 ppt
protozoan parasite	15–35 ppt

QRQ 16. Select which of the following are competing interests and conditions involving water allocation from the Apalachicola-Chattahoochee-Flint River Basin: agricultural water usage, aqueducts, balance of saline water and freshwater for oyster growth, growth of Metro Atlanta, hydroelectric power, Lake Lanier reservoir and recreation area, paper mills, severe droughts in the state of Georgia, spread of disease, steel production

Image Credits

Fig. 21.1: Source: https://commons.wikimedia.org/wiki/File:Eastchina_and_Eastchina_Sea_tmo_2017313_lrg.jpg.

Fig. 21.2: Source: https://commons.wikimedia.org/wiki/File:Gulf_of_Mexico_oil_spill_May_17_cropped.jpg.

Fig. 21.3: Copyright © by Thakur Dalip Singh (CC BY 3.0) at https://commons.wikimedia.org/wiki/File:-Coastal_Pollution_Thakur_Dalip_Singh_(136164175).jpeg.

Fig. 21.4: Copyright © by q phia (CC BY 2.0) at https://commons.wikimedia.org/wiki/File:Plastic_on_corals,_pockets,_wakatobi,_2018_(44900243615).jpg.

LAB #21

Laboratory Exercises

21.5. Laboratory on Corrosion

21.5.1. Preliminary Work

In this laboratory, we examine the interaction of water with metal. From the text's website, view the introductory video; download and open *"Corrosion.nlogo."*

1. **Question 1.** Click the Info tab at the top of the page and read the section "What Is It?"
   a. ________________ Select all that apply to the term *pitted corrosion*: A. commonly causes failure and leakage of pipes, B. corrosion of pits, C. deterioration of metal, D. deterioration of rubber gaskets, E. easier to detect than regular corrosion, F. forms pits, G. intermediate corrosion, H. more difficult to detect than regular corrosion, I. takes place at focal points
   b. __________ Select all that apply to the term *passivation*: A. chemicals, interacting with water components, are deposited as scale along the inner surface of a pipe, B. deterioration of a pipe's protective lining, C. eliminating turbulent water in a pipe, D. process of forming a lining, E. protects a pipe in places from further corrosion
   c. _____ Define *depassivation* using the choices from part b.

2. Read "How It Works."
   a. Give the color (A. black, B. blue, C. brown, D. cyan, E. gray, F. green, G. orange, H. red, I. violet, J. white, K. yellow) of each model component.

      ______ treated water

      ______ corrosive water

      ______ metal

      ______ passive

      ______ corrosive product
   b. _____ What rate(s) does treated water employ, DEPASSIVATION-RATE (D) or PASSIVATION-RATE (P)?

c. ______ What rate(s) does corrosive water employ, DEPASSIVATION-RATE (D) or PASSIVATION-RATE (P)?

d. __________ List all that apply when the HOLE? switch is ON: A. Initially, a corrosive water patch occupies the hole in the passive layer. B. Initially, corrosive water patches are above the passive layer. C. Initially, metal patches are below a passive layer. D. Initially, treated water patches are above a passive layer. E. Model is initialized with a one-patch thick, horizontal passive layer with a hole at the origin. F. Model is initialized with an unbroken, one-patch thick, horizontal passive layer through the origin.

e. __________ List all that apply when the HOLE? switch is OFF. Use choices from part d.

3. Read "How to Use It."

a. ______ Give the value of HOLE? for Scenario 1, OFF (F) or ON (N).

b. ______ Give the value of HOLE? for Scenario 2, OFF (F) or ON (N).

c. ______ For Scenario 1, the system initially has **1.** (C. corrosive, T. treated) water.

d. ______ Write 0.0052 in scientific notation, with one nonzero digit before the decimal point, that the model can use for a rate.

e. ______ In Scenario 2, what is the value in the TICKS-TO-UNTREATED or TICKS-TO-TREATED-AGAIN input box that causes the system to ignore the box? A. −1, B. 0, C. 1, D. OFF, E. ON

f. ______ In Scenario 2, if both TICKS-TO-TREATED-AGAIN and TICKS-TO-UNTREATED input boxes have values, which must be larger? A. TICKS-TO-TREATED-AGAIN, B. TICKS-TO-UNTREATED?

21.5.2. Laboratory Work

4. This question involves Scenario 1, with one hole in a thin, passive layer.

a. ______________________________ Turn HOLE? ON. Make sure PASSIVATION-RATE is 0.001 and DEPASSIVATION-RATE is 1.0E-6, which is $1.0 \times 10^{-6} = 0.000001$. Set the simulation speed to be faster. Observe carefully the animation, noting particularly the appearance of the associated graph when a new hole in the passive layer occurs. Also, at approximately 5,000, 10,000, 15,000, and 20,000 ticks, stop execution by pressing GO and export the interface (found under *File*, *Export*, *Export Interface...*), saving the picture under a name indicating the tick count. Select all statements concerning your observations about the developing pitted corrosion: A. corrosive particles go into the pits, B. eventually, pits have finger-like projections, C. more and more product particles are in the water, D. number of passive particles decreases, E. number of passive particles increases, F. number of product particles in the water decreases, G. pit formed at initial hole is a ragged semicircle, H. pit formed at initial hole is a smooth semicircle, I. pits tend to merge, J. when a new pit forms, the number of product particles rises sharply, K. with each new pit, the number of unprotected metal particles increases slightly and then levels off

b. Decrease PASSIVATION-RATE to $0.0003 = 3.0 \times 10^{-4} = 3.0E\text{-}4$, keeping DEPASSIVATION-RATE = 1.0E-6. Uncheck *view updates* or increase the simulation's speed. At approximately 5,000, 10,000, 15,000, and 20,000 ticks, check *view updates* and export the interface as described in part a. Before restarting the simulation, uncheck *view updates*. Compare the interfaces at corresponding times from this part and part a. For each of the following statements, indicate which simulation applies: A. part a with the larger passivation rate (0.001), B. part b with the smaller passivation rate (0.0003), S. about the same

_______ The pits are smoother and closer to being semicircles.

_______ The number of passive particles is greater.

_______ The number of unprotected metal patches is greater.

c. With PASSIVATION-RATE = 0.0003, decrease DEPASSIVATION-RATE to 0.0000003 = 3.0×10^{-7} = 3.0E-7. Repeat part b. Compare the corresponding interfaces from this part and part b. For each of the following, indicate which simulation applies: B. part b with the larger depassivation rate (1.0E-6), C. part c with the smaller depassivation rate (3.0E-7), S. about the same

_______ The pits are smoother and closer to being semicircles.

_______ The number of product particles is greater.

_______ More pits are formed.

d. Depending on the parameters (passivation rate and depassivation rate), the model generates two kinds of pits, smooth-surfaced brightening pits (S) or rough-surfaced etching pits (R). Categorize the kinds of pits for each part.

_______ Part a

_______ Part b

_______ Part c

e. **1.** _______ **2.** _______ Complete the following statement using depassivation (D) and/or passivation (P): A higher **1.** rate tends to build up the protective passive layer, while a higher **2.** rate tends to break down the protective layer leading to corrosion and seepage of toxic materials into the water.

5. This question involves Scenario 2, which has augmentation of the passive layer before introduction of corrosive water.

a. _______________ Switch HOLE? OFF, so that the thin passive layer is unbroken, and switch TREATED? ON, so that initially the passive layer will build. Set PASSIVATION-RATE to 0.003 and DEPASSIVATION-RATE to 1.0E-4, so that both the processes of passivation and depassivation run faster. To observe a situation in which authorities use treated water and then untreated water, set values for TICKS-TO-UNTREATED 700. Increase the speed of the simulation. After 700 ticks, stop execution by pressing GO. Check all statements that describe the situation: A. corrosive water starts flowing into the system from the top at 700 ticks, B. number of passive particles increases, C. number of product particles increases, D. number of unprotected metal particles increases, E. protective layer has increased about sevenfold, F. protective layer is ragged, G. system has treated water during the first 700 ticks, H. system has untreated water during the first 700 ticks

b. _______ Start the simulation again but stop it as soon as a rupture appears and a pit begins to form. Select all statements that describe the situation: A. once a pit formed, the metal started corroding and releasing corrosion products into the water, B. once untreated water flowed into the system, the number of passive particles decreased in a linear fashion, C. when a pit formed, the number of product and unprotected metal particles increased

c. _______________________ Run the simulation for about another 4,000 ticks before stopping the simulation. Select all statements that describe the situation: A. adjacent pits tend to join, B. after an initial seepage, no additional product particles are released, C. after decreasing somewhat, number of passive particles starts to increase, D. after increasing initially, number of product particles remains fairly constant, E. corrosive water does not seep into the pits, F. corrosive water seeps into the pits, G. many product particles are released into the water, H. new passive particles form in the pits, I. number of passive particles decreases, J. number of product particles increases sharply, K. pits are rough-surfaced etching pits, L. pits are smooth-surfaced brightening pits

d. **1.** ______ **2.** ______ **3.** ______ **4.** ______ **5.** ______ With the initial presence of **1.** (A. treated, B. untreated) water, **2.** (C. corrosion begins, D. the passive layer builds up). After **3.** (A. treated, B. untreated) water is introduced into the system, **4.** (E. corrosion halts, F. the passive layer begins to break down), eventually pits form and **5.** (G. harmful corrosive products, H. passive particles) are released into the water.

REFERENCES

1. Science for Citizens

AAP (American Academy of Pediatrics). "Vaccine Safety: Examine the Evidence." HealthyChildren.org, July 24, 2018. https://www.healthychildren.org/English/safety-prevention/immunizations/Pages/Vaccine-Studies-Examine-the-Evidence.aspx.

———. "Do Vaccines Cause Autism?" The History of Vaccines, January 25, 2018, https://www.historyofvaccines.org/content/articles/do-vaccines-cause-autism.

APA (American Psychiatric Association). *Diagnostic and Statistical Manual of Mental Disorders, Fifth Edition (DSM-5)*. Arlington: American Psychiatric Association, 2013.

Araghi, Sadaf Oliai, Jessica C. Kiefte-de Jong, Suzanne C. van Dijk, Karin M. A. Swart, Hanneke W. van Laarhoven, Natasja M. van Schoor, Lisette C. P. G. M. de Groot et al. "Folic Acid and Vitamin B12 Supplementation and the Risk of Cancer: Long-Term Follow-Up of the B Vitamins for the Prevention of Osteoporotic Fractures (B-PROOF) Trial." *Cancer Epidemiology, Biomarkers, and Prevention* 28, no. 2 (2019): 275–82.

Barrois, M., I. Bieche, S. Mazoyer, M-H. Champème, B. Bressac-de Paillerets, and R. Lidereau. "Real-Time PCR-Based Gene Dosage Assay for Detecting BRCA1 Rearrangements in Breast–Ovarian Cancer Families." *Clinical Genetics* 65, no. 2 (2004): 131–36.

Bastian, Hilda. "Lucy Wills (1888–1964): The Life and Research of an Adventurous Independent Woman." *Journal of the Royal College of Physicians of Edinburgh* 38, no. 1 (2008): 89–91.

Bes-Rastrollo, Maira, Matthias B. Schulze, Miguel Ruiz-Canela, and Miguel A. Martinez-Gonzalez. "Financial Conflicts of Interest and Reporting Bias Regarding The Association Between Sugar-Sweetened Beverages and Weight Gain: A Systematic Review of Systematic Reviews." *PLoS Medicine* 10, no. 12 (2013): e1001578.

Binns, Colin W., Mi Kyung Lee, and Andy H. Lee. "Problems and Prospects: Public Health Regulation of Dietary Supplements." *Annual Review of Public Health* 39 (2018): 403–20.

Brasky, Theodore M., Emily White, and Chi-Ling Chen. "Long-Term, Supplemental, One-Carbon Metabolism–Related Vitamin B Use in Relation to Lung Cancer Risk in the Vitamins and Lifestyle (VITAL) Cohort." *Journal of Clinical Oncology* 35, no. 30 (2017): 3440–48.

Brown, Paul. "Cosmic Rays Are Not Causing Climate Change." *Scientific American*, November 11, 2013, https://www.scientificamerican.com/article/cosmic-rays-not-causing-climate-change.

CDC (Centers for Disease Control and Prevention). "Fast Facts: Cost and Expenditures." Smoking and Tobacco, November 15, 2019, https://www.cdc.gov/tobacco/data_statistics/fact_sheets/fast_facts/index.htm.

———. "Measles (Rubeola)." October 4, 2019, https://www.cdc.gov/measles/index.html.

———. "Smallpox." July 12, 2017, https://www.cdc.gov/smallpox/index.html.

———. "Thimerosal in Vaccines." Vaccine Safety, Common Concerns, October 27, 2015, https://www.cdc.gov/vaccinesafety/concerns/thimerosal/index.html.

Chowdhury, Rajiv, Samantha Warnakula, Setor Kunutsor, Francesca Crowe, Heather A. Ward, Laura Johnson, Oscar H. Franco et al. "Association of Dietary, Circulating, and Supplement Fatty Acids with Coronary Risk: A Systematic Review and Meta-Analysis." *Annals of Internal Medicine* 160, no. 6 (2014): 398–406.

Deer, Brian. "How the Case Against the MMR Vaccine Was Fixed." *British Medical Journal* 342 (2011): c5347.

Dietary Supplement Health and Education Act of 1994, Public Law No. 103–417 (1994), https://ods.od.nih.gov/About/DSHEA_Wording.aspx.

Dow, Caitlin. "How *Industry Influences Nutrition Science*." Interview of Marion Nestle by Bonnie Liebman. Nutritian Action, March 15, 2017. https://www.nutritionaction.com/daily/food-safety/how-industry-influences-nutrition-science/.

Ebbing, Marta, Kaare Harald Bønaa, Ottar Nygård, Egil Arnesen, Per Magne Ueland, Jan Erik Nordrehaug, Knut Rasmussen et al. "Cancer Incidence and Mortality After Treatment with Folic Acid and Vitamin B12." *Journal of the American Medical Association* 302, no. 19 (2009): 2119–26.

Ewald, Ingrid Petroni, Patricia Lisboa Izetti Ribeiro, Edenir Inêz Palmero, Silvia Liliana Cossio, Roberto Giugliani, and Patricia Ashton-Prolla. "Genomic Rearrangements in BRCA1 and BRCA2: A Literature Review." *Genetics and Molecular Biology* 32, no. 3 (2009): 437–46.

Flaherty, Julie. "Special Supplement: 10 Red Flags of Misleading Nutrition Claims." *Health and Nutrition Letter*, Tufts University Updated February 20, 2020, https://www.nutritionletter.tufts.edu/issues/10_15/special-reports/Special-Supplement-10-Red-Flags-of-Misleading-Nutrition-Claims_1652-1.html.

Godlee, Fiona, Jane Smith, and Harvey Marcovitch. "Wakefield's Article Linking MMR Vaccine and Autism Was Fraudulent." *British Medical Journal* 342 (2011): c7452.

Henning, Justine. "Science Is Why My Cancer Diagnosis Isn't a Death Sentence. Today I March for Science." Vox.com, April 22, 2017, https://www.vox.com/first-person/2017/4/21/15374532/breast-cancer-march-for-science.

Howlader N, A. M. Noone, M. Krapcho, D. Miller, K. Bishop, C. L. Kosary, M. Yu et al. (eds). "SEER Cancer Statistics Review, 1975–2014." National Cancer Institute, Bethesda, MD, https://seer.cancer.gov/csr/1975_2014/, based on November 2016 SEER data submission, posted to the SEER web site, April 2017.

Institute of Medicine. Dietary Reference Intakes: Thiamin, Riboflavin, Niacin, Vitamin B6, Folate, Vitamin B12, Pantothenic Acid, Biotin, and Choline. Washington, DC: National Academy Press, 1998.

Kuchenbaecker, Karoline B., John L. Hopper, Daniel R. Barnes, Kelly-Anne Phillips, Thea M. Mooij, Marie-José Roos-Blom, Sarah Jervis et al. "Risks of Breast, Ovarian, and Contralateral Breast Cancer for BRCA1 and CRCA2 Mutation Carriers." *Journal of the American Medial Association* 317, no. 23 (2017): 2402–16.

Lippman, Scott M., Eric A. Klein, Phyllis J. Goodman, M. Scott Lucia, Ian M. Thompson, Leslie G. Ford, Howard L. Parnes et al. "Effect of Selenium and Vitamin E on Risk of Prostate Cancer and Other Cancers: The Selenium and Vitamin E Cancer Prevention Trial (SELECT)." *Journal of the American Medical Association* 301, no. 1 (2009): 39–51.

Maenner, Matthew J., Kelly A. Shaw, Jon Baio, Anita Washington, Mary Patrick, Monica DiRienzo, Deborah L. Christensen et al. "Prevalence of Autism Spectrum Disorder Among Children Aged 8 Years—Autism and Developmental Disabilities Monitoring Network, 11 Sites, United States, 2016." *Surveillance Summaries* 69, no. 4 (2020): 1–12, https://www.cdc.gov/mmwr/volumes/69/ss/ss6904a1.htm?s_cid=ss6904a1_w.

Malhotra, Aseem, Rita F. Redberg, and Pascal Meier. "Saturated Fat Does Not Clog the Arteries: Coronary Heart Disease Is a Chronic Inflammatory Condition, the Risk of Which Can Be Effectively Reduced from Healthy Lifestyle Interventions." *British Journal of Sports Medicine* 51, no. 15 (2017): 1111–12.

Mayo Clinic. "Hormone therapy for breast cancer." Updated February 5, 2019, accessed January 7, 2020. https://www.mayoclinic.org/tests-procedures/hormone-therapy-for-breast-cancer/about/pac-20384943

NASA (National Aeronautics and Space Administration). 2017. "Graphing Global Temperature Trends," updated September 6, 2017, accessed February 6, 2021. https://www.nasa.gov/mission_pages/noaa-n/climate/climate_weather.html/.

———. "What Is the Sun's Role in Climate Change?" *Ask NASA Climate*, September 6, 2019, https://climate.nasa.gov/blog/2910/what-is-the-suns-role-in-climate-change.

NCI (National Cancer Institute). "BRCA Mutations: Cancer Risk and Genetic Testing." Cancer Causes and Preventions, Genetics, January 30, 2018, https://www.cancer.gov/about-cancer/causes-prevention/genetics/brca-fact-sheet.

NIH (National Institute of Health). *"Autism Spectrum Disorder."* Mental Health Information, March 2018, https://www.nimh.nih.gov/health/topics/autism-spectrum-disorders-asd/index.shtml.

———. *"Vitamin B12."* Health Information, updated March 30, 2020, https://ods.od.nih.gov/factsheets/VitaminB12-HealthProfessional.

Oreskes, Naomi. "The Scientific Consensus on Climate Change." *Science* 306, no. 5702 (2004): 1686.

Oreskes, Naomi, and Erik M. Conway. *Merchants of Doubt: How a Handful of Scientists Obscured the Truth on Issues From Tobacco Smoke to Global Warming.* New York: Bloomsbury Publishing, 2010.

PBS (Public Broadcasting Service). "American Experience: The Poison Squad. Dr. Havey Wiley: Father of Pure Food." PBS, Arlington (video), January 16, 2020.

Powell, James Lawrence. 2015. "Climate Scientists Virtually Unanimous: Anthropogenic Global Warming Is True." *Bulletin of Science, Technology & Society* 35, no. 5–6 (2015): 121–24.

Rylaarsdam, Lauren Elizabeth, and Alcia Guemez-Gamboa. "Genetic Causes and Modifiers in Autism Spectrum Disorder." *Frontiers in Cellular Neuroscience* 13 (2019): 385.

Sacks, Frank M., Alice H. Lichtenstein, Jason H. Y. Wu, Lawrence J. Appel, Mark A. Creager, Penny M. Kris-Etherton, Michael Miller et al. "Dietary Fats and Cardiovascular Disease: A Presidential Advisory from the American Heart Association." *Circulation* 136, no. 3 (2017): e1–e23.

Satterstrom, F. Kyle, Jack A. Kosmicki, Jiebiao Wang, Michael S. Breen, Silvia De Rubeis, Joon-Yong An, Minshi Peng et al. "Large-Scale Exome Sequencing Study Implicates Both Developmental and Functional Changes in the Neurobiology of Autism." *Cell* 180, no. 3 (2020): 564–84.

Schroder, Tim. "The Protein Puzzle." *Max Plank Research* 3 (2017): 54–59, https://www.mpg.de/11447687/W003_Biology_medicine_054-059.pdf.

US Food and Drug Administration. "Tainted Products Marketed as Dietary Supplements_CDER." FDA Home, December 19, 2019, https://www.accessdata.fda.gov/scripts/sda/sdNavigation.cfm?sd=tainted_supplements_cder.

Verkaik-Kloosterman, Janneke, S. Marije Seves, and Marga C. Ocké. "Vitamin D Concentrations in Fortified Foods and Dietary Supplements Intended for Infants: Implications for Vitamin D Intake." *Food Chemistry* 221 (2017): 629–35.

Vosoughi, Soroush, Deb Roy, and Sinan Aral. "The Spread of True and False News Online." *Science* 359, no. 6380 (2018): 1146–51.

2. Chemistry for Life—Essentials

Ali, Saba M., and Gil Yosipovitch. "Skin pH: From Basic Science to Basic Skin Care." *Acta Dermato-Venereologica* 93, no. 3 (2013): 261–69.

Baranda, Lourdes, Roberto González-Amaro, Bertha Torres-Alvarez, Carmen Alvarez, Victoria Ramírez. Correlation between pH and Irritant Effect of Cleansers Marketed for Dry Skin. *International Journal of Dermatology* 41, no. 8 (2002): 494–99.

Barrett, Stephen. "A Critical Look at 'Dr.' Robert Young's Theories and Credentials," Quackwatch. July 3, 2017, https://www.quackwatch.org/11Ind/young3.html.

CSPI (Center for Science in the Public Interest). "Alkaline Water—Find or Fraud?" January 16, 2018, https://cspinet.org/tip/alkaline-water—find-or-fraud.

Gorski, David. "More, Please! A Victim of Cancer Quack Robert O. Young Wins a $105 Million Settlement." *Science-Based Medicine,* November 5, 2018, https://sciencebasedmedicine.org/more-please-a-victim-of-cancer-quack-robert-o-young-wins-a-105-million-settlement.

Reddy, Avanija, Don F. Norris, Stephanie S. Momeni, Belinda Waldo, and John D. Ruby. "The pH of Beverages in the United States." *Journal of the American Dental Association* 147, no. 4 (2016): 255–63.

Rubin, Courtney. "All of Those Products Are Making Your Skin Worse." *New York Times,* July 30, 2019.

Villines, Zawn. "Does the Alkaline Diet Work?" *Medical News Today,* January 24, 2019, https://www.medicalnewstoday.com/articles/324271.php.

3. Chemistry for Life—Carbon-Based Molecules

CDC (Centers for Disease Control and Prevention). "Defining Adult Overweight and Obesity." Overweight and Obesity, updated April 3, 2020, accessed November 19, 2019, https://www.cdc.gov/obesity/adult/defining.html.

———. "National Diabetes Statistics Report, 2017: Estimates of Diabetes and Its Burden in the United States." https://www.cdc.gov/diabetes/pdfs/data/statistics/national-diabetes-statistics-report.pdf.

———. "Obesity and Overweight." June 13, 2016. https://www.cdc.gov/nchs/fastats/obesity-overweight.htm.

Desilver, Drew. "What's on Your Table? How America's Diet Has Changed over the Decades." *FactTank,* Pew Research Center, December 13, 2016, https://www.pewresearch.org/fact-tank/2016/12/13/whats-on-your-table-how-americas-diet-has-changed-over-the-decades.

Dow, Caitlin. "Why Extra Weight—Not Just Obesity—Matters." *Nutrition Action Newsletter,* Center for Science in the Public Interest, July 16, 2018.

Fothergill, Erin, Juen Guo, Lilian Howard, Jennifer C. Kerns, Nicolas D. Knuth, Robert Brychta, Kong Y. Chen et al. "Persistent Metabolic Adaptation 6 Years after *The Biggest Loser* Competition." *Obesity* 24, no. 8 (2016): 1612–19.

Harvard Health. "Why the FDA Banned Ephedra." March 2004.

Kuddus, Mohammed. "Potential Applications of Microbial Chitinase: Recent Development." *Biochemical and Cellular Archives* 14, no. 1 (2014): 1–7.

Luzzatto, Lucio. "Sickle Cell Anaemia and Malaria." *Mediterranean Journal of Hematology and Infectious Diseases* 4, no. 1 (2012): https://www.ncbi.nlm.nih.gov/pmc/articles/PMC3499995/pdf/mjhid-4-1-e2012065.pdf.

Mayo Clinic. "Sickle Cell Anemia," updated January 30, 2020, accessed Nov. 19, 2019, https://www.mayoclinic.org/diseases-conditions/sickle-cell-anemia/symptoms-causes/syc-20355876.

NIH (National Institutes of Health). "Barth Syndrome." Genetic and Rare Diseases Information Center. Accessed November 18, 2019, https://rarediseases.info.nih.gov/diseases/5890/barth-syndrome.

______. "Barth Syndrome." Genetics Home Reference. Accessed November 18, 2019, https://ghr.nlm.nih.gov/condition/barth-syndrome.

______. "HBA1 Gene." Genetics Home Reference. Accessed November 19, 2019, https://ghr.nlm.nih.gov/gene/HBA1.

______. "HBB Gene." Genetics Home Reference. Accessed November 19, 2019, https://ghr.nlm.nih.gov/gene/HBB.

______. "Body Mass Index Table." Accessed November 19, 2020, https://www.nhlbi.nih.gov/health/educational/lose_wt/BMI/bmi_tbl.htm

Oyeleye, Ayokunmi, and Yahaya M. Normi. "Chitinase: Diversity, Limitations, and Trends in Engineering for Suitable Applications." *Bioscience Reports* 38, no. 4 (2018): BSR2018032300.

Pew Research Center. "The New Food Fights: US Public Divides over Food Science." Science and Society, December 1, 2016, https://www.pewresearch.org/science/2016/12/01/the-new-food-fights.

Speakman, John R., and Colin Selman. "Physical Activity and Resting Metabolic Rate." *Proceedings of the Nutrition Society* 62, no. 3 (2003): 621–34.

Trexler, Eric T., Abbie E. Smith-Ryan, and Layne E. Norton. "Metabolic Adaptation to Weight Loss: Implications for the Athlete." *Journal of the International Society of Sports Nutrition* 11, no. 1 (2014): 7.

USDA (United States Department of Agriculture). "Food Availability (per Capita) Data System." Economic Research Service. Updated January 9, 2020, accessed November 19, 2019, https://www.ers.usda.gov/data-products/food-availability-per-capita-data-system.

Vichinsky, Elliott. "Sickle Cell Disease and Thalassemia: Disorders of Globin Production." *50 Years of Hematology. Research That Revolutionized Patient Care*, 3–5. American Society of Hematology, 2008, https://www.hematology.org/About/History/50-Years/1534.aspx.

4. Understanding the Importance of Cells

AAAS (American Association for the Advancement of Science). "The Cells in Your Body," *ScienceNetLinks*, accessed November 21, 2019, http://sciencenetlinks.com/student-teacher-sheets/cells-your-body.

Bianconi, Eva, Allison Piovesan, Federica Facchin, Alina Beraudi, Raffaella Casadei, Flavia Frabetti, Lorenza Vitale et al. "An Estimation of the Number of Cells in the Human Body," *Annals of Human Biology* 40, no. 6 (2013): 463–71.

Census of Marine Life. "How Many Species on Earth? About 8.7 Million, New Estimate Says," *Science News*, August 24, 2011, https://www.sciencedaily.com/releases/2011/08/110823180459.htm.

Guzman, Flavio. "The Mechanism of Action of SSRIs," Psychopharmacology Institute (video lecture). Updated April 6, 2020, accessed November 23, 2019, https://psychopharmacologyinstitute.com/section/the-mechanism-of-action-of-ssris-2051-4050.

Locey, Kenneth J., and Jay T. Lennon. "Scaling Laws Predict Global Microbial Diversity." *Proceedings of the National Academy of Sciences* 113, no. 21 (2016): 5970–75.

NIMH (National Institutes of Mental Health). 2018. "Depression," *Mental Health Information*. Accessed November 23, 2019. https://www.nimh.nih.gov/health/topics/depression/index.shtml.

NIMH (National Institutes of Mental Health). 2019. "Major Depression: Prevalence of Major Depressive Episode Among Adults," *Mental Health Information*. Accessed November 23, 2019. https://www.nimh.nih.gov/health/statistics/major-depression.shtml.

Stork, Nigel E. "How Many Species of Insects and Other Terrestrial Arthropods Are There on Earth?" *Annual Review of Entomology* 63 (January 2018): 31–45, https://doi.org/10.1146/annurev-ento-020117-043348.

UNOS (United Network for Organ Sharing). "Organ Transplants in United States Set Sixth Consecutive Record in 2018." *UNOS News*, January 8, 2019, https://unos.org/news/organ-transplants-in-united-states-set-sixth-consecutive-record-in-2018.

5. Inside the Plasma Membrane

Chial, Heidi, and Joanna Craig. "mtDNA and Mitochondrial Diseases." *Nature Education* 1, no. 1 (2008.): 217, https://www.nature.com/scitable/topicpage/mtdna-and-mitochondrial-diseases-903.

Falkenberg, Maria. "Mitochondrial DNA Replication in Mammalian Cells: Overview of the Pathway." *Essays in Biochemistry* 62, no. 3 (2018): 287–96.

NIH (National Institutes of Health). "Mitochondrial DNA." National Library of Medicine. National Institutes of Health. Updated June 23, 2020, accessed August 16, 2019, https://ghr.nlm.nih.gov/mitochondrial-dna.

Sylvester, James E., Nathan Fischel-Ghodsian, Edward B. Mougey, and Thomas W. O'Brien. "Mitochondrial Ribosomal Proteins: Candidate Genes for Mitochondrial Disease." *Genetics in Medicine* 6, no. 2 (2004): 73.

6. Channeling Energy: Photosynthesis

Bright, Monika, Julia Klose, and Andrea D. Nussbaumer. "Giant Tubeworms." *Current Biology* 23, no. 6 (2013): R224–25.

Reich, Peter B., Kerrie M. Sendall, Artur Stefanski, Roy L. Rich, Sarah E. Hobbie, and Rebecca A. Montgomery. "Effects of Climate Warming on Photosynthesis in Boreal Tree Species Depend on Soil Moisture." *Nature* 562, no. 7726 (2018): 263–67.

Schlesinger, William H., Emily S. Bernhardt, E. H. DeLucia, David S. Ellsworth, A. C. Finzi, George R. Hendrey, Kirsten S. Hofmockel et al. "The Duke Forest FACE Experiment: CO2 Enrichment of a Loblolly Pine Forest." In *Managed Ecosystems and CO2: Case Studies, Processes, and Perspectives, edited by J. Nösberger et al.* (Berlin: Springer, 2006), 197–212.

Warren, Jeffrey M., Anna M. Jensen, Belinda E. Medlyn, Richard J. Norby, and David T. Tissue. "Carbon Dioxide Stimulation of Photosynthesis in *Liquidambar styraciflua* Is Not Sustained during a 12-Year Field Experiment." *AoB Plants* 7 (2015).

7. Metabolism of the Cell—Using Energy Constructively

Allen, David Grant, and Sofie Trajanovska. "The Multiple Roles of Phosphate in Muscle Fatigue." *Frontiers in Physiology* 3 (2012): 463.

Baggish, Aaron L., Rory B. Weiner, Gen Kanayama, James I. Hudson, Michael T. Lu, Udo Hoffmann, and Harrison G. Pope, Jr. "Cardiovascular Toxicity of Illicit Anabolic-Androgenic Steroid Use." *Circulation* 135, no. 21 (2017): 1991–2002.

Brewers Association. "National Beer Sales & Production Data." Stats and Data, accessed December 5, 2019, https://www.brewersassociation.org/statistics-and-data/national-beer-stats.

Costa, Daniel P. "Diving Physiology of Marine Vertebrates." In *eLS* (*New York: John Wiley & Sons, 2001*), https://doi.org/10.1002/9780470015902.a0004230.

Costa, Daniel P. and Barry Sinervo. "Field Physiology: Physiological Insights from Animals in Nature." *Annual Review of Physiology* 66, no. 1 (2004): 209–28.

Gallone, Brigida, Jan Steensels, Troels Prahl, Leah Soriaga, Veerle Saels, Beatriz Herrera-Malaver, Adriaan Merlevede et al. "Domestication and Divergence of Saccharomyces Cerevisiae Beer Yeasts." *Cell* 166, no. 6 (2016): 1397–410.

Hindell, Mark A., David J. Slip, Harry R. Burton, and Michael M. Bryden. "Physiological Implications of Continuous, Prolonged, and Deep Dives of the Southern Elephant Seal (*Mirounga leonina*)." *Canadian Journal of Zoology* 70, no. 2 (1992): 370–79.

Hooker, Sascha K., Andreas Fahlman, Michael J. Moore, Natacha Aguilar De Soto, Yara Bernaldo de Quiros, Alf O. Brubakk, Daniel P. Costa et al. "Deadly Diving? Physiological and Behavioural Management of Decompression Stress in Diving Mammals." *Proceedings of the Royal Society B: Biological Sciences* 269, no. 1729 (2011): 1041–50.

Kingma, Boris, and Wouter van Marken Lichtenbelt. "Energy Consumption in Buildings and Female Thermal Demand." *Nature Climate Change* 5, no. 12 (2015): 1054.

Mirceta, Scott, Anthony V. Signore, Jennifer M. Burns, Andrew R. Cossins, Kevin L. Campbell, and Michael Berenbrink. "Evolution of Mammalian Diving Capacity Traced by Myoglobin Net Surface Charge." *Science* 340, no. 6138 (2013): 1234192.

NIH (National Institutes of Health). "Anabolic Steroids," US National Library of Medicine, MedLine Plus, accessed December 1, 2019, https://medlineplus.gov/anabolicsteroids.html.

Pavid, Katie. "Secrets of the Deepest-Diving Whales." Oceans, Natural History Museum, accessed August 13, 2019, https://www.nhm.ac.uk/discover/secrets-of-deep-diving-whales.html.

Starrs, Jenny. "Pushing the Limits: After the Deepest of Breaths, Free Diver Plunges 400 Feet Underwater to Set Two World Records in Three Days." *Washington Post*. May 3, 2016, https://www.washingtonpost.com/news/morning-mix/wp/2016/05/03/free-diver-plunges-400-feet-underwater-to-set-two-world-records-in-three-days.

Westerblad, Håkan, David G. Allen, and Jan Lännergren. "Muscle Fatigue: Lactic Acid or Inorganic Phosphate the Major Cause?" *Physiology* 17, no. 1 (2002): 17–21.

8. Hosting Unwelcome Guests—Noncellular Entities

CDC (Centers for Disease Control and Prevention). "Update: Multistate Outbreak of Monkeypox—Illinois, Indiana, Kansas, Missouri, Ohio, and Wisconsin, 2003." *Morbidity and Mortality Weekly Report* 52, no. 27 2003): 642–46, http://www.cdc.gov/mmwr/preview/mmwrhtml/mm5227a5.htm.

Heymann, David L. "How SARS Was Contained." *The New York Times*, March 14, 2013.

Li, Wenhui, Swee-Kee Wong, Fang Li, Jens H. Kuhn, I-Chueh Huang, Hyeryun Choe, and Michael Farzan. "Animal Origins of the Severe Acute Respiratory Syndrome Coronavirus: Insight from ACE2-S-Protein Interactions." *Journal of Virology* 80, no. 9 (2006): 4211–49, http://jvi.asm.org/content/80/9/4211.full.

Mydans, Seth. "The SARS Epidemic: Containment; How Vietnam Halted SARS and Saved the Life of a Nurse." *The New York Times*, May 7, 2003.

Reilley, Brigg, Michel Van Herp, Dan Sermand, and Nicoletta Dentico. "SARS and Carlo Urbani." *New England Journal of Medicine* 348, no. 20 (2003): 1951–52.

UCMP (University of California Museum of Paleontology). "Tracking SARS Back to Its Source." Understanding Evolution, January 2006, updated July 2013, http://evolution.berkeley.edu/evolibrary/news/060101_batsars.

9. Cell Reproduction and Genetics

Dobnell, Clifford. Anthony van Leeuwenhoek and His Little Animals, p. 113. New York: Russell and Russell.

IUPUI (Indiana University–Purdue University Indianapolis). "Human Chromosomal Disorders." Department of Biology, April 30, 2003, https://nanopdf.com/download/human-chromosomal-disorders_pdf.

Jahovic, Nermina, Esra Güzel, Serap Arbak, and Berrak Ç. Yeğen. "The Healing-Promoting Effect of Saliva on Skin Burn Is Mediated by Epidermal Growth Factor (EGF): Role of the Neutrophils." *Burns* 30, no. 6 (2004): 531–38, https://www.ncbi.nlm.nih.gov/pubmed/15302417.

National Geographic Society. "History of the Cell: Discovering the Cell." May 23, 2019, https://www.nationalgeographic.org/article/history-cell-discovering-cell.

O'Connor, Clare, and Ilona Miko. "Developing the Chromosome Theory." *Nature Education* 1, no. 1 (2008): 44, https://www.nature.com/scitable/topicpage/developing-the-chromosome-theory-164.

PBS. "Human Chromosome 2." Evolution (video), Clear Blue Sky Productions, 2007. https://www.pbs.org/wgbh/evolution/library/07/3/l_073_47.html.

Rejón, Manuel Ruiz. "The Origin of the Human Species: A Chromosome Fusion?" OpenMindBBVA, January 17, 2017, https://www.bbvaopenmind.com/en/science/bioscience/the-origin-of-the-human-species-a-chromosome-fusion.

Schultz, Myron. "Rudolf Virchow." *Emerging Infectious Diseases* 14, no. 9 (2008): 1480, https://www.ncbi.nlm.nih.gov/pmc/articles/PMC2603088.

Theoi Project. "Asklepios." Accessed May 8, 2020, https://www.theoi.com/Ouranios/Asklepios.html.

Woodhouse, Margaret, Diana Burkart-Waco, and Luca Comai. "Polyploidy." *Nature Education* 2, no. 1 (2009): 1, https://www.nature.com/scitable/topicpage/polyploidy-1552814.

10. Fundamental Genetics

American Red Cross. "Facts about Blood and Blood Types." Accessed May 7, 2020, https://www.redcrossblood.org/donate-blood/blood-types.html.

Baranski. "Nikolai Ivanovic Vavilov (1887–1943)." The Embryo Project Encyclopedia, April 15, 2014, https://embryo.asu.edu/pages/nikolai-ivanovic-vavilov-1887-1943.

Biography.com Editors. "Gregor Mendel." Updated June 5, 2020, accessed May 7, 2020, https://www.biography.com/scientist/gregor-mendel.

Children's Wisconsin. "Multifactorial Inheritance." Accessed May 7, 2020, https://chw.org/medical-care/genetics-and-genomics-program/medical-genetics/multifactorial-inheritance.

Digg. "The Most Common Blood Types in the US, Visualized." September 4, 2019, https://digg.com/2019/most-common-blood-types-in-the-us.

Fossella, J. A. "Incomplete Dominance." *Encyclopedia of Genetics* (2001): 1016–17, https://www.sciencedirect.com/topics/agricultural-and-biological-sciences/incomplete-dominance.

Griffiths AJF, Miller JH, Suzuki DT, et al. *An Introduction to Genetic Analysis*. 7th edition. New York: W. H. Freeman; 2000. Penetrance and expressivity. https://www.ncbi.nlm.nih.gov/books/NBK22090/

JoVE. "Polygenic Traits." Classical and Modern Genetics, accessed May 7, 2020, https://www.jove.com/science-education/10778/polygenic-traits.

LeMieux, Julianna. "Calico Cats Are a Walking Genetics Lesson." American Council of Science and Health, July 27, 2016, https://www.acsh.org/news/2016/07/27/calico-cats-are-a-walking-genetics-lesson.

Mayo Clinic. "Hemophilia." Diseases and Conditions, accessed May 7, 2020, https://www.mayoclinic.org/diseases-conditions/hemophilia/symptoms-causes/syc-20373327.

______. "Phenylketonuria (PKU)." Diseases and Conditions, accessed May 7, 2020, https://www.mayoclinic.org/diseases-conditions/phenylketonuria/symptoms-causes/syc-20376302.

Miko, Ilona. "Epistasis: Gene Interaction and Phenotype Effects," *Nature Education* 1, no. 1 (2008): 197, https://www.nature.com/scitable/topicpage/epistasis-gene-interaction-and-phenotype-effects-460.

Native Seeds. "Teosinte," accessed May 7, 2020, https://www.nativeseeds.org/pages/teosinte.

NCBI (National Center for Biotechnology Information). OMIM, accessed May 7, 2020, https://www.ncbi.nlm.nih.gov/omim.

NIH (National Institutes of Health). "X Chromosome." Genetics Home Reference, accessed May 7, 2020, https://ghr.nlm.nih.gov/chromosome/X.

______. "Y Chromosome." Genetics Home Reference, accessed May 7, 2020, https://ghr.nlm.nih.gov/chromosome/Y.

Schulz, Laura C. "The Dutch Hunger Winter and the Developmental Origins of Health and Disease." *Proceedings of the National Academy of Sciences* 107, no. 39 (2010): 16757–58.

Stearns, Frank W. "One Hundred Years of Pleiotrophy: A Retrospective." *Genetics* 186, no. 3 (2010): 767–73, https://www.genetics.org/content/186/3/767.

WIE (What Is Epigenetics). "A Super Brief and Basic Explanation of Epigenetics for Total Beginners." September 1, 2019, https://www.whatisepigenetics.com/what-is-epigenetics.

Zimmer, Carl. "Dutch Genes Still Bear Scars." *New York Times*, January 31, 2018, https://www.nytimes.com/2018/01/31/science/dutch-famine-genes.html

11. Life's Instructions—It's All about DNA

Cobb, Matthew. "What Is Life? The Physicist Who Sparked a Revolution in Biology." *The Guardian*, February 7, 2013, https://www.theguardian.com/science/blog/2013/feb/07/wonders-life-physicist-revolution-biology.

DNA Learning Center. "Friedrich Miescher." DNA From the Beginning, 2011, http://www.dnaftb.org/15/bio.html.

Doessing, Simon, Katja M. Heinemeier, Lars Holm, Abigail L. Mackey, Peter Schjerling, Michael Rennie, Kenneth Smith et al. "Growth Hormone Stimulates the Collagen Synthesis in Human Tendon and Skeletal Muscle without Affecting Myofibrillar Protein Synthesis." *The Journal of Physiology* 588, no. 2 (2010): 341–51.

Greenwood, Michael. "What Are Introns and Exons?" *News Medical Life Sciences*, November 2, 2018, https://www.news-medical.net/life-sciences/What-are-introns-and-exons.aspx.

Lee, Kwanghyuk, Saadullah Khan, Muhammad Ansar, Regie Lyn P. Santos-Cortez, Wasim Ahmad, and Suzanne M. Leal. "A Novel ESRRB Deletion Is a Rare Cause of Autosomal Recessive Nonsyndromic Hearing Impairment among Pakistani Families." *Genetics Research International* (2011).

NIH (National Health Institutes). "The Human Genome Project FAQ." Human Genome Research, February 24, 2020, https://www.genome.gov/human-genome-project/Completion-FAQ.

Pray, Leslie A. "Discovery of DNA Structure and Function: Watson and Crick." *Nature Education* 1, no. 1 (2008): 100, https://www.nature.com/scitable/topicpage/discovery-of-dna-structure-and-function-watson-397.

Sagan, Carl. *Cosmos*. New York: Random House, Inc., 1980.

Science Learning Hub. "Role of Proteins in the Body." June 10, 2011, https://www.sciencelearn.org.nz/resources/209-role-of-proteins-in-the-body.

Scitable. "Protein Function." 2014, accessed May 7, 2020, https://www.nature.com/scitable/topicpage/protein-function-14123348.

12. Biotechnology—Transforming the Future

Azad, Abul Kalam, Latifah Amin, and Nik Marzuki Sidik. "Gene Technology for Papaya Ringspot Virus Disease Management." *The Scientific World Journal* 2014 (2014), https://www.hindawi.com/journals/tswj/2014/768038.

Baeshen, Nabih A., Mohammed N. Baeshen, Abdullah Sheikh, Roop S. Bora, Mohamed Morsi M. Ahmed, Hassan AI Ramadan, Kulvinder Singh Saini et al. "Cell Factories for Insulin Production." *Microbial Cell Factories* 13, no. 1 (2014): 141, https://www.ncbi.nlm.nih.gov/pmc/articles/PMC4203937.

Bier, Ethan, and Elliot Sober, "Gene Editing and the War against Malaria." *American Scientist* 108, no. 3 (May-June 2020): 162–69, https://www.americanscientist.org/article/gene-editing-and-the-war-against-malaria.

Genetics Generation. "Transgenic Organisms." Accessed May 9, 2020, https://knowgenetics.org/transgenic-organisms.

Mukherjee, Siddhartha. *The Gene: An Intimate History*. (New York: Scribner, 2016): 428–36.

Murnaghan, Ian. "Therapeutic Cloning," February 10, 2020, http://www.explorestemcells.co.uk/therapeuticcloning.html.

NAS (National Academy Science). "A Brief History of Malaria." In *Saving Lives, Buying Time: Economics of Malaria Drugs in an Age of Resistance*, edited by Kenneth J. Arrow, Claire Panosian, and Helen Gelband, (Washington, DC :National Academies Press, 2004), https://www.ncbi.nlm.nih.gov/books/NBK215638.

NIH (National Institutes of Health). "Cloning Fact Sheet." National Human Genome Research Institute, March 21, 2017, https://www.genome.gov/about-genomics/fact-sheets/Cloning-Fact-Sheet.

______. "DNA Sequencing Fact Sheet." 2020. National Human Genome Research Institute, accessed February 6, 2021, https://www.genome.gov/about-genomics/fact-sheets/DNA-Sequencing-Fact-Sheet.

______. "FOXP2 Gene." Genetics Home Reference, accessed May 9, 2020, https://ghr.nlm.nih.gov/gene/FOXP2#resourcesfox.

______. "How Does Gene Therapy Work?" Genetics Home Reference, accessed May 9, 2020, https://ghr.nlm.nih.gov/primer/therapy/procedures.

______. "NHGRI History and Timeline of Events." National Human Genome Research Institute, accessed May 9, 2020, https://www.genome.gov/about-nhgri/Brief-History-Timeline.

______. "Polymerase Chain Reaction (PCR) Fact Sheet." National Human Genome Research Institute, June 16, 2015, https://www.genome.gov/about-genomics/fact-sheets/Polymerase-Chain-Reaction-Fact-Sheet.

______. "SSH Gene." Genetics Home Reference, accessed May 9, 2020, https://ghr.nlm.nih.gov/gene/SHH.

Vidyasagar, Aparna. "What Is CRISPR?" *Live Science*, April 21, 2018, https://www.livescience.com/58790-crispr-explained.html.

WHO. "World Malaria Report 2019." December 4, 2019, https://www.who.int/news-room/feature-stories/detail/world-malaria-report-2019.

13. Evolution of Evolutionary Thought

AMNH (American Museum of Natural History). "*A Trip around the World*." Darwin Exhibition, accessed December 29, 2019, https://www.amnh.org/exhibitions/darwin/a-trip-around-the-world.

BBC (British Broadcasting Coporation). "Bird Seduction Techniques," Life Story, September 10, 2015, https://www.youtube.com/watch?v=_H9TyXiXM2k.

Beccaloni, George, and Charles Smith. "*Biography of Wallace*." The Alfred Russel Wallace Website, January 2015, http://wallacefund.info/content/biography-wallace.

Bittel, Jason. "Epic Eel Migration Mapped for the First Time." *National Geographic*, October 27, 2015, https://www.nationalgeographic.com/news/2015/10/151027-american-eel-migration-animal-behavior-oceans-science.

Brent, Peter. 1981. *Charles Darwin: A Man of Enlarged Curiosity*. Harper and Rowe Publishers, New York.

Darwin, Charles. *The Origin of the Species. London:* John Murray, Albemarle Street, 1859.

______. *The Autobiography of Charles Darwin. London: Collins, 1958.*

______. 2005. *The Darwin Compendium*. Barnes and Noble Publishing. USA.

Dobzhansky, Theodosius. "Nothing in Biology Makes Sense Except in Light of Evolution." *The American Biology Teacher* 35, no. 3 (1973): 125-129.

Dugatkin, Lee, and Lyudmila Trut. "How to Tame a Fox and Build a Dog." *American Scientist*, accessed December 29, 2019, https://www.americanscientist.org/article/how-to-tame-a-fox-and-build-a-dog.

Goldman, Jason G. "Man's New Best Friend? A Forgotten Russian Experiment in Fox Domestication." *Scientific American*, September 6, 2010, https://blogs.scientificamerican.com/guest-blog/mans-new-best-friend-a-forgotten-russian-experiment-in-fox-domestication.

———. "What DNA from Pet Foxes Teaches Us about Dogs—and Humans." *National Geographic*, August 6, 2018, https://www.nationalgeographic.com/animals/2018/08/fox-dogs-wild-tame-genetics-study-news.

Huxley, Thomas Henry. *On the Reception of the 'Origin of Species'*. Good Press, 2019.

Jones, Lucy. "A Soviet Scientist Created the Only Tame Foxes in the World." BBC, September 13, 2016, http://www.bbc.com/earth/story/20160912-a-soviet-scientist-created-the-only-tame-foxes-in-the-world.

McCarthy, Eugene M. "Jean-Baptiste Lamarck." Macroevolution.net, accessed December 29, 2019, http://www.macroevolution.net/jean-baptiste-lamarck.html.

———. "Alfred Russel Wallace." Macroevolution.net, accessed December 29, 2019, http://www.macroevolution.net/alfred-russel-wallace.html.

National Geographic. "Alfred Wallace." Resource Library, accessed December 29, 2019, https://www.nationalgeographic.org/encyclopedia/alfred-wallace.

———. "Sep 15, 1835 CE: Darwin Explores Galapagos Islands." Accessed December 29, 2019, https://blog.education.nationalgeographic.org/2016/09/11/this-week-in-geographic-history-september-12-18/.

PBS (Public Broadcasting Service. "Jean Baptiste Lamarck." Evolution Library, accessed December 29, 2019, https://www.pbs.org/wgbh/evolution/library/02/3/l_023_01.html.

Russ, Michael. 1998. *Taking Darwin Seriously.* Prometheus Books, New York

Smithsonian Institution. "Evolution of Whales Animation." Accessed December 17, 2019, https://ocean.si.edu/through-time/ancient-seas/evolution-whales-animation.

Stark, Louisa A. 2009. *Darwin on the Web: Resources for Darwin 200 and Beyond*. CBE Life Sci Educ. 2009 Spring; 8(1): 1–6. doi: 10.1187/cbe.08-12-0078. Accessed December 29, 2019. https://www.ncbi.nlm.nih.gov/pmc/articles/PMC2649646/.

Switek, Brian. "How Did Whales Evolve?" *Smithsonian Magazine*, December 1, 2010, https://www.smithsonianmag.com/science-nature/how-did-whales-evolve-73276956.

Thewissen, J. G. M. *The Walking Whales: From Land to Water in Eight Million Years*. Oakland: University of California Press, 2014.

Thewissen, J. G. M., Lisa Noelle Cooper, John C. George, and Sunil Bajpai. "Land to Water: The Origin of Whales, Dolphins, and Porpoises." *Evolution: Education and Outreach* 2, no. 2 (2009): 272–88.

Thomson, Keith. *The Young Charles Darwin*. New Haven, CT: Yale University Press, 2009.

UCMP (University of California Museum of Paleontology). "Jean-Baptiste Lamarck (1744-1829)." University of California, Berkeley, accessed December 29, 2019, https://ucmp.berkeley.edu/history/lamarck.html.

———. "The Evolution of Whales," Understanding Evolution, accessed December 29, 2019, https://evolution.berkeley.edu/evolibrary/article/evograms_03.

USFW (US Fish & Wildlife Service). "American Eel." Accessed December 29, 2019. https://www.fws.gov/fisheries/fishmigration/american_eel.html.

Van White, John. *The Darwin Experience: The Story of the Man and His Theory of Evolution*. Washington, DC: National Geographic, 2008.

Wyhe, John van, ed. Darwin Online. Accessed December 29, 2019, http://darwin-online.org.uk.

14. Evidence for Evolution

AAAS (American Association for the Advancement of Science). "Mutant Chickens Grow Teeth," *Science*, February 21, 2006, https://www.sciencemag.org/news/2006/02/mutant-chickens-grow-teeth.

Bartfal, David, and Gunit Kang. "The Life History of Archaeopteryx lithographica," Tree of Life Project, accessed January 6, 2020, http://tolweb.org/treehouses/?treehouse_id=5136.

Beans, Laura. "10 Success Stories Thanks to the Endangered Species Act," EcoWatch, December 7, 2013, https://www.ecowatch.com/10-success-stories-thanks-to-the-endangered-species-act-1881837279.html.

Byrnes, Laurel. "Ancient Myths Inspired by Fossils." Biodiversity Heritage Library, October 13, 2015. https://blog.biodiversitylibrary.org/2015/10/ancient-myths-inspired-by-fossils.html.

Center for Biological Diversity. "The Endangered Species Act: A Wild Success." Accessed January 6, 2020, https://www.biologicaldiversity.org/campaigns/esa_wild_success.

______. "The Extinction Crisis." Accessed January 6, 2020, https://www.biologicaldiversity.org/programs/biodiversity/elements_of_biodiversity/extinction_crisis.

Chu, Jennifer. "End-Permian Extinction, Which Wiped Out Most of Earth's Species, Was Instantaneous in Geologic Time," *Phys Org*, September 19, 2018, https://phys.org/news/2018-09-end-permian-extinction-earth-species-instantaneous.html.

Cowen, Richard. "The K-T Extinction." UCMP Berkley, 1999, accessed January 6, 2020, https://ucmp.berkeley.edu/education/events/cowen1b.html.

E. O. Wilson Biodiversity Foundation. "Humans Are Speeding Extinction and Altering the Natural World at an 'Unprecedented' Pace." May 7, 2019, https://eowilsonfoundation.org/humans-are-speeding-extinction-and-altering-the-natural-world-at-an-unprecedented-pace.

Everhart, Mike. "Marine Birds with Teeth in the Late Cretaceous Seas." Updated January 19, 2014, accessed January 11, 2020, http://oceansofkansas.com/Hesperornis.html.

Greshko, Michael, and National Geographic Staff. "What Are Mass Extinctions and What Causes Them?" National Geographic Society, September 26, 2019, https://www.nationalgeographic.com/science/prehistoric-world/mass-extinction.

Hendry, Lisa. "Who Were the Neanderthals?" Natural History Museum, London. May 5, 2018, https://www.nhm.ac.uk/discover/who-were-the-neanderthals.html.

Hickey, Hannah. "What Caused Earth's Biggest Mass Extinction?" Stanford Earth, December 6, 2018, https://earth.stanford.edu/news/what-caused-earths-biggest-mass-extinction#gs.o4m0mj.

Hoffman, Hillel. "The Permian Extinction—When Life Nearly Came to an End." National Geographic, accessed January 6, 2020, https://www.nationalgeographic.com/science/prehistoric-world/permian-extinction.

Longitude Prize. "10 Most Dangerous Antibiotic-Resistant Bacteria," August 26, 2014, https://longitudeprize.org/blog-post/10-most-dangerous-antibiotic-resistant-bacteria.

Mayo Clinic. "Staph Infections." Diseases and Conditions, accessed January 6, 2020, https://www.mayoclinic.org/diseases-conditions/staph-infections/symptoms-causes/syc-20356221.

Mayor, Adrienne. *The First Fossil Hunters: Dinosaurs, Mammoths, and Myth in Greek and Roman Times.* Princeton, NJ: Princeton University Press, 2011.

National Geographic. "The Echoes of Toothed Birds," Science, May 22, 2015, https://www.nationalgeographic.com/science/phenomena/2015/05/22/the-echoes-of-toothed-birds.

———. "Extinction." Resource Library, accessed January 6, 2020, https://www.nationalgeographic.org/encyclopedia/extinction.

Palumbi, Stephen R. "What Can We Do to Reduce Antibiotic Resistance?" *Survival*, accessed January 6, 2020, https://www.pbs.org/wgbh/evolution/survival/enemy/statement_04.html.

Peppe, Daniel J. and Alan L. Deino. "Dating Rocks and Fossils Using Geologic Methods." *The Nature Education Knowledge* 4, no. 10 (2013): 1, https://www.nature.com/scitable/knowledge/library/dating-rocks-and-fossils-using-geologic-methods-107924044.

Pickrell, John. "How Chinese Fossils Are Rewriting the History of Feathered Dinosaurs," *Cosmos: The Science of Everything*, May 29, 2017, https://cosmosmagazine.com/palaeontology/jurassic-flight-school.

Science Daily. "Antibiotic Resistance." Accessed January 6, 2020, https://www.sciencedaily.com/terms/antibiotic_resistance.htm.

Service, Robert. "Fossil Feathers Reveal How Dinosaurs Took Flight," *Science*, January 28, 2019, https://www.sciencemag.org/news/2019/01/fossil-feathers-reveal-how-dinosaurs-took-flight.

University of Chicago. "Meet *Tiktaalik roseae*: An Extraordinary Fish." Accessed January 6, 2020, https://tiktaalik.uchicago.edu/meetTik.html.

USFWS (US Fish and Wildlife Service). "American Alligator *Alligator mississippiensis*." Accessed January 6, 2020, https://www.fws.gov/uploadedFiles/American-Alligator-Fact-Sheet.pdf.

USGS (US Geological Survey) Earthquakes Map. July 4, 2020. https://earthquake.usgs.gov/earthquakes/map/.

Wong, Kate. "The Current Mass Extinction." *Scientific American*, October 30, 2000, https://www.scientificamerican.com/article/the-current-mass-extincti.

World Press. "The 8 Biogeographic Realms and 14 Biomes." Accessed January 6, 2020, https://chandanjani.files.wordpress.com/2012/01/biogeographic-realms.pdf.

15. Fit for Life—Adaptation and Diversity

APF (American Porphyria Foundation). "Variegate Porphyria (VP)." Accessed April 14, 2020, https://porphyriafoundation.org/for-patients/types-of-porphyria/vp.

Bunje, Paul. "The Mollusca: Sea Slugs, Squid, Snails, and Scallops." University of California Museum of Paleontology, 2003, accessed January 5, 2020, https://ucmp.berkeley.edu/taxa/inverts/mollusca/mollusca.php.

Evarts, Holly. "Staying Cool: Saharan Silver Ants." Columbia University Engineering, June 18, 2015, https://engineering.columbia.edu/news/staying-cool-saharan-silver-ants.

Feldberg, Sarah. "Scientists Are Getting Creative to Save This Muppet-Faced, Flightless Parrot." National Geographic, March 5, 2019, https://www.nationalgeographic.com/animals/2019/03/endangered-kapako-breeding-technology.

Hanlon, Roger. "Cephalopod Dynamic Camouflage." *Current Biology* 17, no. 11 (2007): R400–4, https://www.cell.com/current-biology/pdf/S0960-9822(07)01138-4.pdf.

Hanlon, Roger, Chuan-Chin Chiao, Lydia M. Mäthger, A. Barbosa, Kendra C. Buresch, and Charles Chubb. "Cephalopod Dynamic Camouflage: Bridging the Continuum between Background Matching and Disruptive Coloration." *Philosophical Transactions of the Royal Society B: Biological Sciences* 364, no. 1516 (2009): 429–37.

Kegel, Brian L. "Bivalve Biology and Physiology." Accessed January 7, 2020, http://ants.biology.utah.edu/tescbiota/kingdom/animalia/phylum/mollusca/bivalves/Biology.htm.

Lazarus, Sarah. "Can Tech Save the Kakapo, New Zealand's 'Gorgeous, Hilarious' Parrot?" *CNN World*, December 26, 2019, https://www.cnn.com/2019/12/26/world/kakapo-conservation-scn-c2e-intl-hnk/index.html.

Linnaeus' *Philosophia Botanica*. "Linnaeus, Carolus." Oxford/New York: Oxford University Press, 2005.

Mäthger, Lydia M., Eric J. Denton, N. Justin Marshall, and Roger T. Hanlon. "Mechanisms and Behavioural Functions of Structural Coloration in Cephalopods." *Journal of the Royal Society* 6, no. S2 (2009): S149–63.

NIH (National Institutes of Health). "PPOX Gene." Genetics Home Reference, accessed April 14, 2020. https://ghr.nlm.nih.gov/gene/PPOX.

Ogden, Lesley Evans. "Three Tricks Orchids Use to Lure Pollinating Insects." *BBC Earth*, February 4, 2015, http://www.bbc.com/earth/story/20150202-three-ways-orchids-trick-insects.

Redmond, Kendra. "The Hairy Optics of Saharan Silver Ants." *PhysicsCentral*, April 26, 2018, http://www.physicscentral.com/explore/plus/saharan-silver-ants.cfm.

Rudman, W. B. "How Sea Slugs Craw," Sea Slug Forum, Australian Museum, May 23, 2001, http://www.seaslugforum.net/locomotion.htm.

Rutgers University. "Class Cephalopoda." General Biology 102, accessed January 5, 2020, http://bio.rutgers.edu/~gb102/lab_2/307em-cephalo.html.

Shi, Norman Nan, Cheng-Chia Tsai, Fernando Camino, Gary D. Bernard, Nanfang Yu, and Rüdiger Wehner. "Thermal Physiology. Keeping Cool: Enhanced Optical Reflection and Radiative Heat Dissipation in Saharan Silver Ants." *Science* 349, no. 6245 (2015): 298–301.

Sick, Stefanie, Stefan Reinker, Jens Timmer, and Thomas Schlake. "WNT and DKK Determine Hair Follicle Spacing Through a Reaction-Diffusion Mechanism." *Science* 314, no. 5804 (2006): 1447–50.

Sokol, Joshua. "Silver Coat Lets Saharan Ants Withstand Scorching Desert Heat." *NewScientist*, June 18, 2015, https://www.newscientist.com/article/dn27748-silver-coat-lets-saharan-ants-withstand-scorching-desert-heat/#ixzz5vvg7Z4Kn.

Sutherland, Richard L., Lydia M. Mäthger, Roger T. Hanlon, Augustine M. Urbas, and Morley O. Stone. "Cephalopod Coloration Model. I. Squid Chromatophores and Iridophores." *Journal of the Optical Society of America* A 25, no. 3 (2008): 588–99.

Turing, Alan Mathison. "The Chemical Basis of Morphogenesis." *Philosophical Transactions of the Royal Society of London. Series B, Biological Sciences* 237, no. 641 (1952): 37–72.

UFAW (Universities Federation for Animal Welfare). "Genetic Welfare Problems of Companion Animals." Accessed April 14, 2020, https://www.ufaw.org.uk/horses/american-paint-horse---overo-lethal-white-foal-syndrome.

University of Michigan. "Evolution and Natural Selection." October 10, 2010, http://www.globalchange.umich.edu/globalchange1/current/lectures/selection/selection.html.

Valentine, Katie. "The Critically Endangered Kakapo Parrot Is Having One Fantastic Year." *Audubon*, August 5, 2016, https://www.audubon.org/news/the-critically-endangered-kakapo-parrot-having-one-fantastic-year.

Wägele, Heike, and Annette Klussmann-Kolb. "Opisthobranchia (Mollusca, Gastropoda)—More Than Just Slimy Slugs. Shell Reduction and Its Implications on Defense and Foraging." *Frontiers in Zoology* 2, no. 3 (2005): 1–18.

Wehner, R., A. C. Marsh, and S. Wehner. "Desert Ants on a Thermal Tightrope." *Nature* 357, no. 6379 (1992): 586.

Whiteway, Christina. "*Strigops habroptila*." *Animal Diversity Web*, accessed January 05, 2020, https://animaldiversity.org/accounts/Strigops_habroptila.

Wood, James B. "The Cephalopod Page." Accessed January 5, 2020, http://www.thecephalopodpage.org.

WWF-Australia (World Wildlife Fund-Australia). "Great Barrier Reef." Accessed January 5, 2020, https://www.wwf.org.au/what-we-do/oceans/great-barrier-reef#gs.qe2l9s.

Young, Emma. "Deceptive Orchids: Luring Wasps for Pollination." *Australian Geographic*, September 16, 2014, https://www.australiangeographic.com.au/topics/science-environment/2014/09/deceptive-orchids-luring-wasps-for-pollination.

16. Population Ecology—Understanding and Caring for Earth

Brisson, Jacques, and James F. Reynolds. "The Effect of Neighbors on Root Distribution in a Creosote Bush (Larrea tridentata) Population." *Ecology* 75, no. 6 (1994): 1693–702

Butler, Matthew J., Grant Harris, and Bradley N. Strobel. "Influence of Whooping Crane Population Dynamics on Its Recovery and Management." *Biological Conservation* 162 (2013): 89–99.

Butler, Matthew, and Wade Harrell. "Whooping Crane Survey Results: Winter 2017–2018." US Fish and Wildlife Service, 2018, https://www.fws.gov/uploadedFiles/WHCR%20Update%20Winter%202017-2018(1).pdf.

CIA (Central Intelligence Agency). *The World Factbook*. Accessed September 21, 2019, https://www.cia.gov/the-world-factbook.

Della Rocca, Gianni, Carmen Hernando, Javier Madrigal, Roberto Danti, José Moya, Mercedes Guijarro, A. Pecchioli et al. "Possible Land Management Uses of Common Cypress to Reduce Wildfire Initiation Risk: A Laboratory Study." *Journal of Environmental Management* 159 (2015): 68–77.

Della Rocca, Gianni, Roberto Danti, Paolo Raddi, Bernabé Moya, and José Moya. "Implementation of the «Cypress System as a Green Firewall. Project CypFire." *Forêt méditerranéenne* (2014).

Devokaitis, Marc. "Whooping Crane Population Hits Historic High In 2018." All about Birds, December 20, 2018, https://www.allaboutbirds.org/whooping-crane-population-hits-historic-high-in-2018.

Evans, Nicholas P., Thomas K. Bauska, Fernando Gázquez-Sánchez, Mark Brenner, Jason H. Curtis, and David A. Hodell. "Quantification of Drought During the Collapse of the Classic Maya Civilization." *Science* 361, no. 6401 (2018): 498–501.

Ingram, Wendy Marie, Leeanne M. Goodrich, Ellen A. Robey, and Michael B. Eisen. "Mice Infected with Low-Virulence Strains of *Toxoplasma gondii* Lose Their Innate Aversion to Cat Urine, Even After Extensive Parasite Clearance." *PLoS One* 8, no. 9 (2013): e75246.

Larson, Don J., Luke Middle, Henry Vu, Wenhui Zhang, Anthony S. Serianni, John Duman, and Brian M. Barnes. "Wood Frog Adaptations to Overwintering in Alaska: New Limits to Freezing Tolerance." *Journal of Experimental Biology* 217, no. 12 (2014): 2193–200.

Roser, Max, Hannah Ritchie, and Esteban Ortiz-Ospina. World Population Growth. Accessed September 20, 2019, https://ourworldindata.org/world-population-growth.

Stalmans, Marc E., Tara J. Massad, Mike J. S. Peel, Corina E. Tarnita, and Robert M. Pringle. "War-Induced Collapse and Asymmetric Recovery of Large-Mammal Populations in Gorongosa National Park, Mozambique." *PLoS One* 14, no. 3 (2019): e0212864.

Statistics Times. List of Countries by Population. Accessed September 21, 2019, http://statisticstimes.com/demographics/countries-by-population-density.php.

Urbanek, R. P., and J. C. Lewis. "Whooping Crane (*Grus Americana*), version 2." The Birds of North America (ed. A.F. Poole), Cornell Lab of Ornithology, 2015, https://doi.org/10.2173/bna.153.

Vyas, Ajai. "Mechanisms of Host Behavioral Change in *Toxoplasma gondii* Rodent Association." *PLoS Pathogens* 11, no. 7 (2015): e1004935.

World Population Review. "Alaska Population and Population Density," accessed February 6, 2021, http://worldpopulationreview.com/states/Alaska-population.

17. Community Ecology—In This Together

Fowler, Amy E., Tiia Forsström, Mikael von Numers, and Outi Vesakoski. "The North American Mud Crab, *Rhithropanopeus harrisii* (Gould, 1841) in Newly Colonized Northern Baltic Sea: Distribution and Ecology *Aquatic Invasions* 8, no. 1 (2013).

Glenner, Henrik, Jens T. Høeg, Jack J. O'Brien, and Timothy D. Sherman. "Invasive Vermigon Stage in the Parasitic Barnacles *Loxothylacus texanus* and *L. panopaei* (Sacculinidae): Closing of the Rhizocephalan Life-Cycle." *Marine Biology* 136, no. 2 (2000): 249–57.

Hardin, Garrett. "The Competitive Exclusion Principle." *Science* 131, no. 3409 (1960): 1292–97.

HHMI (Howard Hughes Medical Institute). "Purple Sea Star." Keystone Species Sampler, accessed December 28, 2019, https://media.hhmi.org/biointeractive/click/keystone/sea-star.html.

Karageorgi, Marianthi, Simon C. Groen, Fidan Sumbul, Julianne N. Pelaez, Kirsten I. Verster, Jessica M. Aguilar, Amy P. Hastings et al. "Genome Editing Retraces the Evolution of Toxin Resistance in the Monarch Butterfly." *Nature* 574 (2019): 1–4.

Keever, Catherine. "Causes of Succession on Old Fields of the Piedmont, North Carolina." *Ecological Monographs* 20, no. 3 (1950): 229–50.

MacArthur, Robert H. "Population Ecology of Some Warblers of Northeastern Coniferous Forests." *Ecology* 39, no. 4 (1958): 599–619.

Mack, Kevin J., Robert D. Podolsky, Virginia Shervette, Amy E. Fowler, and Dara H. Wilber. "Spatial and Temporal Associations between Native Crabs and the Invading Green Porcelain Crab, Petrolisthes armatus, throughout Its Northernmost Invaded Range." *Estuaries and Coasts* 42, no. 2 (2019): 537–47.

Mueller-Dombois, Dieter, and Hans J. Boehmer. "Origin of the Hawaiian Rainforest and Its Transition States in Long-Term Primary Succession." *Biogeosciences* 10, no. 7 (2013): 5171–82.

North Carolina Wildlife Resources Commission. 2017. "Eastern Cottontail." Accessed December 19, 2019. https://www.ncwildlife.org/Portals/0/Learning/documents/Profiles/Mammals/Eastern_Cottontail_Wildlife_Profile_2017.pdf.

Smithsonian (Smithsonian National Zoo and Conservation Biology Institute). "Poison Frogs." Accessed December 12, 2019.

Wikipedia. "Waterfall Frog." Accessed May 6, 2020.

Zagrobelny, Mika, Søren Bak, Anne Vinther Rasmussen, Bodil Jørgensen, Clas M. Naumann, and Birger Lindberg Møller. "Cyanogenic Glucosides and Plant–Insect Interactions." *Phytochemistry* 65, no. 3 (2004): 293–306.

Zeh, David W., and Jeanne A. Zeh. "On the Function of Harlequin Beetle-Riding in the Pseudoscorpion, *Cordylochernes scorpioides* (Pseudoscorpionida: Chernetidae)." *Journal of Arachnology* 20, no. 1 (1992): 47–51.

18. Ecosystems—Natural Networks

Anderson Lab. "Florida Red Tide Mitigation Using Clay Dispersal." Woods Hole Oceanographic Institute, accessed January 20, 2020, https://www2.whoi.edu/site/andersonlab/current-projects/florida-clay-mitigation.

Carbon Farmer. "The Carbon Farmer, Inc." Accessed May 15, 2020, https://thecarbonfarmer.ca/content/climate-change.

Center for Biological Diversity. "Sea Butterfly/Pteropod (Limacina helicina)." Accessed June 16, 2014, http://www.biologicaldiversity.org/programs/climate_law_institute/the_arctic_meltdown/slideshow_text/sea_butterfly.html.

Costanza, Robert, Rudolf de Groot, Paul Sutton, Sander Van der Ploeg, Sharolyn J. Anderson, Ida Kubiszewski, Stephen Farber et al. "Changes in the Global Value of Ecosystem Services." *Global Environmental Change* 26 (2014): 152–58.

DigDeep (DigDeep and US Water Alliance). "Closing the Water Access Gap in the United States: A National Action Plan." Accessed November 18, 2019, http://uswateralliance.org/sites/uswateralliance.org/files/Closing%20the%20Water%20Access%20Gap%20in%20the%20United%20States_DIGITAL.pdf.

Doney, Scott C., Victoria J. Fabry, Richard A. Feely, and Joan A. Kleypas. "Ocean Acidification: The Other CO_2 Problem." *Marine Science* 1 (2009): 169–92.

Dybas, Cheryl Lyn. "Florida's Indian River Lagoon: An Estuary in Transition." *BioScience* 52, no. 7 (2002): 554–59, https://doi.org/10.1641/0006-3568(2002)052[0555:FSIRLA]2.0.CO;2.

Errera, Reagan M., and Lisa Campbell. "Osmotic Stress Triggers Toxin Production by the Dinoflagellate *Karenia brevis*." *Proceedings of the National Academy of Sciences* 108, no. 26 (2011): 10597–601.

———. "Osmotic Stress Does Trigger Brevetoxin Production in the Dinoflagellate *Karenia brevis*." *Proceedings of the National Academy of Sciences*110, no. 25 (2013): E2255-E2255.

Farwig, Nina, Roland Brandl, Stefen Siemann, Franziska Wiener, and Jörg Müller. "Decomposition Rate of Carrion Is Dependent on Composition, Not Abundance of the Assemblages of Insect Scavengers." *Oecologia* 175, no. 4 (2014): 1291–300, https://www.researchgate.net/profile/Nina_Farwig/publication/262610440_Decomposition_rate_of_carrion_is_dependent_on_composition_not_abundance_of_the_assemblages_of_insect_scavengers/links/00b49538722375af22000000/Decomposition-rate-of-carrion-is-dependent-on-composition-not-abundance-of-the-assemblages-of-insect-scavengers.pdf.

FDEP (Florida Department of Environmental Protection). "Salt Marshes." Updated June 1, 2020, accessed November 9, 2019, https://floridadep.gov/rcp/saltmarshes.

Florida Museum. "Adaptations: Mangroves." South Florida Aquatic Environments, accessed November 9, 2019, https://www.floridamuseum.ufl.edu/southflorida/habitats/mangroves/adaptations.

Florida Museum. "Seagrasses." South Florida Aquatic Environments, accessed November 9, 2019, https://www.floridamuseum.ufl.edu/southflorida/habitats/seagrasses.

Florida State Parks. "Ecology of the Indian River Lagoon." Florida Department of Environmental Protection, accessed November 10, 2019, https://www.floridastateparks.org/learn/ecology-indian-river-lagoon.

GLOBE. "Pools, Fluxes and a Word about Units. Globe Carbon Cycle Project." University of New Hampshire, accessed June 20, 2014, http://globecarboncycle.unh.edu/CarbonPoolsFluxes.shtml.

Google Corporation. "Google's Carbon Offsets: Collaboration and Due Diligence." White Paper, 2011, accessed June 20, 2014, https://static.googleusercontent.com/media/www.google.com/en/us/green/pdfs/google-carbon-offsets.pdf.

Henry, Raymond P., Čedomil Lucu, Horst Onken, and Dirk Weihrauch. "Multiple Functions of the Crustacean Gill: Osmotic/Ionic Regulation, Acid-Base Balance, Ammonia Excretion, and Bioaccumulation of Toxic Metals." *Frontiers in Physiology* 3 (2012): 431.

Henshaw, Patricia C., Robert J. Charlson, and Stephen J. Burges. "Water and the Hydrosphere." In *Earth System Science: From Biogeochemical Cycles to Global Changes*, vol. 72, edited by Michael Jacobson, Robert J. Charlson, Henning Rodhe, and Gordon H. Orians, 109–31. Cambridge, MA: Academic Press, 2000.

IPCC (Intergovernmental Panel on Climate Change). "The Carbon Cycle and the Climate System." Sec. 7.3 Climate Change 2007: Working Group I: The Physical Science Basis. Intergovernmental Panel on Climate Change, Fourth Assessment Report: Climate Change, accessed June 16, 2014, https://www.ipcc.ch/site/assets/uploads/2018/02/ar4-wg1-chapter7-1.pdf.

Lewis, Andrew J. "The Ecology of Carrion Decomposition: Necrophagous Invertebrate Assembly and Microbial Community Metabolic Activity During Decomposition of *Sus scrofa* Carcasses in a Temperate Mid-West Forest." PhD dissertation, University of Dayton, December 2011, https://pdfs.semanticscholar.org/97dd/88fc3588d7ddc28daf90582572c3d5d913ef.pdf.

Mackenzie, Fred T. *Our Changing Planet: An Introduction to Earth System Science and Global Environmental Change*, 2nd edition. Upper Saddle River, NJ: Prentice Hall, 1998.

Motte, Jacob Rhett. "Journey into Wilderness: An Army Surgeon's Account of Life in Camp and Field during the Creek and Seminole Wars 1836–1838." Edited by James F. Sunderman. Gainesville, FL: University of Florida Press, 1953, https://ufdc.ufl.edu/UF00020423/00001.

Navajo Nation Department of Water Resources. "Draft Water Resources Development Strategy for the Navajo Nation." July 2011, http://www.frontiernet.net/~nndwr_wmb/PDF/Reports/DWRReports/DWR2011%20Water%20Resource%20Development%20Strategy%20for%20the%20Navajo%20Nation.pdf.

NOAA (National Oceanic and Atmospheric Administration). "Adaptions to Life in the Estuary." Accessed November 2, 2019, https://oceanservice.noaa.gov/education/tutorial_estuaries/est07_adaptations.html.

Novey, Joelle. "Carbon Offsets Demystified." *Green American* (Feature Article), accessed June 20, 2014, http://www.greenamerica.org/green-living/carbon-offsets-explained.

Parmenter, Robert R., and James A. MacMahon. "Carrion Decomposition and Nutrient Cycling in a Semiarid Shrub–Steppe Ecosystem." *Ecological Monographs* 79, no. 4 (2009): 637–61. http://citeseerx.ist.psu.edu/viewdoc/download?doi=10.1.1.669.8721&rep=rep1&type=pdf.

Pechal, Jennifer L., Tawni L. Crippen, Aaron M. Tarone, Andrew J. Lewis, Jeffery K. Tomberlin, and M. Eric Benbow. "Microbial Community Functional Change during Vertebrate Carrion Decomposition." *PLoS One* 8, no. 11 (2013): e79035.

Perkins, Sid. "Inner Workings: Ramping Up the Fight against Florida's Red Tides." *Proceedings of the National Academy of Sciences* 116, no. 14 (2019): 6510–12.

PMEL (Pacific Marine Environmental Laboratory). "What Is Ocean Acidification?" NOAA/PMEL Carbon Program, accessed June 16, 2014, http://www.pmel.noaa.gov/co2/story/What+is+Ocean+Acidification%3F.

Ransford, Marc. "Ball State Teams with Chevrolet to Go Green: New Carbon Methodology Brings Value to Campus Energy Efforts." Ball State University, February 12, 2014, http://cms.bsu.edu/news/press-center/archives/2014/2/ball-state-partners-with-chevrolet-to-validate-new-methodology-for-verifying-carbon-reduction.

Riebeek, Holli. "The Carbon Cycle," Earth Observatory, NASA. Updated June 16, 2011, accessed January 18, 2020. http://earthobservatory.nasa.gov/Features/CarbonCycle/.

Roach, Mary. *Stiff: The Curious Lives of Human Cadavers*. New York: W. W. Norton & Company, 2003.

Royal Society. *Ocean Acidification Due to Increasing Atmospheric Carbon Dioxide*. London: The Royal Society, 2005.

Schlesinger, W. H. and Emily S. Bernhardt. *Biogeochemistry: An Analysis of Global Change*. Waltham, MA : Academic Press, 2013.

Smithsonian. "Mangroves." Ocean Life, April 2018. https://ocean.si.edu/ocean-life/plants-algae/mangroves.

Sunda, William G., Cheska Burleson, D. Ransom Hardison, Jeanine S. Morey, Zhihong Wang, Jennifer Wolny, Alina A. Corcoran et al. "Osmotic Stress Does Not Trigger Brevetoxin Production in the Dinoflagellate *Karenia brevis*." *Proceedings of the National Academy of Sciences* 110, no. 25 (2013): 10223–28.

Tomberlin, Jeffery K., and Peter H. Adler. "Seasonal Colonization and Decomposition of Rat Carrion in Water and on Land in an Open Field in South Carolina." *Journal of Medical Entomology* 35, no. 5 (1998): 704–9, https://academic.oup.com/jme/article-abstract/35/5/704/2221718?redirectedFrom=fulltext.

Towle, David W., and Dirk Weihrauch. "Osmoregulation by Gills of Euryhaline Crabs: Molecular Analysis of Transporters." *American Zoologist* 41, no. 4 (2001): 770–80.

Weisberg, Robert H., Lianyuan Zheng, Yonggang Liu, Alina A. Corcoran, Chad Lembke, Chuanmin Hu, Jason M. Lenes et al. "*Karenia brevis* Blooms on the West Florida Shelf: A Comparative Study of the Robust 2012 Bloom and the Nearly Null 2013 Event." *Continental Shelf Research* 120 (2016): 106–21.

19. Human Impact—Over Our Heads? Atmospheric Carbon and Climate Change

Boisier, Juan P., Roberto Rondanelli, René D. Garreaud, and Francisca Muñoz. "Anthropogenic and Natural Contributions to the Southeast Pacific Precipitation Decline and Recent Megadrought in Central Chile." *Geophysical Research Letters* 43, no. 1 (2016): 413–21.

Church, J. A. and N.J. White Church, John A., and Neil J. White. "A 20th-Century Acceleration in Global Sea Level Rise." *Geophysical Research Letters* 33, no. 1 (2006). Accessed May 12, 2020. https://agupubs.onlinelibrary.wiley.com/doi/full/10.1029/2005GL024826.

Cooke, Kieran. “Chile Faces ‘Critical Situation’ as Floods and Droughts Batter Country.” *Climate Change News,* March 31, 2015, http://www.climatechangenews.com/2015/03/31/chile-faces-critical-situation-as-floods-and-droughts-batter-country.

Ding, Y. D. J. G., David J. Griggs, Maria Noguer, Paul J. van der Linden, Xiaosu Dai, Kathy Maskell, and C. A. Johnson. *Climate Change 2001: The Scientific Basis.* Vol. 881. Cambridge, UK: Cambridge University Press, 2001.

DOE (US Department of Energy). “How Fossil Fuels Were Formed.” Accessed February 7, 2021, https://www.energy.gov/science-innovation/energy-sources/fossil.

Frich, P., Lisa V. Alexander, P. M. Della-Marta, B. Gleason, M. Haylock, AMG Klein Tank, and Tom Peterson. “Observed Coherent Changes in Climatic Extremes during the Second Half of the Twentieth Center.” *Climate Research* 19, no. 3 (2002): 193-212. Accessed May 2020. https://www.int-res.com/articles/cr2002/19/c019p193.pdf.

Garreaud, René, Camila Alvarez-Garreton, Jonathan Barichivich, Juan Pablo Boisier, Duncan Christie, Mauricio Galleguillos, Carlos LeQuesne et al. “The 2010–2015 Mega Drought in Central Chile: Impacts on Regional Hydroclimate and Vegetation.” *Hydrology and Earth System. Sciences* 21, no. 12 (2017): 6307–27.

Gleason, Karin L., Jay H. Lawrimore, David H. Levinson, Thomas R. Karl, and David J. Karoly. “A Revised US Climate Extremes Index.” *Journal of Climate* 21, no. 10 (2008): 2124-2137. Accessed May 12, 2020. https://journals.ametsoc.org/jcli/article/21/10/2124/31825.

Hansen, James, Makiko Sato, Reto Ruedy, Andrew Lacis, and Valdar Oinas. Global Warming in the Twenty-First Century: An Alternative Scenario. *Proceedings of the National Academy of Sciences* 97, no. 18 (2000): 9875–80.

Harris, Gardiner. “Borrowed Time on Disappearing Land: Facing Rising Seas, Bangladesh Confronts the Consequences of Climate Change.” *New York Times,* March 28, 2014.

Karl, Thomas R., Jerry M. Melillo, and Thomas C. Peterson (eds.). *Global Climate Change Impacts in the United States.* New York: Cambridge University Press, 2009.

Kwok, R., and D. A. Rothrock. “Decline in Arctic Sea Ice Thickness from Submarine and ICESat Records: 1958–2008.” *Geophysical Research Letters* 36, no. 15 (2009). Accessed May 12, 2020. https://agupubs.onlinelibrary.wiley.com/doi/full/10.1029/2009gl039035.

Levitus, Sydney, John I. Antonov, Julian Wang, Thomas L. Delworth, Keith W. Dixon, and Anthony J. Broccoli. “Anthropogenic Warming of Earth’s Climate System.” *Science* 292, no. 5515 (2001): 267-270. Accessed May 12, 2020. https://www.researchgate.net/profile/Thomas_Delworth/publication/12031466_Anthropogenic_Warming_of_Earth%27s_Climate_System/links/00b7d52389be304def000000/Anthropogenic-Warming-of-Earths-Climate-System.pdf.

Levitus, Sydney, John I. Antonov, Timothy P. Boyer, Ricardo A. Locarnini, Hernan E. Garcia, and Alexey V. Mishonov. “Global Ocean Heat Content 1955–2008 in Light of Recently Revealed Instrumentation Problems.” *Geophysical Research Letters* 36, no. 7 (2009). Accessed May 12, 2020. https://agupubs.onlinelibrary.wiley.com/doi/full/10.1029/2008GL037155%4010.1002/%28ISSN%291944-8007.GRL40.

NASA (National Aeronautics and Space Administration). 2020. “Overview: Weather, Global Warming and Climate Change. Global Climate Change. https://climate.nasa.gov/resources/global-warming-vs-climate-change/.

NRC (National Research Council). “Advancing the Science of Climate Change.” Washington, DC: National Academies Press, 2010.

———. "Climate Change: Evidence, Impacts, and Choices." Division on Earth and Life Studies. Washington, DC: National Academies Press, 2013.

Reeves, Philip. "In Chile, Many Regard Climate Change as the Greatest External Threat." *Weekend Edition Sunday*, July 16, 2017, http://www.npr.org/sections/parallels/2017/07/16/536896748/in-chile-many-regard-climate-change-as-the-greatest-external-threat.

Scripps CO_2 Program. "Atmospheric CO_2." Scripps Institution of Oceanography, accessed December 23, 2019, https://scrippsco2.ucsd.edu/research/atmospheric_co2.html.

Vicuña, Sebastian, René D. Garreaud, and James McPhee. "Climate Change Impacts on the Hydrology of a Snowmelt-Driven Basin in Semiarid." *Climatic Change* 105 (2011): 469-488.

WTO (World Trade Organization). "Chile and the WTO." Accessed December 23, 2019, https://www.wto.org/english/thewto_e/countries_e/chile_e.htm.

20. Commandeering the Nitrogen Cycle—Human Transformation of the Biosphere

Almaraz, Maya, Edith Bai, Chao Wang, Justin Trousdell, Stephen Conley, Ian Faloona, and Benjamin Z. Houlton. "Agriculture Is a Major Source of NOx Pollution in California." *Science Advances* 4, no. 1 (2018): eaao3477.

Battye, William, Viney P. Aneja, and William H. Schlesinger. "Is Nitrogen the Next Carbon?" *Earth's Future* 5, no. 9 (2017): 894–904

Bricker, Suzanne B., Ben Longstaff, William Dennison, Adrian Jones, Kate Boicourt, Caroline Wicks, and Joanna Woerner. "Effects of Nutrient Enrichment in the Nation's Estuaries: A Decade of Change." *Harmful Algae* 8, no. 1 (2008): 20–32.

Burkholder, JoAnn M., David A. Tomasko, and Brant W. Touchette. "Seagrasses and Eutrophication." *Journal of Experimental Marine Biology and Ecology* 350, no. 1–2 (2007): 46–72.

CIA. "*The World Factbook Europe: Netherlands.*" Accessed December 2, 2019, https://www.cia.gov/the-world-factbook/countries/netherlands.

DEC (NY State Department of Environmental Conservation). "DEC Discovers Brook Trout in Previously 'Fishless' Adirondack Lake Impacted by Acid Rain: Lake Colden Tributaries Show Signs of Fish and Habitat Recovery." Press Release. 7 October, 2019.

Driscoll, Charles T., Kimberley M. Driscoll, Myron J. Mitchell, and Dudley J. Raynal. "Effects of Acidic Deposition on Forest and Aquatic Ecosystems in New York State." *Environmental Pollution* 123, no. 3 (2003): 327–36.

FAO (Food and Agriculture Organization). Statistical Databases, Rome, 2006.

Fenn, L. B., and L. R. Hossner. "Ammonia Volatilization from Ammonium or Ammonium-Forming Nitrogen Fertilizers." In *Advances in Soil Science*, edited by B. A. Stewart, 123–69. New York: Springer, 1985.

Galloway, James N., John D. Aber, Jan Willem Erisman, Sybil P. Seitzinger, Robert W. Howarth, Ellis B. Cowling, and B. Jack Cosby. "The Nitrogen Cascade." *Bioscience* 53, no. 4 (2003): 341–56.

Galloway, James N., Alan R. Townsend, Jan Willem Erisman, Mateete Bekunda, Zucong Cai, John R. Freney, Luiz A. Martinelli, et al. "Transformation of the Nitrogen Cycle: Recent Trends, Questions, and Potential Solutions." *Science* 320, no. 5878 (2008): 889–92.

Jiang, Mei, and Richard Jagels. "Detection and Quantification of Changes in Membrane-Associated Calcium in Red Spruce Saplings Exposed to Acid Fog." *Tree Physiology* 19, no. 14 (1999): 909–16.

Kim, Youngsug, Bernard A. Engel, Kyoung Jae Lim, Vickie Larson, and Brean Duncan. "Runoff Impacts of Land-Use Change in Indian River Lagoon Watershed." *Journal of Hydrologic Engineering* 7, no. 3 (2002): 245–51.

Lapointe, Brian E., Laura W. Herren, and Bradley J. Bedford. "Effects of Hurricanes, Land Use, and Water Management on Nutrient and Microbial Pollution: St. Lucie Estuary, Southeast Florida." *Journal of Coastal Research* 28, no. 6 (2012): 1345–61.

NASA (National Aeronautics and Space Administration). "The Ozone We Breathe," NASA Earth Observatory, April 19, 2002, https://earthobservatory.nasa.gov/features/OzoneWeBreathe/ozone_we_breathe3.php.

Ogden, Lesley Evans. "The Bittersweet Story of How We Stopped Acid Rain." August 7, 2019, *BBC Future.* https://www.bbc.com/future/article/20190823-can-lessons-from-acid-rain-help-stop-climate-changeRudd, John W. M., Carol Ann Kelly, Vincent St. Louis, Raymond H. Hesslein, Akira Furutani, and Morris H. Holoka. "Microbial Consumption of Nitric and Sulfuric Acids in Acidified North Temperate Lakes 1." *Limnology and Oceanography* 31, no. 6 (1986): 1267–80

Schindler, D. W., K. H. Mills, D. F. Malley, D. L. Findlay, J. A. Shearer, I. J. Davies, M. A. Turner et al. "Long-Term Ecosystem Stress: The Effects of Years of Experimental Acidification on a Small Lake." *Science* 228, no. 4706 (1985): 1395–401

Schlesinger, William H., and Anne E. Hartley. "A Global Budget for Atmospheric NH_3." *Biogeochemistry* 15, no. 3 (1992): 191–201

Solomon, S., D. Qin, M. Manning, Z. Chen, M. Marquis, K. Averyt, M. Tignor, and H. Miller. "IPCC Fourth Assessment Report (AR4)." Intergovernmental Panel on Climate *Change,* 2007.

Statistics Netherlands 1. "Agricultural Export Value over 90 Bn Euros in 2018." Accessed December 2, 2019. https://www.cbs.nl/en-gb/news/2019/03/agricultural-export-value-over-90-bn-euros-in-2018.

______. "National *Cattle Herd Declining Slightly.*" Accessed December 2, 2019, https://www.cbs.nl/en-gb/news/2019/29/national-cattle-herd-declining-slightly.

Stokstad, Erik. "Nitrogen Crisis Threatens Dutch Environment—and Economy." *Science* 366 (2019): 1180–81.

Torsethaugen, Gro, Eva J. Pell, and Sarah M. Assmann. "Ozone Inhibits Guard Cell K+ Channels Implicated in Stomatal Opening." *Proceedings of the National Academy of Sciences* 96, no. 23 (1999): 13577–82

US Census Bureau. "Population and Housing Unit Estimates Tables." Accessed July 4, 2020. https://www.census.gov/programs-surveys/popest/data/tables.html.

Vitousek, Peter M., John D. Aber, Robert W. Howarth, Gene E. Likens, Pamela A. Matson, David W. Schindler, William H. Schlesinger et al. "Human Alteration of the Global Nitrogen Cycle: Sources and consequences," *Ecological Applications* 7, no. 3 (1997): 737–50.

WHO (World Health Organization). "Effects of Ozone on Vegetation: Critical Levels." *Air Quality Guidelines,* 2nd ed., WHO Regional Office for Europe, Copenhagen, Denmark, 2000. http://www.euro.who.int/__data/assets/pdf_file/0006/123099/AQG2ndEd_12ozonelevel.pdf?ua=1.

Wikipedia. "Lake Colden." Accessed December 16, 2019, https://en.wikipedia.org/wiki/Lake_Colden.

21. Human Activity—The Future of Water Not So Clear

Alpert, Alice. "Coral-Current Connections." *Oceanus*, October 20, 2014.

Andrews, Jay D. 1988. "Epizootiology of the Disease Caused by Oyster Pathogen, *Perkinsus marinus*, and Its Effects on the Oyster Industry." *American Fisheries Society Special Publication* 18: 47–63.

AreaVibes. "Flint, MI Employment." Accessed July 4, 2020. https://www.areavibes.com/flint-mi/employment/.

Babcock, Russell C., Jeffrey M. Dambacher, Elisabetta B. Morello, Eva E. Plaganyi, Keith R. Hayes, Hugh PA Sweatman, and Morgan S. Pratchett. "Assessing Different Causes of Crown-of-Thorns Starfish Outbreaks and Appropriate Responses for Management on the Great Barrier Reef." *PLoS One* 11, no. 12 (2016).

Blanchon, Paul, Anton Eisenhauer, Jan Fietzke, and V. Liebetrau. 2009. "Rapid Sea-Level Rise and Reef Back-Stepping at the Close of the Last Interglacial Highstand." *Nature* 458: 881–84.

Braun, Ashley. "Sea-Star Murdering Robots Are Deployed in the Great Barrier Reef." *Smithsonian Magazine* August 31, 2018, https://www.smithsonianmag.com/innovation/sea-star-murdering-robotsa-are-deployed-in-great-barrier-reef-180970177.

Brodie, Jon, Katharina Fabricius, Glenn De'ath, and Ken Okaji. "Are Increased Nutrient Inputs Responsible for More Outbreaks of Crown-of-Thorns Starfish? An Appraisal of the Evidence." *Marine Pollution Bulletin* 51, no. 1–4 (2005): 266–78.

Bruckner, Andrew W. "New Threat to Coral Reefs: Trade in Coral Organisms." *Issues in Science and Technology* 17, no. 1 (Fall 2000): 1–6.

Carlson, Rachel R., Shawna A. Foo, and Gregory P. Asner. "Land Use Impacts on Coral Reef Health: A Ridge-to-Reef Perspective." *Frontiers in Marine Science* 6 (2019): 562.

Castro, Peter and Michael Huber. "Ch. 14, Coral Reefs." In *Marine Biology*, 10th ed. New York: McGraw-Hill Education, 2016.

Census Reporter. "Atlanta Metro Population Estimate." *Census Reporter*, 2018, https://censusreporter.org/profiles/31000US12060-atlanta-sandy-springs-roswell-ga-metro-area.

Coral Reef Alliance. "Saving the Worlds Coral Reefs: Direct Threats." Accessed February 3, 2020, https://coral.org/coral-reefs-101/reef-threats/direct.

Couch, Carol Anne, Evelyn H. Hopkins, and P. Suzanne Hardy. *Influences of Environmental Settings on Aquatic Ecosystems in the Apalachicola-Chattahoochee-Flint River Basin. Water-Resources Investigations, Report 95-4278*. US Department of the Interior, US Geological Survey, 1996.

Denchak, Melissa. "Flint Water Crisis: Everything You Need to Know." Natural Resources Defense Council, November 18, 2018, https://www.nrdc.org/stories/flint-water-crisis-everything-you-need-know.

De Villegas, Rob Diaz. "The Apalachicola Oyster in 2016—A Long Way to Recovery." *WFSU Ecology Blog*, October 27, 2016, https://blog.wfsu.org/blog-coastal-health/?p=9854.

Domínguez, Lucía Molina, Francisco Otero-Ferrer, and Marisol Izquierdo López. "Coral Reefs: Threats and Future Focusing in Over-Fishing, Aquaculture, and Educational Programs." *Management of Natural Resources, Sustainable Development and Ecological Hazards* 99 (2006): 305.

Done, Terence J. "Simulation of Recovery of Pre-Disturbance Size Structure in Populations of Porites spp. Damaged by the Crown of Thorns Starfish *Acanthaster planci*." *Marine Biology* 100, no. 1 (1988): 51–61.

Doney, Scott C., Victoria J. Fabry, Richard A. Feely, and Joan A. Kleypas. "Ocean Acidification: The Other CO_2 Problem." *Marine Science* 1, no. 1 (2009):169–92.

Elliott, Debbie. "Florida Closes Iconic Apalachicola Oyster Fishery," National Public Radio (NPR) Morning Edition, July 22, 2020, https://www.npr.org/2020/07/22/894074674/floridas-oyster-beds-devastated-by-years-of-drought-other-pressures.

Encomio, Vincent. "The Oyster Is Our World at Florida Oceanographic Society." October 1, 2015, https://coastalanglermag.com/the-oyster-is-our-world-at-florida-oceanographic.

EPA (Environmental Protection Agency). "Threats to Coral Reefs." May 4, 2018, https://www.epa.gov/coral-reefs/threats-coral-reefs.

Field, Michael E., Andrea S. Ogston, and Curt D. Storlazzi. 2011. "Rising Sea Level May Cause Decline of Fringing Coral Reefs." *Eos, Transactions American Geophysical Union* 92, no. 33 (2011): 273–74.

Fisch, Nicholas C., and William E. Pine III. "A Complex Relationship between Freshwater Discharge and Oyster Fishery Catch per Unit Effort in Apalachicola Bay, Florida: An Evaluation from 1960 to 2013." *Journal of Shellfish Research* 35, no. 4 (2016): 809–25.

Florida Sea Grant. *Apalachicola Bay Oyster Situation Report*. April 24, 2013, https://www.flseagrant.org/news/2013/04/apalachicola-oyster-report.

Frick Elizabeth A., Gary R. Buell and Evelyn H. Hopkins. "1996 Nutrient Sources and Analysis of Nutrient Water-Quality Data, Apalachicola-Chattahoochee-Flint River Basin, Georgia, Alabama and Florida, 1972-1990." US Geological Survey Water-Resource Investigations Report 96-4101, 1996.

GBRMPA (Great Barrier Reef Marine Park Authority). "Ocean Currents." Accessed February 3, 2020, http://www.gbrmpa.gov.au/our-work/threats-to-the-reef/climate-change/ocean-currents.

Georgia Department of Natural Resources. "River Basin Characteristics. Chattahoochee River Basin Watershed Protection Plan, Section 2." Environmental Protection Division, Georgia Department of Natural Resources, 1997.

Glynn, Peter W. "Coral Reef Bleaching in the 1980s and Possible Connections with Global Warming." *Trends in Ecology & Evolution* 6, no. 6 (1991): 175-179.

Great Barrier Reef Foundation. "Crown-of-Thorns Starfish." Accessed December 12, 2020, https://www.barrierreef.org/the-reef/threats/Crown-of-thorns%20starfish.

Gross, Terry. "Pediatrician Who Exposed Flint Water Crisis Shares Her 'Story of Resistance'." NPR, June 25, 2018.

Harriott, Vicki. "Crown-of-Thorns Starfish on the Great Barrier Reef" Current State of Knowledge. Reef Resilience Network, November 2003, https://www.reefresilience.org/pdf/COTS_Nov2003.pdf.

Havens, Karl, Mike Allen, Traci Irani, Angela Lindsey, Andrew Kane, David Kimbro, Steve Otwell, et al. "Apalachicola Bay Oyster Situation Report." Technical Publication 201. Gainesville, FL: University of Florida Sea Grant, 2013.

Hawkins, Gary. "Georgia-Florida Water-Based Lawsuit." *University of Georgia Extension Blog*, November 5, 2016, https://site.extension.uga.edu/water/2016/11/georgia-florida-water-based-lawsuit/.

Henkel, Timothy P. "Coral Reefs." *Nature Education Knowledge* 3, no. 10 (2010): 12.

Lamb, Joleah B., Bette L. Willis, Evan A. Fiorenza, Courtney S. Couch, Robert Howard, Douglas N. Rader, James D. True et al. "Plastic Waste Associated with Disease on Coral Reefs." *Science* 359, no. 6374 (2018): 460–62

Leitman, Sasha, W. E. Pine, and Gregory Kiker. "Management Options During the 2011–2012 Drought on the Apalachicola River: A Systems Dynamic Model Evaluation. *Environmental Management* 58:193–207.

Lesser, Michael P. "Oxidative Stress Causes Coral Bleaching During Exposure to Elevated Temperatures." *Coral Reefs* 16, no. 3 (1997): 187–92.

Livingston, Robert J. "Climate Change and Coastal Ecosystems: Long-Term Effects of Climate and Nutrient Loading on Trophic Organization." Boca Raton, FL: CRC Press. 2015.

Livingston, Robert J., F. G. Lewis, G. C. Woodsum, X.-F. Niu, B. Galperin, Wenrui Huang, John D. Christensen et al. "Modelling Oyster Population Response to Variation in Freshwater Input." *Estuarine Coastal and Shelf Science* 50, no. 5 (2000): 655–72.

Madin, Kate. "Ocean Acidification: A Risky Shell Game." *Oceanus*, December 4, 2009.

Manganiello, Chris. "The Tri-State Water Wars and Opportunities for Change. History of the Dispute." Chattahoochee River Keeper, July 24, 2019, https://chattahoochee.org/our-work/tri-state-water-conflict.

Nature Conservancy. "Crown of Thorns Starfish." Reef Resilience Network, accessed February 12, 2020, https://reefresilience.org/stressors/predator-outbreaks/crown-of-thorns-starfish.

———. "Sea-Level Rise." Reef Resilience Network, accessed February 3, 2020, https://reefresilience.org/climate-and-ocean-change/sea-level-rise.

NOAA (National Oceanic and Atmospheric Administration). "How Does Overfishing Threaten Coral Reefs?" June 25, 2018, https://oceanservice.noaa.gov/facts/coral-overfishing.html.

Normille, Dennis. "El Nino's Warmth Devastating Reefs Worldwide." *Science Magazine Online*, March 29, 2016, http://www.sciencemag.org/news/2016/03/el-ni-o-s-warmth-devastating-reefs-worldwide.

Oczkowski, Autumn J., F. Graham Lewis, Scott W. Nixon, H. Lee Edmiston, Rebecca S. Robinson, and Jeffrey P. Chanton. "Fresh Water Inflow and Oyster Productivity in Apalachicola Bay, FL (USA)." *Estuaries and Coasts* 34, no. 5 (2011): 993–1005.

Pieper, Kelsey J., Rebekah Martin, Min Tang, LeeAnne Walters, Jeffrey Parks, Siddhartha Roy, Christina Devine et al. "Evaluating Water Lead Levels during the Flint Water Crisis." *Environmental Science & Technology* 52, no. 15 (2018): 8124–32.

Pine, William E., III, Carl J. Walters, Edward V. Camp, Rachel Bouchillon, Robert Ahrens, Leslie Sturmer, and Mark E. Berrigan. "The Curious Case of Eastern Oyster Crassostrea virginica Stock Status in Apalachicola Bay, Florida." *Ecology and Society* 20, no. 3 (2015).

Pollock, F. Joseph, Naohisa Wada, Gergely Torda, Bette L. Willis, and David G. Bourne. "White Syndrome-Affected Corals Have a Distinct Microbiome at Disease Lesion Fronts." *Applied and Environmental Microbiology* 83, no. 2 (2017): e02799-16.

Reilly, Amanda. "Special Master Hands Georgia a Victory in Southeast Water War." *E&E News*, February 15, 2017, https://www.eenews.net/stories/1060050120.

Rivera-Posada, Jairo, Morgan S. Pratchett, Catalina Aguilar, Alexandra Grand, and Ciemon F. Caballes.. "Bile Salts and the Single-Shot Lethal Injection Method for Killing Crown-of-Thorns Sea Stars (Acanthaster planci)." *Ocean and Coastal Management* 102 (2014): 383–90.

Ruckart, Perri Zeitz, Adrienne S. Ettinger, Mona Hanna-Attisha, Nicole Jones, Stephanie I. Davis, and Patrick N. Breysse. "The Flint Water Crisis: A Coordinated Public Health Emergency Response and Recovery Initiative." *Journal of Public Health Management and Practice* 25, no. S1 (2019): S84.

Schultz, David. "Was Flint's Deadly Legionnaires' Epidemic Caused by Low Chlorine Levels in the Water Supply?" *Science*, February 5, 2018, https://www.sciencemag.org/news/2018/02/was-flint-s-deadly-legionnaires-epidemic-caused-low-chlorine-levels-water-supply.

Smith, Lindsey. *Not Safe to Drink*. Michigan Radio (documentary), National Public Radio. Aired December 16, 2015.

Smithsonian Ocean Portal Team. "Corals and Coral Reefs," reviewed by Nancy Knowlton. Smithsonian Ocean Portal. April 2018, http://ocean.si.edu/corals-and-coral-reefs.

Torrice, Michael. "How Lead Ended Up in Flint's Tap Water." *Chemical and Engineering News* 94, no. 7 (2016), https://cen.acs.org/articles/94/i7/Lead-Ended-Flints-Tap-Water.html.

Twichell, Dave, L. Edmiston, Brian Andrews, W. Stevenson, J. Donoghue, R. Poore, and Lisa Osterman. "Geologic Controls on the Recent Evolution of Oyster Reefs in Apalachicola Bay and St. George Sound, Florida." *Estuarine, Coastal and Shelf Science* 88 (2010): 385–94, http://dx.doi.org/10.1016/j.ecss.2010.04.019.

US Census Bureau. "Flint City, Michigan (2019)." QuickFacts. accessed July 4, 2020, https://data.census.gov/cedsci/all?q=Flint.

Uthicke, Sven, Murray Logan, M. Liddy, David Francis, Natasha Hardy, and Miles Lamare. "Climate Change as an Unexpected Co-Factor Promoting Coral-Eating Seastar (Acanthaster planci) Outbreaks." *Scientific Reports* 5 (2015): 8402.

Ward, Carlton. "An Oysterman Hero in Apalachicola." Voices Florida Wildlife Corridor, National Geographic Explorers, March 10, 2015, https://blog.nationalgeographic.org/2015/03/10/an-oysterman-hero-in-apalachicola/.

Wikipedia. "Flint Water Crisis." Accessed February 1, 2020. https://en.wikipedia.org/wiki/Flint_water_crisis.

Zahran, Sammy, Shawn P. McElmurry, Paul E. Kilgore, David Mushinski, Jack Press, Nancy G. Love, Richard C. Sadler et al. "Assessment of the Legionnaires' Disease Outbreak in Flint, Michigan." *Proceedings of the National Academy of Sciences* 115, no. 8 (2018): E1730–39.

INDEX

C

F

G

H

I

J

K

L

M

Q

R

S

T

U

V

W

X

Y

Z

www.ingramcontent.com/pod-product-compliance
Ingram Content Group UK Ltd.
Pitfield, Milton Keynes, MK11 3LW, UK
UKHW052006041026
15590UKWH00008B/2188

9 781793 51379